THE
BIRDS
OF
ARIZONA

Masked Bobwhite — *William J. Schaldach*

THE BIRDS of ARIZONA

by

ALLAN PHILLIPS
JOE MARSHALL
GALE MONSON

THE UNIVERSITY OF ARIZONA PRESS
TUCSON   1964

Copyright© 1964

The Board of Regents

of the Universities

and State College

of Arizona.

All rights reserved.

Library of Congress

Card Catalog Number

64-17265

PREFACE

JUST twenty-four years ago I first visited Tucson, Arizona, as part of an informal program of seeing as many species of American birds as I could in a twelve-month period. I was at that time a vice president of Bankers Trust Company in New York, as well as president of the National Audubon Society; and my associates had libelously suggested that my field trips were planned primarily to see the largest possible number of birds rather than because of the urgent needs for banking service on the part of our customers.

At any rate, with the cooperation of many local bird people, I managed to record, in the year 1939, 497 species of birds (not including subspecies), which was a record that stood for 13 years in the amateur field. In 1954 Roger Peterson and James Fisher became a little tired of having this record put up to them, and set out on a formal expedition to surpass my record. My record had been strictly confined to the continental area of the United States, but their list started in Newfoundland and ended in Alaska. I may say that I encouraged and helped to finance their trip to show that there were no hard feelings. They wired me at a certain point in their tour, saying that they had passed the 500 mark. Later on they wrote a book, *Wild America*, giving a fascinating description of their trip, and were kind enough to dedicate the book to Guy Emerson.

I mention this at some length to indicate why I first came to Tucson. I had a letter of introduction to Prof. Charles T. Vorhies who, in addition to being a distinguished professor of biology at the University of Arizona, was also a pioneer in bird-watching in that area. He had developed several of his students as excellent field men. These men were always available to go into the field with bird people from other states.

When I had arrived in town and had gone on field trips with growing enthusiasm with different men on different days, Dr. Vorhies seemed to think that I was sufficiently equipped to rate an expedition with his number one field expert and friend, Dr. Allan Phillips. When I first went out with Dr. Phillips, it was a red-letter day in my experience. I found him to be a man of deep culture in the science of ornithology, with enormous persistence and tireless energy. I soon learned that he was to make a life-work of the systematics of Arizona ornithology. He is singularly well equipped to do this job thoroughly, possessing as he does not only the laboratory talents necessary for that work, but also to a high degree the field skills, including an extraordinary eye and ear.

I returned many times to Arizona and made memorable visits to the many beautiful mountains for which Tucson is the point of departure. And on these trips we often discussed the absence of an outstanding book on the birds of Arizona, and in our talks it was more or less assumed that a book by Allan Phillips would be produced to fill that gap. If Dr. Vorhies had lived, it would have been natural to have collaboration between Allan Phillips and himself.

As the years passed by I'm afraid I became rather a nag in my relations with Allan Phillips because as I traveled around the country I was so often asked why there was not an adequate book on Arizona birds. Arizona was an area of pilgrimage, sharing the top place in numbers and beauty of species with Texas, California and Florida.

The answer was that Phillips was a perfectionist, and whenever he started on a first paragraph of a BIRDS OF ARIZONA he became appalled at the volume and variety of information which he lacked. So that first paragraph was never finished.

As the years went on he became fascinated by certain phases of bird life in Mexico, as practically all of the bird people in the border states do sooner or later. He was welcomed by the people at the University of Mexico and took up his residence in Mexico City in 1957. I kept up a correspondence with him urging the need for a book on Arizona birds but had almost given up hope when some years ago, on a visit to Tucson, I became acquainted with Dr. Joe Marshall, who had become professor of biology succeeding Phillips. I have found by long experience that most professors of biology are unusual people. They certainly are never content with excelling in one talent. This is particularly true of Joe Marshall, who in addition to having a highly developed taxonomic skill, and easily one of the most highly developed field skills in the United States, is a distinguished harpsichordist, and also has built two complete harpsichords with his own hands.

One evening after a day in the Chiricahua Mountains with Marshall and Bill George, one of his most brilliant pupils, we were sitting around reminiscing about the day's adventures, when I had what later turned out to be an inspiration. I suggested to Marshall that he collaborate with Phillips in getting out a first-class book on the birds of Arizona. Marshall did not jump up and down with enthusiasm over my idea, but I could see that he was definitely interested. The trouble was that he had a full teaching schedule at the University, and was in the midst of a field study of the Abert's Towhee and of other bird studies in neighboring Sonora.

The next day I saw Dr. Richard Harvill, president of the University, whom I knew as a man of imagination and action. As soon as he heard the facts, he gave his enthusiastic approval. He immediately sent for Marshall and asked him to undertake the job. It was arranged that Marshall would have a substantial time free from academic duties to work on the book.

Phillips gave his approval, but was still the perfectionist. He realized that his field studies had not covered the important Colorado River Valley and southwestern Arizona. Gale Monson of the U. S. Fish and Wildlife Service had done thorough work in that and other parts of the state for many years, and he very kindly agreed to put his long experience, judgment, and fine records at the disposal of the project.

Meanwhile, Marshall went to Mexico and started work with Phillips. It soon became evident that the material available was unusually thorough, particularly on the technical side. So after full discussion with the officers of the University Press, it was agreed that they would bring out immediately the *Annotated Checklist of the Birds of Arizona,* by Gale Monson and Allan R. Phillips (University of Arizona Press, 1964), which had been pending for over a year. This Checklist contains scientific data on each of the 423 full species of Arizona birds. The issuance of this book has made available the scientific data for a leading paragraph which heads each species account in THE BIRDS OF ARIZONA.

THE BIRDS OF ARIZONA is illustrated with sixty-four reproductions in color. We were specially fortunate in obtaining the twelve "sketches" as a gift to Allan Phillips from the eminent bird painter, George Miksch Sutton. They were done by the artist on a trip to the Santa Rita Mountains. The fifty-one color photographs by Eliot Porter add greatly to the distinction of the book.

I believe these two books, put out by the University of Arizona and its Press under the able direction of Dr. Jack L. Cross, will put Arizona in the forefront of American ornithological literature. A number of state bird books have appeared during the past fifty years. One of the pioneer works in this field was, and still remains, among the most distinguished. It is *The Birds of Massachusetts and Other New England States,* in three volumes, about 1450 pages, by Edward Howe Forbush, illustrated by color plates from drawings by Louis Agassiz Fuertes, perhaps the best bird painter since Audubon. It was published in 1925–1929, under the auspices of the Massachusetts Department of Agriculture, and issued "by the Authority of the Legislature." It has long been a collector's item.

It is hoped that THE BIRDS OF ARIZONA will win favor not only with Arizona residents but also with the thousands of bird-watchers throughout the nation who will inevitably come to Arizona to see the birds of incomparable beauty and interest which are found within the borders of the state.

Martha's Vineyard
August 17, 1964

Guy Emerson

CONTENTS

	Page
Abbreviations used in the text	viii
Introduction	ix
Historic Changes and Conservation of Arizona Habitats	xiii
Map of Arizona State Highway System	xviii
George Miksch Sutton Field Sketches	fp 76
Eliot Porter Photographs	fp 140
Index	213

THE BIRDS OF ARIZONA

Family	Page
Loons *Gaviidae*	1
Grebes *Podicipedidae*	1
Tropic Birds *Phaethontidae*	3
Pelicans *Pelecanidae*	3
Boobies and Gannets *Sulidae*	4
Cormorants *Phalacrocoracidae*	4
Anhingas *Anhingidae*	4
Frigate Birds *Fregatidae*	4
Herons and Bitterns *Ardeidae*	5
Storks and Jabirus *Ciconiidae*	7
Ibises and Spoonbills *Threskiornithidae*	8
Swans, Geese, and Ducks *Anatidae*	8
American Vultures *Cathartidae*	18
Hawks, Old World Vultures, and Harriers *Accipitridae*	19
Ospreys *Pandionidae*	25
Caracaras and Falcons *Falconidae*	25
Grouse and Ptarmigan *Tetraonidae*	27
Quail *Phasianidae*	28
Turkeys *Meleagrididae*	30
Cranes *Gruidae*	30
Rails, Gallinules, and Coots *Rallidae*	30
Plovers *Charadriidae*	32
Snipe and Sandpipers *Scolopacidae*	33
Avocets and Stilts *Recurvirostridae*	36
Phalaropes *Phalaropodidae*	37
Jaegers and Skuas *Stercorariidae*	37
Gulls and Terns *Laridae*	37
Pigeons and Doves *Columbidae*	40
Parrots *Psittacidae*	44
Cuckoos, Roadrunners, and Anis *Cuculidae*	45
Barn Owls *Tytonidae*	46
Typical Owls *Strigidae*	46
Goatsuckers *Caprimulgidae*	55
Swifts *Apodidae*	58
Hummingbirds *Trochilidae*	59
Trogons *Trogonidae*	66
Kingfishers *Alcedinidae*	67
Woodpeckers *Picidae*	68
Cotingas *Cotingidae*	76
Tyrant Flycatchers *Tyrannidae*	77
Larks *Alaudidae*	93
Swallows *Hirundinidae*	95
Crows and Jays *Corvidae*	102
Titmice, Bushtits, and Verdins *Paridae*	109
Nuthatches *Sittidae*	113
Creepers *Certhiidae*	115
Ouzels *Cinclidae*	116
Wrens *Troglodytidae*	117
Mockingbirds and Thrashers *Mimidae*	120
Thrushes *Turdidae*	126
Old-World Warblers, Gnatcatchers, and Kinglets *Sylviidae*	133
Wagtails and Pipits *Motacillidae*	137
Waxwings *Bombycillidae*	138
Silky Flycatchers *Ptilogonatidae*	139
Shrikes *Laniidae*	140
Starlings *Sturnidae*	141
Vireos *Vireonidae*	142
Wood Warblers *Parulidae*	146
Weaver Finches *Ploceidae*	162
Meadowlarks, Blackbirds, and Orioles *Icteridae*	162
Tanagers *Thraupidae*	174
Grosbeaks, Finches, Sparrows, and Buntings *Fringillidae*	176

ABBREVIATIONS

of museums referred to in the text

AMNH	American Museum of Natural History, New York
ARP	Allan R. Phillips Collection, Mexico (collected by Phillips unless indicated otherwise)
ARIZ	University of Arizona, Tucson
ASU	Arizona State University, Tempe
CAS	California Academy of Sciences, San Francisco
CLM	Specimens formerly in Cleveland Museum
CM	Carnegie Museum, Pittsburgh
CU	Cornell University, Ithaca, N.Y.
F	Chicago Natural History Museum
GCN	Grand Canyon National Park, Arizona
GM	Gale Monson Collection (now at ARIZ)
JSW	Johnson-Simpson-Werner Collection (at ASU)
KANU	University of Kansas Museum of Natural History, Lawrence
LA	Dickey Collection, University of California, Los Angeles
LAM	Los Angeles County Museum
LLH	Lyndon L. Hargrave Collection, Globe, Ariz.
MCZ	Museum of Comparative Zoology at Harvard College, Cambridge, Mass.
MICH	University of Michigan Museum of Zoology, Ann Arbor
MIN	University of Minnesota, Minneapolis
MNA	Museum of Northern Arizona, Flagstaff
MVZ	Museum of Vertebrate Zoology, University of California, Berkeley
PH	Philadelphia Academy of Natural Sciences
PX	Phoenix College
RSC	Richard S. Crossin Collection, Tucson
SBM	Santa Barbara Museum (formerly)
SD	Natural History Museum, San Diego
SHL	Seymour H. Levy Collection, Tucson
SWAC	Southwest Archeological Center, Globe, Ariz.
SWRS	Southwest Research Station, Portal, Ariz.
US	United States National Museum, Washington, D.C. (includes Fish and Wildlife Service Collection)
UT	University of Utah, Salt Lake City
WF	Western Foundation of Vertebrate Zoology, Los Angeles
WGG	Wm. G. George Collection, New York City
WJS	William J. Sheffler Collection, Los Angeles (being transferred to Louisiana State University)

(See also Introduction page x column 1)

INTRODUCTION

OUR AIM is to tell exactly where and when each kind of Arizona bird can be found and to remark what is interesting about it in Arizona. We try to present ornithology as an engaging pursuit full of absorbing problems, not as a static discipline with everything settled by pompous dicta of the experts. Original information that cannot be found in other books is emphasized. On almost every page we seek to entice the attention of the amateur ornithologist toward the biological problems that birds so superbly illuminate, in hopes that he will be encouraged to contribute to their solution.

This book contains the original thinking of Allan Phillips. He is responsible for the scientific names and classification used. Largely from Phillips' manuscripts and files, Monson wrote the summarizing paragraph which opens each species account, and Marshall wrote the rest — much of it during conferences with Phillips — except for those families, genera, and species marked in the text as having been prepared by Phillips himself. (Two of these, the Bobwhite and Orange-crowned Warbler, were originally written in 1955 for his projected fuller account of Arizona ornithology.) This is Phillips' Birds of Arizona "as told to" Marshall and Monson. To it, Monson adds his field observations from southwestern Arizona, and Marshall his from the southeastern mountains and on night birds.

Every species of bird in Arizona has its own special requirements for making a living, and is unique in its distribution and status. Therefore we cannot present it by means of a standard formula. The distribution maps show this, for no two necessarily portray the same kind of information. Rather, each focuses attention upon a particular aspect of the biology of a species. Birds that occur all over the state are not mapped. A species that winters and migrates throughout Arizona, but which has a restricted breeding range, will have only the latter mapped. Therefore, do not expect the map to tell you everything about the bird. Subspecies symbols represent *specimens*. With these admonitions in mind, the reader should find the maps with their legends to be self-explanatory.

Though splendidly varied in topography, Arizona may conveniently be divided into a smaller northeastern part above the Mogollon Rim (including at its western edge the Grand Canyon region) and a larger southwestern part below the Rim. For convenience we refer to these as "northern Arizona" and "southern Arizona," respectively. A severe shortage of data from northern Arizona, however, occasionally forces us to cite migration dates from nearby areas rather than from northern Arizona proper.

Most Arizona land birds choose their habitat according to an instinctive awareness of the shape of the vegetation. A migrating Grace's Warbler, arriving here for the summer, will choose pine forest; but a Black-throated Gray Warbler will seek oak or juniper-piñon woodland. Though they feed on the same sorts of insects, the first is equipped by instinct to gather this food among pine needles, the second among oak or juniper leaves. A convenient way to state their resulting distributions is through the use of terms for Life Zones. These are an Arizona tradition, for it was here, on the San Francisco Peaks, that C. Hart Merriam first recognized and named them. A Life Zone is a band around a mountain of vegetation whose shape, or "life form," is visibly distinct to man and bird alike. Presumably, a given climate and soil will eventually confer the same life form of plants to various diverse spots on the earth's surface, even though the species of plants will be different. The Life Zones in Arizona are the hot Lower Sonoran Zone of desert vegetation; the foothill and mesa Upper Sonoran Zone of oak woodland or piñon-juniper woods, grassland with yucca, chaparral, or sagebrush in the north; the Transition Zone of ponderosa pine forest; the Canadian Zone of fir-aspen forest; the timberline Hudsonian Zone of spruce forest; and the Arctic-Alpine Zone of sedge meadows or rocks above timberline. The last three are sometimes lumped into the "boreal province," within which the Canadian and Hudsonian Zones may be termed "boreal forests."

All of the 424 species of birds which are represented from Arizona by museum specimens are given a designating serial number for Arizona — not the American Ornithologists' Union or A.O.U. number which is used for cataloguing collections of eggs. Unnumbered species are enclosed in brackets; their presumed occurrence in the state is not yet substantiated by a specimen. They are the "hypotheticals," some of which are represented by specimens erroneously attributed to Arizona.

An example of Phillips' shorthand for citing evidence for the occurrence of a bird at a particular time and place might be *"January 15, 1939* (near Flagstaff — Hargrave, MNA)." This would mean that on January 15, 1939, Lyndon L. Hargrave collected this bird near Flagstaff, and that it is a scientific specimen in the Museum of Northern Arizona. *All dates in italics, then, refer to museum specimens.* Dates in ordinary type are not substantiated by preserved specimens. Abbreviations for museums, as recommended in the "Index Internationalis Herbariorum," are listed on page *viii*.

Two additional abbreviations are "BNav" for the publication "Birds of the Navajo Country" (Woodbury and Russell, Bull. Univ. Utah 35, 1945: 1-157) and "FW files" for card files on distribution and migration, and collectors' notebooks, all at the Fish and Wildlife Service, U. S. National Museum, Washington, D.C., or at the Patuxent Research Refuge, Laurel, Md.

References to "Brown" are to Arizona's pioneer resident naturalist and Territorial Prison warden, Herbert Brown. The "Mormon Lake" we refer to is that one located southeast of Flagstaff, Coconino County. The names "Pinaleno" and "Graham" Mountains, and "Mount Graham," all refer to the same range in Graham County southwest of Safford.

The following subjective designations of abundance are used: *abundant* = in numbers; *common* = always to be seen, but not in large numbers; *fairly common* = very small numbers or not always seen; *uncommon* = seldom seen, but not a surprise; *rare* = always a surprise, but not out of normal range; *casual* = out of normal range; *accidental* = far from normal range and not to be expected again.

Subspecies (races) teach us more about migration than any other source of information. Since Phillips' study in Arizona has been slanted principally at migration, races receive much attention in this book. He and Hargrave at first banded wintering Dark-eyed Juncos at Flagstaff for years, noting a turnover of individuals during weather changes, and obtaining bountiful evidence that they revisit the same spot in successive winters. Nevertheless, banding, in the vast reaches of our state, has yielded practically no knowledge of actual migratory paths, since the chance of recovering a marked bird along such a path is infinitesimal. But races constitute whole *populations* which are "marked" by their peculiarities of color, size, and proportions. By carefully identifying a bird to race, we can tell from which general breeding area of the species it originated, just as surely as if it were banded. From your own experience with White-crowned Sparrows, think how rewarding your observations become through your ability easily to distinguish the two races at your garden feeding station. In fall and spring you find the black-lored Rocky Mountain subspecies; whereas your overwintering flock is of the Alaskan form, with its white lores and smaller yellowish bill. Although most subspecies are not identifiable in the field, nevertheless, they also provide this kind of information when trapped, netted, or preserved as museum specimens. Monson's valuable collection is small, but it contains a large proportion of new state records. Phillips' collection is perhaps the finest regional collection in existence. Each specimen has been taken either because it is in perfect new fall plumage or because it helps answer a particular scientific question.

In our references to size differences between races, the *wing* measurement is always of the *chord*.

Let us now examine the conclusions from all these accumulated data. Lying between major migratory "flyways" to the west and east, Arizona receives stragglers from both sides of the continent. These boost the roster of kinds of birds to an impressive total, exceeded by few states.

There have been dramatic changes in the native bird-life of Arizona during the present century. We have lost the Aplomado Falcon and the Magpie, but have unfortunately gained the English Sparrow and the Starling. We have witnessed an invasion of tropical birds from Mexico; these have become established first along the southeastern border of the state. They are in order of their appearance the Bronzed Cowbird, Black Vulture, Boat-tailed Grackle, Rose-throated Becard, Violet-crowned Hummingbird, Black-bellied Tree Duck, Preste-me-tu-cuchillo or Ridgway's Whip-poor-will, and Thick-billed Kingbird.

The Anna's Hummingbird, Say's Phoebe, Phainopepla, Red Crossbill, and Cassin's Sparrow are found to defy the ordinary laws of migratory behavior. Unexpected migrations are performed by the Flammulated Screech-Owl, Elf Owl, Scrub Jay, Bridled Titmouse, and White-breasted Nuthatch. Mountain birds wintering in the Arizona lowlands often prove not to have descended from nearby peaks; rather, they have come from far to the north or northwest. The earlier spring arrival of a species in the southwest than in the southeast of Arizona is explained by the discovery that the earlier birds are of races bound for the Pacific Coast, and that few of these migrate as far east as the San Pedro Valley.

Several eastern birds, which have invaded the west in Recent geologic time, are still accustomed to retrace that invasion path when they migrate. Thus in fall the Catbird, Veery, and Red-eyed Vireo fly far eastward before turning south, so that they do not normally cross southern Arizona at all.

In their spring migration, segments of some species "leap-frog" over each other, as do the Yellow Warblers. The southernmost breeding population arrives first and establishes itself on territories. Then migrants bound for more northerly stations pass through, and finally the northernmost races appear and are still migrating in early June. The Western Flycatcher, Western Tanager, and Common Grosbeak have such protracted periods of migration that the last north-bound individuals have scarcely passed in June before the first returning "fall" migrants appear in early July!

You will notice in many of our species accounts that the Arizona race is "larger, paler, and grayer" than are the populations of the same species elsewhere. Thus numerous kinds of birds respond alike, in their racial evolution, to the sunny Arizona (or Great Basin) environment.

"Phillips' Law" is a generalization about geographic variations in size, derived from Phillips' comparisons of Arizona subspecies with those over the rest of North America. From north to south there is commonly an increase in size of birds that live at increasing altitudes in the mountains southward; examples include the Goshawk, Western Flycatcher, Violet-green Swallow, Common Crow, Hermit Thrush, Ruby-crowned Kinglet, Water Pipit, Solitary and Warbling Vireos, Yellow-rumped Warbler, Red Crossbill, Pine Siskin, and Brown-eyed Junco. The opposite trend is shown in those birds that continue to inhabit the *lowlands* southwardly, where they become *smaller* in the desert. The Common Screech-Owl, Flicker, Cliff Swallow, Common Raven, and Song Sparrow are striking illustrations, plus of course several birds in eastern North America.

We are grateful to Mrs. Elsie R. Marshall and the Señoritas Ma. Cristina López Viadero, Ma. Antonieta Martín, and Rosalinda Villegas for typing this manuscript. We thank the many museum curators and owners of private collections for their innumerable courtesies and loans of specimens. The contributions by a large number of field observers, especially Mr. Lyndon L. Hargrave, are gratefully acknowledged. They are credited throughout the text, but we express here our deep indebtedness to Anders H. and Anne Anderson, James T. Bialac, the late Herbert Brandt, Mont A. Cazier, Harry L. and Ruth Crockett, Richard S. Crossin, Robert W. Dickerman, William George, the late M. French Gilman, the late Ludlow Griscom, Vic H. Housholder, E. C. Jacot, R. Roy Johnson, Albert J. and Louise Levine, Seymour H. and Jim Levy, Abe S. Margolin, Loye H. Miller, the late J. A. Munro, William Musgrove, Mary Jane Nichols, Ivan Peters, Warren M. Pulich, Mrs. Elizabeth Rigby, W. J. Sheffler, James M. Simpson, J. O. Stevenson, the late D. D. Stone, the late A. J. van Rossem, the late Charles T. Vorhies, James R. Werner, Milton A. Wetherill, and Lewis D. Yaeger.

We thank Dr. George Miksch Sutton for the twelve sketches painted during his trip to the Santa Rita Mountains in 1940. These he intentionally left unfinished, to preserve their charm and authenticity as on-the-spot field sketches. We are grateful to Mr. William J. Schaldach for painting our frontispiece; and last, but far from least, to Dr. Eliot Porter for his many fine action photographs in color.

In compiling his records and tables, Phillips had much help from Mrs. L. C. Daves, and William X. and the late Alma J. Foerster. Bibliographic assistance was received from A. H. Anderson, R. S. Palmer, and the late H. S. Swarth. Old manuscripts of J. G. Cooper, P. L. Jouy, and E. A. Mearns were examined through the courtesy of Dr. Herbert Friedmann, then of the United States National Museum; and of C. F. Batchelder and J. C. Cahoon through Dr. R. A. Paynter, Jr., of the Museum of Comparative Zoology at Harvard College. Notes of the late Messrs. Herbert Brown, W. L. Dawson, and Frank Stephens were made available by their widows.

Finally we must stress that, despite our own efforts and the truly prodigious amount of help received from so many generous and talented persons, the bird-life of large sections of Arizona — including many promising habitats — remains virtually *unexplored* today. Even in the best-known parts of the state, our records are undoubtedly incomplete. We can only hope, therefore, that this book will be helpful in exposing what is *not* known, and needs to be found out by those fortunate enough to study birds in Arizona.

Tucson, Arizona
February, 1964

Joe Marshall

Garden Canyon, Huachuca Mountains. (Fig. 1)

Chiricahua National Monument, Chiricahua Mountains. Within the United States, this area holds many birds found only in southeastern Arizona and southwestern New Mexico.

HISTORIC CHANGES AND CONSERVATION OF ARIZONA HABITATS

by Allan Phillips and Gale Monson

SINCE THIS IS THE FIRST full treatment of the bird life of Arizona, it is important to present as accurate a picture as possible of the *original* condition of Arizona as a home for birds. The ornithologist visiting the state today can get but a hazy picture of what Arizona was like when such pioneer naturalists as Woodhouse, Kennerly, Cooper, Coues, Bendire, Henshaw, Scott, Mearns, and Herbert Brown explored it. While a large book could be written on this subject, it is possible to present here only the highlights. This is best done by considering selected important habitats.

Grasslands

Originally, large parts of eastern and central Arizona were covered by extensive stands of perennial grass. This supported herds of Antelope, Prairie Dog towns, and, in season, large numbers of characteristic grassland birds such as Horned Larks, Baird's Sparrows, and Longspurs. Locally, the southeastern grasslands also had such birds as Aplomado Falcon, Masked Bobwhite, Scaled Quail, and Botteri's Sparrow. This abundant pasturage naturally attracted the attention of stockmen, who turned great numbers of cattle onto the open range in the 1870's and 1880's. It soon became evident that the grassland could not support indefinite numbers of livestock, and many died during dry years. The effect on the land and its bird inhabitants was profound. The most grass-dependent species, the Masked Bobwhite, disappeared almost immediately. Botteri's Sparrow survived only in isolated colonies in grass too tough for cows to chew. Horned Larks, on the other hand, probably increased at first, since they prefer low, open grass.

Grazing pressure continued to be abusive for many years. The grasses were weakened, and woody shrubs, which had been unable to compete with healthy and normal grasses, took root and spread out from their original precarious footholds along streams and in rocky places. Some claim that this invasion of shrubs came about by the prevention of range fires — a claim which we must frankly question. Similarly, the many cows turned loose in the mountains produced a thickening of the woody vegetation there, due to the destruction of the grasses and the consequent loss of soil and ground moisture. This resulted eventually in a considerable reduction in the open grass-loving birds of the southern mountains (such as Buff-breasted Flycatchers and Western Bluebirds) and the nearly complete extermination in some areas of the Mearns' Quail. On the other hand, it probably increased the numbers of Scrub Jays, Bush-tits, and Rufous-sided Towhees, though of course accurate numerical data are lacking. In a parallel manner on the plains below, the spread of mesquite enabled the Verdin and Lucy's Warbler to inhabit areas not previously available to them.

A further result of this grazing was the opening of deep channels and rocky gullies along the washes below the mountains. Originally, the healthy grasses were able to hold back and absorb most of the water. Figure 1 shows the bed of the wash draining Garden Canyon in the Huachuca Mountains in 1950. The Army officers at Fort Huachuca had for years prevented the overgrazing of this region. As a result, notice that the enormous drainage basin has absorbed all the summer downpours without damage to the soil; no gully has formed, nor was silt carried into the San Pedro River below. No other such example can be found today in Arizona!

The Lower Colorado Valley

Originally, the Colorado River got its name from the Spanish *colorado,* meaning "red." This color came from the erosion of mountains and badlands in the Rocky Mountain region. The river reached a high and muddy crest in late spring with the melting of those distant snows. Receding, it annually left large backwater

Such birds as the Broad-tailed Hummingbird, Williamson's Sapsucker, Red-breasted Nuthatch, Warbling Vireo, Yellow-rumped Warbler, and Brown-eyed Junco are common in summer in the fir-aspen forests of northern Arizona.

Saguaros and desert shrubs in the foothills of the Santa Catalina Mountains, home for White-winged Doves, Elf Owls, Verdins, Curve-billed Thrashers, and Black-throated Sparrows.

areas and silt flats, creating excellent habitat for many species of birds, especially waterbirds. Along its valley in many places grew fine woodlands of cottonwood with some willows and a light understory of mesquite and arrowweed. These cottonwoods thronged with Arizona Crested-Flycatchers, Bell's Vireos, Lucy's and Yellow Warblers, and Summer Tanagers. There were local colonies of Flickers, Traill's Flycatchers, and Bewick's Wrens.

The construction of Hoover Dam in the 1930's changed all this. The river became a steady, clear-flowing stream that no longer annually overflowed its banks to create lagoons and silt flats. The building of this and other dams produced large lakes of clear, open water that drowned much excellent bird habitat. Most of the surviving river-bottom habitat has been cleared, leveled, and converted to farmlands. On the other hand, for some years the drowned tree and brush tops at the head of some of the reservoirs provided satisfactory substitute nesting sites for large waterbirds. Perhaps nowhere else in Arizona have the changes been more dramatic.

Other River Valleys

Except for the Salt River, the history of other Arizona marsh and water habitats can be summarized in one sad word: destruction. Their desiccation was accomplished in many ways: ditching, pumping, diversion of water, destruction of woody vegetation, and above all by the incessant overgrazing of the entire drainage basins, permitting the rains to run off in short, disastrous flash floods that trenched the floors of the canyons, washes, and valleys. This of course caused long-stored waters to vanish down the river-beds. In some less-frequented areas, the marshes disappeared before any naturalist had visited them. Such places included Marsh Pass near Kayenta, and the extensive marshes of the Little Colorado River and the San Simon Valley. The bordering fringes of cottonwoods and willows (and, in the south, mesquites) have been decimated by cutting, clearing, and the lowering of the water table. Areas of once-abundant artesian water like the San Pedro and Sulphur Springs Valleys must now pump water for farming purposes from far underground.

Sahuarita Butte bottom-land (Fig. 2).

Particularly dramatic have been the changes along the valley of the Santa Cruz. This river originally flowed north to the San Xavier Indian Reservation, sank underground, and reappeared. It then flowed into Silver Lake, a pleasant cottonwood-shaded dam pond where persons from old Tucson could pass the time in boating and fishing. This was about at the present 29th Street in the southwestern part of Tucson. Another dam formed a small lake, Warner's Lake, just west of town. Its water was used to run a mill. The river then flowed northward another nine miles and sank into the desert. It was not connected with the Gila River on the surface. During the early severe overgrazing, and extreme drought of 1892, conditions deteriorated so badly as to produce a raging flood that cut through and destroyed the dam at Silver Lake. The river became a continuous channeled affair without permanent bodies of water or marshes marking its course. Above Tucson the Papagos annually constructed an earthen dam with which to irrigate their fields near San Xavier Mission. During the 1920's, this was replaced by a supposedly superior concrete dam, "Indian Dam," which promptly silted full. The Santa Cruz, however, continued to flow below the dam and was diverted for irrigation. This flow finally ceased about 1945.

Prior to World War II, the river at Sahuarita Butte (between Indian Dam and San Xavier Mission) was a paradise for birds. There were fine groves of cottonwoods, and, in the more open areas, thickets of batamote on the sandy bottoms back of the shallow channel itself. Along the irrigation ditch and elsewhere grew tree tobacco. Above these bottoms was a dense but not tall mesquite thicket, of a general level of fifteen feet or more, with much understory of vines and flowers in the summer rainy season. This abutted against Sahuarita Butte with its palo verdes and sahuaros (Figure 2).

No book on Arizona birds would be complete without a mention of the Tucson mesquite forest. This famous bosque of great mesquite trees, all of them forty feet or more high, occupied the river valley on the San Xavier Indian Reservation before it became channeled. The trees were cut down for firewood in the early decades of the Twentieth Century, and would probably have died in any case with the drastic lowering of the water table. Figure 3 shows a horizontal mesquite root abruptly turning down, searching for water. The oldest part of the root is now above the level of the ground. Figure 4 shows one of the few remaining tall mesquites, long-dead, still to be seen in 1948. Today in this area it is difficult to find even a single healthy cottonwood along the Santa Cruz.

Lakes

Most lakes in Arizona have their own peculiar individual histories. The largest natural body of water, at times, has been Mormon Lake southeast of Flagstaff. This was originally a moist meadow with a small sink in the middle. When cattle were introduced, this promptly silted shut, producing a large lake. Mearns,

Erosion in Tucson mesquite forest (Fig. 3).

A mesquite survivor (Fig. 4).

in 1887, stated that this was a most congenial home for American Bitterns. The implication is that there were extensive stands of tule or cattails. Presumably, the lake continued to grow. In 1933, it was a large body of open water, with a rocky east shore and rimmed by mud elsewhere. Low herbs and grasses surrounded it on most shores, but at the southwest part a few stands of cattails persisted. Shortly after World War II, the lake dried up and since then it has never provided good bird habitat. No bittern in its right mind would be found there today.

Today, our open water habitats are primarily manmade lakes, some of them very extensive. These can provide excellent habitat when water levels remain constant enough to permit the growth of vegetation. Often, however, they are drastically lowered seasonally, resulting in barren, unsightly shores, devoid of biological value. The smaller represos, or tanks, are usually maintained for the watering of cattle. They cannot be recommended to the visiting naturalist.

History is never a closed subject. The human population of Arizona continues to increase at an appalling rate. If the state is to remain an attraction to naturalists, it is well-nigh past time that action be taken to *preserve* some of its natural beauties. Let us hope that the data presented in this book do not represent the obituary of some of our most interesting birds. Rather, may these birds continue to find shelter and safety in a green Arizona!

Saguaro and desert scrub (cholla, ocotillo, bur-sage).

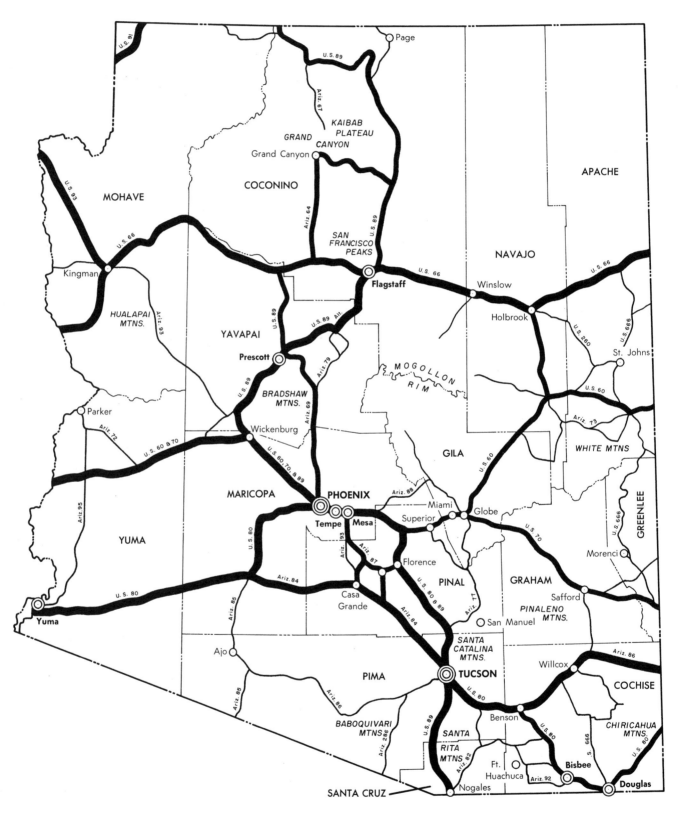

Map of Arizona State Highway System

THE
BIRDS
OF
ARIZONA

THE BIRDS OF ARIZONA

LOONS *GAVIIDAE*

Loons are large waterfowl which feed under water on fish. As known in Arizona, they are silent birds found singly upon a lake or where they have fallen on the desert during migration. They prefer open water where they stay well out from shore, and they dive with a forward leap from the surface. Loons, also known as divers, can swim long distances under water without surfacing. Heavy-bodied and short-necked, they are built like cormorants but are whitish beneath and have the bill pointed. Practically all Arizona examples are in dull winter plumage, including the strays that linger into summer. The three front toes of a loon are united by a web, the feet are placed far aft, and a tail of sorts is present. They cannot walk on nor take off from land.

1. COMMON LOON *Gavia immer* (Brünnich)

Rare transient throughout Arizona, except on Colorado River lakes, where it is sometimes common in April and has been found every month of the year, including one at Havasu Lake, *July 28, 1954* (Monson, MNA).

The Common Loon might retain some of the white spots from a previous plumage, and this would have led Herbert Brown to think that a mounted bird in the Arizona State Museum was a Red-throated Loon. The bill is larger in the Common Loon than in the other loons. The period when transients are seen away from the Colorado River is late October to November (*November 5, 1959* at Globe — Pat Kelley and Norman Messinger, SWAC) and April to May.

2. ARCTIC LOON; PACIFIC LOON *Gavia arctica* (Linnaeus)

Very rare fall and early winter visitant; five specimens, north and east to Camp Verde (Mearns, AMNH) and Tucson (H. Brown, ARIZ). Has been seen near Yuma April 19, 1941 (Lee W. Arnold) and on Havasu Lake in spring and August 13, 1954 (Monson).

In its dull fall plumage, this loon appears to be a smaller edition of the Common Loon, but a sight identification is practically impossible without the latter for comparison. The fall records, starting in August and September, are notably early as compared with the Common Loon.

[RED-THROATED LOON *Gavia stellata* (Pontoppidan)]

[Hypothetical. One near Fort Mohave, November 21, 1947 (Pulich) and one on Havasu Lake June 8 and 15, 1948 (Monson). We have been unable to verify a supposed specimen taken many years ago (Scott, Auk 3, 1886: 383); and sight records for the Papago Indian Reservation and from near Tuba City are unsubstantiated.]

GREBES *PODICIPEDIDAE*

Grebes, also popularly known as helldivers, have lobed toes and no tail. The only other birds in Arizona which have lobes fringing the toes instead of webs for propulsion through the water are the American Coot and phalaropes. The absence of a tail is unique and diagnostic, though hard to discern in the water. The feet are placed far back, and these birds cannot walk on land. Several species are found commonly in Arizona, diving in clear water for fish and possibly for plants. But the stomach of a grebe is usually full only of feathers — for reasons known only to grebes! Grebes dive from the surface with an upward leap, or they merely submerge gradually, to swim around with only a part of the head out. When alarmed they dive and stay down or out-of-sight for long periods. From ducks and loons they are distinguished by thinner necks and generally by shorter or thinner bills. Juvenile grebes are boldly striped.

3. HORNED GREBE *Podiceps auritus* (Linnaeus)

Rare transient along Colorado River; several sight records but only one specimen, lower Havasu Lake, *October 27, 1952* (GM). Not yet certainly identified elsewhere in state; at least one alleged specimen proves to be *P. caspicus,* and others have been lost.

Very similar to the more numerous Eared Grebe, the Horned Grebe is somewhat more purely white on cheeks and underparts and differs in the slightly deeper bill. The scarcity of records for the state may reflect only the extreme difficulty of distinguishing the Horned Grebe in its winter plumage, for even specimens in the hand have been confounded with the Eared Grebe by leading ornithologists. In spring plumage, however, the Horned Grebe is easily identified by its red neck and upward-directed ear-tufts.

4. EARED GREBE *Podiceps caspicus* (Hablizl)

Fairly common to common transient statewide. Winters in small numbers in the lower Colorado and Salt River valleys and occasionally elsewhere, and a few may sometimes be found in summer. Nests on lakes in San Francisco and White Mountains regions, and possibly in Verde Valley (Peck's Lake — Mearns, ms.).

The Eared Grebe may be distinguished at all times from the Pied-billed Grebe, the other common small grebe of Arizona, by its thin bill with straight edges. The iris is red. Nests are floating platforms of vegetation in colonies on mountain lakes. Only one brood per year of two to three, or rarely four, eggs is laid in late June by Arizona birds, though elsewhere clutches are larger.

In the Salt River Valley the migration periods are from mid-March to May and from September to early November. On the Colorado River this grebe occurs more or less numerously (flocks up to 2000) from late August or early September through early December. A few are present through winter, especially on Havasu Lake; in spring it is rather uncommon except for the first half of May. Generally one to five non-breeding birds may be found on either Imperial or Havasu Lake Refuges from mid-May to late August.

There was an unusually large migratory flight in the fall of 1956 of birds evidently headed for the Gulf of California, including a flock of 16 at Kinsley's (October 7, 1956 — Phillips, J. T. Bialac) and numerous examples at many of the southern Arizona tanks. Seymour Levy writes: "Our dog brought in a live Eared Grebe on *September 24, 1956* at our house on the east slope of the Tucson Mountains. We found several other carcasses in this area during the next few weeks. Also at that time there appeared articles and photographs in local newspapers of Eared Grebes picked up in and around Tucson. Apparently a large flight was struck with some catastrophe on the night of September 24th which caused the birds to alight on the desert." A similar but much later catastrophe to the north of Arizona occurred on a December 19th (Cottam, Condor 31, 1929: 80).

5. LEAST GREBE; MEXICAN GREBE *Podiceps dominicus* (Linnaeus)

A straggler in recent years along the south edge of Arizona; recorded from Quitovaquito in Organ Pipe Cactus National Monument, *April 28, 1939* (Huey, SD); near Tucson, December 28, 1941 and *September 26, 1943* (Wm. X. Foerster and Hargrave, ARP); and 10 miles north of Sasabe *January 14, 1958* (Levy, US). There is an old sight record for Camp Verde (Mearns, ms.). Nested at California end of Imperial Dam (West Pond) in *1946* (McMurry and Monson, Condor 49, 1947; 125-126).

The Least Grebe in immature plumage resembles the Eared Grebe but is darker on the cheeks and much smaller — about the size of a tail-less Robin. The adult is entirely dark, mostly grayish, with black around the bill and a yellow eye. It was first recorded for the state by Laurence M. Huey, who states (Trans. San Diego Soc. Nat. Hist. 9, 1942: 363): "A single specimen was collected on April 28 from the small pond at Quitobaquita and constitutes one of the few Arizona records for the species. When first seen this tiny swimmer ducked into the sheltering branches of a partly submerged willow tree, and at the same moment issued a rather loud alarm call."

A life history of this little grebe in Cuba is reported, with illustrations, by Alfred O. Gross (Auk 66, 1949: 42-52). The Arizona specimens have been identified as the small race *P. d. bangsi* (van Rossem and Hachisuka) of western Mexico.

6. WESTERN GREBE *Aechmophorus occidentalis* (Lawrence)

More or less common transient and winter resident, rare to uncommon in summer, on lakes of the Colorado River; very unusual migrant elsewhere, unrecorded on the high plateaux.

The Western Grebe is sometimes known as the Swan Grebe because of its long slender neck and white underside. At all times of the year the front part of the neck, throat and cheeks are white. The bill is long and thin; the crown and upper parts are blackish, strongly demarcated against the pure white underparts—in other words, a very striking bird. The iris is red. These grebes tend to occur in loose flocks in winter and they remain well out in the open water.

On the lower Colorado River, Monson has numerous summer records of non-breeding birds, including a flock of eight on Havasu Lake, July 16, 1954. The maximum seen by him in one day was 181 on March 27, 1954 at the same lake, where there were unusually large numbers of wintering birds from late December, 1953, through mid-April, 1954. The few records away from the Colorado River are from August to December.

The race in Arizona (according to Dickerman, Condor 65, 1963: 66-67) is the large northern *Ae. o. occidentalis* (Lawrence).

7. PIED-BILLED GREBE *Podilymbus podiceps* (Linnaeus)

Nests scatteringly in marshy lakes throughout state, casually even in late winter in the southeast, but commonly only in lower Colorado Valley and in high mountain country; winters in most ice-free waters.

The Pied-billed Grebe, also known as the Dabchick, has a short thick bill which abruptly tapers; this and the heavy head distinguish it from other grebes. In winter the color is similar to the Eared Grebe but not so white. The Eared Grebe is dark gray above and white beneath; whereas the Pied-billed Grebe is grayish brown above and on the neck, with only the belly white. Adults in summer have a black throat and a black ring around the bill. Their hollow cuckoo-like breeding call is heard along the lower Colorado River as early as mid-December.

The Dabchick is solitary, and it may be found both in deep water and along the shore. The nest, though resembling that of the Eared Grebe, is not in colonies and is usually placed more in concealment among tules.

A most amazing thing about this grebe was its nesting at Fort Huachuca, where normally it only winters. Here Brandt and Phillips saw large young with their parents by April 6-8, 1950! However, the potential nesting season is a long one, and small young have been seen in late March and May on the lower Colorado River (Monson), May in the Salt River Valley (Vic H. Housholder), to early September (not fully grown, apparently begging from an adult —Phillips and Louise Levine) at Love Lake near Shumway.

The few individuals seen during the summer in south-central Arizona may be belated migrants or may belong to a sparse non-breeding population. Northeast of the Mogollon Rim, where the bird does not winter, presumed migrants have been recorded from latter August to early December. There are no satisfactory spring arrival dates for that region. Migration in the south is during April and May, and again in latter August through October.

TROPIC-BIRDS
PHAETHONTIDAE

8. RED-BILLED TROPIC-BIRD *Phaëthon aethereus* Linnaeus

Accidental straggler; exhausted specimens have been found near Phoenix and in Apache Pass, near Dos Cabezas, Cochise County, *April 10, 1905* (AMNH) and *September 15, 1927* (ARIZ), respectively (Vorhies, Condor 36, 1934: 85-86).

They belong to the race *mesonauta* Peters, which breeds on islands from the Gulf of California to northern South America.

PELICANS *PELECANIDAE*

Pelicans belong to a group of water birds, the order Pelecaniformes, which on our Arizona list includes those species from the tropic-birds through the Anhinga and the frigates. These dive for their food of fishes, and they all possess a web which encloses the four toes within one continuous membrane. Also, all but the tropic-birds have a bare, distinctively colored pouch at the base of the mandible. The pelican family includes the following two species which are about the largest birds in Arizona; their pouch is huge, occupying the throat and the entire distance between the long mandibles.

9. WHITE PELICAN *Pelecanus erythrorhynchos* Gmelin

Regular transient, also wintering and summering in smaller numbers, along the lower Colorado River. Elsewhere occurs irregularly, subject to great fluctuations over the years but recently in small numbers or as singles. No verified breeding record.

The White Pelican is an enormous and very distinctive bird, all white except for black major wing feathers (the primaries and secondaries). On the water, with wings folded, it appears entirely white, with a very long bill. If it happens to be on a sand bar, the legs are seen to be short. In flight the great expanse of wings, not much tail, feet not trailing behind the tail, neck pulled in, and habit of soaring in flocks are diagnostic. The birds are completely silent and they hunt commonly by swimming in a line while scooping up fish from near the surface. They never dive from the air, and accordingly they do not have the wish-bone (clavicles) united to the sternum as do Brown Pelicans.

The probable sources of our White Pelicans are the breeding colonies at Great Salt Lake (cf. Behle, Condor 37, 1935: 24-35), Yellowstone Park, and Pyramid Lake, Nevada. For the latter, a long and rather tragic record has been kept, starting in 1924 (Marshall and Giles, Condor 55, 1953: 105-116). A bird banded at Yellowstone Park was recovered at Imperial Refuge near Yuma (Monson).

Until 1913, the White Pelican was found principally along the lower Colorado River, where it was common in September and April and wintered in the vicinity of Yuma. Coincident with an apparent decrease in the Yuma region, the birds appeared as first scarce (Willett, FW files) and then in the 1930's as common winter residents on the newly formed Roosevelt Lake. Also in the 1930's they were common at Picacho Reservoir, near Coolidge, where they are now rare. Elsewhere away from the Colorado River, this pelican is known only as a very uncommon transient, principally in April.

10. BROWN PELICAN *Pelecanus occidentalis* Linnaeus

A few enter the state in summer and fall, rarely staying until winter or even spring; most records are for the Colorado Valley, but stragglers reach most of the state, exceptionally even the northeast (Tolani Lakes, July 1937 – Paul Phillips *fide* Monson; Flagstaff, *August 3, 1954*–MNA).

In Arizona the Brown Pelican feeds like a White Pelican; but in its normal environment of surf, it catches fish by diving from the air. It is distinctly brown at any age, but there is no record of an adult in Arizona, and it goes without saying that the bird does not breed here.

Prior to 1930, there were only two records for Arizona, and these were for the southeast (Law, Condor 26, 1924: 153; Bruner, Condor 28, 1926: 232). Later appearances began on the Colorado River, extended over the rest of the state, and have continued to the present. The highlight is an invasion in June and July of 1936 (Vorhies and Phillips, Condor 39, 1937: 175).

Specimens from Arizona belong to the race *californicus* Ridgway, which nests on islands along the Pacific Coast from California to Nayarit and in the Gulf of California.

BOOBIES AND GANNETS
SULIDAE

11. BLUE-FOOTED BOOBY   *Sula nebouxii* Milne-Edwards

Rare late summer and fall straggler to lower Colorado Valley, at least in 1953 and 1954, north to Havasu Lake (*September 19, 1953*–GM); also found once at Phoenix, *July 29, 1953* (ARP).

12. BROWN BOOBY; BREWSTER'S BOOBY   *Sula leucogaster* (Boddaert)

Rare late summer and fall straggler to lower Colorado Valley, north to Havasu Lake (*September 5, 1953*–GM). One present continuously from early September 1958 to October 1960 at Martinez Lake about seven miles above Imperial Dam (Monson). Colorado River specimens have been identified as the race *brewsteri* Goss, which breeds in the Gulf of California.

CORMORANTS
PHALACROCORACIDAE

13. DOUBLE-CRESTED CORMORANT; FARALLON CORMORANT   *Phalacrocorax auritus* (Lesson)

Common transient and winter resident in lower Colorado Valley, where it breeds locally. Occurs in smaller numbers on Lakes Mohave and Mead, on larger lakes of the Salt River, and rarely elsewhere in the state, where chiefly a transient.

Cormorants are goose-like in form, especially when in flight. A migrating flock goes up a river valley in a perfect "V." Here the resemblance to geese ends, for cormorants are dark all over, and they swim low in open water, with the bill elevated above the horizontal. Young birds are light on the breast and chest. The two species in Arizona, differing mainly in size, are impossible to distinguish afield; but the Double-crested Cormorant is the only one which nests here. The Double-crested Cormorant is the larger of the two, and its gular pouch is plain yellow. The nests of sticks are in colonies and placed upon drowned trees and bushes.

The periods of migration seem to be March to April and late October to November; however, there are scattered occurrences away from the breeding areas in other months including mid-summer. This cormorant is a rare transient on the lakes and ponds of the Mogollon Plateau (Lake Mary, September 29, 1926 – Griscom; Lakeside, April 17, 1952–Phillips and Dickerman). About Tucson these birds were rather common from *March 29* to June 20, *1897* (Brown, ms.) though few have been seen since. The race occurring in Arizona should be *albociliatus* Ridgway, but the identification would have to be made from specimens with nuptial plumes, of which we have none.

14. OLIVACEOUS CORMORANT; MEXICAN CORMORANT; NEOTROPICAL CORMORANT   *Phalacrocorax olivaceus* (Humboldt)

Casual in the region of Nogales. Two specimens, *March 10, 1961* (Levy, ARIZ, US).

Now that this cormorant's presence in Arizona is attested by specimens, we can admit our suspicion that other sight records have pertained to it. The two species are really hard to tell apart, because they differ mainly in size. The Olivaceous Cormorant is the smaller, and has a relatively smaller bill. With a telescope, a shiny black adult in breeding plumage can be distinguished by the narrow white border of the lower edge of the yellow gular pouch. These cormorants sit in the sun on rocks or pilings, often with the wings held out, as all cormorants and Anhingas like to do.

Cormorants, mergansers, Ospreys, and Belted Kingfishers are detested and shot whenever seen by fish-and-game rangers and owners of ponds and lakes used for fishing. Thus it was that the two specimens, seen alive at Canoa Ranch by Seymour Levy and later by Marshall and his class from the University of Arizona, were found dead by Levy at Arivaca Junction (Kinsley's), a few miles away.

ANHINGAS   ANHINGIDAE

15. ANHINGA; WATER-TURKEY   *Anhinga anhinga* (Linnaeus)

Accidental; only one certain record, from Tucson, *September 12, 1893* (Brown, Auk 23, 1906: 217-218). Old sight records at Yuma, where it perhaps occurred about 1900, and at California end of Laguna Dam, February 9, 1913, and more recently, but none substantiated.

The swimming Anhinga is a ludicrous sight when its body is submerged and only its long snake-like head and neck show. The neck lurches fore and aft as if it were a lever for propulsion – as in the old-fashioned railway hand-car. Anhingas sit around higher in trees than cormorants do. They like to soar and to fly with the head held down.

The one shot at Tucson long ago and measured by Herbert Brown, was probably not saved. Its measurements indicate that the race was *leucogaster* (Vieillot), lending little support to van Rossem's separation (Annals and Mag. Nat. Hist. 4, 1939: 439) of western birds on the basis of allegedly smaller size.

FRIGATE-BIRDS   FREGATIDAE

FRIGATE-BIRD; MAN-O'-WAR BIRD   *Fregata sp.*

[Hypothetical. Seen at Havasu Lake August 13, 1954, by Monson and September 8, 1955, by C. R. Darling; at Tumacacori National Monument on June 18, 1953, by Sallie Pierce Van Valkenburgh; and at Picacho Reservoir by Seymour Levy on August 29, 1962. Both the latter were suspected of being *F. minor* because of their buffy coloration.]

HERONS AND BITTERNS
ARDEIDAE

This family and the next two belong to the order Ciconiiformes, composed of long-legged, long-necked long-billed wading birds. The toes also are long, with no webs. When they get well underway in flight, herons curl the neck and pull in the head; ibises do not. Herons are less noisy than cranes, with which they may easily be confused. The latter are scarcer and more local (see Gruidae). Ciconiiforms feed by standing extremely still in or by shallow water and spearing or probing for small animal life. An exception is the Reddish Egret and, at times, the Snowy Egret, which Phillips has seen rushing about "pretty fast," catching fish. Sometimes Great Blue Herons, Common Egrets, and ibises feed in alfalfa and other plowed fields. Except for the bitterns, most of these birds spend much time perched in trees. A. J. Meyerriecks has written "Comparative breeding behavior of four species of North American herons" (Publ. Nuttall Ornith. Club no. 2, 1960).

16. GREAT BLUE HERON; *Ardea herodias*
TREGANZA'S HERON   Linnaeus

Found at any season, but in small numbers, throughout the state wherever there is much open water; nesting in scattered small heronries and singly.

A large heron, the Great Blue stands three or four feet tall and is blue-grey. The crown of immatures and the side of the adult's head are blackish. *Herodias* is conspecific with the "Great White Heron" and perhaps with the Common Heron of Eurasia, *A. cinerea* Linnaeus. There is a deep-pitched, rough call. The Great Blue Heron is the most frequently observed large water bird in Arizona. A life history reference is "The Great Blue Heron: behavior at the nest" by W. P. and B. D. Cottrille (Univ. of Mich. Misc. Publ. no. 102, 1958; 1-15). Juvenile birds, like most young herons, have a proclivity for wandering northwards after leaving the nest.

A gloomy reminder of what happens to habitat for waterbirds is Phillips' observation near Tucson of a Great Blue Heron which was attracted to an old tire, a small part of which was sticking above the water in a small ditch. The heron came in, circled, lit, and speared the tire. Edouard Jacot never found fish in their stomachs in Arizona — just small turtles and mammals. The bulky nests of sticks are placed high in trees, often, or formerly, in small colonies. But the species is not strictly colonial, and probably the majority of Arizona birds now nest in isolated pairs, though such nests are hard to find to prove this. For instance, a colony of about 60 nests was found on the Gila River near Avondale on April 13, 1930 (Harry L. and Ruth Crockett). The trees are now gone and the heronry as well, and it is now believed that only scattered nests occur in the Salt River Valley.

In summer the Great Blue is fairly common at ponds and streams almost throughout Arizona, but is rather rare in the higher parts of the White Mountains region. Most of the birds seem to be non-breeders, as known nesting colonies are few. In winter it is usually much scarcer on and northeast of the Mogollon Plateau, but remained in the Transition Zone to November 27, 1936 (Black River-Phillips), and winters in low Transition and below, where open water permits it to, as at Springerville, McNary, Lakeside, and Show Low (Jacot, Phillips). While there is no proof of wintering farther north, it has been noted as late as November 29 and 30, 1941, at Kayenta (Wetherill, ms.), and December 7, 1936, near Tuba City (Monson).

Known breeding stations are at various points along the Colorado River up to the Boulder Dam Region (Grater), at Roosevelt Lake, and are scattered along the Gila River and its tributaries in southern and central Arizona, including Eagle Creek (Jacot, Hargrave). Nelson was told that in 1909 two pairs nested at Tuba City; and Ligon (FW files) reports it "nesting" near St. Johns, without details. In late August and early September, 1933, Phillips saw what appeared to be a family of five young at Mormon Lake, but as they were full-grown there is no certainty that they had been raised there. The same applies to "four young" at Stoneman Lake, June 15, 1931 (Crocketts). Absolute proof that the species breeds anywhere in northern Arizona, on or north of the Mogollon Plateau, is therefore lacking.

Great Blue Herons differ racially in how light or dark they are. The light race, *Ardea herodias treganzai* Court, breeds in the Great Basin, Arizona, and on into Mexico. An immature wanderer from the desert at Tucson, *June 22, 1953* (ARP) is so extremely light that it is identified as *A. h. sancti-lucae* Thayer and Bangs from southern Sonora and Baja California. This is quite probable, as immature herons do wander and move north. The darker race, *A. h. herodias* (including *hyperonca* Oberholser) is distributed over most of the continent from California to the East, but well north of Arizona. It winters in Arizona (Levy, Condor 63, 1961: 98; identified by Burleigh).

17. GREEN HERON   *Ardea virescens*
Linnaeus

Breeds uncommonly along streams, particularly in willow areas, south and west of the Mogollon Plateau; found sparingly in winter in same areas, except norththernmost parts, chiefly in recent years (since 1940).

The Green Heron, or Shitepoke, is one of the most frequently seen herons in Arizona — always singly. The only common dark small heron in Arizona, it appears more dark gray than green, with a rich brown neck in the adult. The neck of the immature is dark, streaked with white. The iris is golden. Their subdued colors fit well with the trees in which Green Herons perch very still, with the neck drawn in so as to appear stocky. The voice is a nasal *quaak*. Nests are in living trees above or near water. Green Herons commonly fish close to or among bushes along the shore. When they fly a short distance, as to another tree, they keep the neck stretched out and raised, and the crown feathers spread. The bird is tame and tends to flush close.

Prior to the 1940's the Green Heron was common in the breeding range outlined above, from late March or early April to September or early October. There were only a few observations in winter, such as that of Vorhies *et al.* at Tucson on December 24, 1929 (Condor 37, 1935: 244). Beginning in late 1942, winter records became commonplace: four or five birds at Picacho Reservoir, December, 1942 (H. C. Lockett, *fide* Hargrave); one near Tucson, December 22, 1942 (Foersters) and January 10, 1943 (E. C. Jacot, Bird Club); and many others in succeeding winters. But in the Big Sandy Valley this species has still not been seen later than September 21 (1917, Goldman), and in fact it seems to move out usually by mid-August (Phillips). Its winter range has not yet engulfed that region.

Geographic variation in the Green Heron involves the darkness of the chestnut neck and possibly also size. The lighter, somewhat larger race *anthonyi* (Mearns) breeds in Arizona and probably forms the bulk of the wintering (or now partly resident?) population. A lookout should be kept for the darkest winter specimens, which might prove to be from the east or south, belonging to the typical race, *virescens* (Linnaeus). Such a candidate is a specimen from Tucson, *January 27, 1945* (ARP), which however is not small, and the neck of which, being telescoped, gives an effect of being darker than it really is.

The Ardeidae seem overly split, and we prefer a retreat to some of the generic standards of Linnaeus and Ridgway.

18. COMMON EGRET; AMERICAN EGRET; GREAT EGRET
Ardea alba Linnaeus

Common resident, breeding in scattered colonies along lower Colorado River; fairly common winter resident at Picacho Reservoir. Elsewhere found in very small transient numbers, a few wintering south of Mogollon Plateau. Reports of its nesting on or north of Mogollon Plateau, or in central Arizona, are not based on valid evidence. May be found in central Arizona all summer, except perhaps end of June and first three weeks of July.

The Common Egret is a large heron almost the size of the Great Blue, standing about three feet tall, with a heavy yellow bill and blackish feet. In spring this heron develops the long aigrettes which were favorite women's wear in the 19th century. These are very delicate and beautiful plumes, which cost untold numbers of egrets their lives during the years of the millinery feather trade. The Common Egret breeds in colonies, but at other seasons is solitary. The stick nests in Arizona are built in drowned bushes out in the water (Monson, Auk 65, 1948: 603-604), or in leafy trees near the edge.

19. SNOWY EGRET; SNOWY HERON; BREWSTER'S EGRET
Ardea thula Molina

Found year-long in Colorado Valley, especially common as a transient and in recent years as a winter resident; has nested in the Topock Swamp and on California side of river at Taylor Lake, 17 miles above Imperial Dam. Elsewhere a frequent migrant, in most places in small flocks or singly. Occasional in central and southeastern Arizona all summer, except perhaps end of June and early July. Only one winter record east of lower Gila Valley and Picacho Reservoir: Benson, January 31, 1940 (Monson, Condor 44, 1942: 222).

The Snowy Egret is a delicate white heron, smaller and more graceful than the Common Egret. It is somewhat commoner in most parts of Arizona, especially as a transient, and differs further in a thin bill which is dark (when not coated with mud). The legs are blackish in adults but greenish in immatures, so that the distinctive yellow toes do not contrast sharply in young birds as they do in adults. The habits and nesting are similar to those of the Common Egret.

The race in Arizona is *brewsteri* (Thayer and Bangs), a rather unsatisfactory subspecies (A. M. Bailey, Auk 45, 1928: 430-440).

20. LITTLE BLUE HERON
Ardea caerulea Linnaeus

Casual; one immature wounded at Camp Verde, July 27, 1885 (Mearns, ms.), and an adult taken near Phoenix, *May 27 1958* (J. M. Simpson, JSW). A sight record for Lake Mead in Nevada is dubious. The closest breeding colony is near Guaymas, Sonora. We place this heron between the Snowy Egret and the Louisiana Heron, with both of which it has hybridized.

21. LOUISIANA HERON; TRICOLORED HERON
Ardea tricolor Müller

Very rare fall straggler in southwestern and central Arizona; has been found north to Camp Verde, *September 24, 1884* (Mearns, AMNH). Other records are: Phoenix, August 27, 1932 (immature captured — Housholder; sketch and photograph examined) and August 20 to September 4, 1956 (Simpson, Werner *et al.*); Picacho Reservoir, September 23, 1943 (Cater) and on August 29, 1946 (Vorhies). The report of one near St. Thomas, Nevada, November 21 to December 8, 1938 (Grater, Condor 41, 1939:121) is so unseasonal as to require specimen substantiation. The nearest breeding colony is at Guaymas, Sonora.

22. REDDISH EGRET
Dichromanassa rufescens (Gmelin)

Rare or casual in late summer: one specimen from near Camp Verde, *August 27, 1886* (Mearns, AMNH). Also, one seen repeatedly along Colorado River above Imperial Dam, September 1954 to March 1955; one on California side of Havasu Lake, *September 9, 1954* (MVZ); and one about 30 river miles above Imperial Dam, September 2, 1960 (all by Monson).

Mearns' specimen belongs to the typical race, *rufescens* (Parkes, Condor 52, 1950: 91). The nearest suspected breeding colony is at Tóbari Bay, southern Sonora. *Dichromanassa*, with its very long tarsus and manner of hunting by running around, seems to be a distinctive genus.

23. BLACK-CROWNED NIGHT HERON
Nycticorax nycticorax (Linnaeus)

Resident in Colorado Valley, where it breeds; frequently seen elsewhere as transient, occasionally in waterless situations, or as winter visitant in southern and central Arizona. Bred formerly on Salt River and Verde Rivers, but present status there uncertain; may breed near Springerville.

Although often seen singly, Black-crowned Night Herons nest in colonies and commonly migrate in flocks. During the day they roost in dense leafy trees or stands of cattail (*Typha*), whence they sally forth at dusk with loud *quahk* calls from which they receive their name, *Nycticorax,* meaning "raven of the night." The adult is a handsome gray heron, white below, with a relatively short bill and jet black crown and back. In flight, however, it gives a clear gray appearance. Immatures are streaky brown and white birds much resembling the more secretive American Bittern, but without dark neck stripes or wing-tips. At any considerable distance they appear uniform brownish. The Night Heron is a stocky bird. Arizona nesting colonies are usually in dead trees over water; like those of the Great Blue Heron, they are small.

The only definite breeding colonies known away from the Colorado River are those formerly at Roosevelt Lake and in the Verde Valley (Mearns, Auk 7, 1890: 51). Breeding at Springerville is indicated by Seymour Levy's having seen about 30 in reeds of Becker's Lake in latter June, 1957. Migration periods are March through May and August to October, complicated by midsummer records of transients as well as the normal winter pattern of occurence. There are several such summer records from southeastern Arizona, and one near Kayenta, *July 14, 1934* (Wetherill, MNA). An excellent life history study was made by Gross (Auk 40, 1923: 1-30 and 191-214).

24. YELLOW-CROWNED NIGHT HERON *Nycticorax violaceus* (Linnaeus)

Accidental: one immature found at Sullivan Lake, north of Prescott, *October 14, 1951* (ARP).

This bird was quietly perched on the foot of the dam in a rocky gorge. It attracted Phillips' attention by its peculiar misshapen bill — often the sign of a hybrid, but not borne out by its plumage. It is identified as the race *bancrofti* Huey which is apparently resident as far north as Tepopa Bay, central Sonora.

25. LEAST BITTERN *Ixobrychus exilis* (Gmelin)

Uncommon but found year-round, principally in late spring and summer, in lower Colorado Valley; rare in summer in marshy places of central Arizona, where it has bred. Elsewhere rare transient or wanderer and no winter records except at Picacho Reservoir, Pinal County, in 1942-43 (December to January, "several" — Housholder).

The Least Bittern is a miniature cinnamon edition of the American Bittern. The two have similar habits, but the Least is more silent. The male is black-backed; the female is dark chocolate on the back. A conspicuous buff patch shows on the wing in flight. This is the smallest of the herons, the most difficult to observe, and the least known.

The Least Bittern has been recorded as far east in Arizona as Tuba City (July 9, 1936 — Phillips) and San Bernardino Ranch (*May 22, 1948,* mentioned by H. Brandt — Hargrave, ARP). The former breeding spot in central Arizona was a marsh near Phoenix where Vic H. Housholder *et al.* saw six birds on June 20, 1943, and where a juvenile was taken *June 25, 1955* (Werner, JSW). The marsh has since dried up.

Additional records of occasional birds away from the lower Colorado River are from Tucson (Herbert Brown, ARIZ), Camp Verde (Mearns), Kingman (*fide* Musgrove), and St. George, Utah (*fide* Hardy), all in late April, May and September. Arizona specimens belong to the large western race *hesperis* Dickey and van Rossem.

26. AMERICAN BITTERN *Botaurus lentiginosus* (Rackett)

Uncommon from late August to May in cattail areas in lower Colorado Valley; in some years absent after January; rare elsewhere and chiefly a transient. Before 1915 nested at lakes on Mogollon Plateau; no winter records for that region or northward.

Mearns, in 1887, considered that this species "breeds commonly in suitable places throughout" the San Francisco Mountain region. At the end of May he found it "especially abundant at Mormon Lake, where it finds a most congenial home" in the tules which it requires. Since then, however, Mormon Lake has become unsuitable for bitterns. In the White Mountains region, a female with ovary "slightly enlarged" was taken *June 16, 1915,* at Big Lake (Goldman, Condor 28, 1926: 162). While it is thus likely that bitterns once nested at the lakes of the Mogollon Plateau, it is extremely doubtful that they do so any longer. American Bitterns are famous for their pumping noise — the "love song" — almost never given off the breeding ground. An inhabitant of the densest cattails, which it actually simulates through the blackish parallel stripes down its neck, this bittern stays hidden for all its activities, including nesting, in cattails above the water. These cattails are trampled into piles where the birds roost.

The American Bittern is chiefly known as a rare transient in Arizona, principally from April to mid-May and September to early October. Extreme fall records are July 31 (1933, near Tucson — A. H. Anderson) and August 23 (1950 at Ehrenberg, Colorado River — Pulich) to *November 12 (1935* at Springerville — Jacot, ARP). It has remained to *May 26 (1893* near Tucson — Price and Wilbur, formerly Stanford Univ. Mus.) and even June 7 (1930, one "pumping" near Prescott—Harry L. and Ruth Crockett).

STORKS AND JABIRUS
CICONIIDAE

27. WOOD IBIS *Mycteria americana* (Linnaeus)

Frequent summer visitant (non-breeding) along lower Colorado and Gila Rivers from late June to early October; uncommon elsewhere in southern part of state, and rare on and north of Mogollon Plateau (St. Johns,

August 29, 1934 — Stevenson, Condor 39, 1937: 87). No winter records since 1903 (when Herbert Brown saw flocks on February 2 and March 5 near Yuma), and no spring record since 1910 (Grinnell, Univ. Calif. Publ. Zool. 12, 1914: 116) earlier than May 15 (1955, seventy flying north over Tucson — Mr. and Mrs. James M. Gates).

The Wood Ibis is the only United States stork. Its body is all white and the wing-tips are black. The head, bare in the adult, is fuzzy brown in immatures. These ibises are gregarious; they like to soar in flocks, like big white vultures. The bill is heavy and curves downward slightly at the tip. In flight the long bill and neck stick out conspicuously in front, and the legs similarly project behind. Supposed young seen in Arizona were probably just egrets feeding in their company. It is likely that the Wood Ibis occurred throughout the year formerly on the lower Colorado River, though Herbert Brown's records cover only February to April.

IBISES AND SPOONBILLS
THRESKIORNITHIDAE

28. WHITE-FACED IBIS; *Plegadis chihi*
GLOSSY IBIS; (Vieillot)
WHITE-FACED GLOSSY IBIS

Fairly common migrant, perhaps less common in northeast; a few (rarely up to about 50) may be found throughout the summer, but no positive breeding record. Rare in winter in southern Arizona.

At any distance White-faced Ibises look all black, but if the sun hits them just right, the adults reflect bronze and green, young have the head and neck lightly streaked with paler, in a blurry pattern. The "white face" is inconspicuous. The bill is slender and regularly decurved. These gregarious birds are frequently quite tame. What they are doing in summer remains a mystery; odd birds show up at different points, but there is no reason to believe they are nesting. The migratory flock flies high and fast. Its members appear very long and slender and black, with the neck held at a downward angle and the long legs trailing.

The transient periods are from mid-April to the first of June and early or middle August to late September. In 1887 Mearns (Auk 7, 1890: 51) found "large flocks" at Mormon Lake, May 31, and thought that they were "probably on their breeding ground." But there are numerous June records of single non-breeding birds at scattered points: Joseph City and San Bernardino Ranch (Harry L. and Ruth Crockett in 1930 and 1934, respectively), Picacho Reservoir (common in 1952 — Phillips, L. K. Sowls, and D. D. Stone), and others found dead (ARP) to mention just a few. There are three winter records, all in December: the Colorado River (Pulich, ARP; Monson), and "Tombstone" in 1908 (Willard, whose "Tombstone" locality included the whole San Pedro Valley and Huachuca Mountains region). There is also a record of four at Mittry Lake, Yuma, November 11, 1942 (Monson, Foersters).

[**WHITE IBIS** *Eudocimus albus* (Linnaeus)]

[One reported as seen at Palo Verde, Colorado River, California, by Dr. John Hornung in "March, 1914" (Lincoln, Condor, 25, 1923: 181). The date was more likely in early February, when Hornung took the Beautiful Buntings (cf. Daggett, Condor, 16, 1914: 260). An adult seen at Martinez Lake above Imperial Dam on the Colorado River, April 4, 1962, was photographed the following day (Monson).]

[**SCARLET IBIS** *Eudocimus ruber* (Linnaeus)]

[Flock of seven or eight seen by Herbert Brown (Auk 16, 1899: 270) on the Rillito near Tucson, September 17, 1890. Unfortunately none was taken, but this sight record by an experienced collector seems to be one of the most satisfactory of the few North American records of this striking species. Brown had specimens of *Plegadis* in his collection at the time of the observation.]

29. ROSEATE SPOONBILL *Ajaia ajaja* (Linnaeus)

Casual in southern Arizona from Mittry Lake above Yuma (Monson, Condor 46, 1944: 19) east to near Phoenix; records extend from May to November, including one at Quitovaquito, Organ Pipe Cactus National Monument, May 30, 1951 (S. Gallizioli, L. Lawson). Specimen evidence of occurrence rests upon a photograph of a mounted bird shot near Beardsley, northwest of Phoenix, in the *summer* (*June* or *July*) of *1942* (fide W. M. Pulich and Mr. and Mrs. Crockett; see Monson, *op. cit.* 19-20).

The Roseate Spoonbill has a soft pink appearance. The bill's odd shape is not appreciated in life, for it appears to be straight. This spoonbill is a rather uncommon summer visitor to Volcano Lake, Baja California (see Bancroft, Condor 24, 1922: 98); and apparently it ranges up the Colorado River as far as Palo Verde, California, in July and August (Grinnell, Bryant, and Storer, *Game birds of California,* 1918, p. 263).

SWANS, GEESE, AND DUCKS
ANATIDAE

Anseriforms are known to everyone. Gregarious, they fly in a *V* (geese) or in rows. When not disturbed they sit in full view on the shore. We follow the outlines of Delacour and Mayr's classification (Wilson Bull. 57, 1945: 3-55) based not only on shape of bill and hind toe, but on biological and behavioral characters and the juvenal plumage as well. Important books on this group are:

J. C. Phillips, *A natural history of the ducks.* 4 vols. Houghton Mifflin, Boston, 1922–26.

Nagamichi Kuroda, *A bibliography of the duck tribe.* Tokyo, 1942 (with supplements in later numbers of Tori).

F. H. Kortright, *The ducks, geese and swans of North America*. Amer. Wildlife Inst., Washington, D.C., 1942.

H. A. Hochbaum, *Travels and traditions of waterfowl*. Univ. Minnesota Press, Minneapolis, 1955.

L. K. Sowls, *Prairie ducks*. Wildlife Mgmt. Inst., Washington, D.C., 1955.

J. Delacour, *The waterfowl of the world*, 4 vols. Country Life Ltd., London, 1954, 1956, 1959, 1964.

Female ducks sometimes steal broods of ducklings which may be of another species. Therefore the mere presence of ducklings following a duck is not proof of her having nested.

This family is conveniently divided into distinct sub-groups: the swans, geese, tree ducks, surface-feeding ducks (genus *Anas,* puddle ducks), perching ducks, diving ducks, mergansers, and stiff-tailed ducks. Further remarks which apply to all the members of such a group, consisting of identification or points of interest in Arizona, are inserted just before the Mallard, at the start of the diving ducks (genus *Aythya*), and at the start of the mergansers. The details of distribution in Arizona are rather monotonously similar in all these anatids, which is of interest in itself. Most species are found all year-round in the state. But this is not by any means equivalent to the status of resident. Each kind migrates south into Arizona in the fall, some in large numbers, to spend varying lengths of time before continuing into Mexico, or, in some species, to winter in the state with only part of the population going on to Mexico. Most species stay on the lakes of the Mogollon Plateau and northeastward into December or January, moving on south only when these lakes become iced over. Such northeastern birds perform therefore a migratory movement, though this involves a long delay en route. In mild winters they may of course remain the whole season in the northeast, and then be classed as winter residents. Then there is a migration northward in the spring. Finally, as we have seen in the grebes, non-breeding individuals remain here and there through the summer. Thus you can appreciate the difficulty of ascribing a definite span to the migration. This can be understood only by noting the bulk arrivals and departures, rather than by paying attention to the bewildering first and last dates. These complications are heightened by the ability of some geese, crippled in the hunting season, to remain and breed, and by the impossibility of separating some ducks in the field (female and eclipse Blue-winged and Cinnamon Teal, Greater and Lesser Scaup).

Digressions of peculiar interest and appropriateness to Arizona are added to the account of the White-fronted Goose (a taxonomist's nightmare), Mallard (evolution), and the Hooded Merganser (a lesson in museum curating). The first two are by Phillips, the last by Marshall.

SWANS, GEESE, AND TREE DUCKS *ANSERINAE* Subfamily

30. WHISTLING SWAN *Olor columbianus* (Ord)

Winters uncommonly in lower Colorado Valley; now rare elsewhere.

This swan is a magnificent great white bird with a long slender neck. It is considerably larger than the Snow Goose and has no black in the wings. The head and neck of the immature are washed with brown. The bill in native American swans is black; the Whistling Swan often has a yellow spot on the lores, not likely to be noticed. Swans are rare in Arizona and are completely protected by law, yet hunters frequently shoot them "by mistake." Feeding is on vegetable matter by tipping-up. The separation from the genus *Cygnus* is due to the trachea's entering the keel of the sternum, which thereby becomes hollowed and double-walled.

Originally, Whistling Swans were probably winter residents along the rivers of southern and western Arizona in small numbers; they are now regular only near Yuma. Emory (*Notes of a military reconn. . . .*, 1848, p. 92) camped on November 19, 1846, at some pools in the old bed of the Gila River northwest of Mohawk; here "all night the swan, brant, and geese were passing." At Fort Mohave, Cooper (*cf.* Baird, Brewer, and Ridgway, *Water birds N. Am.* 1, 1884: 431 and ms.) found a flock wintering in 1860–1861, but he could not determine the species. Herbert Brown (ms.) was told of two birds shot from a small flock near Tucson, winter of 1873; and swans were reported as "found sparingly every year, and some were killed" at Camp Verde prior to 1884 (*fide* Mearns, ms.). Although in none of these reports was the species critically determined, they doubtless all apply to the Whistling Swan, not the Trumpeter. It is highly doubtful that the Trumpeter (*Olor buccinator*) has ever occurred south of Oregon; it is a strictly northern species, and in particular Phillips doubts the 1931 record for New Mexico (Merrill, Auk. 1932: 460), which is unconfirmed. All specimens from Arizona have proved to be *columbianus*.

Since 1885, Whistling Swans have occurred in Arizona, away from Yuma, chiefly as uncommon and rather irregular late fall transients, with dates falling between early October and early January. There is only one spring record: desert tank half-way between Flagstaff and Winslow, May 13, 1934 (Hargrave, Condor 37, 1935: 178). Evidence of a tendency to remain through the winter in places other than the Yuma area is furnished by an individual seen at Canoa Ranch (near Continental) from January 1 to February 24, 1956 (Phillips *et al.*, Lanyon and class, respectively). Normal wintering span on the Colorado River near Yuma is from mid- or late November to February (Monson).

31. CANADA GOOSE; HONKER *Branta canadensis* (Linnaeus)

Winters more or less commonly in southwestern and central Arizona and along the Virgin River; uncommon migrant in northern Arizona.

The Honker is a big goose, dark above, light below in Arizona birds, with a black head and neck. A conspicuous white patch on the lower throat, extending up onto the rear of the cheeks, relieves this black. The call is a ringing *quarump*. These geese rest during the day out on open water or sand bars and go to land to graze mostly at night.

The wintering period along the lower Colorado River varies from mid-October or early November to late February or late March (Monson). A flock of six was seen near Topock on September 12, 1951 (Monson). In central Arizona at Gila Bend, Arlington, the Salt River Valley, and Roosevelt Lake these geese winter from November to April, but are rare after February (Johnson, Simpson and Werner). At Roosevelt Lake they were commoner in 1939-1940 than during the preceding 25 years (J. J. Lane, *fide* Hargrave); over 1000 were there on February 4, 1940 (Hargrave), and in recent years more than 2000 have occurred (Monson). The Honker remains on the lakes of the Mogollon Plateau until at least January (San Francisco Mountain region, January 1934 — *fide* Hargrave), and if it is unable to winter there, it certainly remains in the valleys immediately below, as Mearns secured specimens in the Verde Valley on a number of dates from *December 18* to *March 4* (of the race *moffitti*, AMNH), and saw them there up to April 10, 1887 (Mearns, ms.), and they occur regularly in parts of the lower Verde Valley at the present time. Records in northern Arizona are from September 11, 1947 (Fredonia — W. M. Pulich) to January. One of the extremely few spring records in the north is a flock of 45 honking as they flew over Flagstaff on March 10, 1955 (H. S. Colton, M. A. Wetherill).

This species is now far less abundant than it was originally. Emory (*Notes of a military reconn. . . .*, 1848: 78-92) saw "flights of geese" on the Gila River near or southeast of the mouth of Mineral Creek, east of Florence, November 7, 1846. North-northeast of Aztec, November 17, the Gila River bottoms were "alive with flocks of white brant . . . geese and ducks"; and northwest of Mohawk, November 19, "all night the swan, brant, and geese were passing."

Most of the foregoing applies to the race *moffitti* Aldrich. It is pale in color and lacks a white crescent at the base of the neck. The sides of the broad bill are parallel. It now nests in the Great Basin area, including the Bear River Refuge, Utah; it is the form most likely to be seen in Arizona.

A smaller race, of the same pale coloration, is *parvipes* (Cassin), which also winters in Arizona. It used to be known in the literature of the southwest as "Hutchins' Goose," and more recently as "Lesser Canada Goose." In Arizona it is much less common than *moffitti*; it breeds in interior Canada and eastern Alaska. Definite records include: Camp Verde, *October 24, 1887* and *January 14, 1886* (Mearns, AMNH); and Picacho Reservoir, *November 9, 1936* (Thornton Gambrell and Vorhies, ARIZ). Sight records of medium-sized Canada Geese extend from October 6, 1943 (Picacho Reservoir — Hargrave) to April 15, 1935 (Lyman Dam south of St. Johns — Housholder). This is in all probability the goose responsible for the reports of the Brant (*Branta bernicla hrota*) in the inland southwest, as Coues evidently concluded as long ago as 1866 (*op. cit.*, 1866: 98) when he omitted the latter — his own published record — from his list of Arizona birds.

The extraordinary racial variation in size, shape of bill, and color, together with extremely local and interdigitated breeding ranges shown by Honkers and Cackling Geese, make this group one of the most interesting among all birds. However, it is difficult to persuade hunters to let you skin their trophy, and scientific study-skins of the adult geese necessary for racial determination are scarce in collections of Arizona birds. Try to secure at least the head with one wing and the tail, and determine the sex. Adults are larger than immatures, and the immature male is larger than the adult female. The adult is distinguished (in order of importance) by its very broad, square contour feathers; by its complete tips of the rectrices (immatures broken out in a notch); and by the glossy black of the flight feathers. There are undoubtedly more kinds in Arizona than those we list here, including *Branta canadensis leucopareia* (Brandt), which has been taken in Sonora (US). It has a slender bill which is tapered, as viewed from above.

32. CACKLING GOOSE   *Branta hutchinsii* (Richardson)

Casual at Topock, Colorado River, *November 14, 1953* (GM). This goose has the same color pattern as the Honker, but is small and stubby-billed. Since there are small races of Honkers in Arizona, this form must be determined by measurements.

The wing of our only preserved specimen is 330 mm., which fits the race *minima* Ridgway, nesting along the coast of western Alaska. Another was taken at Topock on December 15, 1949, and photographed alongside a presumed *moffitti*, but could not be secured from the hunter; its wing measured 355, its bill 27 mm. (Monson). This photograph was submitted to Capt. Delacour, who writes that "There is no doubt at all" that the smaller bird is *minima*.

[BRANT OR BLACK BRANT?   *Branta* sp.]
[One examined by Jacot and Vorhies at Picacho Reservoir about November, 1941 or 1942. It was muddy and bedraggled and was not saved. Old reports (Coues, Ibis 1, 1865: 538) and recent ones from Lake Mead (Grater, Condor 41, 1939: 221) are unlikely and unsubstantiated.]

33. WHITE-FRONTED GOOSE   *Anser albifrons* (Scopoli)

Now rare to uncommon; formerly a common fall transient in the Colorado Valley and in smaller numbers elsewhere. Winter and even spring records are not numerous and are chiefly from central-southern Arizona.

The White-fronted Goose, also known as the Specklebelly, is similar to the Canada Goose but as its popular name implies, adults have black marks on the belly. The head and neck, however, are brown like the back. The white forehead, present only in adults, is unlikely to attract attention. Like the snow goose, this is sometimes seen on smaller bodies of water than would attract Canada Geese. The migrations of the White-front are unusually early for a goose.

Although Coues (Proc. Acad. Nat. Sci. Phila. 1866: 98) considered them "abundant" on the Colorado River in September, 1865, and some of the geese mentioned by early explorers may have been this species, White-fronted Geese are now rare in Arizona. There are nine specimens preserved; these and various sight records fall in the southwestern half of the state, except for Mormon Lake (September 16, 1940 — Phillips) and Parks Lake, San Simon Valley, southeastern Graham County (Vorhies and Taylor, Condor 37, 1935: 175). Fall transients occur in September and October. Winter records extend through to February along the lower Colorado River, consisting mainly of one to four birds (Monson); north to Topock, *January 4, 1948* (Pulich, ARP). "Spring" records are from late January to May! These include the arrival of a flock of 56 at Ferguson Lake on the California side above Imperial Dam, January 24, 1943 (Monson); seven at Martinez Lake above Imperial Dam, January 25, 1960 (Monson), and a flock of 100 near Gadsden on February 15, 1948 (*fide* W. M. Pulich, who saw 24 there on the next day). May records are near St. Thomas, Nevada on May 6, 1938 (Grater, Condor 41, 1939: 30); and one near Topock, May 10, 1947 (Monson).

The scientific name of a bird, even of a subspecies, is determined quite simply. The earliest name published with a description that fits the bird, will be used. Be amazed then at how weirdly complicated the application of this simple rule can become — in this case for the races of the circumpolar White-fronted Goose. They differ from each other in size, though not as profoundly as do races of the Canada Goose. Swarth and Bryant discovered a new large wintering goose in California (which came to be called the Tule Goose in the 1931 A.O.U. Check-list). They looked around for a name, and found that *gambelli* Hartlaub had been applied to a big goose. Its type locality was "Texas et du sud de l'Amerique du nord"; the type specimen was taken early in the 19th Century and it undoubtedly did not come from that part of the West, which had not yet been explored. The fact is that no large White-fronted Goose occurs anywhere near "Texas and of the south of the North America." Kuroda measured the types and found measurements so much smaller than Hartlaub's that he thought he had the wrong birds. However, W. E. Clyde Todd has worked out the breeding range and taxonomy in North America; he finds that the migration of the big California goose is exactly parallel to that of Ross' Goose: from central northern Canada to Sutter and Solano counties of central California. Thus it appears certain that Hartlaub could not have had this race, and that the birds examined by Kuroda in Berlin were the same ones described so inaccurately by Hartlaub. Therefore, the choice appears to be whether (1) to accept the name *gambelli* for these Texas birds, in which case the accurately but subsequently named *frontalis* Baird would fall as a synonym; or (2) to reject Hartlaub's name as fictitious, a case somewhat parallel to Audubon's "Washington Eagle" or giant "bird of Washington" (see Mengel, Wilson Bull. 65, 1953: 150). Phillips favors (2), the suppression of a name based on such a faulty description. In either case, the large goose wintering in California remains innominate.

The one in Arizona is smaller than that wintering in California, but a little larger than the European race, *albifrons;* it should continue to be called *frontalis*. The Tucson specimen identified by Swarth himself as the large Tule Goose (Vorhies, Condor 37, 1935: 244; ARIZ) proves upon reexamination to be the normal adult of the common race, *frontalis*.

34. SNOW GOOSE; *Anser caerulescens* BLUE GOOSE (Linnaeus)

Formerly common winter visitant to Colorado and Gila Valleys, now reduced in numbers; still winters on lower Colorado River, and in small numbers in Salt River Valley east of Roosevelt Lake. Elsewhere an uncommon, almost rare, and local fall migrant.

The Snow Goose occurs in Arizona almost invariably in the snow white plumage with black wings and tail. It is distinguished from all other similar white and black waterbirds by its short bill and short feet, besides being considerably smaller than a pelican. A distinctive stripe along the sides of the bill at its base is called the "grinning patch." There is a second plumage, the blue color phase, which some consider to be a separate species, the "Blue Goose." See, however, Cooch (Auk 78, 1961: 72-89).

This goose also appears to have been common originally, at least in southwestern Arizona in the fall. During Emory's trip down the Gila River, he found the river bottoms northeast of Aztec "alive with flights of white brant (wing tipped with black), geese, and ducks," on November 17, 1846 (*Notes of a military reconn. . . .*, 1848. 91). Coues (Proc. Acad. Nat. Sci. Phila. 1866: 98) called it "common" on the Colorado River in September 1865. The wintering period along the lower Colorado River is from late September to mid-March (Monson), and in the Salt River Valley from mid-October to mid-January (Simpson, Werner). There are only two winter records outside of this range: bottom of Grand Canyon (January 3 to 9, 1937 — Victor Veatch) and St. Johns (one shot January 11, 1951 — Housholder). Migrants have been seen in scattered localities over the state from early October (McKee) to November and in late February and March.

The foregoing account applies to the white Snow Goose phase. There are three valid Arizona records of the Blue Goose phase at Topock (January 20–31, 1950 — Crocketts, Monson; and another December 29, 1953 and during January 1954 — Monson); and Benson (*November 11, 1961* –LLH).

35. ROSS' GOOSE *Anser rossii* Cassin

Casual. Five records: one near Camp Verde (*October 24, 1887*—Mearns, AMNH); one near Topock (*December 10, 1948* — Pulich, LLH); one adult shot below Laguna Dam (about the end of *1959* — mounted specimen in Fish and Wildlife Service Museum at Yuma); four adults at Martinez Lake above Imperial Dam (November 5 to at least December 31, 1960 — photographed by Monson), and one adult at same place (February 5 to March 18, 1961 — Monson).

Ross' Goose is a miniature Snow Goose without the "grinning patch."

36. BLACK- *Dendrocygna autumnalis* BELLIED TREE DUCK (Linnaeus)

Of sporadic and rare occurrence in summer in southeastern and central Arizona; has nested near Tucson (pair with 16 newly-hatched young, September 13, 1949 — Phillips), Hereford, San Pedro Valley, and Nogales. Two families were wintering near Nogales in December, 1960 (Bill Harrison).

Tree Ducks are goose-like — with a long neck. They sometimes perch in trees. In contrast to the Fulvous Tree Duck, almost all the records of the Black-bellied

Tree Duck are from recent years. Its head and neck are plain grayish brown. In adults the belly is black and the bill is red. These birds are not yet numerous enough to attract attention by flying around in large flocks at night calling a piping *pichichi,* which is the common name in Mexico. No nest has yet been found in Arizona, but the distinctively colored ducklings have been seen, at irregular intervals.

The most recent such family was a pair with eight downy young, found by John Stair six miles northeast of Nogales on August 9, 1960. There are several occurrences of these ducks in June to early September from 1953 to the present at Phoenix (aviary escapes?), Hereford, Tucson, and Nogales. Before 1949, there were only two records: two females at a pond near Tucson, *June 14, 1942* (Mr. and Mrs. W. X. Foerster, Vorhies; ARP and ARIZ); and six killed out of flock of eight on the Santa Cruz River, two miles south of Tucson, on May 5, 1899, one of them examined by Brown (Auk 23, 1906: 218; ms.).

37. FULVOUS TREE DUCK — *Dendrocygna bicolor* (Vieillot)

Now rare from spring to early winter in southern and western Arizona; recent records are almost entirely of singles or pairs. Formerly a fairly common winter resident at Yuma. No breeding record. This species used to be the more numerous Tree Duck in Arizona, having been fairly common at Yuma (*1899-1901* — Herbert Brown, ARIZ) and occasional farther east. The long neck is brown with a patch of streaked dark and light corrugations behind the cheek. Otherwise the bird is dark above and with the pale brown extending over the belly.

One of the old records is a pair "about twenty miles from Fort Whipple," near Prescott, November 1864 (Coues, Proc. Acad. Nat. Sci. Phila. 1866: 98). A flock of six was seen in *May 1956* near Peoria (Simpson and Werner, Condor 60, 1958: 68). This duck has been found as far east in the state as Picacho Reservoir (*November 2, 1958* — SHL; and June 20, 1952 — Phillips, L. K. Sowls) and the Altar Valley (*June 4, 1892* — Wm. Reed and Herbert Brown, ARIZ). The largest flocks seen in recent years were both at Martinez Lake above Yuma: 15 on April 24, 1956 and 27 on November 16, 1961 (both photographed — Monson).

OTHER DUCKS AND MERGANSERS — *ANATINAE* Subfamily

PUDDLE DUCKS OR DABBLING DUCKS, INCLUDING TEAL — Genus *ANAS*

In this group hybridization is relatively frequent, resulting in some odd combinations of color traits. These ducks nest out on the shore, often far from water, and only the female visits the nest. Most species have prominent colors in the spread wing. They rise almost vertically from the water.

38. MALLARD, INCLUDING MEXICAN DUCK — *Anas platyrhynchos* Linnaeus

Common transient and winter resident wherever there is open water. Breeds on suitable ponds throughout Arizona (except the southeast), including high mountain lakes in the north, and locally in the lower Colorado Valley, and formerly near Phoenix. This applies to the typical Mallard, *Anas p. platyrhynchos.* A female-plumaged race, *Anas platyrhynchos novimexicana* Huber (*Anas diazi* Ridgway, the Mexican Duck of the A.O.U. Checklist) is found locally in summer, at least, in extreme southeast; definitely known to nest at San Bernardino Ranch east of Douglas, and in San Simon Cienega (New Mexican part). Straggler west to Salt River (one mounted bird, about *1943,* from Stewart Mountain Dam, examined at Buckhorn Hot Springs near Mesa).

This is the common domestic duck in the wild state. The wild male in spring and summer has the blackish-green of the head separated from the chestnut color of the breast by a narrow white collar. This coloration is found almost throughout Arizona. But in the extreme southeast the so-called "Mexican" or "New Mexican" duck is a slightly darker race in which the male's plumage differs little or not at all from the brown-streaked female.

In the Mallard and most ducks, *Ortstreue* (faithfulness to the place of first breeding — perhaps stronger than attachment to place of hatching) has to compete with faithfulness to the mate, in migration. Now in geese, we have unerring attachment to both, as well as mating for life, which results in inbreeding within a family flock or colony, and this has been a famous example of the formation of a lot of local races. More remarkably, the varieties persist even when the nesting habitat is destroyed and the colony or subspecies has to move (see maps in Delacour, *Waterfowl of the world* vol. 1, showing former and present breeding areas, especially in races of the Canada Goose). In marked contrast are the ducks, which do not stay together on the wintering grounds the way geese do, but form great mixed companies there, of diverse origins, within which the pair is formed each winter. In fact, the families break up before fall migration, the young migrating earlier than the adults, at least in the Mallard. Here again we have a conflict between family ties and the migratory instinct, and migration wins in the fall. This is even more pronounced in some birds like the Cliff and Barn Swallows and House Martin, where the parents, at migration time, simply fly off and leave their late broods in the nest! In spring the drake follows his winter-acquired mate to her breeding ground, resulting in almost world-wide mixing of stocks and we find world-wide species with no races or at most a weakly differentiated one in each hemisphere. Tree Ducks have virtually no races at all.

Doubtless the Mallard is actually intermediate in these respects between ducks and geese. In fall and winter it behaves as do the other ducks. But the fact that it nevertheless has well-marked races is highly suggestive that the mixed pairs formed on the wintering grounds do not migrate together and that *Ortstreue* takes precedence. Furthermore, the hen-feathered populations, except for the Black

Duck, are much more sedentary than is the sexually dimorphic and widespread common Mallard, and this of course contributes to the maintenance of the subspecies. Obviously a male Mallard who pairs with a female Mottled Duck (Florida Duck) does not stay there to breed with her. Thus, over the world, there are a couple of dozen more or less sedentary hen-feathered races and only two sexually dimorphic ones (*platyrhynchos*, migratory, holarctic in distribution except eastern North America; and *conboschas*, sedentary, of Greenland). To derive *platyrhynchos* as a single dominant mutation toward the much more attractive (to the females) drake type of Mallard, which swamped the Greenland population via Iceland, is much simpler and more logical a presumed course of evolution than to force dozens of other races each to make the same mutation away from male plumage. Furthermore, it is a generally accepted principle of evolution that the more advanced types are found on the larger land masses, particularly of the northern hemisphere, whereas the more primitive types remain in the islands of the Pacific. This works for Mallards if we regard the dull female-plumaged types as primitive. They occupy such remote spots as New Zealand, the Marianas Islands, and Laysan. And despite all the arguments for rapid evolution in tiny populations, surely the large ones on great continental areas have a better chance of coming up with favorable mutations.

In Arizona, spring migration of the Mallard lasts into May; fall migration is well underway in August. "Hundreds" were already present at Aguirre Lake, north of Sasabe, by August 25, 1923 (Goldman, FW files). The wintering period in the Salt River Valley (where the Mallard no longer breeds) is usually from mid-August to early or middle May, and of course this includes migrants; the peak in numbers there is from mid-November to the middle of February (Johnson, Simpson, Werner).

A common summer resident on the Mogollon Plateau, the Mallard also breeds farther northeast at Kayenta (Wetherill, Woodbury, Russell) and probably in the Lukachukai Mountains (*June 25, 1927*, Wheatfield Creek — Trapier, US). Former nesting in the Salt River Valley is indicated by a female with nine young one-third grown on the Gila River south of Goodyear, June 23, 1924 (Housholder); two nests at Papago Park in 1936 (Gus Evers); and reported nesting near Buckeye prior to 1948 (J. Field). Occasional birds, probably non-breeders, linger into the summer in southern Arizona (Anderson, ms.).

39. GADWALL *Anas strepera* Linnaeus

Fairly common transient, and winters along low river valleys. Nests on high mountain lakes, and locally and irregularly in very small numbers in the lower Colorado Valley.

The Gadwall, or Gray Duck, is a blotchy grayish-brown duck with white lower surface, grayer head, and a small white patch at the hind part of the secondaries, which is most visible in flight. Like all dabbling ducks it secures its food beneath the surface by tipping-up.

The Gadwall is now a rather uncommon summer resident on the lakes of the San Francisco Mountain region, where Mearns considered it common in 1887. Dates there extend from late April through October; whereas the wintering period in the Salt River Valley to the south is chiefly from mid-December through March. In some years this is one of the most abundant ducks wintering along the Colorado River.

[EUROPEAN WIDGEON *Anas penelope* Linnaeus]

[Hypothetical. A male seen at Topock, December 18, 1947 (Monson and Pulich).]

40. AMERICAN WIDGEON; BALDPATE *Anas americana* Gmelin

Fairly common migrant, wintering on open waters; non-breeding birds may be seen in summer. Prior to 1929 nested at least occasionally on lakes of the Mogollon Plateau.

The Widgeon is a brownish duck whose white underparts contrast with a cinnamon wash along the sides of the chest. Spring and summer males have the side of the neck a delicate grayish-green color, speckled with whitish, and the top of the head white, whence the oft-used name, "Baldpate."

Mearns (Auk 7, 1890: 50) shot "a number" in late May, 1887, at Mormon Lake, where they were "doubtless breeding." But they are now rare in the San Francisco Mountain region in summer, and the only recent breeding record is for Ashurst Lake, September 5, 1929 (F. M. Murphy). The winter range includes the Mogollon Plateau and the northeastern section of the state in mild years (a number killed south of St. Johns at Salada Slough, December 1956 to January 1957 — Housholder). In southern Arizona Baldpates are common from late September through April; flocks up to 200 are not rare.

41. PINTAIL; SPRIG *Anas acuta* Linnaeus

Abundant to common fall migrant, wintering on open waters; rather uncommon spring migrant. Nests on lakes of the San Francisco Mountain region. Non-breeding birds may be seen elsewhere in summer.

The slender neck of this duck is intermediate in length between that of the tree ducks and of the other puddle ducks. The female is very plain — speckly brown and white without prominent wing markings. The male has a rather pointed tail. In spring and summer he shows a handsome chocolate-brown head and neck, relieved by a white stripe down the side of the neck. This is the most abundant Arizona duck. It has increased greatly as a summering bird during the present century — probably during the past forty years.

The Pintail winters in mild seasons in parts of northeastern Arizona, as evidenced by observations near Kayenta (December 1941 — Wetherill, ms.), Joseph City and Ganado Lake, Apache County (February 7 and 10, 1938 — Monson). Fall migration is mainly in August and September; it reaches a high peak in the north in late August. In the south the birds usually arrive in August, are common by September, and remain to late March. Large numbers go to Mexico via the Colorado River Valley in fall.

[COMMON TEAL *Anas crecca* Linnaeus]

[Hypothetical. A male seen at Pichacho Reservoir, Pinal County, January 18, 1953 (Johnson, Margolin, Simpson). Possibly conspecific with *A. carolinensis*.]

42. GREEN-WINGED TEAL — *Anas carolinensis* Gmelin

Abundant to common transient, and winter resident on open waters. Nests rarely on lakes of Mogollon Plateau.

The Green-winged Teal is the smallest of Arizona ducks (if not of all ducks) and the one most commonly seen in early winter, particularly on small ponds and irrigation ditches. The female and eclipse plumages are speckly-brown with a patch on the rear of the folded wing of a green color so dark as to appear black. The male in spring and summer has a chocolate-brown head relieved by a green stripe through the eye, and there is a white crescent just forward of the gray sides. Perhaps these teal do not really fly faster than other ducks, but their small size makes them appear to. The Green-winged Teal is common in southern Arizona from mid-August to mid-April.

43. BLUE-WINGED TEAL — *Anas discors* Linnaeus

Spring transient throughout state, numbers and dates variable from year to year. Fall status uncertain, but recorded to at least *October 31 (1936* — ARP, Crescent Lake, White Mountains) in northern Arizona and November 17 along lower Colorado River; teal with blue wings are common fall migrants statewide. One casual winter record, pair above Imperial Dam, December 23, 1960 (G. Duncan, Gallizioli). Sometimes seen in summer, even in south, and quite possibly nests on Mogollon Plateau.

The Blue-winged Teal is at most times the rarest of the three teal in Arizona. The female and eclipse plumages cannot be safely distinguished in the field from the Cinnamon Teal; in fact, the determination of museum specimens is difficult. The best (and nearly the only) character in the hand is the slightly smaller bill of the Blue-winged Teal, which does not tend to broaden toward the tip. The male in spring has a dark gray head with a white crescent before the eye, in sharp contrast; and the body has no bright cinnamon hue.

At least in spring, therefore, something definite can be concluded about this species: it migrates north later than does *cyanoptera,* and in much smaller numbers, particularly in western Arizona. Yet recent studies (Hargrave, Levy) show a considerable flight in southeastern and central Arizona in *October,* when the Cinnamon Teal is apparently very rare or absent from Arizona (LLH, ARP, SHL)! The latest specimen is *December 14 (1961* — LLH).

44. CINNAMON TEAL — *Anas cyanoptera* Vieillot

Common migrant in spring, and probably in early fall, although status uncertain then. Small numbers may occasionally winter in extreme southwest (Monson). Breeds commonly in the lakes of the Mogollon Plateau. Non-breeding birds sometimes fairly numerous in Colorado Valley in summer.

The Cinnamon Teal is one of the commonest ducks in Arizona during the spring migration, and it is one of the handsomest. The male at that time has the entire head, neck and underparts a deep cinnamon-brown hue. The female is just another of those little speckly-brown ducks. Both sexes, like the Blue-winged Teal and the Shoveler, have a large area of light blue on the forepart of the wing. As in all other ducks, the male in late summer and throughout the fall wears an eclipse plumage similar to that of the female. All three teal are found along the margins of lakes and in small bodies of water. The Cinnamon Teal is one of the more numerous ducks breeding in northern Arizona.

The spring migration is very early; in southern Arizona it regularly commences in January and lasts into early May. Northern Arizona migration is from March through May. We lack October and November records, but males were taken *December 4* and *14* in the San Pedro Valley area (*1959, 1961* — LLH).

45. SHOVELER; SPOONBILL — *Anas clypeata* Linnaeus

Common transient, wintering on ice-free waters. Non-breeding birds or transients may be seen during summer. No positive nesting record.

The Spoonbill is so-named for the broad swollen front of the bill. In plumage, the female and eclipse male resemble Blue-winged and Cinnamon Teal. But the male in spring is a striking bird with dark green head and deep cinnamon breast set off by white belly and neck, which gives in flight the appearance of alternate black and white areas. Also in flight the broad bill gives the bird a front-heavy appearance.

Mearns (Auk 7, 1890: 51) found this species very abundant on the lakes of the Mogollon Plateau in the San Francisco Mountain region in May and (early?) June, 1887; but in recent years the only possible indication of breeding is a doubtfully identified nest at Big Lake, White Mountains (Goldman, Condor 28, 1926: 162). The large influx into southern Arizona begins by late August and this duck is common through the end of March, in flocks up to 150 (Salt River Valley — Simpson and Werner). Northern Arizona migration dates are March to May and mid-August to December; as with most of the ducks considered so far, numbers may remain on the lakes of the high mountains and northeastern section of the state until ice forms.

46. WOOD DUCK — *Aix sponsa* (Linnaeus)

Very rare, found almost throughout state, mostly in fall and early winter; one summer record, a male near Topock, June 7, 1946 (Monson). Southernmost records are at Tucson (including *January 17, 1953* — Dickerman, CU), and Arivaca (male, February 23, 1953 — Dickerman).

The Wood Duck cannot be described; its unbelievable colors require a painting, such as the one by Allan Brooks in the *National Geographic Book of Birds.* It has a tufted head and the female's eye is circled by a tear-drop light patch. In Arizona this species does not nest, and therefore it will not be seen flying around the trees looking for holes.

DIVING DUCKS

These ducks, on our list from the Canvasback to the Bufflehead, are chunkier and shorter-necked than puddle ducks. They generally stay in deeper water, where they dive for submerged vegetation and invertebrates. In flying from the water they patter along the surface. They lack brightly colored markings in the wings, which are either solidly dark or relieved only by white or gray. Predominantly northern in breeding distribution, they are not common in Arizona during summer; nevertheless, several species nest in the White Mountains and northward. Color patterns are much the same in several species, and these ducks are not easy to identify.

47. CANVASBACK *Aythya valisineria* (Wilson)

Rather uncommon migrant, and winters in some numbers locally on open waters; sometimes numerous in fall on mountain lakes. There are a few summer records.

Compared with the other species of *Aythya,* the profile of the head and bill of the Canvasback is linear, angular, and awkward-looking; the neck and bill are a little longer. The back is very white and both sexes are lighter-backed than the Redhead. The Canvasback is famous for the quality of its flesh when fed on eel-grass. It winters in considerable numbers on certain ponds south of Tucson, but it is very rare anywhere in Arizona in summer. An exceptionally fine book is Hochbaum's *The Canvasback on a prairie marsh* (1944, American Wildlife Institute, Washington, D. C.).

Excluding some midsummer records (Werner, Simpson), southern Arizona dates fall between late September and early May — chiefly from mid-October to mid-April. A straggler remained at Kinsley's Arivaca Junction in 1952 to May 30 (Phillips, Yaeger) and June 1 (Andersons). Northern Arizona migrations are in April to early May and October to early December, complicated by overwintering in favorable years (15 at Lakeside, January 21, 1936 — Jacot; Ganado Lake and near Holbrook in February, 1938 — Monson). The species was also seen on Mormon Lake, August 5, 1933 (Phillips). No young have yet been taken in Arizona.

48. REDHEAD *Aythya americana* (Eyton)

Fairly common transient. An occasional bird winters in southern Arizona, especially in lower Colorado Valley. Nests locally in the Colorado Valley (Monson) and on the Mogollon Plateau (Goldman, Condor 28, 1926: 162). Non-breeding birds may be seen during summer. Seymour Levy saw two females with four young each at Picacho Reservoir, August 13, 1958, and another brood the following year.

The head and bill of the Redhead make a profile full of graceful curves, which include the short neck. The head is deep reddish-brown in the male, while the female has no red tinge at all, is very plain, and lacks the white marks that distinguish the females of some other diving ducks. This and the Ruddy are the only diving ducks in southwest Arizona in summer, and are the only ones known to nest below the Mogollon Rim.

The period of fall migration in southern Arizona is late September into December. Spring numbers seem much smaller, and records span the months of January to May. It is not at all unusual to see singles, pairs, or small flocks throughout the winter. There has evidently been some confusion between the Redhead and the Canvasback in Arizona, unless the discrepancies between the notes of different early observers are due to fluctuations from year to year.

49. RING-NECKED DUCK *Aythya collaris* (Donovan)

Rather uncommon migrant, sometimes common in fall; locally common in winter on open waters. Fairly common as a nesting bird locally in the White Mountains. Non-breeding birds may be seen rarely elsewhere in summer.

The Ring-necked Duck, also known as the Ring-bill, from the pale bluish ring around the middle of the black bill, appears like a darker, more elegantly trimmed Scaup, but the male has a vertical white crescent at the front of its gray sides. The whitish area just behind the bill of the female (sometimes lacking) grades gradually into the dark head. In both sexes the wing stripe is gray, rather than the white of the scaups. The breeding colony in the White Mountains is isolated by a considerable distance from the next closest nesting birds, in central Colorado.

In the Salt River Valley, the Ring-necked Duck is one of the later ducks to arrive for the winter; it is there chiefly from mid-November through March (Simpson, Werner). Near its breeding range in the White Mountains it has been recorded from April 12 (1937, near McNary — Phillips) to *December 4, (1936,* Lakeside — Phillips and Correia, ARP, from a flock of about 20). Away from this region, dates range from September 3 (1953, near Shumway — Louise Levine and Phillips) and September 26 (1949, north of Prescott — Phillips) to May 12 (1949, at Kinsley's — Mr. and Mrs. Erle D. Morton) with casuals on June 5 (1960, California side of Imperial Refuge, two pairs — Monson) and June 30 (1959, Kinsley's — Phillips).

50. LESSER SCAUP; BLUEBILL *Aythya affinis* (Eyton)

Common to abundant migrant, especially in fall; winters occasionally wherever there is open water, even on the Mogollon Plateau lakes. Occasional non-breeding birds may be seen throughout the summer.

The male Lesser Scaup is a white-looking duck with a black head and neck. The female has a dark chocolate-brown head with a sharply defined white area at the base of the bill. This is probably the most numerous diving duck in Arizona at most times and places. There is a white stripe across the spread wing in flight.

The principal migration periods along the Colorado River are mid-October to late November and early March to mid-May; it winters there in considerable numbers where it finds food conditions good (Monson).

51. GREATER SCAUP   *Aythya marila* (Linnaeus)

Probably more common and widespread than the three verified records would indicate: one from near Clarkdale, *December 8, 1887* (Mearns, CLM); another near Arivaca, *December 3, 1961* (SHL); and remains found at Picacho Reservoir, *January 31, 1953* (Dickerman, ARP). Also one reported from Lower Colorado River (in Baja California), *February 7, 1928* (Lamb, *fide* Grinnell, U. C. Publ. Zool. 32, 1928: 76).

The Greater Scaup is not to be identified by sight alone in Arizona, because of the difficulty of detecting it among flocks of Lesser Scaup. It is doubtless taken sometimes by hunters.

52. SURF SCOTER   *Melanitta perspicillata* (Linnaeus)

Casual; two specimens — one from Hillside, west of Prescott, *October 20, 1929* (Jacot, ARIZ) and one from Havasu Lake, *October 23, 1949* (GM). Also fall sight records, one each from Topock and Havasu Lake, and two near Yuma (all by Monson); and one male, March 14, 1948 at Canoa Ranch, southern Pima County (A. H. Anderson).

This is a black sea-duck with small white head spots.

52a. WHITE-WINGED SCOTER   *Melanitta deglandi* (Bonaparte)

One, Ashurst Lake near Flagstaff, *November 16, 1963* (Charles Biller, SHL). Also, one below Supai, December 20, 1955 (E. Tad Nichols).

53. OLDSQUAW   *Clangula hyemalis* (Linnaeus)

Rare fall and winter visitant, recorded south to Tucson (late December, 1949 to *January 2, 1950* — Tucson Bird Club, ARP), Arlington (*February 14, 1953* — Wendell G. Swank and Wesley Fleming, ARIZ), Bill Williams Arm of Havasu Lake (*November 10, 1953* — Monson, US), and Martinez Lake above Yuma (November 7, 1959 — Monson).

The Tucson bird, collected by Phillips, was swimming by itself in open water of a small pond at the sewage plant.

54. COMMON GOLDENEYE; AMERICAN GOLDENEYE   *Bucephala clangula* (Linnaeus)

Rare to uncommon winter visitant on unfrozen lakes, chiefly of central and western Arizona; sometimes locally numerous along Colorado River (including more than 140 in Parker Dam vicinity, February 9, 1951 — Monson). Fall migrant on Mogollon Plateau and north. Individuals or pairs may remain until May.

The Common Goldeneye, or Whistler, is another flashy black and white duck, easily identified in the male by a round white spot between the eye and the bill. The female's entire head is deep brown, sharply delineated against the whitish neck; thus it resembles in color a female merganser, from which it is distinguished by the short chunky bill and neck.

Goldeneyes are hardy northern birds, here near their southern limits, and are accordingly the last to arrive of our waterfowl in the autumn; migrants on the lakes south of Flagstaff are found principally in *November* (and *December 2, 1934* — Murphy, MNA).

[BARROW'S GOLDENEYE   *Bucephala islandica* (Gmelin)]

[Hypothetical. An old specimen supposedly taken in Phoenix on *March 30* (year?), formerly in Breninger's collection (*fide* FW files). We have not been able to locate it for re-examination. Mr. E. R. Blake writes that it is not at Chicago Natural History Museum.]

55. BUFFLEHEAD   *Bucephala albeola* (Linnaeus)

Still a frequent transient; in winter fairly common in Colorado Valley but much scarcer eastward, especially in recent years. Casual near Phoenix, June 20, 1943 (Housholder).

The Bufflehead, or Butterball, is a teal-sized little duck of a handsome black and white pattern in the male; the black top and forepart of the head is relieved by white spots on the lower rear of the head. The browner female has a smaller white spot just below and behind the eye.

In the 1930's one or more Buffleheads were apt to be seen on almost any little pond. But in recent years their occurrence has been rare and irregular, chiefly as transients, except in certain favored spots. They have never been seen here in large numbers, as are Scaups and Canvasbacks.

MERGANSERS   Genus *MERGUS*

Mergansers and smews are fish-eating ducks provided with a slender saw-toothed bill for catching and holding fish. They have a moderately long and slender neck, and the general coloration at any distance appears black or brown and white, thus resembling a Goldeneye. In Arizona they nest on cliffs; elsewhere in cliffs or holes in trees. Two or three females may join together a great brood of young.

56. HOODED MERGANSER   *Mergus cucullatus* Linnaeus

Winter resident in very small numbers, principally in lower Colorado Valley above Parker Dam, from late October to early April. One summer record, if correctly labeled: Santa Cruz River near Calabasas, *June 1, 1890* (H. Brown, ARIZ).

The Hooded Merganser is the smallest and rarest merganser in Arizona. The male is similar in general color to the Bufflehead, but it has a flat instead of puffy crest. The female shares the crested outline of the male but has a plain grayish head.

Unfortunately, the entire Herbert Brown collection (now ARIZ), was re-labeled by Dr. Byron Cummings and Mrs. Wheeler at the Arizona State Museum. New, pretty, printed labels, of a paper that has since turned yellow and brittle, were substituted for the original labels of Brown's birds, which were collected mostly from 1884 to 1903. Written-out months, when occurring, were transformed into the intolerable numbered months. Also, many of the scientific study skins were exhibited in glass cases. The fading and irreparable loss of original field data which resulted are the worst things that can happen to scientific specimens outside of being eaten by insects. In the present case, Brown may have labeled his pair of mergansers "1–6–90," meaning January 6; simple copying in the European system of date-writing would thus change a winter date to the present summer one.

Mearns' "large numbers" of female Hooded Mergansers at Peck's Lake in the winter of 1887–1888 (ms.) refer of course to the Common Merganser. His only sight records and manuscript determinations of specimens that cannot be accepted are mergansers, sandpipers, and *Myiarchus* and *Empidonax* flycatchers. Otherwise the data of this greatest of early Arizona ornithologists are eminently reliable.

57. COMMON MERGANSER; AMERICAN MERGANSER *Mergus merganser* Linnaeus

Common winter resident on larger bodies of open water, uncommon elsewhere. Uncommon fall transient on lakes of Flagstaff area. Breeds in small numbers along streams below the Mogollon Plateau and White Mountains. Occasional individuals, probably cripples, may remain summer-long in the Colorado Valley.

The Common Merganser is the only merganser nesting in Arizona, where it uses cliffs. It is the species likely to be seen on running water. The male has a blackish-green head, smooth and uncrested. The entire posterior underparts are white. The female has a crested brown head which is sharply delineated against the white neck. Though fairly widespread, this is never a very numerous duck in Arizona, except on the Colorado and formerly the Salt and Verde Rivers.

This species appears to have reached a peak at Roosevelt Lake in the winters of 1915–1916 and 1916–1917, when it was common (Willett, FW files). Perhaps this corresponded with a peak in the population or availability of certain fishes there. Phillips' doctoral dissertation of 1946 treated this merganser as "a fairly common resident of the streams flowing south and west from the Mogollon Plateau in the Verde Valley and White Mountains regions." A recent nesting record from that area is fifteen young seen, about three days old, at the mouth of Sycamore Creek on the upper Verde River, May 2, 1954 (M. A. Wetherill).

58. RED-BREASTED MERGANSER *Mergus serrator* Linnaeus

Irregular transient in Colorado Valley, sometimes common; elsewhere decidedly uncommon in recent years. Occasional individuals may remain winter- or summer-long in the Colorado Valley. Lingering spring migrants include 8 at Topock, June 24, 1949, and 22 near Imperial Dam, June 1, 1958 (Monson). Wintered abundantly at one time on Roosevelt Lake, and a few formerly wintered at Lakeside, Navajo County, on the Mogollon Plateau.

The Red-breasted Merganser is very similar to the Common Merganser in appearance, but the spring male has a cinnamon chest band and a crested rear of the head. The female differs from the female Common (American) in having the brown of the neck paler at the lower end and more gradually merging into the whitish below. Exhausted individuals occasionally alight on the desert during their April migration. One such was very fat (ARP)! There sometimes are large flocks on the Colorado River during the April and October migrations.

This species, like the last, reached a peak at Roosevelt Lake in the winters of 1915–1916 and 1916–1917 (Willett, FW files), though it was evidently a little earlier in doing so, as "at least" 800 were seen December 24, 1914. In the two succeeding winters it was thought to be less common than the Common Merganser. Of the two species together, hundreds and at times thousands were seen in 1915–1916 and in November 1917.

Red-breasted Mergansers were seen at Springerville and St. Johns, November 11-19, 1935, and at Lakeside on January 21, 1936 (all by Edouard Jacot); and four were taken at Benson, *November 29* and *December 4, 1954* (E. E. Jans and Hargrave, LLH). These data point to a small late fall flight through eastern Arizona, besides the larger April one in the west.

STIFF-TAILS

59. RUDDY DUCK *Oxyura jamaicensis* (Gmelin)

Common transient in lower Colorado Valley, less common in eastern Arizona. Winters in small numbers almost wherever there is open water. Breeds on Mogollon Plateau, locally along Colorado River, rarely elsewhere (Salt River Valley – Simpson and Werner).

This, our only stiff-tailed duck, is one of the commonest and tamest ducks of Arizona. It is a dumpy short-necked bird with the tail cocked up in the air and unusually conspicuous for the tail of a duck. The dark crown is more or less (depending on the season and sex) contrasted with light cheeks. Summer males show the handsome ruddy color on the back. Ruddies often swim near shore when undisturbed, but on heavily hunted lakes they congregate and remain in the middle, without flying about as other ducks do. They dive for food. A nest at Mormon Lake was on a platform of cattails within the cattail growth, resembling that of a Coot. The Ruddy is the most widespread duck in Arizona in summer.

AMERICAN VULTURES
CATHARTIDAE

Vultures are nature's garbage department — avian Digby O'Dells. Extraordinary vision enables them to spy dead animals of any kind from great distances while soaring high in the air. They seem to be able to gorge themselves tremendously so as to live several days between autopsies. They are long-lived birds with small clutches and few young at a time, as is natural in a bird with few enemies, and depend upon fortuitous meals. Male, female and young alike share the funereal costume, but the downy chicks are white or whitish-brown. They rarely bathe, but sun themselves much, and their aroma protects them from most enemies. They gather in large roosts at night and are often in flocks during the day. In the early morning they sit around at or near the roost, sunning themselves until air currents become suitable. The nests, of which there are remarkably few discoveries, are in caves, hollow logs, ledges on cliffs, mine shafts, and rock heaps on the desert. These birds lack vocal chords and a voice, maintaining a silence befitting their profession. There are no talons as in hawks; despite their large size, therefore, vultures can carry nothing away in their feet.

60. TURKEY VULTURE *Cathartes aura* (Linnaeus)

Common summer resident, except in extensive forested areas where it is rather uncommon or locally rare. Winters in agricultural areas in Colorado Valley north to at least Topock, less commonly in Gila Bend and Phoenix areas, Sells and Sasabe. (A few records of stragglers at Tucson in November — Phillips, Marshall).

One of the most ubiquitous summer birds of Arizona, the Turkey Vulture appears all black, with paler gray undersides of the flight feathers. The wings are held partly bent in a characteristic dihedral above the back. The head of adults is red like that of a Turkey, but the immatures have a dusky head which has caused them to be mistaken for Black Vultures. Turkey Vultures soar effortlessly without moving the wings, as long as rising currents of warm air last. Stager (ms.) discovered that the Turkey Vulture can find food hidden under dense trees and bushes by scent, thus leading the Black Vultures to the carcass. He demonstrated that these two species differ in the development of the olfactory regions of the brain, by which he could infer a marked difference in their sense of smell, as proven by their behavior. Both of course have astonishing visual powers — probably the best developed of any land animal.

Hanna (Condor 63, 1961: 419) reports nesting at Picacho in late April and May. Copulation has been observed in February on the floor of the desert (J. Stair).

Migration takes place by day in spectacular flocks which pause to tower upward in a giant spiral. This makes it easy to define the periods of migration and of summer residency from hundreds of records. At Tucson a single bird or two may be seen early in March, but the big flocks arrive in the second or third week. The summer resident flocks at Tucson, with large roosts in cottonwoods at the San Xavier Indian Reservation and in eucalyptus at "Desert Treasures," remain until the middle of October; migrating flocks can be seen in southern Arizona from the middle of September through about October 20th. Northern Arizona records, northeast of the Mogollon Plateau, comprise a lesser span from late March to early October, but this differs much less than the usual discrepancy between northern and southern periods of most migratory birds in Arizona, since these large birds probably move rather far in a single day's travel.

The winter distribution is spotty, particularly along the international boundary with Sonora, where the species has been seen in late November and December north of Sasabe (1948 — Phillips *et al.*), and near Sells in February (1962 — Marshall), but not at Tumacacori, Nogales, or the Ajo Mountains.

Three races of the Turkey Vulture might pass through Arizona: *septentrionalis* Wied of the northeastern United States is large, with a wing chord of about 530 mm. or more; the northwestern *teter* Friedmann is smaller but with a tail presumably equally long (usually over 250 mm.); whereas the southern *aura* (Linnaeus) is small and has the tail usually less than 250 mm. Except for one *teter* at Yuma (*March 6, 1903* — ARIZ), only *aura* has so far been identified among Arizona specimens, even as far north as Snowflake (ARP). Specimens from the extreme north appear to be larger, but their tails are too worn for racial determination (MNA, GCN). Their wings measure from 480 to 510 mm. It would be well, therefore, to save the wing and tail of Turkey Vultures found dead along the highway, or at least to measure them, in order to determine the source of our migrants. Measurements of the two sexes are the same, but it is possible that the tail of immatures is longer than that of adults, as in the Red-tailed Hawk.

61. BLACK VULTURE *Coragyps atratus* (Bechstein)

Recently a resident of the Santa Cruz Valley above Picacho Peak. Probably resident on the Papago Indian Reservation west (irregularly?) to the Organ Pipe Cactus National Monument. Occasionally seen along Gila River north to Palo Verde (Werner *et al.*) and Gila River Indian Reservation areas (Housholder, Phillips, Foersters). Recorded casually from Wickenburg (approximately 50 on October 5, 1944 — A. J. van Rossem), north of Oracle Junction in Pinal County (Housholder), and extreme southeastern Arizona (Phillips). Supposed records for Tonto Basin, Verde Valley, and north are not considered authentic.

The Black Vulture throughout life has a black head with a thinner and longer bill than a Turkey Vulture's, as seen when perched close-up. In the air, the wings are shorter, broader, and more rounded; they are held out straight when soaring and are entirely black except for a white patch near the tip. The tail does not project far behind the wings. When there are few rising air currents, the Black Vulture flies with several rapid flaps and a glide, rather like an *Accipiter* but much less graceful and powerful. Although a former slaughterhouse on the outskirts of Tucson used to attract them, these vultures do not frequent cities the way they do in Mexico. Here they are distinctly local, as contrasted to being ubiquitous farther south.

The first Black Vultures seen in Arizona were two birds at Indian Oasis, south of Sells, June 8 and 10, 1920, by M. French Gilman (ms.). "At least a dozen" were seen 12 miles south of Tucson, May 7, 1922, by Kimball (Condor 25, 1923: 109). The next records were from Tucson, April 20 and November 28, 1928. Soon after this they became of regular occurrence. Taylor and Vorhies (Condor 35, 1933: 205-206) summarized the early records and implied that the birds appeared first as winter residents. If there was any seasonal movement, however, it favored a summer arrival. But by 1928 they were permanently resident. The first actual specimens were taken by Vorhies near Sells, *March 21, 1933* (ARIZ); and the first summer specimen, by Phillips near Tucson, *July 19, 1936* (MNA). The first large flocks (of more than 25 or 30 individuals) were seen at Sahuarita, December 14, 1933 (*fide* Taylor, FW files), and south of Tucson, January 26 and 27, 1934 (Vorhies, ms.). The records along the Gila River began in 1947, with flocks having been seen in November, December, February, and May from Chandler and Palo Verde.

The supposed occurrence of this species in the Tonto Basin ("several seen May, 1890, by Dr. A. K. Fisher" — Cooke, Auk 31, 1914: 403) is extremely doubtful. Dr. Fisher informed Phillips that he was not in Arizona before 1892. According to the FW files, whence came this information, the birds were seen by someone named Jones, whose record reached the files via Fisher's ms. notes. The record is so much at variance with the known distribution of this species at that early date that it must be rejected. So must those from the Grand Canyon (Behle and McKee, Auk 60, 1943: 278), which in turn leaned heavily on the Tonto report.

It is not likely that the Black Vulture performs any regular migrations; the larger concentrations in winter may be drawn to the most favorable feeding ground from a considerable area of residency. Nesting has not yet been observed in Arizona.

62. CALIFORNIA CONDOR *Gymnogyps californianus* (Shaw)

A few sight records, principally in the 1880's, from southeast to northwest (and southwestern Utah). Most of the few bird bones (not petrified) recently recovered from caves in the Grand Canyon are of this species.

The California Condor formerly occurred in Arizona, but possibly only as a winter visitor by the time settlement began. No specimen is extant, aside from the prehistoric bones. Further indication of its occurrence in Arizona is its breeding in Texas 1500 to 3000 years ago (Wetmore and Friedmann, Condor 35, 1933: 37-38). The latest date is apparently about 1924 near Williams (one seen by E. C. Jacot — ms.).

Other records are Colorado River at Fort Yuma, California, "individuals observed" September 1865 (Coues, Proc. Acad. Nat. Sci. Phila. 1866: 42); Pierce's Ferry, on the Colorado River, March 1881 (*fide* Herbert Brown, Auk 16, 1899: 272); Cave Creek, Chiricahua Mountains, March 7, 1881 (Frank Stephens, *fide* Brewster, Bull. Nuttall Orn. Club 8, 1883: 31); Santa Catalina Mountains about 1885 (*fide* Rhoads, Proc. Acad. Nat. Sci. Phila. 1892: 114); between Ash Creek and Bumblebee, one feeding on a dead horse, March 26, 1885 (Mearns, ms.); and Yuma County prior to 1935 (*fide* Harris, Condor 43, 1941: 43).

Of the above records, none is very doubtful except Coues' and possibly Harris' and Stephens', on which last Brewster unfortunately gives no details. Yet so careful an ornithologist as Brewster would hardly have mentioned it had it not been satisfactory.

HAWKS, FALCONS
Suborder *FALCONES*

Eagles, hawks, and falcons have powerful sharp claws for catching live prey, and a hooked beak for tearing it as it is held down with the feet. These birds occur in Arizona in a bewildering array of species and color phases, some of which are virtually impossible to identify in life. The best that can be done in some cases is to determine the genus to which the individual belongs. This is particularly true of those which are black with broad wings and short tail. Road kills should be investigated if a safe place can be found to park beside the highway; a wing and tail saved from the rare wintering species will be appreciated by Arizona museums.

All these hawks have white, downy young which develop slowly. The eggs are incubated from the first; therefore the young in a nest are of graduated sizes. The females of most species are bigger than the males, and immatures of many have a longer tail than does the adult.

HAWKS, OLD WORLD VULTURES AND HARRIERS   *ACCIPITRIDAE*

WHITE-TAILED KITES
ELANINAE Subfamily

[WHITE-TAILED KITE   *Elanus caeruleus* (Desfontaines)]

[Hypothetical. Phillips could find no specimen in the Pember Museum at Granville, New York, to substantiate the occurrence at Gila Bend. There are some recent sight records.]

BIRD HAWKS *ACCIPITRINAE*
Subfamily

These short-winged, long-tailed hawks hide in dense trees, then surprise and chase their prey through the foliage in a short, rapid dash — often of amazing dexterity.

63. GOSHAWK; AMERICAN GOSHAWK   *Accipiter gentilis* (Linnaeus)

Rare to uncommon resident of the mountains of eastern and central Arizona and the Kaibab Plateau. In winter it occurs rarely well away from the pine forests and irregularly so on the desert, even reaching the Colorado Valley during major flights (Palo Verde, California, *1916* — Wiley, *fide* Grinnell, Condor 19, 1917: 70).

The Goshawk is the largest and most powerful of the bird-hunting hawks. Its use in falconry in the middle ages was therefore restricted to the nobility, whence the name *gentilis*. Due to its preference for high mountains and its relatively non-migratory nature, it is the least seen of our three accipiters. It is also the only one which is ever wholly gray (the adult). The immature differs from immature *cooperi* in its white superciliary stripe, and larger size in the female. The male Goshawk is about the same size as the female Cooper's. Band-tailed Pigeons are an item in the food.

Goshawks of northern and central Arizona belong to the small, pale race *atricapillus* (Wilson). A notable invasion was that of winter 1916–1917, when they evidently occurred in all parts of the state, including the Chiricahua Mountains (common, specimen preserved by Austin Smith, AMNH). In the winter of 1942–1943 there was another lowland invasion. At Tucson alone, Goshawks were seen on eight occasions from November 7 through March 17 (Harold Webster, Hargrave, and the Foersters). In no other winters have more than one or two individuals been seen in the lowlands.

Accipiter gentilis apache van Rossem is the resident race of the mountains of southeastern Arizona, where it is rare. These birds are large, with heavy tarsi, and the adults are dark, almost blackish on the back. In addition to the specimens from the Chiricahua Mountains mentioned in the original description (Proc. Biol. Soc. Wash. 51, 1938: 99), the following which we have examined belong to *apache*: Santa Catalina Mountains, *February 12, 1953* (R. W. Dickerman, CU, adult female); Huachuca Mountains, *February 24, 1923* and *November 15, 1932* (Jacot, ARIZ, females, adult and immature respectively), and an immature *September 21, 1949* (ARP). Though hard to find because of its retiring habits, this Goshawk has been seen rather regularly in summer in the Santa Catalinas (Marshall and class); in the Santa Ritas (van Rossem and Loye Miller); in the Huachucas (adults at nest, 1951, Charles Wallmo); and in the Chiricahuas (nest June 16 to July 26, 1956 — observed by J. W. Hardy).

64. SHARP-SHINNED HAWK *Accipiter striatus* Vieillot

Uncommon summer resident of the mountains of eastern Arizona. Winters commonly in southern and western Arizona, uncommonly on Mogollon Plateau and farther north where it is mainly a transient.

The male Sharp-shinned Hawk is about the size of a Mourning Dove, whereas the larger female appears equal to the male *cooperii*. Sharp-shins differ from the latter in that the tail is square, not rounded, at the tip. If seen from above, the crown of the adult is slaty blue, concolor with the back. Immatures usually have the ventral streaks less blackish than those of *cooperii*.

The food consists of small birds, usually of sparrow size or less, which the Sharp-shinned Hawk catches by amazing bursts of speed and twisting flight through the trees. On the north slope of the White Mountains in winter, one Sharp-shinned Hawk was feeding upon the body of another, still warm (Ligon, FW files)!

The wintering period, during which the Sharp-shinned Hawk is common in the main valleys of southern Arizona, is chiefly from October through April, although migrants can be seen in September (Salome, *September 7, 1941* — WJS) and in May (north end of Hualapai Mountains, where it does not breed, *May 3, 1938* — Huey, SD). Mearns took specimens at Camp Verde (where the "general migration begins in August" — ms.) from *September 13, 1887* to *May 1, 1884* (*fide* Fisher, U. S. Dept. Agric., Div Orn. and Mam, Bull. 3, 1893: 37); these dates are probably fairly representative of its stay in the Lower Sonoran Zone, though not extreme. As in all the hawks which are hard to identify, it is futile to attempt a more accurate estimation of the migration periods on the basis of sight records.

Though common in summer in the Pinal Mountains (1951 — Marshall), this hawk is extremely rare in the high mountains to the south, with only three nesting records (Santa Rita Mountains — S. Levy; Huachuca Mountains, nests, Willard and Bent, and two specimens — Breninger, F.; and Brandt, Univ. Cincinnati). These tend toward the northern race, *velox* (Wilson), to which the entire foregoing account applies.

Immediately across the boundary, a relatively unbarred race, *suttoni* van Rossem, breeds in the forested San Luis Mountains of the Chihuahua-Sonora boundary (Mearns, US) and the Ajos Mountains of Sonora (Marshall, WJS). Unless specimens of it turn up from the southern Arizona mountains or Animas Mountains of New Mexico, it should be taken off the A.O.U. Check-list, because the San Luis Mountains do not afford suitable habitat for it at the border of New Mexico, where the mountain is low and supports no forest.

65. COOPER'S HAWK *Accipiter cooperii* (Bonaparte)

Nests scatteringly throughout state, but rare in the west. Winters commonly in southern and western Arizona except in higher parts of mountains. Transient throughout.

This is a medium or crow-sized accipiter, a larger edition of the Sharp-shinned Hawk. The tip of the tail is definitely rounded. The adult has a blackish cap and reddish underparts barred with white, while the young are brown with streaked underparts. Although this hawk is more widely distributed than the Sharp-shinned in summer, it is never common. It feeds mostly on Robin-to Flicker-sized birds and is not above snatching chicks in the barnyard. The nest is in tall sycamores, alders, or cottonwoods; females have been observed incubating in late April and early May, in some years before the leaves are on the nest trees (Marshall).

Northern Arizona records are from March 9 and April 18 (1936, near Flagstaff — Phillips) to October 15 and 16 (1931, at Oraibi and Shungopovi — Hargrave). Cooper's Hawk is rather common everwhere during the migration in September. It is a fairly common summer resident in tall trees from the Lower Sonoran to the Transition Zone. Apparently it breeds locally as high as aspens of the Canadian Zone (8800 feet, west side of Escudilla Peak, White Mountains, *June 12, 1934* — Stevenson, MVZ). This should at least dispatch the temptation to identify accipiters by altitude!

HAWKS AND EAGLES

BUTEONINAE
Subfamily

These are the soaring hawks with long, broad wings and short, fanned-out tail. They stoop from high in the air to grasp their prey, usually small mammals and snakes, at the ground. All the buteos scream or whistle near the nest; their voices are distinguishable as to species, with practice. A completely black phase occurs in several species.

66. RED-TAILED HAWK — *Buteo jamaicensis* (Gmelin)

Common resident almost statewide. Winters to upper part of Transition Zone and possibly higher.

In Arizona, any single or paired broad-winged hawk is likely to be this species — it does not gather into flocks as does *swainsoni*. It occurs in a wide variety of plumages, many of which do not have a red tail (for detailed descriptions see Friedmann, Bull. U. S. Nat. Mus. 50, pt. XI, 1950: 237-269). The color of the tail is partly a matter of age and partly a matter of race. In any plumage except the most melanistic you see a contrast between the dark tips of the primaries and their light bases, also between the dark sides of the head and a light chest; the belly may be dark also. The white sides of the scapulars show in immatures, giving them some light mottling on the back. Even the experts see hawks they cannot identify, so don't worry. The greatest failing of beginners is to insist upon making the immature Red-tail into some other species; they cannot cope with its cross-banded, brownish-gray tail, which is longer than that of the adult. Later they become familiar with the patterns on the face and underwing, which are common to all the ages, phases, and races of the Red-tailed Hawk.

The Red-tailed is unsuspicious for a hawk, and many are shot by the ignorant, who suppose they are all killers of chickens. Common nest sites are crotches of tall trees or saguaros, although in some extreme desert areas in southeastern California the birds are reduced to nesting in the base of low shrubs like the ocotillo. Young hatch around April 9–12 on deserts in the Salt River Valley region (Mr. and Mrs. Crockett).

As you can infer from the status of "common resident" and "winters," mentioned in the first paragraph, more Red-tailed Hawks are here in winter than in summer. Many of these wintering birds are of the melanistic phase, which is rare in the resident population. Around the farms of the San Xavier Indian Reservation, as many as three have been seen at a time (Marshall). Individuals post themselves day after day in the same places, and they give us an idea of the migration periods of some northern population of which they are a color phase. Outside dates at the San Xavier Reservation (1954 through 1962) are October 12, 1958 to March 26, 1958 and April 13, 1957 (all by Marshall).

The heavily marked western race *calurus* (Cassin) winters throughout Arizona and is also the resident form over most of Arizona. Most of the wintering black individuals also belong to it. Towards the southeastern corner of the state, it can be seen that more and more of the resident Red-tails are pale and lack any marks on the whitish underparts. They approach the race *fuertesi* Sutton and Van Tyne, specimens of which have been identified by Phillips within a winter series from the Chiricahua Mountains region (AMNH).

Kriderii Hoopes, a northern Plains race (if not an individual variant) with a very pale to white tail is known in Arizona only as a rare or casual fall transient on the Mogollon Plateau: about 12 seen and one taken at Doney Park, northeast of Flagstaff, *October 12, 1931* (Hargrave, MNA).

Buteo jamaicensis harlani (Audubon) is another northern race or variant in which the tail of the adult is irregularly clouded or streaked with whitish and chocolate color. It has been recorded twice in Arizona: 10 miles south of St. David, November 22, 1959 to *December 4, 1959* (Levy, US); and 15 miles southeast of Phoenix, *January 10, 1962* (Housholder, US, identified by Alexander Wetmore). The uncertainty concerning these last two non-red-tailed forms — in this species and genus known to be so highly polymorphic in color — concerns whether or not they occupy a breeding range of their own, for it seems more likely that they are just variant individuals within the northern populations of *calurus* and *borealis*. For the evidence of free crossing, but with different conclusions, see Taverner (Condor 38, 1936: 66-71).

67. RED-SHOULDERED HAWK; RED-BELLIED HAWK — *Buteo lineatus* (Gmelin)

One specimen: Little Colorado River near Holbrook, *December 5, 1853* (formerly US, figured in Baird *et al., Birds of North America*, 1860, plate 3). A few subsequent sight records from northeastern Arizona.

The Red-shouldered Hawk is thought to be rare but probably regular, at least in winter, in northeastern Arizona. Possibly this is a relic population.

Mr. J. O. Brew, Director of the Awatovi Expedition of the Peabody Museum, Harvard University, wrote Phillips that "a small number of bones in the debris accumulated at Awatovi during the 15th, 16th, and 17th centuries" were identified as *Buteo lineatus* by the late Dr. Glover M. Allen.

On *December 5* (not November 17), *1853*, Kennerly and Möllhausen took their specimen on the Little Colorado River at the mouth of Leroux Wash, two miles west of Holbrook. This, with the rest of their collection, was sent to Washington, and doubtless arrived there in 1854. The birds were determined by the best ornithologists of the country, and there is no reason to question their identifications. In fact, this particular specimen is doubtless the basis of Cassin's assignment of *elegans* to "New Mexico" in his original description of this western race. It should, therefore, be considered a cotype, though not the lectotype. Its present whereabouts are unfortunately unknown, but it is figured, perhaps not wholly accurately, in the atlas to Baird, Cassin, and Lawrence's book, mentioned above.

The most definite of the few scant reports are entrance to Cataract Canyon, west of Grand Canyon Village, September 22, 1926 (Griscom, ms.); east of Kayenta, December 7–11, 1937 (Hargrave and Wetherill, ms.); and California bank of Colorado River near Yuma, February 16, 1962 (Marshall *et al.*)

68. BROAD-WINGED HAWK *Buteo platypterus* (Vieillot)

Casual; one specimen, Southwestern Research Station, Chiricahua Mountains, *September 22, 1956* (J. G. Anderson and Ellen Ordway, SWRS).

69. SWAINSON'S HAWK *Buteo swainsoni* Bonaparte

Common summer resident of grassy plains of eastern Arizona; also, but sparingly, to central and central-southern Arizona, and possibly to eastern Mohave County. A migrant state-wide except in forests, but rare in western Arizona. Casual in winter, most reports dubious and no specimens between October and March. An old breeding record near Yuma (Brown, Auk 20, 1903: 44) is not considered authentic, nor do we believe it a "common migrant," starting in February, in adjacent parts of Nevada (Gullion *et al.,* Condor 61, 1959: 278-279).

This is the hawk apt to be seen in the largest numbers in Arizona during the migrations. Also it is the commonest large hawk in the open grasslands of eastern Arizona. It avoids rough, wooded terrain and nests rather low in yuccas or mesquites. The Swainson's Hawk is distinguished by entirely dark remiges and a wing slightly more pointed than other buteos. Except in the very blackest color phases, the forward undersurface of the wing (under wing coverts) is pale and unmarked. In the hand, only the three outermost primaries are incised, whereas the outer four of the Red-tailed Hawk are definitely emarginate.

The migration of Swainson's Hawk occurs chiefly in April and in late August to September, when large flocks appear in the southeast. These sometimes pause to feed upon caterpillars and grasshoppers on the ground. Extreme specimen dates for occurrence in Arizona are *March 18, 1933* near Sells (R. Jenks, ARIZ) to *October 18, 1915* at Tucson (Ivan Peters Collection). The breeding range includes the high grasslands of the White Mountains. Nests have been found westward only as far as Sells (S. H. Levy). The only authenticated records of sizeable groups in the western part of the state are a flock of 42 in a Bermuda grass field near Roll, Yuma County, July 14, 1951 (Gallizoili), and a spring flight *April 7, 1939* at Bouse, Yuma County, from which five specimens were taken: two males and three females, all melanistic or rufous-bellied (WJS)! Since the Swainson's Hawk winters in Argentina, we must regard the winter records in Arizona as pertaining not to normal, healthy birds. For instance, one reported by Mearns (ms.) at Cherry Creek settlement, November 28, 1884, was "very tame" — probably ill.

70. ZONE-TAILED HAWK *Buteo albonotatus* Kaup

Fairly common summer resident of mountains in northwestern, central, and southeastern Arizona; has also nested in the Santa Cruz and Rillito valleys, on the Gila River, and near mouth of Bill Williams River (Monson; skins and eggs, WJS); formerly commoner near Tucson. May possibly nest irregularly west to Organ Pipe Cactus National Monument, where individuals seen May 9 and 13, 1939 (Huey). Occurs rarely in migration west to Colorado River. A very few winter records for Tucson region (Phillips; Weisner and Taylor, US), some old ones from Yuma (*January 23, 1902* — H. Brown, ARIZ).

The Zone-tailed Hawk is a Turkey Vulture turned hawk — only the close observer will notice the one or two white bands (dark gray as seen from above) across the tail. If the bird is close, the feathered dark head can be seen, and of course Turkey Vultures do not scream. However, the general appearance, shape, and flight are almost identical in these two birds. Zone-tailed Hawks nest in tall trees along the main rivers and canyons but do much of their soaring over the higher mountains. Although the species is rather local, the nest sites are quite permanent; some have acquired a degree of fame, such as the one near Fort Lowell whence Major Bendire was chased by Apaches (with the egg in his mouth — US), and the one in the Santa Rita Mountains where Allan Brooks' lovely picture was painted (Condor 50, 1948: 5) and where an overenthusiastic photographer broke his back in falling from the nest tree.

Migration is in April and September; the nest in the Santa Ritas is occupied by late April each year. The northernmost records in Arizona are the Juniper and Hualapai Mountains (Phillips); near Valentine, Mohave County (Phillips); Sierra Ancha (R. R. Johnson *et al.*); and Pinaleno Mountains (Monson; Marshall). Outside dates of birds presumed not to be wintering are March 19 (1953, in suburban Tucson — Bruce Cole) to September 20 (1961, two at Santa Cruz River above Tucson — Marshall and James Ambrose) and September 21 (1953, near Tucson — Phillips).

The Zone-tailed Hawk, now limited in nesting mostly to the tall riparian trees of the Upper Sonoran Zone, formerly bred fairly commonly in pure Lower Sonoran Zone about Tucson, where streams supported heavy tree growth, now cut down. Early workers (Bendire, Stephens, Mearns) found it breeding there on the Santa Cruz and Rillito Rivers. Herbert Brown took a pair in the Tortolita Mountains, north of Tucson, *May 4, 1898* (ARIZ). Since then, there have been some single birds sighted in summer, but the only known nesting in this Lower Sonoran area was a pair and nest in the remnants of the mesquite forest, in 1917 (Dawson, ms.; *cf.* Jour. Mus. Comp. Oology 2, 1921: 35.)

This species and the Mexican Black Hawk have frequently been confused, although they were correctly interpreted long ago by Mearns (Auk 3, 1886: 60-69) and Bendire (Life Hist. N. Amer. Birds 1, 1892: 232). Price's widely quoted report from the lower Colorado Valley probably refers actually to the Harris' Hawk. We do not know the basis for van Rossem's allusion (Trans. San Diego Soc. Nat. Hist. 8, 1936: 127) to "several" records from that region. While it is unlikely that the few Colorado River wintering birds might come from Baja California, the several San Diego, California, birds are more possibly so derived.

71. WHITE-TAILED HAWK *Buteo albicaudatus* Vieillot

Casual; one or two specimens from Phoenix, *1899* (Breninger, MNA); nest recorded (egg taken, Breninger) from between Florence and Red Rock, 1897; one seen repeatedly near Marinette, Maricopa County, winter of 1954–55 (Abe S. Margolin *et al.*), and three the same winter west of Gila Bend (Housholder and H. Yost).

72. ROUGH-LEGGED HAWK; AMERICAN ROUGH-LEGGED HAWK   *Buteo lagopus* (Pontoppidan)

Rare in winter, but probably of statewide occurrence, except in southwest. Specimens south to Sulphur Springs Valley at Sonora border (winter of *1953–54* near Douglas — ARP, and near Pearce — LLH), where it may occur regularly.

These very unsuspicious hawks are never so common in Arizona as they are in northeastern New Mexico, but they seem to occur regularly in the extreme southeast and presumably farther north. Major recent flights were the three winters from 1953 to 1956, with several sightings and two specimens — one a mummy (LLH) and one scraped with loving care from the highway near Douglas (ARP), not quite close enough to the International Boundary to qualify also as a record for Sonora! Arizona birds are mostly immatures. All that Phillips has recognized as this species have been in the light phase with marked contrasts: light head and chest, blackish belly, white base and dark tip of the tail.

73. FERRUGINOUS HAWK; FERRUGINOUS ROUGH-LEG   *Buteo regalis* (Gray)

Uncommon, but widely distributed, summer resident of grassy plains of northern Arizona, and locally and irregularly in southeastern Arizona. Fairly common in winter in north and southeast, relatively rare elsewhere.

This is another variable hawk — from nearly white to blackish, showing white bases of the primaries and a tail which is either white or whitish, sometimes clouded near the end, without abrupt contrast, and never really dark. The head is pale in the light phase, and the only dark markings in the abdominal region are the flags (feathers of the thighs). The rusty or ferruginous tints are not obvious. This is a large sedate hawk, not given to hovering or to tilting the wings above the back; it behaves much like the Red-tail. The "rough leg" refers to the tarsus being feathered all the way down to the base of the toes. The Ferruginous Hawk is a bird primarily of open grassy plains. It and the Red-tail have a reputation in Arizona for perching low near a working badger in order to catch what prey slips past the digging mammal.

Edouard Jacot found the first Arizona nest of the Ferruginous Hawk on April 15, 1926, near Prescott (Condor 36, 1934: 84-85). He writes (ms.) that in the Lonesome Valley, near Prescott, the prairie dog was still common up to about 1943, and that it furnished 35% of the food of this hawk. Other items were rabbits, squirrels, rats, and grasshoppers (twice). Additional summer (mostly May) localities are somewhere in the Boulder Dam Region (Grater — so that it might possibly breed above Pierce Ferry), Hyde Park and west of Seligman (Phillips), Peach Springs Station (Monson and Phillips), north fork of White River (Huey), and west side of the Chiricahua Mountains (*May 13 to June 10, 1915* — van Rossem, LA). Outside of the breeding range the Ferruginous Hawk has been recorded from *October 23, 1932*, near Phoenix (Housholder, KANU) to *April 25, 1937* on the Salt River (G. Gillespie, F).

74. GRAY HAWK; MEXICAN GOSHAWK   *Buteo nitidus* (Latham)

Formerly a fairly common summer resident along wooded streams in southeast, now probably restricted to Santa Cruz River above Tucson (where irregular) and the Sonoita Creek — Nogales area. One old winter sight record (south of Tucson — H. E. Weisner). No authentic record as far north as the Gila River.

The Gray Hawk has only two plumages; the adult is gray, without any brown tinges, solid above and barred below with white. The tail is evenly barred with black and white. The immature is a brown, streaked affair like any other immature hawk. There are no erythristic or melanistic color phases to complicate the picture. The feet and tarsi are large and yellow, and there are no flags. This hawk is the smallest broad-winged hawk normally occurring in southern Arizona except for some of the accipiters. Its tail is slightly longer in proportion than those of other buteos, but not to the extent of the accipiters'. It is one of the noisiest hawks near the nest, which is along one of the major rivers, and from which it does not wander widely. It appears to require considerable stands of cottonwoods and willows for its nesting and hunting. The food in Arizona consists mainly of the large lizard, *Sceloporus clarkii*.

The Gray Hawk spends the period of April to September in Arizona; extreme dates are March 11 (1939, near Nogales — Fred Dille) to October 2 (1940, south of Tucson — Monson). It ranges west to Arivaca (Breninger, AMNH; Levy); north to lower Aravaipa Creek (in Pinal County — Phillips); and east to Hereford (about April 20, 1953 — *fide* Ed Lehner), San Bernardino Ranch (April 20, 1948 — Brandt and Hargrave), Chiricahua Mountains (Law, Condor 31, 1929: 219), and near Fort Bayard, New Mexico (Stephens). Its principal breeding range in our state has been, and still is, the Santa Cruz River drainage. This river, in its Arizona portion, ceased to be a permanent stream in about 1948; and the Gray Hawk had already abandoned the section through the San Xavier Indian Reservation. But after a decade's absence, it was seen again in May, 1958 (Marshall), and nested in the cottonwoods there in July, 1959 (W. G. George). In 1963 pairs appeared in unexpected places: Rillito Creek (Stensrude, Marshall and class), Babocomari Ranch and the San Pedro River generally (Short).

75. HARRIS' HAWK   *Parabuteo unicinctus* (Temminck)

Locally resident along Colorado River north to Topock; eastward to Superior, east of Florence, and Tucson; and southward to Mexican boundary in saguaros on Papago Indian Reservation and Organ Pipe Cactus National Monument. Rarely seen in migratory flocks, usually with Swainson's Hawks, in Tucson region and westward. Has occurred to the upper San Pedro Valley (*1890* — H. Brown, ARIZ), casually to Verde Valley (*1886* — Mearns, FW files; *1963* — Short). Found by Mearns in extreme southeastern Arizona (San Bernardino Ranch, *August 3, 1892* — US).

This is a buteo-like hawk with dark body and a broad white band across the center of the tail. There is also a narrow white tip on the tail in unworn plumage. Of our three always-black hawks with white bands on the tail, this is the only species that winters in any numbers. Harris' Hawk is distinguished from other similar dark hawks by the chestnut patches on the legs and basal parts of the wing, which are usually easily seen at any moderate distance. Immatures have some light streaking beneath, as is true of most young black hawks. The diet is unusually varied, including small birds and wounded waterfowl such as Coots.

There are several reports of numbers of Harris' Hawks in migratory flocks of Swainson's Hawks. Most of them are from southeastern California, but there is one as far east as Tucson in May. It is difficult to conceive whither such large numbers of Harris' Hawks might be going, since the northern limit of the breeding range is in the vicinity of Needles! It would therefore be desirable that a specimen be captured to verify these reports, which must pertain to the darker and redder plumages of *swainsoni*.

The breeding habitat of this hawk in Arizona is as variable as its food: saguaro deserts from Florence to the southwestward, and also the drowned trees of the Colorado River above Topock. Yet about Tucson the bird is very local — in a mesquite area 12 miles south of Tucson, and the saguaros and paloverdes of the Catalina Foothills Estates, where they have been since about 1955 (Mary Jane Nichols).

76. BLACK HAWK; MEXICAN BLACK HAWK *Buteogallus anthracinus* (Deppe)

Regular summer resident along permanent streams in southeastern and central Arizona as far northwest as the Big Sandy drainage (Phillips); also one-time sight records from Cataract Canyon southwest of Grand Canyon (Jackson, FW files) and near Parker (Monson). Most winter sight records are not considered authentic. There is no factual basis for the statement "possibly resident" (A.O.U. Check-list).

The Mexican Black Hawk is a characteristic hawk of running water, but will often soar high above the creeks and rivers. The single broad tail band (in addition to the narrow white tip) is white as seen both from above and below; it appears to arise at the level of the rear end of the very long secondaries. This broadened wing, its flat soaring position, and an occasional hint of whitish on the primaries gives the bird a resemblance to the Black Vulture, in contrast to the Zone-tailed Hawk's definite mimicry of the Turkey Vulture. Phillips observed one at La Casita, just south of Nogales, in late April; it carried a green branch, gave its piping scream in the air, then dove straight down to the nest. As seen at close range the Black Hawk is extraordinarily beautiful: the cere and feet are bright yellow, and the back is suffused with a powder-gray bloom.

The Black Hawk is in Arizona from mid-March to early October; migrants may be seen away from known nesting sites in March and September. We have long repudiated all winter records because of our not having seen the bird at that season in such well-frequented breeding spots as Sonoita Creek. It is therefore somewhat embarrassing to take the author's Olympian privilege of accepting our own: Kinsley's Lake at Arivaca Junction, January 17, 1963 (Marshall and John Schaefer)! The requirement for permanent streams makes this hawk commonest all along the base of the Mogollon Rim.

77. GOLDEN EAGLE *Aquila chrysaëtos* (Linnaeus)

Sparingly distributed throughout state in all mountain areas; somewhat more common in the lower mountains in winter, when it also occurs along the Colorado River. Virtually absent in summer, after breeding, in some desert regions (*viz.*, lower Little Colorado Valley), from early May to August or mid-September.

The largest Arizona bird of prey, the Golden Eagle feeds on rabbits and ground-squirrels. Formerly it doubtless took prairie dogs also. In the distance it resembles a Turkey Vulture, but it flies with the wings flat; these are long, straight and broad. Also the eagle's long neck and rather long tail give it a distinctive outline in flight. The adult appears entirely dark. Younger birds show a variable amount of white in the base of the flight feathers. These partly white feathers are important in the religious ceremonies of the Hopi Indians who were accustomed to keeping the young eagles, obtained from the nesting cliffs, in captivity so that they might be plucked. Normally the Golden Eagle nests in a pothole of a cliff at an inaccessible height, and with a good view. The species is holarctic (all around the northern hemisphere) and has been much discussed. Its fondness for carrion often gets it into traps. While not common, it is much the commoner of the two Arizona eagles, with the possible exception of certain winter flights of Bald Eagles at the lakes of the Mogollon Plateau.

In the hand, the Golden Eagle is distinguished from the Bald Eagle by its completely feathered tarsus, as in the rough-legged hawks. Eagles are really just outsized hawks, with a voice similar to the buteos'.

The Golden Eagle has decreased in size since the Pleistocene, according to a study by Hildegard Howard ("An ancestral Golden Eagle raises a question in taxonomy," Auk 64, 1947: 287-291). This eagle is probably as common in Arizona as in any part of the world, and it even inhabits the extremely arid mountains of Yuma County, such as the Kofas and Sierra Pintas (Monson).

78. BALD EAGLE *Haliaeetus leucocephalus* (Linnaeus)

Not uncommon resident about lakes and streams of the White Mountain region, rarer west down to Salt River and to Flagstaff region; said to leave after nesting, at least on Salt River (Hargrave). Transient in northern mountains and Kaibab Plateau. Winter records mainly from the Flagstaff and Colorado River regions. Rare in southeast.

The Bald Eagle, our national emblem, is protected by law, but is by no means a common bird in Arizona, nor is it ever likely to increase. This is due to its fondness for a diet of fish, though it will also take crippled waterfowl. In general appearance the young Bald Eagle greatly resembles the adult Golden Eagle, though it lacks the golden tinge on the nape. The adult, however, has snowy white head and tail, whence the name, bald. Nests in Arizona are always on the tops of buttes or chimneys of rock.

Occasional flights have been observed in winter on the lakes south of Flagstaff. Some former nests were at Bartlett Dam on the Verde River (Werner), Sahuaro Lake from 1930–1936 (Jas. J. Lane), and Mulehoof Bend of the Salt River in 1935 (Housholder — this nest said by Indians to have been used for many years; birds seen again there in 1944 and 1949).

There are no breeding specimens from Arizona, but the few wintering birds that have been measured (mounted, unsexed, 5-6 in various shops) seem to be too large for the typical race, and may therefore belong to *alascanus* Townsend.

HARRIERS *CIRCINAE* Subfamily

79. MARSH HAWK *Circus cyaneus* (Linnaeus)

Common winter resident of well-vegetated open areas below Transition Zone, except in northeast where it is mainly a transient. Before 1890 nested in parts of eastern Arizona, and may still do so west of Holbrook, but no positive record. Summer reports along North Rim of Grand Canyon probably do not indicate breeding. One seen at Topock June 12 and July 25, 1950 (Monson).

The Marsh Hawk or Harrier is perhaps the most distinctive of all our hawks because of its habit of hunting low and flashing a conspicuous white rump before a long tail. The wings are flapped regularly and are often held at a considerable angle over the back. At close range the peculiar owl-like facial ruff is visible. No other Arizona hawk presents the combination of such a long tail and long but not sharply-pointed wings. Marsh Hawks feed principally on rodents and small birds found in heavy grass or weed cover in open terrain. The adult male is of a lovely gull-like color with black wing-tips, and is the palest of Arizona hawks. Females and immatures are cinnamon-brown below and dark brown above, setting off the white rump to advantage.

The period of winter residency in southern Arizona is chiefly from September to April; in northern Arizona the Marsh Hawk becomes common in the last half of August and remains to early May. While it may not winter on the high prairies of the Transition Zone on the Mogollon Plateau, it remains there through November (White Mountains and San Francisco Peaks — Phillips). Between April and September there are numerous reports for early arrivals and late departures which grade into casual midsummer records.

Bendire (Life Hist. N. Amer. Birds 1, 1892: 185) "found a nest near the laguna," the old sink of the Santa Cruz River nine miles northwest of Tucson, doubtless in 1872; and in the late 1880's Swinburne, who worked about Springerville, wrote to Bendire (*loc. cit.*) "that in Arizona the usual number of eggs found by him is two or three."

OSPREYS *PANDIONIDAE*

80. OSPREY; *Pandion haliaëtus*
FISH HAWK (Linnaeus)

Nests locally along streams below Mogollon Rim, also one possible nesting record from Mohave Lake area (1950). Found rarely in summer on Mogollon Plateau, and along Colorado River and upper Salt and Verde Rivers. May occur almost anywhere in migration. Winters sparingly along lower Colorado River (Pulich, Monson), occasionally in the Salt (Margolin) and Gila Valleys (Housholder), and formerly at Roosevelt Lake.

The Osprey, or Fish Hawk, as the name implies, is a very rare bird in Arizona, where fish are scarce. Along our rivers it may be expected in season, however, usually soaring high in the air. Snowy white underparts and the long wings bent in the middle (wrist joint), with a black mark at the bend, are unmistakable. The nests are placed in tall trees and are occupied for many years. The Osprey is a cosmopolitan species, placed in a distinct family. The claws are round in cross-section and the scales on the soles of the toes are equipped with spines for grasping slippery fish. How this magnificent hawk can survive in Arizona is a marvel, considering that even fish-and-game rangers are instructed to shoot them on sight. Man cannot tolerate an animal that is a better fisherman than he is!

Migrations are in April and September. Ospreys wintered at Roosevelt Lake in 1915–1917, perhaps coincident with a peak in the fish population there. Nests, active from April to June, have been found in such places as Granite Reef Dam (R. Roy Johnson); Whiteriver (Phillips, S. Leamann Green, and Correia, ARP); three miles east of McNary (Ralph M. Brown); Black River Canyon (Chas. W. Quaintance); and Buffalo Creek south of Springerville (Stevenson, ARP).

CARACARAS AND FALCONS *FALCONIDAE*

81. CARACARA; *Caracara cheriway*
AUDUBON'S CARACARA (Jacquin)

Resident in small numbers on the Papago Indian Reservation; casual east to Santa Cruz Valley, west to Gila Bend and extreme southeastern Yuma County. Old records extend to Santa Cruz and Salt River Valleys, Oracle, and Yuma. A recent record near Pearce, Sulphur Springs Valley, is for May 1, 1954 (Hargrave).

The Caracara is a falcon turned vulture. The head is only partially bare, unlike true vultures. It is easily recognized at a distance by the broadly white base of the tail and a large white patch toward the tip of the wing, contrasting with the blackish body and rest of wing. The primaries curve down peculiarly in flight. Close-up, the red skin near the eye and the white cheeks and throat contrast strongly with the dark crown and breast. This all applies to the adults; young have brown instead of black. Caracaras are usually very tame, and their habit of sitting on telephone poles, in areas devoid of large cacti, probably led to their extermination south of Tucson by thoughtless gunners. They drum with the bill, like a slowly-turned wooden noise-maker (E. Tinkham).

The Caracara was formerly a resident about Tucson, ranging thence west to the lower Colorado Valley and north to the Salt River, where it was probably never common, but was taken *November 19, 1886* (Mearns, AMNH). It was found east to Oracle (Rhoads, Proc. Acad. Nat. Sci. Phila. 1892: 115), but as Scott did not report it there in his extensive residence, it was probably of irregular occurrence. The last nest found near Tucson was in 1889 (Brown, *fide* Bendire, Life Hist. N. Amer. Birds 1: 317), and at about that time the species began to decline. The last summer records for the vicinity of Tucson were in 1917 (Dawson, Jour. Mus. Comp. Oology 2, 1921: 32-34), though it persisted in the Santa Cruz Valley farther south until at least *August 19, 1930* (Gorsuch, ARIZ). In the lower Colorado Valley, the last definite records seem to be *January 15, 1905*, at Yuma (Herbert Brown, ARIZ) and extreme northeastern Baja California, *March 15, 1928* – Chester C. Lamb, MVZ).

Returning again to the opening paragraph, let us document those statements not yet covered above. In 1960, Seymour H. Levy found two nests on the Papago Indian Reservation. The first was being built on March 20; it contained three nestlings by May 24. The second nest, of another pair, had two young out on June 9. Publishing these facts (Auk 78, 1961: 99) together with the exact localities resulted in plunder, much to Levy's chagrin, and he has not found a trace of the birds since. A recent record from the Santa Cruz Valley is of an adult near Continental on January 10, 1963 (Marshall and class). The Gila Bend record is of two seen in January, 1956 by Monson. Monson also encountered the bird at Monreal Well, extreme southeastern Yuma County, April 17, 1959.

FALCONS Genus *FALCO*

Falcons are distinguished by sharply-pointed wings, extremely rapid flight in chase, and a fairly long tail. They resemble in flight a purposeful nighthawk. The Arizona species are of small to moderate size, never as big as most broad-winged hawks. In the hand, they show a toothed bill. Their hunting, at least that of the larger forms and the Merlin, is one of the most splendid of all natural sights.

82. PRAIRIE FALCON   *Falco mexicanus* Schlegel

Scarce resident statewide, also winter visitant to southern and western Arizona. Formerly commoner.

The Prairie Falcon is of the moderate-sized class, about the size of a Cooper's Hawk and pale in color (contrasting in the immatures with dark axillars). The general color above is a pale sandy brown and the malar area is crossed with a thin dark moustache. The flight is lighter and more maneuverable than that of the Peregrine, which permits it to snatch a ground squirrel from the entrance of its hole. (Falconers state that the Peregrine would never risk such a dangerous stoop to the ground.) It also pursues doves during their evening flights. The Prairie Falcon is a shy bird which does not make itself at home in towns, as do other falcons (except that Marshall once saw it plucking a bird while soaring over Park Avenue, Tucson). The nest is on a ledge of a commanding cliff, which has brought it into competition with the formidable Peregrine.

Though more often seen in winter than summer, this is partly due to a greater feeding range at that season. The only obvious migration performed seems to be a withdrawal in winter from the Transition Zone and above, where it is recorded as yet only from June to September. Nests have been found south to the Huachuca Mountains. Peregrines replaced Prairie Falcons about 1940 at Peña Blanca (W. J. Sheffler), and between 1939 and 1947 in the Ajo Mountains (Huey, Phillips).

83. PEREGRINE FALCON; *Falco peregrinus* DUCK HAWK   Tunstall

Nests rarely at cliffs throughout state, even at some distance from water. Found statewide in migrations; winters occasionally along lower Colorado River, in central Arizona, and the Santa Cruz Valley.

The Peregrine is another of the medium falcons, about the same size as the Prairie Falcon, but of a more solid build. It has a broad black moustache and dark upper parts; these are slate blue in the adults and brownish-black in the immatures. Exclusively a bird-hunter, the Peregrine can knock a teal out of the air with its closed fist. It is widespread over the world and nests in cliffs. In Arizona it has apparently increased as a nesting bird since 1939, at the expense of *mexicanus*.

The migratory periods seem to be prolonged: March through April and from late August into November. The regular wintering grounds are the Colorado River north to Topock (Monson, Pulich, E. S. Stergios), and the Salt River Valley (Johnson, Simpson, Werner).

Near Tucson single birds were seen December 2, 1956 and January 9, 1960. But in the winter of 1960–1961 several were repeatedly seen at the San Xavier Indian Reservation from December 15, 1960 to January 5, 1961 (all by Marshall).

84. APLOMADO FALCON *Falco femoralis* Temminck

Before 1890 a fairly common summer (or permanent?) resident in the southeast; since then virtually extinct in state, with only three exact records: near Tucson *prior to 1910* (Lusk, *fide* Visher, Auk 27, 1910: 281); near McNeal, Cochise County, November 13, 1939 (Monson); and St. David, San Pedro River, October 7, 1940 (Huey, ms.).

This third medium-sized falcon is distinguished by the black breast contrasting with the white throat. As far as we have observed, it is a sluggish bird; the few seen by Phillips in Mexico were just sitting. They spend much time perched on a prominent lookout in open country. Though found by most of the early explorers in or near southern Arizona, the Aplomado Falcon disappeared abruptly, for unknown reasons, leaving its seasonal status here in considerable doubt.

Our principal source on the bird is Lt. H. C. Benson, whose work on the nesting, etc., at Fort Huachuca was reported by Major Bendire (Proc. U. S. Nat. Mus. 10, 1887: 552). The birds were found on the grass-yucca plains of the southeast. Benson found five nests near Fort Huachuca in 1887, but the species must have decreased sharply soon afterward.

Though usually considered a summer visitor only, this species probably wintered in small numbers. Heermann's specimen was taken in late February or early March, and Benson saw it "as late as January" and found young in the nest as early as April 25, 1887, so it can hardly have absented itself from Arizona for any length of time. But there is no certain February record; J. C. Cahoon, who was at Fort Huachuca in February 1887 never saw this falcon at any time; and Willard's list of hawks observed in the San Pedro Valley on February 13, 1910 (Condor 12, 1910: 110) fails to inspire confidence.

85. PIGEON HAWK; MERLIN *Falco columbarius* Linnaeus

Uncommon to rare transient and winter resident virtually statewide. Summer reports unsubstantiated.

The Merlin is a small falcon — a miniature Peregrine, lacking the heavy moustache. It is scarcely larger than the Sparrow Hawk, but with much heavier, more purposeful flight. In coloration it differs in lacking any trace of a moustache and in being uniform blue or brown (not rufous) above; in having the whole tail dark with narrow crossbars; and in having the underparts sharply streaked with dusky. It feeds on small birds, but the Tucson Bird Club, while assembling at the University of Arizona Campus for the start of a field trip, was surprised to see Flicker feathers raining down from an ash, where a Merlin sat plucking it. There used to be (1930's to 1950's) Merlins on the campus every winter; they fed chiefly on the flights of House Finches and English Sparrows which arrived in late afternoon to roost.

Merlins inhabit Arizona principally from late September through April. Some very extreme records are August 25 in the Salt River Valley (Simpson, Werner) and September 7 (1951 in the Bill Williams Delta — Monson) to May 3 (1937 northeast of Flagstaff — Wm. X. Foerster).

Arizona specimens of the Merlin are about equally divided between the medium dark northwest race, *bendirei* Swann, and the extremely pale *richardsonii* Ridgway from the Prairie Provinces. None examined so far is anywhere near dark enough to be the eastern nominate race or the black form from coastal British Columbia, known as *suck-* *leyi* Ridgway. A female *richardsonii*, banded at Rosebud, Alberta on July 6, 1931, was collected by C. T. Vorhies at Tucson on *January 28, 1933* (ARIZ).

86. SPARROW HAWK *Falco sparverius* Linnaeus

Resident and transient, abundant in some areas in migration; uncommon to rare in winter in the higher mountains, and in summer in western Arizona.

This is the smallest and also the commonest hawk in Arizona. It is easily recognized by its unique black-and-white head pattern. Like the Marsh Hawk and Merlin, the Sparrow Hawk, or American Kestrel, is sexually dimorphic: the male has the tail uninterruptedly rufous down to the broad black subterminal band, has the wing mostly blue, and differs elsewhere from the browner female. Unlike most hawks, which have streaked young, Sparrow Hawk juveniles resemble the adults of their respective sex. We have seen the great migrations of this species mainly in Arizona. Numbers can be seen in the grassy parks of the ponderosa pine belt in August, and up to 35 were counted at two adjacent irrigated alfalfa fields near Tucson in April. The flight is light and airy, and the birds often hover at one spot while the wings are merely vibrated. Principally they feed on grasshoppers and other insects, but those in the city fill in their diet with small birds. The Sparrow Hawk frequents parks, campuses, and vacant lots in town, and nests in holes in tall trees, usually excavated by Flickers.

Since the sexes are so easily distinguished in the field, observers should work out the average differences in time of migration and relative numbers between the males and the females. Migration takes place generally from the end of March through the first twenty days of April, and again from August into October. Perhaps the start of the migration is represented by a group of ten, including immatures, seen on some flats near Whiteriver, *July 20, 1937* (S. Leamann Green, ARP). One of the immatures was collected and was very fat.

The Sparrow Hawks of Arizona all belong to the typical race except for the small form, *peninsularis* Mearns of Baja California and Sonora, which either occurs at or influences the population around Yuma. The problem is that there are no breeding specimens — taken when the birds first begin to explore nesting holes, but before they wear off the wing-tips. The larger race occurs all over Arizona in winter (including Yuma — ARIZ); it is responsible for the flights that go through on migration.

GROUSE AND PTARMIGAN
TETRAONIDAE

87. BLUE GROUSE; DUSKY GROUSE *Dendragapus obscurus* (Say)

Fairly common resident of the White Mountains region in boreal zones; less common along the North Rim of the Grand Canyon and in the Chuska Mountains of northern Apache County (Paul Phillips, Auk 54, 1937: 203-204).

The Blue Grouse is a chicken-sized dark gray bird with a fairly long tail, restricted to fir and spruce forests. It feeds upon the ground, but when flushed often flies up to perch stupidly among the lower branches. It is generally solitary and silent, but like all grouse, the male emits a prominent love note in spring. The peculiar altitudinal movements of this grouse — upward in winter and downward in summer — occurring farther north, are not yet known to take place in Arizona.

Blue Grouse were apparently never common on the San Francisco Peaks, and the last record known was of a hen with young at the north base about the year 1910 (*fide* Hargrave).

[SAGE GROUSE   *Centrocercus urophasianus* (Bonaparte)]

[Hypothetical. One seen near Nixon Spring in the Mt. Trumbull region, July 29, 1937 (Leroy Arnold).]

QUAIL   *PHASIANIDAE*
Genus *COLINUS*
by Allan Phillips

88. BOBWHITE; MASKED BOBWHITE   *Colinus virginianus* (Linnaeus)

Extinct in Arizona for many years; before 1890 common in tall grass plains from Baboquívari Mountains east to upper Santa Cruz Valley. Grazed out of existence by early 1900's; attempted reintroductions have all failed owing to destruction of grass. Records for Huachuca and Whetstone Mountains are not well-founded, though repeated in A.O.U. Check-list.

Description: Length 10 inches. Male of local race, *ridgwayi*: above mixed brown, black, and pale buff. Face and throat black, sometimes intermixed with some white. Rest of underparts rich chestnut brown. Female and young: no solid black or chestnut. Throat whitish to pale buffy; rest of underparts buffy barred with black.

Field marks: An ordinary-looking brown quail from above, without a crest, and given to rising suddenly from nearly underfoot. Under such circumstances it cannot be distinguished from the Mearns' Quail where these two occur together. On a bush or by a trail, the black-and-ruddy underparts of the male are distinctive. Females look like a female Mearns' Quail with boldly barred underparts, but are less pinkish, without the pompadour on the back of the head.

Voice: A soft, almost whistled *hoo-ee*. Song of male much louder, a clear rising whistle, *bob-white*.

Status: Central part of southern border, from east slope of Baboquívari Mountains east to upper Santa Cruz Valley and perhaps San Rafael Valley. Now extinct in Arizona. Up to 1885 a resident locally in tall grass, bordered by mesquite. Most of the birds disappeared promptly about then with the general abuse and overgrazing of southern Arizona, and the last specimens were taken for Herbert Brown (ARIZ, MNA) at Calabasas, *December 29, 1897*. A few birds may have lingered on for 20 or 30 years in grassy pockets of the Pajaritos Mountains or Canelo Hills.

Attempts at reintroduction have been unsuccessful, as there is no ungrazed grassland within the former range in Arizona.

Subspecies: The Arizona race was the famous, "almost mythical," *Colinus virginianus ridgwayi* Brewster, the Masked Bobwhite.

Virtually our entire knowledge of Arizona's most famous bird we owe to Herbert Brown, and no one interested in the birds of Arizona should fail to read his "Conditions governing bird life in Arizona" and "Masked Bob-white (*Colinus ridgwayi*)" (Auk. 17: 31-34, 1900, and 21: 209-213, 1904). The fame of the Masked Bobwhite rests on its handsome colors, the rumors of its existence before its actual description, its very limited range which was completely restricted to Arizona and our sister state of Sonora, Mexico, and the very few men who ever saw the bird before it was so promptly grazed out of existence. From time immemorial it had thrived in the prosperous grasslands of the border; but it died off almost instantly at the demise of its home with the coming of the great herds and their owners. Let those who really wish to conserve our wild heritage ponder well the lesson!

If there is one point on which I dare differ with so great an authority, it is in the wide range that Brown accords the Masked Bobwhite to the east. Inexperienced persons probably confused it with the Mearns' Quail. Brown himself failed to find it on the Babocomari in 1881, and notes that Bendire did not see it at "Camp Buchanan," later Fort Crittenden — nor, it should be added, did Henshaw's party nor the Stephens-Nelson party. Benson spent several years in the Huachuca Mountains in the 1880's, but he and Cahoon did not obtain it until they arrived as far south as Bacuachi, Sonora. (Nor is it mentioned in Cahoon's very sketchy, hasty notes farther north in Sonora.) To my mind, there is no good evidence that Bobwhites ever inhabited the high grasslands above the level of common occurrence of mesquite; that is, the Sonoita Plains, Huachuca or upper Baboquívari Mountains, or the Cananea region of Sonora. These remarks are not intended to reflect on the value and accuracy of Brown's own work. Let them serve as a warning, however, against further waste of funds in vain attempts to introduce the Masked Bobwhite in high grasslands where it never thrived. Ours is not the hardy "old New England Bobwhite"!

Virtually nothing is known of the life history of the Masked Bobwhite. An egg was allegedly taken from the oviduct of a hen by Brown in the "spring of 1890," according to Bendire; but Brown himself later wrote "it was never my good fortune to see an egg of this bird." He examined a robbed nest, but unfortunately does not mention the date. Judging by the few specimens of young birds I have seen from Arizona and Sonora, and the general pattern of nesting of most birds of the grasslands of western Mexico, nesting is likely to have occurred chiefly, at least in most years, in August, when the grass was most lush and animal life at its peak.

With such flimsy guesswork we must close the account of the first bird wiped from the face of Arizona. Efforts are now being made by Seymour Levy and Lewis Wayne Walker to propagate the birds reared by Stokely Ligon, in the hopes of restocking suitable grasslands, if such can be found or grown.

89. SCALED QUAIL; *Callipepla squamata*
COTTON-TOP (Vigors)

Common resident of grassy plains in southeastern and south-central Arizona, west and northwest to Baboquívari Mountains, south-central Pinal County, and east of San Carlos. Occurs locally in upper Little Colorado drainage, perhaps by introduction. Hybrids with *gambelii* have been found occasionally in recent years. Three such specimens (ARIZ) are from the west side of Antelope Peak southwest of Winkelman, from near Oracle, and the west base of the Santa Rita Mountains.

The Scaled Quail, or Cotton-top, belongs to a group of running quail which find safety by their strong legs, and fly only when hard-pressed. These species hybridize; probably the genera *Callipepla* and *Oreortyx* should be united. The Scaled Quail derives its names from the curved black margins of the chest feathers, which are not easily noticed in life; and from the short, bushy, white crest. It is much the plainest, most uniformly colored, of Arizona quail. The male and female are essentially the same in coloration. It is chiefly an inhabitant of grassland, and as a result of the destruction of the grass, it has been greatly reduced or locally exterminated in many parts of the state. On the other hand, the Gambel's Quail increases in these same areas as the grass is replaced by mesquite bushes and cholla cactus.

An island of grass on the top of Black Mountain supports Scaled Quail near Tucson (Keith Justice, ARIZ); doubtless from this colony came a lone male who used to follow along behind a large covey of *gambelii* on the San Xavier Indian Reservation in 1955 (Marshall). An occasional variant shows the chestnut belly characteristic of the Texas race (Sasabe – W. J. Schaldach, Jr., ARIZ). Nevertheless, the Arizona population belongs in *pallida* Brewster, the northern pale race.

90. GAMBEL'S QUAIL; *Callipepla gambelii*
DESERT QUAIL (Gambel)

Abundant resident in all areas where mesquite occurs; locally higher (along foot of Mogollon Plateau). Native occurrence in northern Arizona not substantiated. Also introduced in various places in the northern part of the state, for the most part unsuccessfully.

Gambel's Quail is a handsome running quail whose slender dark top-knot often curves forward so that the tip is directly above the bill. The body is gray, elegantly variegated with the white, chestnut, and buff; the throat and belly of the male are black. It is similar to, and sometimes erroneously called, the California Quail, a species which does not occur in Arizona. It is the common quail of southern and western Arizona.

George O. Hand was a pioneer from Oneida County, New York, whose interests, as an enlisted man in Company "G," First Infantry, California Volunteers, were whiskey, women, and whiskey. The few comments in his diary on the natural history of the country therefore show that it must have been amazing, for him to notice it. He states he was a good shot. An entry for 1862 is the following: "July 30, – Left Kinnion's Station [on or near the Gila River east of Oatman Flat] at twenty five minutes till two and travelled eight miles. Struck a bend in the river at 4 a.m., had a bath, filled our canteens, and started again. All along this day's march the quail were astonishing; big flocks of them two hundred yards long. I really think there were millions of them in each flock. If I were to tell my old friends in California that, they would say that I had lost my senses, and would not believe me. Came to camp, Gila Bend, about eight in the morning."

Although Gambel's Quail gather to drink at ponds, one covey will not ordinarily cross the territory of another in order to reach surface water (Hungerford). Distribution on the desert is independent of such water, and the birds fill their needs by appropriate selection of greens (Hungerford, Condor 64, 1962: 213-219).

The paler nominate race of Gambel's Quail occurs throughout the species' range in Arizona, except in the extreme east, as at Whiteriver. There, and possibly also in the vicinity of the Chiricahua Mountains, resides the darker, browner, and more richly colored *fulvipectus* (Nelson) (see Phillips, Anales Inst. Biol. México 29, 1958: 361-374).

91. MEARNS' QUAIL;
FOOL QUAIL; *Cyrtonyx montezumae*
HARLEQUIN QUAIL (Vigors)

Uncommon to fairly common resident of grassy open woods of the mountains of southeastern and central Arizona, west to the Baboquívari Mountains and north to the Pinal and White Mountains regions; formerly ranged sparingly to Flagstaff and Prescott areas. Found principally in Upper Sonoran Zone, but has occurred casually to Hudsonian Zone in summer (Baldy Peak, White Mountains, 10,000 feet, June 1, 1934 – Housholder)!

This quail in general rather resembles a Bobwhite but is larger. It lies low in the grass until almost stepped on whereupon it bursts into the air with a startling roar of wings. It has strong feet and claws specialized for digging out bulbs of nut-grasses.

Mearns' Quail has become reduced in some areas with the decrease of grass. Nesting takes place in the late summer rainy season (Wallmo, Condor 56, 1954: 125-128). In W. E. D. Scott's time this quail must have been numerous in the Santa Catalina Mountains, but now, with the increase of brush, it is very rare there; nevertheless it can be seen with fair regularity at lower Molino Basin (John Tramontano).

[RING-NECKED PHEASANT *Phasianus colchicus* Linnaeus]

[Has been introduced repeatedly in agricultural areas, without success; it persists only by continued artificial replenishment. The statement (A.O.U. Check-list) that it is established in southeastern Arizona is without any basis known to us.]

[CHUKAR *Alectoris* sp.]

[Has been introduced in many parts of the state, without any degree of success except possibly on Mingus Mountain and Kanab Creek.]

TURKEYS *MELEAGRIDIDAE*

92. TURKEY *Meleagris gallopavo* Linnaeus

In early days an abundant resident of nearly all forests except on Hualapai Mountains and Kaibab Plateau, descending to some of the valleys in winter. By 1930 shot out except in San Francisco and White Mountains regions. Since then restocked over most of former range and elsewhere with some considerable success, especially in southeastern mountains and on Kaibab Plateau. We regard as doubtful the statement that it formerly nested in a swamp at Calabasas.

The wild counterpart of the familiar domestic Turkey needs no description. Its dusk flight up into a tall tree to roost is impressive. Turkeys have been familiar inhabitants of Arizona since pre-Columbian times, when the Indians held them in domestication. Arizona is one of the few states that can boast wild turkeys of pure prehistoric lineage.

These are of the race *merriami* Nelson, the northwestern form characterized by cinnamon tips of the rectrices and upper tail coverts, which end in a buff margin. This is the race which has always occupied the large, continuous, Mogollon habitat. Also it is the stock which has in the last decade or so been used to repopulate the mountains of southeastern Arizona and the Hualapais. The white-tipped race, *mexicana* Gould, may originally have lived in these southern scattered peaks. It has been identified not far to the southeast in Chihuahua (Leopold, MVZ). But it is difficult to reconstruct the history from available specimens, which are either old and faded, or taken in summer and faded, or are possible reintroductions.

According to A. Starker Leopold, Turkeys can be aged and sexed as follows: the adult has a uniform series of big, purple secondary coverts; the male has broad rectrices, and his contour feathers terminate in a velvet, black bar; whereas the female's smaller rectrices are narrow and her body feathers have a light fringe distal to the velvet black bar.

The former range of the Turkey in Arizona resembled that of the Mearns' Quail and it extended west to the Baboquívari Mountains (Mearns, ms.). Decrease was marked by the 1880's, and the last records for southern Arizona are 1906 in the Chiricahua Mountains (*fide* Goldman, FW files) and 1907 in the Santa Catalina Mountains (*fide* Visher, Auk 27, 1910: 281). It was still present in 1916–1917 in the Juniper, Mingus, and Chuska Mountains and the Sierra Ancha; and in 1913 in the Mazatzal Mountains (all *fide* Taylor, Goldman, and Willett — FW files).

CRANES *GRUIDAE*

93. SANDHILL CRANE; LITTLE BROWN CRANE *Grus canadensis* (Linnaeus)

Formerly common; now distinctly uncommon and irregular winter resident in irrigated tracts of southern Arizona. Prior to 1936, a few northern Arizona records, including summer occurrences. Virtually unknown as a migrant in recent years. "Abundant" below Yuma, April 9, 1862 (George Hand).

Though popularly often confused with herons, the true cranes are very different, both in anatomy and behavior. Our only species, the Sandhill Crane, is a uniformly gray-brown bird. At close range the adults show a prominent red forehead of naked skin above a bill which is shorter than in herons. The ringing trumpeting is of a more musical quality than in herons. In Arizona, cranes usually occur in flocks. They remain in places with an ample view where they cannot be surprised, such as sandbars in rivers or alfalfa fields. With their long legs and necks, they stand chest high to a man.

Cranes were formerly abundant transients and winter residents along the lower Gila and Colorado River valleys, and fairly common transients generally in September, October, and late February to early April. Some recent winter records from west of Gila Bend are fifteen on December 18, 1949 (L. D. Yaeger); 85 on February 17, 1950, over 200 on February 18, 1952 (Housholder); and eighteen on February 4, 1956 (Monson *et al.*). Near Parker on February 28, 1961, L. D. Hatch saw 210 cranes.

Specimens are too few to permit more than a guess at the status of the two races of Sandhill Crane in Arizona. The smaller, typical subspecies of the north is represented by an individual examined by Holt (FW files), which had been shot at the foot of the Pinaleno Mountains in the spring of 1910; and by two birds shot at about the same time near Tucson (Slonaker, Condor 14, 1912: 154). The large form, *tabida* (Peters), of more southern distribution, was formerly an uncommon summer resident at Mormon Lake (1886 — Mearns, Auk 7, 1890: 51) and on the Apache Indian Reservation at Reservation Ranch, White Mountains region (1910 — Leopold, *fide* Bailey, *Birds of New Mexico*, 1928, p. 239). The only specimen was taken by Möllhausen on the Colorado River, "California," in *February* (no. 11927, US). Due to the serious reduction of *tabida*, it is likely that most records since the turn of the century pertain to *canadensis*.

RAILS, GALLINULES AND COOTS *RALLIDAE*

These birds are usually found slinking or swimming among dense cattails and tules; only one species, the Coot, commonly ventures into open water. Nests are platforms of tules hidden among these stands, and the precocial chicks are all black. Rails and Coots walk and swim with the head bobbing back and forth.

94. CLAPPER RAIL *Rallus longirostris* Boddaert

Summer resident (irregular?) in some alkaline cattail marshes along lower Colorado River, north to Bill Williams Delta. No authentic winter records.

The Clapper Rail is the largest of Arizona true rails (excluding the all-dark Coots and Gallinules, which are about the same size but which have much shorter bills). It is closely restricted to the Colorado River in Arizona, where it is seldom seen. Its colors are similar to those of the next species.

Although the nest has not yet been found in Arizona, a nest with seven slightly incubated eggs was collected near the Salton Sea, California, *May 12, 1940* (E. E. Sechrist, J. B. Dixon Collection at Palomar Junior College). Monson has observed this rail 19 times in all, during the summer period from May 8 (1961) to September 16 (1960), along the river from Yuma to Parker (GM) and the Bill Williams Delta (GM). On July 17, 1948, he saw an adult accompanied by three young at least two weeks old. The absence of winter records, from October through April, makes one wonder if this population could be migratory, although the species is so secretive that it would be hard to find in winter.

95. VIRGINIA RAIL *Rallus limicola* Vieillot

Summer resident of marshes in the White Mountains region down to Upper Sonoran Zone, and probably breeds in certain Maricopa County marshes. Locally a common migrant, at least along Colorado River, and formerly at Tucson. Rare to uncommon during winter, when it has been found as high as McNary, Apache County.

The Virginia Rail is a slim marsh bird a little larger than a Robin, with long legs and toes, and a long decurved bill which is orange at the base. The general colors are dark brown to dusky above and whitish to pale cinnamon-vinaceous below. The flanks are dusky, barred rather narrowly with whitish.

Migrations are mostly in September and April; but they are protracted and also complicated by summer records in the Salt River Valley (observed commonly in 1955 — Simpson, Werner) and wide winter occurrence: Colorado River, Yuma region north to Topock (Monson *et al.*); near Laveen, southwest of Phoenix, *December 22, 1919* (E. P. Walker, US); San Pedro River east of the Santa Catalina Mountains, January 28, 1886 (Scott, Auk 3, 1886: 385-386) McNary, January, 1950 (H. Tvedt); and Concho, in snow and ice on January 10, 1959 (S. H. Levy). Areas of summer residency are Big Lake (Goldman); North Fork of White River (Huey); Becker's Lake, Springerville (female with "ovary enlarged," *July 14, 1934* — J. O. Stevenson, MVZ); and Pachete Ranch, 26 miles east of Fort Apache (Jenks and Jacot, ARP). One was observed at a represo in the Growler Valley, extreme western Pima County, on August 24, 1961 (Monson).

96. SORA; CAROLINA RAIL *Porzana carolina* (Linnaeus)

Breeds commonly at marshes in northern Arizona, also probably in central Arizona. Common transient virtually statewide. Winters commonly at unfrozen marshes; recorded twice from frozen marshes near Lakeside (*December, January, 1936* — Phillips and Jacot, respectively, ARP).

The Sora is somewhat smaller than a Robin and has a short, straight bill which is brownish. Immatures are chiefly brown, paler below; adults are a handsome gray below with a black patch around the bill. This is the commonest rail in Arizona and the only one, exclusive of Coots and Gallinules, likely to be seen outside of a cattail patch — though at no great distance. Like other rails its presence is most readily detected by tossing a stone into the cattails and listening for the clear, descending whistles, which almost run together into a whinny. Migration seems to be in September and May, principally.

In Central Arizona, specimens have been taken in the Salt River Valley in *July* and *August* (*1958* — JSW), including a male with enlarged gonads; 10 or more were heard on June 20, 1952 at Picacho Reservoir (L. K. Sowls and Phillips); and there are other June and July observations at Blue Point and Chandler (L. D. Yaeger and Phillips; Housholder, respectively). Thus the Sora must breed commonly there.

97. YELLOW RAIL *Coturnicops noveboracensis* (Gmelin)

One caught alive near Sacaton, Pinal County, *March 28, 1909* (Gilman, Condor 12, 1910: 46).

The Yellow Rail is very similar to a young Sora but is smaller, with a white patch on the spread wing. Like the Black Rail it behaves more like a mouse than a bird, and it can rarely be flushed.

[BLACK RAIL *Laterallus jamaicensis* (Gmelin)]

[Hypothetical. Sight records by competent collectors in central-southern Arizona: Tucson (Stephens, April 23, 1881) and near Casa Grande (D. D. Stone), but they were unable to capture the birds.]

98. PURPLE GALLINULE *Porphyrula martinica* (Linnaeus)

Casual summer visitant (*July* to *September*) to the Tucson area, where there are several records (Brown, Vorhies, ARIZ); also one record for western Santa Cruz County, Montana Lake, shot by J. S. Andrews near Oro Blanco, *August 2, 1909* (Brown, ARIZ). One caught alive in Arizona near Rodeo, New Mexico, June, 1935 (Ralph Morrow). Occurrence elsewhere doubtful. There was a flight to Tucson in *July, 1951*, when a few were found exhausted or hit by cars, and brought in by boys to Phillips (ARP, two preserved).

This bird is a very deep purple on the head and

underparts; the back is greenish. Otherwise it is much like the Common Gallinule.

99. COMMON GALLINULE; FLORIDA GALLINULE; BLACK GALLINULE   *Gallinula chloropus* (Linnaeus)

Common to fairly common resident locally across the southern and western parts of the state where cattails or tules are associated with shallow streams, ponds, or canals; somewhat less common in winter. Occasional migrant in less favorable spots.

The nearly cosmopolitan Common Gallinule is a solidly slate bird with a short bill which is red in the breeding season but dull olive in winter. Except for the bill, it appears much like a slender, nervous Coot, but has a white line along the flanks. In behavior it is intermediate between the other rails and the Coot, since it habitually swims, but seldom goes far from the nearest cattail patch. In England it is known as the Waterhen, and is the subject of an important book by Eliot Howard, *A Waterhen's World* (1940).

This species has decreased with the destruction of the marshes in southern Arizona. Although found by Breninger in some numbers on the Pima Indian Reservation in September 1900, Gilman (ms). saw none during his residence at Sacaton from 1907 to 1915. There were still ten in a marsh near Tempe in about *1933* (Yaeger, MNA), where the vegetation was destroyed before 1940. Around Tucson there are no summer or breeding records since the turn of the century, although seen as late as May 24, 1933 (Anders H. Anderson). Transient in April and late August to early September.

100. AMERICAN COOT; MUD-HEN   *Fulica americana* (Gmelin)

Common summer resident, and common to abundant migrant, of water areas with cattail-tule margins. Winters commonly wherever open water is present, and abundantly along lower Colorado River.

Coots swim out from shore, effectively propelled by their lobed toes (constructed like those of grebes). They are totally blackish except for the short white bill and the white under tail coverts. They utter staccato exclamations, fight a lot, and paddle the water in taking off. Their behavior, territorial strife, and seasonal growth of the white forehead shield have been studied in California by Gordon Gullion (series of paper in The Condor, Wilson Bull., and Jour. Wildlife Mgmt. 1950–1952). Coots are loath to fly, and when lakes are heavily hunted, they remain in a mass far out on the open water, often accompanied by Ruddy Ducks. One of the most widespread and common water-birds of Arizona, they graze, dive, and tip-up for their food.

The principal months of migration in Arizona seem to be March and October. Along the lower Colorado River Coots breed, and winter in great numbers which fall off in January and February. The normal number of eggs laid on the Mogollon Plateau is from five to twelve. Coots remain in the north until driven out by winter ice. Unlike most other Arizona water birds, they are almost never found on open represos or livestock watering holes.

SANDPIPERS AND GULLS
CHARADRIIFORMES

Very few charadriiforms nest in Arizona, but they are common migrants here as in practically every place in the world. Birds of fabled powers of flight, some species perform the longest total migrations known as well as the longest non-stop flights across oceans. Most have sharply pointed wings and are colored in black, brown, or gray, with much white.

PLOVERS   *CHARADRIIDAE*

Plovers are short-billed feeders along beaches and mudflats, usually solidly brown above and white below with more or less of a black collar. But two species rare in Arizona, the Golden and Black-bellied Plovers, are entirely black on the underparts in summer plumage. Plovers resemble sandpipers but have a bigger head and shorter bill.

101. SEMIPALMATED PLOVER; RINGED PLOVER   *Charadrius semipalmatus* Bonaparte

Uncommon transient along lower Colorado River, elsewhere rather rare; hardly any records in north and east.

This plover is small, with dark brown underparts and one complete dusky collar. It is usually found singly or in two's in Arizona.

This species is doubtfully conspecific with the Ringed Plover of the Old World, *C. hiaticula* Linnaeus. Transients appear in April to May and July to October. A northern record is Mormon Lake, *August 31, 1934* (Russell, MNA).

102. SNOWY PLOVER   *Charadrius alexandrinus* Linnaeus

Fairly common transient along lower Colorado River. Two wintered in Bill Williams Delta, 1952–1953, and two above Imperial Dam, 1960 to 1961 (all by Monson). Eastward, only one record, near Peoria, *September 10, 1957* (Johnson, JSW).

The Snowy Plover is similar to the Semipalmated, but is a very pale sandy-brown above. The periods of migration along the lower Colorado River are late March to early June and the last days of June to mid-November.

103. KILLDEER   *Charadrius vociferus* Linnaeus

The ubiquitous water-bird of the state, found year-long anywhere at open water. Abundant at times in migration, and in irrigated areas in winter.

This is the only plover known to nest in Arizona, and the only one with two black collars. As it takes flight, the rufous rump and long tail are prominently displayed. The common name is taken from its call, often heard in the evening, and its loudness gives us the scientific name, *vociferus*. Not restricted to lakes and ponds, Killdeers are sometimes seen at small trickles of water on the desert, or soggy meadows among the pines. Nesting begins in March; there can be at least two broods in southern Arizona. The migration periods are March and October, and some Killdeers remain through the winter in the north.

104. MOUNTAIN PLOVER — *Eupoda montanus* (Townsend)

Mostly a rare bird; in fall in small numbers in the western part of the state; and as wintering flocks on barren desert flats and fallow fields in the Florence, Phoenix, Yuma and Kingman regions. Very rare farther east, where there are no recent records. Several flocks seen northeast of Springerville in August, 1914 (Ligon, FW files), suggesting possible breeding; otherwise, occurs in Arizona from September to April.

This is a bird never seen in mountains, for it prefers the most open, barren, desert flats, where the flocking birds run around like Horned Larks. They have very plain colors: brown above, whitish below, and no black.

105. AMERICAN GOLDEN PLOVER — *Pluvialis dominicus* (Müller)

One record: near Peoria, Maricopa County, *May 15, 1953* (Johnson and Simpson, PX). The race is *C. d. dominicus*.

106. BLACK-BELLIED PLOVER — *Pluvialis squatarolus* (Linnaeus)

Rare transient probably statewide, with most records from the lower Colorado Valley and none from northeast as yet. One winter record, at Yuma, December 30, 1940 (Lee Arnold). Migrants are seen in April, May, and August to October.

The Black-bellied Plover's plain winter dress is different from that in spring and summer, when it is one of the few birds that has solid black underparts. The change in plumage takes place while the plovers are migrating. Two seen along the Arizona shore of Havasu Lake on August 11, 1951 were still in full summer plumage. This, the earliest record of fall migrants in Arizona, indicates that adults enter the state as early, if not earlier, than do immatures. Arizona records thus far consist only of singles or two's. Any large, stocky shorebird with a Plover bill should be carefully checked in fall to make certain it is not the much rarer American Golden Plover.

[RUDDY TURNSTONE *Arenaria interpres* (Linnaeus)]
[There are two records for the California side of Havasu Lake: *September 16, 1952* (GM) and August 21, 1953 (Monson).]

[BLACK TURNSTONE *Arenaria melanocephala* (Vigors)]
[One seen on the California side of Havasu Lake, May 21, 1948 (Monson).]

SNIPE AND SANDPIPERS
SCOLOPACIDAE

Though resembling plovers, the birds in this family have longer, more slender bills, as well as longer necks and feet for wading. In Arizona they are mostly migrants, although some spend the winter. They generally fly in compact flocks and utter distinctive, shrill calls.

107. COMMON SNIPE; WILSON'S SNIPE — *Capella gallinago* (Linnaeus)

Fairly common to common transient throughout state, wintering generally at unfrozen waters. Breeds near Springerville, where "winnowing" was heard in early July and a young bird, not fully grown, was taken *July 5, 1936* (F. G. Watson, ARP).

Known among hunters as the Jacksnipe, this holarctic species, unlike the other sandpipers, finds safety by crouching in low vegetation where its marvelously concealing color pattern — in browns, buff and black — serves it best. When approached too closely it flies, uttering a harsh *scape*. The long bill, held vertically, probes deep in the mud for food. This remarkable organ can be moved at the tip, where also there are sense receptors by which, perhaps, it can detect worms.

Except for the breeding area at Springerville, the period spent in Arizona is mostly from mid-September to early April. Outside records are from *mid-August* (Mearns, US) to May (A. H. Anderson, etc.).

108. LONG-BILLED CURLEW — *Numenius americanus* Bechstein

Uncommon transient statewide, more common in lower Gila and Salt River Valleys and along lower Colorado River. Very rare in winter in south, and even in the bottom of the Grand Canyon; occasional along lower Colorado River in summer, when it has also been found in the north.

The Long-billed Curlew is a chicken-sized shorebird with an enormously elongated, decurved bill. It is colored chiefly in shades of cinnamon, with no contrasting pattern anywhere. It inhabits open fields, grasslands, and edges of lakes. Migrations in Arizona are from mid-March to early May and from the last days in June to early October.

There are winter records as follows: bottom of Grand Canyon, January 11, 1937 (Victor Veatch); McNeal, Sulphur Springs Valley, 4 seen December 29, 1951 (Gallizioli); and along the Sonora boundary between the Baboquívari Mountains and Yuma in *January* and *February 1894* (Mearns and Holzner, US). Summer stations are the Colorado Valley (frequent records of singles or two's, 1943 to 1961 —

Monson); near Big Lake, White Mountains, June 25, 1915 (Jackson, *fide* Goldman, Condor 28, 1926: 164); and near the south rim of the Grand Canyon, July 1928 and June 12, 1929 (McKee). The species may have bred south to the Grand Canyon in former times, therefore. Two races of the Long-billed Curlew are currently recognized by differences in bill length (shorter in the north), and both have been identified among Arizona specimens, even within the same flock. However, there is so much overlap in this extremely variable member that such identifications are no better than a guess as to where our migrants originate!

109. WHIMBREL; HUDSONIAN CURLEW *Numenius phaeopus* (Linnaeus)

Very rare along lower Colorado River in September; one specimen, Bill Williams Delta, *September 17, 1946* (Monson, US), plus three sight records (Monson). No verifiable report elsewhere.

The Hudsonian Curlew is the smaller of our two species, without the cinnamon, and with a decidedly patterned head. It is chiefly marine-littoral in its distribution.

110. UPLAND PLOVER; BARTRAMIAN SANDPIPER *Bartramia longicauda* (Bechstein)

A few transient records in August, early September, and May from Cochise, Graham, and Pima Counties, the last on *May 8, 1887* (Brown, ARIZ). One seen on California side of Havasu Lake, September 11, 1952 (Monson).

The Upland Plover is a very short-billed, streaky-brown sandpiper usually found in open grassland. It has longer legs and a more slender neck than the Killdeer, which is about the same size.

111. SPOTTED SANDPIPER *Actitis macularia* (Linnaeus)

Breeds along streams and lakes of Mogollon Plateau and just below, rarer northward. Not on San Francisco Mountain. Common transient statewide, and winters more or less commonly along Colorado River and in central Arizona, north irregularly to Camp Verde.

This close relative of (if not conspecific with) the Common Sandpiper of Eurasia is recognizable at a distance by the constant up-and-down teetering of the rear part of the body. It flies low over the water, with the primaries arched down, and displaying a white streak along the base of the secondaries. The call is a shrill *peet-weet*. For a sandpiper, the neck is rather short; spots appear on the underparts only in the summer (alternate) plumage. The Spotted Sandpiper is not found in flocks. Its nest is in grass back from the shore of a stream or lake.

The migration periods are in April and May and from July to September; migrants have also been found in early June (Tucson – A. H. Anderson; Walker Lake near San Francisco Peaks, *June 5, 1887* – Mearns, FW files). While migrants are passing northward, summer residents are beginning breeding activity in spring: ovum 4 mm., Eagle Creek, *April 27, 1934* (Quaintance, ARIZ); sexual displays, Camp Verde, May 12, 1962 (Marshall and Ambrose). Thus date is no criterion of breeding status, and it is unlikely that the species breeds anywhere in the San Francisco Peaks area or Kaibab Plateau, on the available small puddles. Similarly, June and July birds along the Colorado River have not been known to nest.

112. SOLITARY SANDPIPER *Tringa solitaria* Wilson

Fairly common transient statewide, except in Yuma area, where it is uncommon; spring records scarce in west and north. One seen near Yuma in winter, December 26, 1950 (J. A. Munro).

As the name states, this species does not occur in flocks. Usually we see a single bird at the shaded edge of some quiet pool. As it flies up it flashes the broadly white sides of the tail. This sandpiper uses the abandoned nests of other birds, in trees; but it does not nest in Arizona.

Migrations are in April and early May and from July through September. There are some October records, such as *October 10, 1892*, San Pedro River at the Sonora border (Mearns and Holzner, US).

Robert T. Orr and Phillips have re-examined the available specimens from Arizona and find them all to be the larger northwestern race, *cinnamomea* (Brewster), including Laurence Huey's from Quitovaquita (SD). Additionally, *cinnamomea* as compared with the typical southern subspecies, is usually freckled with white on the outer primary, has a back of different color in spring plumage, and is more cinnamon (on the dorsal light spots) in immature plumage. In many species, the northern race will migrate farther south in winter than does the more southern race. If this is true in *Tringa solitaria*, then any birds wintering so far north as Arizona should be the smaller *solitaria*, of which so far we have identified none from the state.

113. WILLET *Catoptrophorus semipalmatus* (Gmelin)

Fairly common migrant statewide, particularly in fall and along the Colorado River. Occasional, even in flocks, along Colorado River in June (Monson).

The Willet is a pigeon-sized bird which seems of ordinary gray color at rest. But on taking flight it reveals a bold pattern of black wings crossed by a broad white stripe. Migration is in late April to middle May and July to early September, with some October records by Coues, Mearns, and Phillips.

The early records cited by Swarth (Pacific Coast Avifauna no. 10, 1914: 20) have been confused. Bischoff's specimen was really taken in Nevada, and we see no reason to believe that Woodhouse saw this species west of Zuni, New Mexico (see Sitgreave's *Report of an Expedition down the Zuni and Colorado Rivers*, 1853: 34-40). Thus Coues' observation at Prescott, October 18, 1864 (Proc. Acad. Nat. Sci. Phila. 1866: 97) seems to be the only definite Arizona record prior to the 1880's. (This is not intended to imply any change of status at that period.)

114. GREATER YELLOWLEGS *Totanus melanoleucus* (Gmelin)

Rather common migrant in small numbers statewide; winters along Colorado River and in south, when rare except in the Yuma and possibly lower Gila River regions. Sometimes occurs in June (Peck's Lake, Clarkdale — Housholder *et al.*; Topock and Bill Williams Delta — Monson).

The Greater Yellowlegs is smaller than the Willet. It is dark brownish above, white below, with white rump and yellow legs. Migrations are from March to early May and from July to October.

115. LESSER YELLOWLEGS *Totanus flavipes* (Gmelin)

Common fall transient in north and (formerly only?) extreme southeast. Rather uncommon migrant over rest of Arizona, particularly scarce westward.

This is a smaller edition of the last, hardly bigger than a Solitary Sandpiper. Fall migration is late July to early October; at this season the Lesser Yellowlegs occurs in fairly large flocks in northern Arizona. In spring it is less common but more widespread, in April to mid-May.

Mearns and Holzner found Lesser Yellowlegs commonly in *August, 1892* at the San Bernardino River, near the International Boundary in the southeast (*fide* Lincoln, Condor 29, 1927: 165).

116. KNOT *Calidris canutus* (Linnaeus)

Casual; one at Topock *July 23, 1952* (GM), and one about three miles above Imperial Dam, October 2, 1959. Also, one on California side of Havasu Lake August 9, 1950 (all by Monson).

117. PECTORAL SANDPIPER *Erolia melanotos* (Vieillot)

Uncommon fall migrant, recorded chiefly in the lower Colorado Valley. One winter record: Martinez Lake above Imperial Dam, December 30, 1957 (Bialac, R. R. Johnson and Monson). Unrecorded as yet in the northeast.

The various species of small gregarious shore-haunting sandpipers are collectively known as "peeps." They commonly flock together in migration. The Pectoral Sandpiper is the largest of them — almost as large as a Solitary Sandpiper. Above it is dark rusty-brown; the chest is streaked with blackish on a buffy ground. The species has been found in Arizona between late August and early October, except for the one winter observation mentioned above.

118. BAIRD'S SANDPIPER *Erolia bairdii* (Coues)

Fall transient, uncommon in west to abundant in east and north. Spring records, cited in A.O.U. Check-list, are based on specimens of *Ereunetes mauri* (Mearns, AMNH).

In size, this is about half-way between the Pectoral and Least Sandpipers but it is a pale buffy-brown above and the chest band is also pale: brownish streaks on a whitish background. Baird's differs notably in its choice of habitat, preferring open ponds — perhaps slightly saline — in the most barren environment, such as the desert flats of northeastern Arizona.

Fall migration is from July through September. The only valid record which even approaches winter is *November 18, 1936* at Crescent Lake, White Mountains (ARP); but there is no reason to believe that the bird winters anywhere in or near Arizona.

119. LEAST SANDPIPER *Erolia minutilla* (Vieillot)

Common to abundant transient statewide, especially in fall; winters more or less commonly from western to central Arizona.

Although as the name states, the Least is the smallest of our sandpipers, the difference between it and the two species of *Ereunetes* is slight. It is best identified at close range by the greenish, not black, legs. Coloration resembles that of the Pectoral Sandpiper; in the hand, the toes show no webbing at the base. This is our most numerous "peep." Recorded here from July through May, it is abundant in migration from the middle of August to mid-October and from early April to early May.

120. DUNLIN; RED-BACKED SANDPIPER *Erolia alpina* (Linnaeus)

Fairly common transient and rare winter visitant in western Arizona, scarcer to east and unrecorded in the north.

The Dunlin is a middle-sized "peep" with a long and somewhat decurved bill. The handsome spring plumage with the red back and black belly is rarely seen in Arizona.

Migrants cross the state in April to early May and late September to November. Birds have wintered in 1952–1953 at the mouth of the Bill Williams River and in 1956–1957 near Imperial Dam (both by Monson); and in 1958–1959 at Picacho Reservoir (Levy).

Geographic variation in color is evident only in the breeding plumage; our one such specimen (Bialac, ARIZ) was identified as the race *pacifica* (Coues) by W. E. Clyde Todd. This is a long-billed form, which breeds in central and southern Alaska (Todd, Jour. Wash. Acad. Sci. 43, 1953: 85-88).

121. SHORT-BILLED DOWITCHER *Limnodromus griseus* (Gmelin)

Rare or casual transient. Only one definite record, an immature near Springerville, *September 21, 1937* (ARP). Possibly not as scarce as this absence of specimens implies, but certainly far outnumbered by the following species. There is no basis for statement (A.O.U. Check-list) that it winters all through the southwestern United States.

122. LONG-BILLED DOWITCHER *Limnodromus scolopaceus* (Say)

Common to fairly common migrant statewide; winters sparingly in southern Arizona and along the Colorado River, generally in flocks. Rare in late June and early July, but no specimens between *May 15* (ARIZ) and *August 13* (UT).

Dowitchers are shorebirds which resemble the Snipe in size and long bill. But they wade out in the open water. The upperparts appear rather uniform brownish or grayish until the bird flies. Then a white stripe is revealed down the middle of the back. Migration is principally from March to May and August to October.

The Long-billed Dowitcher winters along the Colorado River to north of Topock, where 37 were seen on December 19, 1947 (Pulich). Winter specimens away from the Colorado River are *January 1, 1940* at Tucson (GM) and *January 31, 1953* at Picacho Reservoir (ARP). The only definite record cited by Swarth (Pac. Coast Avif. 10, 1914: 19) for Arizona really pertains to New Mexico and is definitely stated by Henshaw to be *scolopaceus*, not *griseus*. Our data fully support van Rossem's discovery (Occ. Papers Mus. Zool. La. State Univ. 21, 1945: 85) that *scolopaceus* is the predominant fresh-water dowitcher, at least in the southwest.

123. STILT SANDPIPER *Micropalama himantopus* (Bonaparte)

Unknown except from the Salt River Valley near Phoenix, where it is an uncommon transient; once found in numbers on *April 25, 1933* (Yaeger, ARIZ, MNA). Also noted there in *September* (ARP), *October* (JSW), and near Palo Verde, *June 12, 1955* (PX).

The Stilt Sandpiper is a bird of about the build of a Lesser Yellowlegs but without yellow on the legs, and with a definite white superciliary stripe. Unlike the yellowlegs, it has two different plumages a year; in summer it is cinnamon underneath.

124. SEMIPALMATED SANDPIPER *Ereunetes pusillus* (Linnaeus)

Known definitely from one specimen: near Sasabe, *April 23, 1957* (Levy, US). Statement (A.O.U. Checklist) that it had been found as a rare transient, both spring and fall, in the intermountain region rests largely or wholly on misidentification.

The Semipalmated Sandpiper differs from the Western only in shorter bill (sex for sex), which in the hand is seen to be slightly spatulate at the tip. This sandpiper remains grayish-brown throughout the year. Were it easier to identify, it might prove to be less rare than the one specimen for Arizona indicates.

125. WESTERN SANDPIPER *Ereunetes mauri* (Cabanis)

Common transient statewide, especially in fall. Probably winters rarely and irregularly.

The Western Sandpiper is a trifle larger than the Least Sandpiper, from which the female, especially, differs in having a distinctly larger bill. The legs are blackish and in the hand show small webs between the front toes. In winter plumage the chest is entirely white. At this time the upperparts resemble those of the Semipalmated, but in spring they become prominently mixed with rich rusty. It is remarkable that this sandpiper is so rare in Arizona in winter, since it winters abundantly much farther north along the Pacific Coast.

Most of the specimens fall within the migration periods of *April 9* (Vorhies, ARIZ) to *May 3* (Randolph Jenks, ARIZ); and from *August 2* (Jacot, ARIZ) to *September 11* (Mrs. C. T. Vorhies, ARIZ). The only winter specimens are *December 22, 1958* at Picacho Reservoir (Levy, US) and *February 15, 1958* at Peoria (Johnson, JSW).

126. MARBLED GODWIT *Limosa fedoa* (Linnaeus)

Rare to uncommon spring transient and fairly common fall transient, with most records from the lower Colorado River and the lakes of the Flagstaff region. One winter record: San Pedro River east of Santa Catalina Mountains, January 27, 1886 (Scott, Auk 3, 1886: 386).

This is a large shorebird very much like a Long-billed Curlew, but its practically straight bill turns slightly upward.

Migration is mostly in *April* (JSW) to *May* (MNA), and July (photograph, Big Lake, White Mountains — Mr. and Mrs. Crockett) to August with some records as late as October (Bialac, Werner, *et al.*).

127. SANDERLING *Crocethia alba* (Pallas)

Rare to uncommon transient along the lower Colorado River.

The Sanderling is a nearly cosmopolitan sandpiper distinguished in its fall plumage by being almost wholly white. It is about the size of a Spotted Sandpiper but is shorter-billed. In the hand it is seen to have only three toes, like a plover.

The only Arizona spcimen was taken above Parker Dam, *September 18, 1947* (GM).

AVOCETS AND STILTS
RECURVIROSTRIDAE

These are large black-and-white shorebirds — long-billed and long-legged — which wade up to their bellies.

128. AMERICAN AVOCET *Recurvirostra americana* (Gmelin)

Fairly common migrant statewide; sometimes occurs in large flocks in fall along lower Colorado River, where one occasionally winters (H. Brown, ARIZ; Monson). Occasional in late June in central and western Arizona (Vorhies; Monson), but not breeding.

In the Avocet, the body is white and the black is restricted to the tail and parts of the wing. Unique among Arizona birds, its long thin bill is decidedly curved upward. The head and neck are washed with soft cinnamon in spring plumage. Large flocks are occasionally seen, but the bird is generally not as common in Arizona as is the Stilt. Migrations are mainly in April and May and from August to October.

129. BLACK-NECKED STILT — *Himantopus mexicanus* (Müller)

Breeds locally along lower Colorado River (Monson) and probably at Picacho Reservoir (Phillips, Sowls, and D. D. Stone); formerly nested on San Pedro River. Otherwise a fairly common migrant in southern and western Arizona, common in fall along the Colorado River. Three winter records: north of Aztec on Gila River, December 11, 1921 (Housholder); west of Gila Bend, *February 17, 1940* (ARP), and near Wikieup, *November 2, 1960* (Musgrove).

The Black-necked Stilt, nearly all black above, has long red legs nearly as long as the body and neck combined. The bill is thin and straight.

The San Pedro River nesting record is of eggs taken May 30, 1901 (in LAM according to Bent, *fide* FW files). Migration is principally in April, May, August and September.

PHALAROPES
PHALAROPODIDAE

These are small sandpipers with lobed toes that go out swimming in the open water. They spin like tops as they stir up food objects with their feet. The female is larger, and in summer her alternate plumage is much brighter-colored than the male's. She does the courting and goes off, leaving the eggs in his care. Probably the same behavior occurs in other shorebirds but is not as well-known in them because of the impossibility of distinguishing the sexes in the field.

130. RED PHALAROPE — *Phalaropus fulicarius* (Linnaeus)

Rare fall transient, recorded mainly from along lower Colorado River. One winter record: tank south of Sierrita Mountains, Pima County, *January 3, 1959* (Levy, US).

This is the rarest of the phalaropes in Arizona, except possibly in late fall. The bill is relatively short and thick — shorter than the head. Late September and October is the time when this bird is most likely to be seen in Arizona.

Specimens have been taken from *July 19* (*1938* at Tonalea — Woodbury and Russell, UT) to *November 21* (*1958* near Tempe — Jim West, JSW).

131. WILSON'S PHALAROPE — *Steganopus tricolor* Vieillot

Common to fairly common transient statewide, an occasional bird lingering through June. Birds seen after mid-June, including at least 105 at Picacho Reservoir, *June 20, 1952* (Phillips, Sowls, and Stone, ARP), and at least 140 at Imperial Wildlife Refuge, June 27, 1961 (Monson), may be returning fall migrants.

Wilson's Phalarope is the commonest phalarope in Arizona. The bill is needle-like and longer than the head. The female in spring has a chestnut stripe down the side of the neck, and a white crown and foreneck. The migration periods would seem to be from *late April* (Hargrave, GCN) to *early June* (LLH), and again from *later June* (as above; specimen from Benson, LLH; and three specimens from Big Lake in the White Mountains — Goldman, Condor 28, 1926: 164) to *September* (Jacot, ARIZ) and October 1 (1926, at Mormon Lake — Griscom, ms.).

132. NORTHERN PHALAROPE — *Lobipes lobatus* (Linnaeus)

Common fall transient in northern and western Arizona only, scarce some years in the west; rare in spring, and in the southeast.

This species has the bill intermediate between the other two phalaropes: needle-like, but only as long as the head. The spring female's neck is almost entirely dark, of blackish and deep chestnut. In Arizona the Northern Phalarope occurs chiefly on the larger bodies of water.

Spring migrants have been taken only in *May*, such as one on the south rim of the Grand Canyon, *May 11, 1903* (Fuertes, CU). In fall, the birds appear from the *middle of August* through *September* (Jacot and Brown, respectively, ARIZ).

JAEGERS AND SKUAS
STERCORARIIDAE

These are big, dark, parasitic gulls, of accidental occurrence in Arizona.

133. POMARINE JAEGER — *Stercorarius pomarinus* (Temminck)

Two records: near Flagstaff, *late October* or *early November, 1927* (Dean Eldredge, MNA), and lower Havasu Lake, *September 26, 1950* (Monson, US).

134. PARASITIC JAEGER — *Stercorarius parasiticus* (Linnaeus)

Two records (Monson, US): Bill Williams Delta, *October 13, 1947*, and lower Havasu Lake, *September 19, 1953*. The latter immature may be *Stercorarius longicaudus* Vieillot, Long-tailed Jaeger.

GULLS AND TERNS *LARIDAE*

Gulls are thick-billed scavengers; terns have pointed bills and dive for fish from the air. Both are white or pale gray, with black around the tips of the wings. Some species are darker, but all those normally found in Arizona are white below as adults (except for the adult

Black Tern in summer). Terns habitually fly with the bill pointing down toward the water, unlike any other bird. Gulls mostly have even tails, whereas those of terns are deeply forked; most terns are smaller than gulls.

135. GLAUCOUS-WINGED GULL — *Larus glaucescens* Naumann

Two records of immatures (GM): lower Havasu Lake, *February 24, 1954,* and Colorado River about eight miles above Imperial Dam, *November 17, 1956.*

This differs from the other largest Arizona gulls in the absence of black on its wing-tips, which are no darker than the rest of the body, even in the immature.

136. WESTERN GULL — *Larus occidentalis* Audubon

One immature of the typical race: Parker Dam, *December 12, 1946* (GM). Some of the large immature Gulls occasionally wintering along the Colorado River may be Westerns, but most are probably Herring Gulls.

The Western Gull is darker above than other Arizona gulls, besides being one of the largest. The blackish-backed adult has not been found in Arizona. Identification of almost all the immature gulls is extremely difficult and usually requires a specimen.

137. HERRING GULL — *Larus argentatus* (Pontoppidan)

Uncommon winter visitant along lower Colorado River, especially Havasu Lake; known elsewhere only from near Tucson (two specimens). *Larus thayeri*, Thayer's Gull, is included here, though some consider it a distinct species (one record, Havasu Lake, *December 13, 1946* — ARP).

This is larger than the Ring-billed Gull and lacks the bill mark and tail pattern of that species, for the tail of the immature Herring Gull is wholly dark.

The four specimens from Arizona are all in immature plumage. The example of the race or species *thayeri* W. S. Brooks, characterized by pale wing-tips, was identified both by A. J. van Rossem and Allen Duvall. The other three (ARIZ) are *smithsonianus* Coues: near Tucson, *November 16, 1894* and Yuma, *February 12, 1903,* both collected by Herbert Brown; and in Tucson, *November 17, 1933,* found dead three or four days by Phillips, identified by Joseph Grinnell, and now missing.

138. CALIFORNIA GULL — *Larus californicus* Lawrence

Fairly common to uncommon migrant, chiefly in spring, along the lower Colorado River; elsewhere only one valid record, Long Lake, Coconino County, *November 20, 1932* (Hargrave, MNA). One summer record: lower Havasu Lake, *July 28, 1954* (GM).

The adult's mantle is darker gray than that of either the Herring or Ring-billed Gulls, between which it is intermediate in size. The immature plumage is dark. These gulls nest due north of Arizona, most of them migrating west to the Pacific Coast thereafter. Presumably in migrating back north from the Gulf of California, they fly more directly home, up the Colorado River (if not arriving overland from the San Diego district).

139. RING-BILLED GULL — *Larus delawarensis* Ord

Common to abundant transient and winter visitor along the Colorado River, fairly common on Salt River lakes. Also a common transient on larger lakes of the Flagstaff and St. Johns regions, but rather rare elsewhere.

The Ring-billed Gull is the commonest of Arizona gulls. The adult has a black ring around the center of the yellowish bill. Immatures have the rear half of the tail dusky, contrasting with the white base.

This species has been found in northern Arizona from *April* (near Kayenta — Wetherill, MNA) to early June (Mormon Lake — Hudspeth) and from August (Mormon Lake — Phillips, H. I. Cone) to *November* (at St. Johns — Jacot, ARP). In the Salt River Valley it is recorded from August through May (Simpson, Werner); and along the Colorado River year-long, with numbers greatly reduced from May to August (Monson). There are some December records from south-central Arizona: Aguirre Lake, north of Sasabe (Phillips *et al.*); near the Quinlan Mountains, west of the Altar Valley (Mr. and Mrs. Foerster); and Benson (Lloyd Rullman and Hargrave, LLH).

140. LAUGHING GULL — *Larus atricilla* Linnaeus

Only one record, near Imperial Dam, *September 3, 1960* (GM). A dubious sight record by Coues from the Colorado River.

This and the succeeding two species of black-headed gulls are all smaller than those previously discussed, being hardly larger than terns. In winter the black of the head is reduced to a spot behind the eye. The slender, pointed wings give these species an airy flight, like that of a nighthawk.

141. FRANKLIN'S GULL — *Larus pipixcan* Wagler

Rather rare migrant along the Colorado River; also record of two near Joseph City, April 26, 1948 (Phillips). Only two specimens: Imperial Dam, *October 11, 1956* (GM), and Colorado River about 15 miles above Imperial Dam, *October 24, 1959* (Marshall, ARIZ).

142. BONAPARTE'S GULL — *Larus philadelphia* (Ord)

Uncommon transient and rare winter visitor along the lower Colorado River; scattered records in spring and fall from northern Arizona, and five records from southern Arizona (including one in winter at Tucson, found recently dead, *January 2, 1950* — LLH).

Bonaparte's Gull is distinguished from the other small, black-headed gulls by a white line extending along the front of the wing, in the primaries. It is the least rare of this group in Arizona.

Four specimens from the Salt River Valley and northern Arizona were taken in *April* (JSW), *May* (two, MNA), and *October* (MNA).

[HEERMANN'S GULL   *Larus heermanni* Cassin]

[Hypothetical. One record, Colorado River about 30 miles above Yuma, November 13, 1955 (Monson). Other sight records, such as one from the California side of Havasu Lake, July 13, 1948, should be disregarded.

Heermann's Gull reverses the usual pattern, for the head is light and the body dark gray; it is virtually unknown away from the ocean.]

143. SABINE'S GULL   *Xema sabini* (Sabine)

Rare fall transient along Colorado River (not "accidental" as termed by A.O.U. Check-list); also flock of seven at Martinez Lake above Yuma, April 13, 1956 (Monson), plus one April 27, 1956 (Harold Irby), and one seen east of "new" Tacna, Yuma County, April 16, 1960 (Monson). Elsewhere in state, only a few fall records, from Tucson (ARIZ), Mormon Lake (MNA), Grand Canyon (Behle and McKee), and Kingman, *October 1, 1959* (Musgrove).

This is still smaller than the above group, and has a peculiar wing pattern: there is a black triangular tip and much white in the secondaries. It is the only gull with a forked tail, but the fork is not deep. It is primarily a pelagic species. Arizona specimens have been determined as the nominate race by J. W. Aldrich and A. Duvall.

Monson's fall specimens (GM) and sight records for Colorado Valley are *September 8*, October 3, and October 9, *1948; September 23* (three birds) and October 2, 1949; October 14, 1950; September 27, 1952; September 30, 1955 (photographed); and *November 4, 1960*.

144. GULL-BILLED TERN   *Gelochelidon nilotica* (Gmelin)

One specimen from two seen, Colorado River about 33 miles above Imperial Dam, *May 24, 1959* (GM).

The Gull-billed Tern's bill is black and heavier than in other terns, thickening to an angle near the tip of the mandible, just like a gull's. But the fishing habits and airy flights are those of a tern.

The Arizona specimen, identified as the race *vanrossemi* Bancroft, is probably a straggler from the only known breeding colony in the western United States — that at the Salton Sea, discovered by A. J. van Rossem. This bird is also found in the Old World, as indeed are most species of Arizona terns.

145. FORSTER'S TERN   *Sterna forsteri* Nuttall

Common to fairly common transient in lower Colorado Valley; elsewhere rare.

Forster's Tern is the commonest of the white terns in Arizona. The bill in breeding adults is supposed to be orange, but all Arizona birds seen by Phillips have been black-billed! Though the immature is distinguished from the next by lacking a dark patch in the wing, adults can only be told safely in the hand, by the outer tail feather. This is dark on the inner web and white on the outer web, instead of the reverse as in other terns.

The migration periods along the lower Colorado River are late April to early June, and early July to mid-October (Monson); specimens there are from *May* (MVZ) to early *June* (ARIZ) and from *July 7, 1948* (GM) at Topock through *August* (ARIZ). Eastern records are from the fall migration, in *August* (ARP) and *September* (AMNH, MNA) at Becker's Lake, Camp Verde, and Mormon Lake, respectively, except for one at Tanque Verde, Tucson on *May 8, 1950* (Marshall, WF).

146. COMMON TERN   *Sterna hirundo* Linnaeus

Uncommon fall transient along lower Colorado River and in northern Arizona; also a few fall records from elsewhere in the state, including one near Peoria, *October 5, 1956* (Simpson and Werner, PX).

Due to difficulty of separating this species from the foregoing, it could occur more often than suspected; it is the more common of the two species in fall at Mormon Lake.

Other specimens are from Havasu Lake (GM), near Prescott (Jacot, ARIZ), Mormon Lake (Perkins, MNA), Becker's Lake at Springerville (Stevenson, ARP) — all in *September* — and the San Pedro River probably near Aravaipa Creek on *October 3, 1873* (Henshaw, US; date given erroneously in various publications).

147. LEAST TERN   *Sterna albifrons* Pallas

Casual: one specimen, lower Havasu Lake, *June 18, 1953* (GM). Also, flock of four or five at Mormon Lake, September 4 to 9, 1933 (Phillips), and one above Imperial Dam, July 30, 1959 (Monson). A published record for the White Mountains (Big Lake) really pertains to *Chelidonias niger* (E. A. Goldman, ms.; see Condor 28, 1926: 161).

This nearly cosmopolitan bird, smallest of the terns, is not much larger than a swallow. In breeding adults the white forehead is distinctly set off against the black crown.

The specimen should be called the race *antillarum* (Lesson) inasmuch as Burleigh and Lowery (Occ. Papers Mus. Zool. La. State Univ. 10, 1942: 175-177) have shown the alleged California race to be inseparable from *antillarum* of the Caribbean-Atlantic coasts.

148. CASPIAN TERN   *Hydroprogne caspia* (Pallas)

Fairly common transient along lower Colorado River; elsewhere, about ten seen at Sullivan Lake, north of Prescott, April 30, 1936 (Vorhies), and September to October sight records from Mormon and Long Lakes in Flagstaff area (Griscom and Hargrave, respectively).

The Caspian Tern is a gull-sized tern with a heavy

red bill. Migration along the lower Colorado River is in late *March* (US) to June, and mid-*August* (Brown, ARIZ) to late October (Monson).

149. BLACK TERN   *Chelidonias niger* (Linnaeus)

Common transient along Colorado River, and fairly common over rest of state, especially in fall. No positive breeding record, though observations span the summer.

The Black Tern is the most common tern in eastern Arizona. It is hardly bigger than the Least. In all plumages the back and mantle are dark slaty — unlike any other tern in Arizona. It also differs in its more irregular, nighthawk-like flight.

Migration is principally in May and early June and in late July to September along the Colorado River. Two pairs at Becker's Lake, Springerville, on June 15, 1957, may have been breeding (Levy).

PIGEONS AND DOVES
COLUMBIDAE

These are the only birds that feed their young on milk, which is produced in the lining of the crop. The bill has a saddle-shaped appearance, with a swollen, fleshy operculum roofing the nostril. Doves and pigeons are the same, the names being applied merely to birds of different sizes. The head is small, legs are short, wings are rather large and powerful, and the tail has a characteristic shape for each genus. These are exclusively fruit- and seed-eaters. The two white eggs (in some species, one) can be seen through the thin bottom of the flimsy nest. The male incubates during the day; the female at night. Arizona has more kinds of pigeons than any other comparable area in the United States save Texas: we have a large one, two of middle-size, and two of small size. Pigeons and doves must drink twice a day, and for this reason the evening flights of Mourning and White-winged Doves in Arizona are often immense and spectacular. Unlike most birds, they drink by sucking up the water.

150. BAND-TAILED PIGEON   *Columba fasciata* Say

Common summer resident of mountains, from northwestern to southeastern Arizona. Rare and irregular in winter in southeast. Casual transient on desert, straggling to Big Sandy Valley (ARP), Ajo (ARP) and Growler (R. D. Johnson, GM) Mountains, and even Yuma (ARIZ). Statement that it winters at Prescott and on Verde River (A.O.U. Check-list) not based on any valid record known to us.

The same size as the Rock Dove (domestic pigeon), the Band-tailed Pigeon is distinguished by its dark rump and by the pattern of its tail, which shows a pale terminal third preceded by a slightly darker band. (Domestic pigeons in any case seldom range far from buildings.) Close up, the Band-tailed Pigeon shows a narrow white

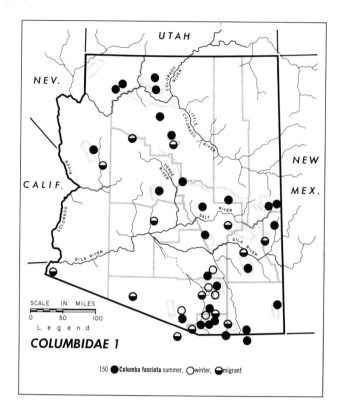

COLUMBIDAE 1

150 ●*Columba fasciata* summer, ○winter, ◐migrant

crescent on the nape. The voice is a deep owl-like double hoot.

The Band-tailed Pigeon is especially common in the oak-juniper-manzanita association and in Canadian Zone fir forests; but the birds nesting in forest generally forage elsewhere — in open meadows or lower down among the oaks. Where acorns or boreal zone meadows are not available, as in pure stands of yellow pine or of piñon and juniper, pigeons are rare or unknown. The distribution is thus very spotty (see accompanying map). Evidence of occurrence on Navajo Mountain, Utah, is not considered quite conclusive, and there are no records for northeastern Arizona. During migration this species occurs rarely in small numbers in the pines near Flagstaff and even along Lower Sonoran Zone rivers, in April to early May, and in October to early November — represented on the map by half-filled circles.

Local movements and abundance, as well as seasonal occurrence, are correlated with the acorn, manzanita, elderberry, madrone, and other fruit crops. For instance, in June 1951, the birds were eating tender young spruce cones at the summit of Heliograph Peak 10,000 feet, Pinaleno Mountains (Marshall). In summer they regularly descend streams of the upper Santa Cruz River drainage to feed on elderberries, as represented by dots (which do not imply nesting) near Tumacacori Mission (L. D. Yaeger) and lower Sonoita Creek.

The southernmost dot, however, does represent a nest: in the Mule Mountains above Bisbee with an egg on July 10, 1939 (Monson). At least in some years, a few pigeons remain in winter in the mountains of central and southeastern Arizona, and two from Gardner Canyon, Santa Rita Mountains, *March 7, 1959* (Hungerford, ARIZ), may have been wintering birds. If not, then they indicate that spring arrival varies with climate and food no less than does fall departure. Usually these are about the first of May and end of October, respectively. But in 1936 there were still nests in fall: a nestling in pinfeathers October 24 near Nutrioso (*fide* Phillips) and a nest with an egg "in which incubation was just starting" on *October 25* near Red Mountain, west of the Blue River, Greenlee County (WJS). Many nests there had small young at the time (W. J. Sheffler). Most extreme dates for nesting are discredited, particularly those of spring, for the Band-tailed Pigeon does not usually begin nesting, at least in the high mountains of Arizona, until the first of July. But both the migrations and nestings of this pigeon appear to be quite irregular and unpredictable. Nests are often, but not always, in small loose colonies. Arizona nests seem invariably to contain only one egg; whether more than one brood is raised by the same adults is uncertain. Arizona specimens, including the straggler from Yuma, all belong to the subspecies *fasciata* Say. It is the relatively pale and large Mexican mainland race. It agrees with other northern populations in its black-tipped yellow bill and white posterior underparts.

151. WHITE-WINGED DOVE *Zenaida asiatica* (Linnaeus)

Abundant summer resident in southern and central Arizona, and along Colorado River below Davis Dam. Mostly rare and irregular in winter. Has probably increased since 1870's, but much nesting habitat is now threatened.

This is a big dove, nearly of pigeon size, whose long broad tail has white corners. The white on the wing coverts is conspicuous both while perched and in flight. The call is a loud "Who cooks for you?" and the song goes on from there. The White-wing is one of the most popular game birds in Arizona, but it is not too satisfactory, for it continues nesting right up to the time of its fall departure for Mexico. Alcorn *et al.* at Saguaro National Monument (*Science* 133, 1961: 1594-1595) prove this dove to be an important cross-pollinator of the saguaro! Other studies of the dove are by Neff (*Jour. Wildlife Mgmt.* 4, 1940: 117-127, 279-290) and Lee Arnold (*The western white-winged dove in Arizona,* Ariz. Game and Fish Commission, 1934: 1-103).

The summer distribution is almost throughout the Lower Sonoran Zone west and south of the Mogollon Rim, where it is abundant. The northernmost nesting in Arizona is a nest with eggs found in Truxton Wash, between Valentine and Hackberry, Mohave County by Wm. G. George on June 2, 1958. It is less common, and local, in the more open oak groves of Upper Sonoran Zone foothills such as at Oracle and in Madera Canyon of the Santa Rita Mountains. A few individuals remain nearly every winter in such places as the University of Arizona campus, Phoenix, and base of the Baboquívari Mountains (RSC); but a flock of about ninety spent the winter of 1959–1960 northeast of Phoenix (Steve Gallizioli). Another unusual concentration was ten to twenty birds at Kinsley's (Arivaca Junction) on October 11, 1954 (Phillips), a date when all summer birds should be gone.

Migration is obscured by the wintering birds, and many March records are probably of wintering birds that were overlooked before they began calling. Usually the first arrivals appear between April 10 and 25, and the species becomes abundant at the beginning of May and remains so to early September. By September 15 the bulk has departed, and the remaining few leave by early October. The nesting season is so long-drawn-out that some wintering birds may have eggs before the migrants return for the summer (incubating near Tucson, April 13 — Swarth, *Birds of the Papago Saguaro National Monument,* 1920: 42), while others have young in the nest as late as September 15 (*fide* Neff, *op. cit.*: 285). By this time nearly all their companions have left for Mexico. Two broods seem to be the usual number, though three might be raised where the birds are left undisturbed.

There is little doubt that the White-wing has increased its range and abundance to the northwest, but the details are confused owing to faulty early records. For instance, Abert's bird mislabelled from New Mexico probably came from southern Arizona in the 1840's; Coues' "tolerably common" birds at Fort Whipple, near Prescott, in 1864–1865 were two caged juveniles. But Frank Stephens called them only "not uncommon" at Tucson in May and June 1881, and saw them only "occasionally" from there west along the Gila; he was genuinely surprised to find them "actually common" at Yuma.

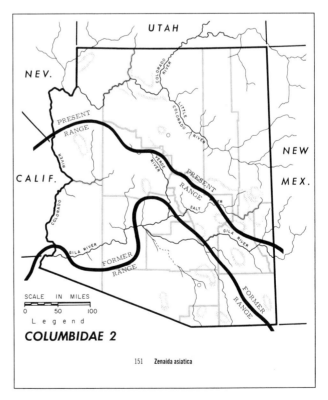

COLUMBIDAE 2

151 *Zenaida asiatica*

Mearns secured neither skins nor eggs at Camp Verde in 1884–1888, and saw a single bird only (spring of 1884, ms.). But he took numbers on a southward trip, the northernmost being near Bumble Bee in the Agua Fria Valley, much closer to the saguaro country. By 1916 the White-wing was abundant at Camp Verde (Taylor and Jackson, FW files), and still is to this day (1962, Ambrose and Marshall); and it ranges up the Verde River to the mouth of Sycamore Creek (1954 – M. A. Wetherill). More than four were heard at and opposite Fort Mohave, and about 22 at Date Creek in May, 1950, by Phillips, where they had not been found in the 1860's by Cooper and by Coues. Even the trembling Coues could hardly have overlooked or misidentified so conspicuous and noisy a bird! It is a safe guess, then, that the White-wing has spread in comparatively recent times northward to Safford and the Gila River generally, to the Verde Valley, and up the Colorado River from Yuma to Needles.

The principal nesting environment of this dove is the dense forest or *bosque* of mesquite which grows in broad riverbottoms, where there is open country or farmland nearby for feeding. Thus there are tremendous numbers nesting in the mesquite bosque on the San Xavier Indian Reservation, surrounded as it is by cultivated fields. It is difficult to make tape recordings of other birds there due to the great volume of sound issuing from the White-wings, and sometimes the brooding doves are so numerous that when a company of them is startled into flight, eggs splash down around the incautious intruder. Similarly, Wetmore (Condor 22, 1920: 140-146) found scattered pairs and colonies up to at least 2000 pairs in mesquite thickets at Arlington in June, 1919, and he believed that reduction in numbers would be inevitable with the cutting of the "mesquite *montes*." This also is in a farming area. In contrast, we find around 1962, that White-wings are few in the splendid mesquite forest along the San Pedro River, especially near Mammoth, where there is still little cultivation. A little to the north, at Feldman, they were abundant in the 1940's (Monson, Phillips).

Thus an important part of the life cycle of the Whitewing is bound up with the mesquite, *Prosopis juliflora*; if you remove it from the riverbottoms, you lose the great nesting concentrations of the dove. Mesquite is a picturesque, fitting, and even a valuable tree in its natural habitat of the bottomland. Not only does it provide an ideal nesting environment for White-winged, Mourning, and Ground Doves and many other birds such as the Gambel's Quail, Screech-Owl, Crissal Thrasher, Least Vireo, and Abert's Towhee, but it makes excellent charcoal, provides a sweet bean which is a food for both man and beast, holds the soil, and in olden times made great beams for buildings. Original stands of large trees, composing a forest beneath which a car could be driven in any direction, were still to be seen in the 1950's along the upper Santa Cruz River, in that part which flows on the Sonora side of the International Boundary, where it is still a permanent stream! Hatred of the mesquite should be directed solely to its occurrence outside of its natural habitat — on the bajadas and plains which used to be desert grassland. Here the mesquite, growing only to large bush size, is an unwanted invader, arriving in retribution for man's insensate total destruction of our local native grasses. (Presumably in Indian times, before the influx of the great Spanish herds, mesquite and cholla were kept in check by seasonal fires in the luxuriant grasslands — opinion of Marshall, not shared by Phillips and Monson). But it is intolerable from the standpoint of soil, water, and wildlife conservation, to see the splendid mesquite *bosques* removed from the riverbottoms where they belong!

Where mesquite has been drowned by the creation of dams, as along the Colorado River, or removed, a most undesirable alien tree, the tamarisk, enters in such abominable density as to make the riparian vegetation impenetrable. Great sums are spent in attempts to control it. Since it is non-native, it constitutes a sterile environment, and is not attractive to birds. The only good thing that can be said for it is that it is hospitable to nesting White-wings. About 125,000 pairs were estimated to have bred in 7000 acres of *Tamarix* thickets along the Gila River from the Salt River mouth to Gillespie Dam in 1959 (Harley Shaw and Jas. Jett). This is of course no argument for replacing *Prosopis* by *Tamarix;* the latter is merely better than no vegetation at all.

152. MOURNING DOVE *Zenaidura macroura* (Linnaeus)

Common resident in valleys of southern, central, and most of western Arizona; abundant in farmlands. Common summer resident in the north, where it is generally rare and local in winter.

This bird, familiar over almost the entire continent from Canada to Panama, is found in truly great numbers in Arizona. Only slightly smaller than the Whitewing, it has plain wings and a long, pointed tail, which gives a characteristic stream-lined shape, distinctive at a great distance. The white in the tail is limited to the sides of the base. The Mourning Dove is an important gamebird here as elsewhere. In the Flagstaff region, where there are almost no trees except ponderosa pine, about half the nests are placed on the ground. A bird banded there in summer was shot near Sásabe, Sonora. The report of two birds banded in Ohio, which were shot in Arizona, is erroneous. The Mourning Dove gets on the highway in late afternoon to pick up gravel.

The summer habitat of the Mourning Dove is in the Sonoran Zones and open parts of the Transition and lower Canadian Zone. Only the densest forest and brush and the most barren deserts are avoided. According to J. M. Simpson and J. R. Werner, the numbers in the Salt River Valley increase by mid-September, providing a dense population through April. They estimated

10,000 birds on a ten-acre plot near Palo Verde on October 16, 1955. For this dove is migratory, at least in the north, where it occurs mostly from the middle of April to the middle of September, except for birds wintering locally along the Mogollon Plateau during some winters, such as *1936* (Jacot, Phillips, and Correia, ARP). Then they occurred at several localities near the Rim, such as Springerville and Lakeside; but they were absent farther north at St. Johns and Holbrook (Jacot). Most of these northern winter records are from farms, and there is a possibility that the species winters in the Hopi farmlands.

Though some pairs are nesting in every month of the year on the Colorado and Salt Rivers, the normal span for eggs is from April to August; males begin their display flights on New Year's day at Tucson. Mr. and Mrs. Morton have observed one nest of three young, and four broods by one pair in only part of the nesting season at Tucson. Some nesting doves banded at Tucson in late June, 1963, bore juvenal wing feathers indicating they had been raised in about February of the same year (L. H. Blankenship and H. Irby)!

153. COMMON GROUND-DOVE *Columbina passerina* (Linnaeus)

Resident in better-watered valleys of southern and central Arizona, especially common in summer. Occasionally seen to north, even to Flagstaff (*October 30, 1931* — Hargrave, MNA) and Grand Canyon Village (October 22–23, 1930 — C. A. Bryant); also in southern Nevada (Hardy, Gullion). Rare or absent some winters. Occasionally wanders into foothills near resident range.

The Ground Dove is a small chunky dove not much larger than a sparrow. The tail is broad and short, appearing mostly black, and is without any white. The bird flies up with a loud clatter, displaying a surprising chestnut underside of the wing. This is the only Arizona dove which has pronounced sexual dimorphism: the male is pinkish anteriorly, in contrast with a bluish crown; whereas the female is mostly dull brown. The call is a monotonous single *coo,* slightly rising — a depressing sound in the stifling midday heat.

The accompanying map shows the range, which seems to be increasing, of the Ground Dove in Arizona. The habitat is in the Lower Sonoran Zone of river bottomlands, where there are mesquite woods (or tamarisk) and farms with brushy or weedy borders. Nests have been found at Tucson from May (Phillips) to September (Bendire); the preferred site is an elder tree within the mesquite *bosque.* In western Arizona nesting may start much earlier, by February. Phillips believes that the species is partly migratory; most of the birds leave from early September to late October and return from mid-March to late May, based on observations at Tucson in the 1940's and early 1950's. Also he regards the various stragglers into higher Life Zones and northern localities (marked by a half-filled circle) as a product of the migratory movement, since most of these records are in the fall. On the other hand, Marshall's intensive observations at the mesquite *bosque* on the San Xavier Indian Reservation from 1957 to 1963 show that the Ground

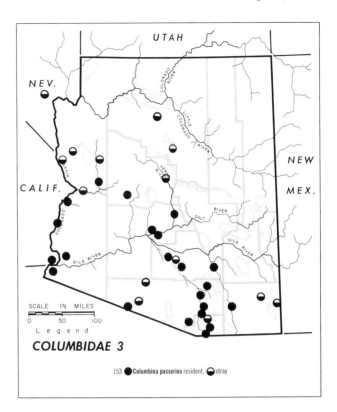

COLUMBIDAE 3

153 ● *Columbina passerina* resident, ◒ stray

Dove is equally common there summer and winter (as it is at Yuma also — Monson). There are even flocks formed in winter, and aggregations up to 25 birds occur in elders beside water holes.

154. INCA DOVE *Scardafella inca* (Lesson)

Common resident of cities and towns of central and central-southern Arizona. Resident at Parker. Also has been found at Bisbee, Yuma, and the California side of Parker Dam, where not yet permanently established. Has straggled to various mountains in southeast. Probably absent from state prior to 1870.

This is a pale grayish dove fairly similar to the Ground Dove but with a long slim tail, bordered with white at the sides. It has a similar flash of chestnut on the underside of the clattering wings. The call is a dreary "no hope." In the hand, the male is pinker on the chest than is the female. The tail is raised high when these doves quarrel among themselves. In fall and winter Inca Doves often gather to roost in sociable little pyramids among the branches. The young, on leaving the nest, also show this huddling instinct and repeatedly clamber onto their parents' backs until they can force their way down to roost between them. In building, the male brings a twig, alights on the back of the female, and passes it down to her while she is sitting on the new platform. This is one of the most friendly and delightful of Arizona birds, which even a shut-in can enjoy. Casting its lot with man, it is confined to settlements. It has the longest breeding season of any Arizona bird, with four

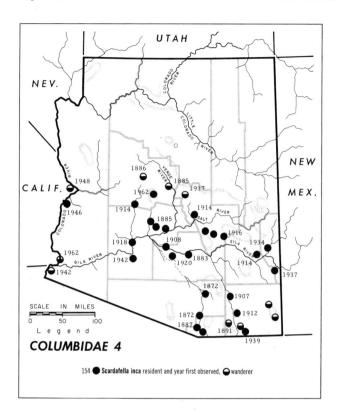

154 ● **Scardafella inca** resident and year first observed, ◐ wanderer

COLUMBIDAE 4

or five broods from January to November, at least in warm winters. The extraordinary behavior rituals of this little bird have been studied by Johnston (Condor 62, 1960: 7-24), and its nesting at Tucson by the Andersons (Condor 50, 1948: 152-154).

While occasional birds appear in unexpected places in September, October and late April (half-filled circles on accompanying map), the species is not truly migratory, nor was it ever migratory, for after its first appearance, in winter, it was breeding the following June. None of the early explorers found this dove in Arizona. Bendire (*Life Hist. N. Am. Bds.* 1, 1892: 153) "first met with this species in February, 1872," at what later became Fort Lowell, near Tucson, where he had arrived four months earlier. He believed "a few pairs bred" near there in 1872, but found his only nest near Tubac, Santa Cruz Valley, June 6, 1872. Possibly he didn't work the ranches and town, and the fort itself was too new for the establishment of doves. At any rate, Inca Doves were rare that year (cf. Coues, Am. Nat. 7, 1873: 321, 323, footnotes). Henshaw never met with this dove, even on his visit to Fort Lowell in September, 1874. Scott (Auk 3, 1886: 321), however, soon found it "common" at Florence and Tucson, presumably in 1882–1883. In the *spring of 1885* Mearns found it already well established at Phoenix, and in *October* he secured three at Camp Verde (AMNH and ms. — where it has not been found since). Other early arrivals are at Calabasas in *1887* (F), Benson in 1907 (V. Bailey, FW files), and Sacaton in 1908–1910 ("common," Gilman, Condor 13, 1911: 55-56). At Tombstone it was not found nesting up to 1911 (see Willard, Condor 14, 1912: 59 and Sloanaker, Wilson Bull. 25, 1913: 195), but it probably arrived that fall, since Willard (FW files) reported it as first seen January 30 and February 3, 1912, and before departing from Arizona in 1916 found it nesting "repeatedly" there. By 1913–1914,

Biological Survey parties found the dove at Wickenburg, Safford, and Solomon, and in 1916 at San Carlos, but not at Camp Verde, whence we know of no recent records.

PARROTS   PSITTACIDAE

Parrots, which include love birds, lories, cockatoos, budgies, and macaws, use their short, thick bills and feet both for climbing about branches and for handling and eating fruits and seeds. Their deafening calls are devoid of musical quality. Most American parrots are green-bodied, and they apparently mate for life soon after attaining their full growth, for they are usually paired even within the great flocks. They gather to roost in immense dormitories. In flight, parrots are recognizable at a distance because their wing-tips are moved only downward from the horizontal.

155. THICK-BILLED PARROT   *Rhynchopsitta pachyrhyncha* (Swainson)

Formerly erratic visitant, mainly in winter, to mountains of southeastern Arizona. The last reliable reports were in 1922 and 1935, and the last major flight in *1917–18;* rumors to 1945. Ranged north centuries ago to the Verde Valley.

Most parrots are tropical, but the Thick-billed Parrot inhabits highland pine forests, where others would die of the cold. It is almost entirely green, with a red forehead. The tail is long and pointed, and the bird is about the size of a Peregrine Falcon. Its screeches are perhaps the most horrible of all parrots'! When not nesting, it is highly gregarious in its wanderings in search of new pine-cone crops; this has in the past brought it as a visitor to Arizona, from the Sierra Madre Occidental, about 100 miles southeast of us, along the Sonora-Chihuahua border. It has never nested in the United States, for in the past it utilized for its nests the holes of the Imperial Ivory-billed Woodpecker, now almost extinct. (The legal protection long afforded this bird in Arizona, while showing the right spirit, would appear rather pointless.)

On its visits to Arizona, the Thick-billed Parrot has sometimes been abundant, and it has largely been limited to stands of pine, mostly of ponderosa pine. Seasonal occurrence varied; usually it came to the Chiricahua Mountains, occasionally also to nearly all the higher mountains east and south of the Santa Cruz and Gila Rivers. Many records for this region are given by Wetmore (Condor 37, 1935: 18-21). Formerly, or casually, it ranged north to Beaver Creek, near Camp Verde (May 5, 1583 — Espejo Expedition; see Colton, Mus. N. Ariz. Mus. Notes 2 [9], 1930: 1-4; Bartlett, New Mex. Hist. Rev. January, 1942: 21-36 and map; and Wetmore, Condor 33, 1931: 35). Archaeological evidence indicates that about 1200 A.D. it may have occurred naturally in the San Francisco Mountains region, if this species was not an object of trade as were the Macaws (see Hargrave, Condor 41, 1939: 208).

Principal invasion years in recent times were 1904 and 1917–1918. Seldom if ever has an ornithologist been on the spot, so nearly all records are by ranchers. The following is a compendium of all recent reports, notably those given to Wetmore, Goldman (FW files), and Phillips:

Chiricahua Mountains: several invasions prior to 1900, not dated (probably including May, 1886, when Shufeldt received reports of them in southern New Mexico); 1898; *"about the middle of June,"* 1900; 1902 (?); *April 26, 1904* (?? — FW files); August 26–29 (or 30), 1904; 1906; August 1908; July *1917* to *March 27, 1918;* April to June, 1920; summer of 1922.

Dragoon Mountains: August to September 1917; summer of 1918.

Pinaleno (Graham) Mountains: sometime prior to 1917.

Galiuro Mountains: May to August 1918.

Santa Catalina Mountains: late summer, 1922.

Whetstone Mountains: probably 1904 (see Smith, Condor 9, 1907: 104).

Huachuca Mountains: August 1904; August 1908.

Patagonia Mountains: August 1904; August 1908; September 1917.

Santa Rita Mountains: about early November 1936.

Because of the rapid destruction of the pine forests of Sonora and Chihuahua, this parrot will probably never be seen again in Arizona.

CUCKOOS, ROADRUNNERS, AND ANIS *CUCULIDAE*

Like the parrots and woodpeckers, these birds have toes number 2 and 3 in front and 1 and 4 pointed back, thus giving the Roadrunner a unique track, like an X, imprinted on desert washes. Most cuckoos except the Ani have moderately slender and decurved bills. The tail is always long. Arizona cuckoos utter series of coos, but not the two-note call of the European Cuckoo and cuckoo clock. Our birds build their own nests and rear their own eggs, unlike certain parasitic species elsewhere.

156. YELLOW-BILLED CUCKOO *Coccyzus americanus* (Linnaeus)

Breeds along main rivers of Sonoran Zones, chiefly of southern and central Arizona; rare transient on desert and in towns.

This is a long slim plain brown bird with white on the underparts and tail corners. In flight it shows rufous under the pointed wing. The base of the mandible is yellow in adults. Short-legged and arboreal, the Yellow-billed Cuckoo is seldom seen near the ground. It moves quietly among the leaves and is famous as a consumer of tent caterpillars — too hairy an item to appeal to most palates. Unfortunately, the Yellow-billed Cuckoo is our latest arrival in spring, giving the caterpillars such a head start that the willows and cottonwoods of the Santa Cruz River near Tucson are often completely denuded (after which they put out a new set of leaves).

The extraordinary vocalizations of this cuckoo consist of an explosive chortling call and a song of 10 strangely inflected *teoo's,* becoming fainter through the measured series. The nest is a thin platform of twigs, and the eggs are plain blue. The amazing thing about the cuckoo is that its principal spring flight in the west seems to occur at the end of June!

The Yellow-billed Cuckoo is a fairly common summer resident, from the first week of June through September, in willow-cottonwood and dense mesquite associations throughout Arizona. Due to the general lack of such vegetation in northeastern Arizona, it is necessarily very local there. Migrants or wanderers are still abroad in non-breeding habitats, such as cities, deserts, and foothills in July as well as June, but of course these are very rare and unusual. For instance, a male flew into a window at Boulder Beach, Nevada, on *July 11, 1950* (Nora Poyser and Grater, collection of Lake Mead Nat. Rec. Area); and one was at Snowflake from July 28 to August 13, 1953 (Albert J. and Louise Levine). Further enumeration of all the exceptional specimens and sightings, such as Mearns' claims for early May at Camp Verde; a lone bird at the Wm. H. Woodins' residence on Sabino Creek, Tucson, October 18, 1957 (Marshall); etc., would merely obscure the fact that the species regularly stays its short allotted time, nests in the summer rainy season when caterpillars are most abundant, and generally escapes notice in the dense foliage because of its sneaky behavior.

157. ROADRUNNER *Geococcyx californianus* (Lesson)

Common resident in Sonoran Zones (chiefly Lower) of southern, central, and western Arizona; scarce in north-central part, and rare in northeast, where still no nesting records. Wanderers appear in high mountains occasionally.

The Roadrunner is perhaps the most distinctive and well-known of Arizona birds. Something like a pheasant in size and shape, it is much more agile and diverting. Though essentially streaky black, gray and white, it has a colorful patch of bare skin behind the eye, and in the hand it shows a surprising variety of iridescent colors. Roadrunners are solitary and silent except in the nesting season. Then they emit a deliberate series of soft descending groans, as well as a rapid drumming of the mandibles. They sometimes fly or climb into a tree, particularly for singing, and the nests are built in small trees, bushes, or cacti. But most of their lives are spent on the ground. Incubation seems erratic, and the young are therefore of varied sizes and of short life-expectancy.

The Roadrunner is the subject of many legends in the Southwest, concerning chiefly the rattlesnake. However, its principal diet consists of lizards and grasshoppers, or when these are scarce, other insects such as stinkbugs. It will occasionally ambush and eat English Sparrows and mice (Bruce Cole; Laurence Huey — movies taken at Willow Beach, and Jour. Ariz. Acad. Sci. 2, 1963: 110). Actually it will probably take any live animal food it can get, big or small. Roadrunners show great ingenuity in getting this food; their antics are well worth watching, including the courtship, which has never been adequately seen or described, but which may consist of some sort of side-stepping dance.

Not only is the Roadrunner common on the open desert and in the more open woods and brush of the Upper Sonoran Zone, but it also inhabits farmlands and urban areas. On and north of the Mogollon Plateau it is an uncommon wanderer at almost any season, with a concentration of records along the Little Colorado River, especially near Holbrook (also to the South Fork, above Springerville — Wm. B. Heed), indicating that it must breed in this region of junipers and grass. This is substantiated by a female "with egg ready to be laid in a day or two" 20 miles east of St. Johns, *July 30, 1937* (H. H. Poor, ARP). Thus far no wanderers have been seen above the Transition Zone.

158. GROOVE-BILLED ANI — *Crotophaga sulcirostris* Swainson

Casual fall straggler into southeastern Arizona; only one presently confirmable record, Pinery Canyon, Chiricahua Mountains, on or about *October 16, 1928* (Frank Hands, ARIZ), though two others are cited by the A.O.U. Check-list.

The Ani violates almost everything we have said to characterize cuckoos, above. It looks like a large blackbird with a grotesquely arched bill and a tail that appears ready to fall off. It does not normally come farther north than the Rio Yaqui, Sonora, where it is a summer resident, only. There you should go to hear its liquid call and to watch it feeding under the cows, on insects scared up by the animal's feet.

The reason that the record for Fort Huachuca in May 1888 cannot be confirmed is that the California Academy of Sciences and all its papers and specimens burned up in the San Francisco fire. The sight record by Walsh (Auk 50, 1933: 124) turns out to have been on the desert at Picacho (Walsh, *in litt.*). There are two other possibly valid sight records, for about 1925 and 1947 (*fide* Phillips, near Tucson and Benson, respectively).

After the above was written, a Groove-billed Ani obliged the Tucson Audubon Society by appearing at Sabino Canyon Dam, Tucson, on October 19, 1963 (Fern Tainter *et al.*).

OWLS — STRIGIFORMES

These birds of prey have forward-directed eyes, the retinas of which are richly or exclusively made up of cones, for nocturnal vision. The extremely sensitive ear has a large opening, shielded by plumes of loose texture to admit the sound waves, which are additionally concentrated by papery feathers of the facial disc, around each ear. The softness of the plumage is accentuated by a nap of velvet on the flight feathers, which makes the wing-strokes of most species noiseless. The outer (fourth) toe is reversible; the owl may perch with two toes in front and two behind, but when pouncing on its prey it swings the outer toes to the side so that the four claws are equidistant — making an effective trap. Cryptic coloration is developed to an extreme of beauty and protectiveness in owls, in order that they may blend with the bark of trees on which they roost. This is achieved through intricate patterns of fine black pencillings together with bars, spots, or hash-marks in lovely rusty, tawny, ochraceous, gray and brown colors. Arizona species of owls are best identified by their voices; in each species the female's is about a fifth higher in pitch than that of the male, despite her superior size.

BARN OWLS — TYTONIDAE
159. BARN OWL — *Tyto alba* (Scopoli)

Fairly common resident throughout most of the open Sonoran Zones except in the deserts of the southwest. Apparently local, but ranging into lower Transition Zone, in the northeast; no winter records north of Needles (California), Salome, and Camp Verde.

This is the lightest in color of all Arizona owls. Individuals vary in the tone of golden-tawny, but most appear white on the entire under surface. The two facial discs surrounding the relatively small, dark eyes, give a heart-shaped outline resulting in the common name of "monkey-faced owl." It was with the Barn Owl that Roger Payne proved in a series of classic, simple experiments that owls can and do catch living prey in total darkness, by means of hearing. His pet Barn Owl was able with unerring accuracy to catch a mouse at the first swoop, as soon as the mouse rustled the leaves on the floor of a blacked-out barn. With one ear plugged with soggy Kleenex, the owl would miss by about a foot on the first flight, but would catch the mouse on a second try (Natural History 72, 1958: 316-323). The call of the Barn Owl is a horrid, rasping or snoring screech. In Arizona the owl has been found roosting in old mines, wells, buildings, large cottonwoods and willows, palm tree rows in towns, and stacks of baled hay. Nests have been found in mines, buildings and holes in banks. Cavities of the large riparian trees are probably also used here. The Barn Owl is much less common in Arizona than in other parts of the world, and fortunate is the community that can boast a family of these rat-catchers in a local steeple.

The few records for the northeast include Springerville (a primary found by Stevenson, ARP), two young photographed up at the edge of the pines at Nutrioso (about August 20, 1935 or 1936, *fide* Phillips), and near Holbrook, July 3, 1949 (Phillips).

TYPICAL OWLS — STRIGIDAE
SCREECH-OWLS; SCOPS-OWLS — Genus *OTUS*

by Joe Marshall

Screech-Owls do not screech. If our naming of birds were consistent, "Screech-Owl" would have to apply to the Barn Owl instead, and the genus *Otus* could bear the term Scops-Owls, after the widespread Old World species. In either case there is considerable merit in retaining a group name, for these owls all look so much alike that some records, even by experts, have to be entered as *"Otus* sp." or Screech-Owl in the generic sense. The A.O.U. Check-list Committee wants shorter names for our three Arizona species: Screech-Owl, Whiskered Owl, and Flammulated Owl instead of Common Screech-Owl, Spotted Screech-Owl, and Flammulated Screech-Owl. The new names are handy and they avoid giving the mistaken impression that these are subspecies of a single species, but they leave us with no common denominator for use as a group designation.

The following points will help to characterize and distinguish Screech-Owls as they occur in Arizona, where they are all of gray color through parallel evolution. Hence they are best identified by voice: a single hoot for the Flammulated, speeded-up trill by the Common, and slowed-down or syncopated trill by the Spotted.

Screech-Owls, like the Elf Owl, eat mostly insects, scorpions, centipedes, and millipedes. They are very abundant, but they pass unnoticed except when there is a general outbreak of hooting — usually on nights of bright moon in early summer and fall. Then several pairs can be heard from a single spot. This is in marked contrast to the carnivorous Pygmy-Owls, each pair of which has a territory about a mile long. Thus there is a correlation between type of food and size of foraging area. There is furthermore a relation among the insectivorous owls between body size and territory size (or spacing apart of the pairs in the population). The Elf Owl and the Flammulated Screech-Owl are the smallest and closest together; the Spotted Screech-Owl is next larger and of medium territory size; the Common Screech-Owl is the largest and most widely spaced — at perhaps 300 yards between pairs in woodlands generally (but 100 yards in mesquite *bosque*).

There are habitat differences which serve nearly to separate the three species of Screech-Owls by altitude. The Flammulated Screech-Owl occupies coniferous forest at the higher elevations; the Spotted Screech-Owl prefers dense woods of sycamores and oaks, or oaks mixed with pines at middle altitudes, and the Common Screech-Owl takes the open woods of oaks, junipers, piñons, cottonwoods, mesquites, or saguaros in the foothills, valleys, and deserts. There is of course some overlap. In a few places, of around 5500 feet altitude in the Huachuca and Chiricahua Mountains, all three species have been found together, along with the Mountain Pygmy-Owl and the Elf Owl. The Common Screech-Owl and Flammulated Screech-Owl are found all over the state, whereas the Spotted Screech-Owl is limited to southeastern Arizona; the first and last-named are permanently resident, but the Flammulated Screech-Owl, like the Elf Owl, is migratory and occurs in Arizona only during the summer.

In Arizona, the three species of Screech-Owls are alike in being predominately gray and in having bold, short, black streaks on the underparts, these streaks crossed with two or three narrower bars. The resulting pattern suggests large ants crawling in rows up the owl's flanks. Unlike the eastern form of the Common Screech-Owl, the ear-tufts of the three Arizona species are not evident; they can be seen only on a roosting bird which changes its shape in day-time by erecting the tufts, while flattening down the rest of the feathers of the head and body. The Flammulated Screech-Owl has brown eyes and naked toes, but the other two, with yellow eyes and feathered toes, cannot be distinguished from each other in the field except by voice. In the hand, the inner web of the outermost primary is crossed by several light bars in the Common Screech-Owl but is solidly dark in the Spotted Screech-Owl. The feet of the former are large and the plumage pattern is fine, whereas the feet of the Spotted Screech-Owl are small and its pattern is coarser, so as to suggest spotting. Elsewhere in their range, both species are colored quite differently, but in Arizona their striking similarity — each species achieving here the extreme for grayness — is a beautiful example of convergent evolution.

160. COMMON SCREECH-OWL; "SCREECH OWL"; MEXICAN SCREECH-OWL; SAGUARO SCREECH OWL *Otus asio* (Linnaeus)

Common to abundant resident throughout open woods of the Sonoran Zones except in northeast, where it is scarce; even occurs in residential and farmland areas.

In Arizona the Common Screech-Owl inhabits all sorts of open woodlands of the deserts, valleys and foothills. These woods include riparian cottonwoods and willows; mesquites and saguaros of the deserts; and oaks, piñons and junipers of the Upper Sonoran Zone. This owl follows junipers on up into the pine forest above Eagar, and at the other altitudinal extreme is found abundantly in willow and tamarisk thickets of the Colorado River near Yuma. Most of these situations have in common a wide spacing of the trees, so that the owls may hunt by swooping down from a branch to bare or grassy open spaces among the trees.

The principal calls of the Common Screech-Owl in Arizona, as in the western United States generally, are first a series of notes at accelerating tempo, which Ralph Hoffmann likened to "a ball bounding more and more rapidly over a frozen surface" (*Birds of the Pacific States*, 1927); and second, a "double trill," consisting of a short trill followed immediately by a longer one. These two calls, uttered at uniform and appropriate pitch by both sexes, are generally delivered as a duet by members of a pair. They function the same as the equivalent calls of eastern birds: respectively the descending whinny and the long single trill. Despite the evident homology between the vocalizations east and west, they are as different as the calls of most full species in the

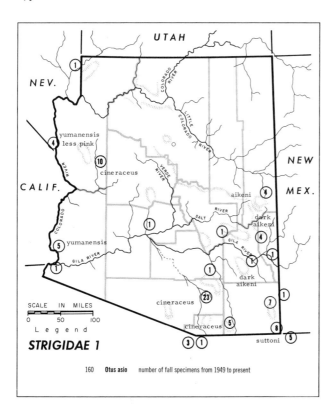

genus, and we must inquire if eastern and western Common Screech-Owls are really the same species. Since all kinds of Screech-Owls look so much alike, they must have to recognize each other primarily by voice, which must to them be a crucial species character; different voice should mark a different species. Absolute proof of such a contention is not forthcoming, although a search for it has produced some startling observations and exciting experiences. Marshall was nearly run down by a horse which galloped up and answered the taped eastern whinny with its identical but very loud real whinny; the tape, kindly provided by the Cornell Laboratory of Ornithology, was being played in an unsuccessful effort to find eastern Common Screech-Owls at Colorado Springs.

Phillips has for years looked for morphological characters which could be used to distinguish eastern from western birds in addition to the vocal differences and the possession by the eastern birds of a red color phase. Marshall found such differences in the color of the bill and in pattern, summarized thus: eastern populations have the whinny and long trill, a yellow or greenish bill, a red color phase, and a pattern of markings in which the cross-bars are emphasized at the expense of the shaft streaks both above and below. Western populations have the "bouncing ball" song and double trill, black bill, no red phase, and a pattern emphasizing the linear shaft streaks. Accordingly, he designated the western forms as the species *Otus kennicottii* (Elliot), and so labelled numerous specimens in museums. But exceptions loomed: first, the birds of typical eastern voice and bill color of the Rio Grande Valley of Texas (race *mccallii*) lack the red phase and have a linear pattern; second, the birds of typical western voice and color pattern from the Puget Sound area (race *kennicottii*) have light bills and a fully developed red phase (8 specimens, Vancouver Island, F, MCZ). Finally, in the *summer of 1962,* a mixed pair was found at the Big Bend of the Rio Grande in Texas (specimens together with tape recordings, Marshall, ARIZ). This pair had successfully reared young; furthermore they indulged in perfectly normal duets. The male (western — *suttoni*) would sing his bouncing ball song or the double trill, and the female (eastern — *mccallii*) would chime in with the typical eastern long trill and she once flew to him and did some billing with him and what appeared by flashlight to be mutual preening of the head. Marshall interprets this to mean that the trill songs of eastern and western Common Screech-Owls are still sufficiently similar to permit normal pair formation between the two kinds of birds; whether this actually happens very frequently remains to be seen, but for the moment he favors retaining both under the one species name, *Otus asio*.

The Common Screech-Owl provides one of the most splendid examples among all North American birds of extreme geographic variation in size and cryptic coloration, by which each population is adapted to its local environment. In the United States the general trends are large size to the north, small to the south; dark, rich coloration in humid regions, pale in arid; coarse markings to the north, finer to the south; increased rufous coloration and possession of a red color phase to the east. Miller and Miller (Condor 53, 1951: 161-177) studied the desert races, including those of Arizona, and clearly established the cryptic function of the complex color pattern which is a means of escape from predation. Screech-Owls prefer camouflaged immobility to flight when encountered by an enemy, and on the day-time roost they distort their outline, elevate the ear tufts, and screen the eyes by spreading their loral plumes to the side. Thus they resemble an extension of the branches upon which they sit. Desert races of Arizona, with their fine black pencillings, are the palest and grayest of all western forms of the Common Screech-Owl; the Millers find a close correlation between the overall effect of their plumage and the color and texture of the bark and foliage in desert vegetation. They also explained the increased size northward (and, in Arizona, to higher elevations eastward) as an adaptation for better heat conservation in the cold.

Despite the magnitude of the geographic differences among Common Screech-Owls, it is hard to apply scientific names for races because their characteristics everywhere grade smoothly and gradually from one area to the next. This "clinal variation" follows from the continuous lowland distribution of the Common Screech-Owl; it may be contrasted with "insular variation" shown by birds of discontinuous distribution such as on mountain tops or islands. In the latter, variation is also discontinuous, and geographic races are easy to characterize; in the former, there is more area occupied by the intermediates than by the distinctive populations, and racial "boundaries" are arbitrarily placed at the half-way point along the cline. In Arizona Common Screech-Owls there are two clines emanating from the northeast, where the birds are predominantly large, clear pale gray, boldly and coarsely marked with black. Towards the southwest, associated with a steady lowering of altitude, the birds become gradually smaller and more finely marked, with a pinkish wash pervading the back and the ventral crossbars becoming fine and indistinct (not pure black) in the birds of the lower Colorado Valley. The second cline, from northeast to southeast, is decrease in size, a progressive darkening of the dorsal ground color to blackish gray, and an increase in density of the black markings, culminating in the dark birds of Guadalupe Canyon, Cochise County.

Populations a hundred miles apart can be distinguished from each other, on the average, along these clines; but to name them all drives the subspecies concept to absurdity. Indeed, numerous races have been named, which can apply to Arizona birds. They are, in order from large to small, *aikeni* (Brewster), *mychophilus* Oberholser, *cineraceus* (Ridgway), *suttoni* Moore, *gilmani* Swarth, and *yumanensis* Miller and Miller. In the past some trouble has been caused by the marked post-mortem color change from gray to brown which shows up even after five years in museum specimens. There is a commensurate change in life, shown by summer birds whose fall gray has turned to brown of varying degrees depending on exposure to sun on their roosts. Actually, all freshly molted, recently taken fall birds from Arizona are clear, cold, pure gray. Those in the southwest have an additional suffusion of pinkish-sandy, but they are still gray, not brown. Accordingly, we shall diagnose races in Arizona just from our recent fall collections, the numbers and localities of which are shown on the accompanying map. We have no evidence for any migration or movement in these owls, for no extreme specimen is more different than the next adjacent race. The most extreme such specimen is a female from the Big Sandy Valley, *October 3, 1949* (ARP); it is the size (wing 168 mm.) and color of *inyoensis* Grinnell of Nevada, but is regarded as a variant of the Big Sandy population, since an occasional specimen even from Tucson shows similar color (but not the large size). (Incidentally, *inyoensis* is identical in size and dorsal coloration with topotypical *aikeni* from Colorado Springs, Colorado; it has the ventral color pattern of *yumanensis*, however, which provides it with a distinct combination of racial traits. It might be expected in extreme northwestern Arizona.)

A most conservative treatment would admit but two subspecies to Arizona: *aikeni* on the Mogollon Plateau, and *yumanensis* in the lower Colorado Valley; everything else is intermediate in varying degrees, depending on proximity to one or the other of the extreme populations. A traditional arrangement admits four, however, in order to be in keeping with the rest of this book, and to recognize the blackness of birds in the southeast and the fair stability of intermediate traits over a large area of central Arizona. Such an alignment follows, as shown on the map: populations not named can be assumed to be intergrades or are represented by too few specimens for allocation to one or another race.

Otus asio aikeni (Brewster) is the large pale gray race with bold coarse black markings, found on the Mogollon Plateau. Arizona birds are only a trifle darker gray and less buff than topotypes from Fountain Creek, Colorado Springs, Colorado (specimens together with tape recordings, proving that *aikeni* is a western race — Marshall, ARIZ). Wing chord of males averages 164 mm., of females, 167 mm. Birds from the south slope of the White Mountains, like those from Silver City, New Mexico (Zimmerman, WF), are of equally large size but are darker gray on the back, approaching the blackness of *suttoni*.

Otus asio suttoni Moore occurs in Guadalupe Canyon at the extreme southeastern corner of Arizona. This is the blackest of all races of the Common Screech-Owl; its dorsal ground color is blackish-gray, and it is boldly and densely marked with black both above and below. Size is fairly small, with wings of males averaging about 152 mm., of females about 157 mm. We have some hesitancy in ascribing this race to Arizona, in view of the paler coloration of birds to the south in Chihuahua (N. Johnson, MVZ) and Durango (Batty, AMNH). Elsewhere, however, on the southern and eastern portions of the Mexican Plateau, in the Big Bend of Texas (Marshall, ARIZ), and at Las Cruces, New Mexico (Ohmart, ARIZ), the birds are very black indeed, and it appears that this trait has a westward extension across southern New Mexico to Guadalupe Canyon, Arizona, and that it involves about half the specimens from the Chiricahua Mountains, Arizona. (The rest of the Chiricahua birds are paler, like *cineraceus*.)

Otus asio cineraceus (Ridgway) is a pale pure gray race like *aikeni*, from which it differs in smaller size and finer black markings. Its traits are best exemplified in the large series from Tucson, where wings of males average 152 mm., of females, 156 mm. Birds from the Huachuca Mountains, the type locality, are larger and somewhat more coarsely marked, approaching *aikeni*. But in general, this race is well characterized in a broad band extending diagonally across the state from the Big Sandy River to the Huachuca Mountains. Based on too few specimens, it appears that this small, light race extends up the Gila River past Safford, thus penetrating the domain of larger, darker birds in the uplands on either side of the river (Mammoth, Aravaipa Creek, Pinaleno Mountains, Granville, Chiricahua Mountains — ARIZ, WJS).

Otus asio yumanensis Miller and Miller occurs along the Lower Colorado Valley and is the same size as *cineraceus* from Tucson. It is a very pale gray dorsally, suffused with pinkish-sandy, and the markings are very narrow. The crossbars of the ventral surface are fine, brownish, diffuse, dense, and are less conspicuous than in any other race of the species.

161. SPOTTED SCREECH-OWL; WHISKERED OWL *Otus trichopsis* (Wagler)

Resident in the southeast, where abundant in some of the border mountain ranges in heavy Upper Sonoran Zone woodlands.

The voice, ecology, similarity to *Otus asio*, and means by which specimens can be distinguished are all treated above under the genus *Otus*. The distribution of this abundant owl is shown on the accompanying map.

Its occurrence on the south slope of the Santa Catalina Mountains deserves comment, for it is rare there in contrast to its abundance in the pine-oak woods of the northeast slope, down the Control Road. This is above the range of the Common Screech-Owl, which is in the more open oak woods from the Control Mine on down to Peppersauce Canyon and Oracle. Strangely, Scott took a Spotted Screech-Owl at Peppersauce Canyon in *1885* (AMNH), where we have always found only *Otus asio*. But to return to the south slope: we had never found Spotted Screech-Owls there except for one lone male, completely surrounded by territories of Common Screech-Owls, in scattered oaks of middle Sabino Canyon at about 4000 feet, *May 24, 1951* (Marshall, WJS). We had done considerable work at night, espe-

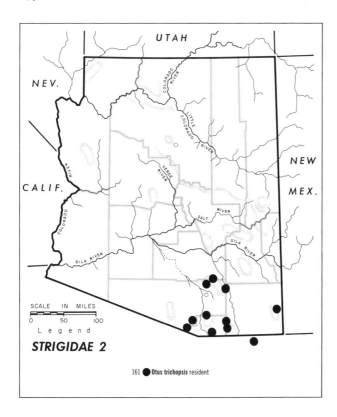

161 ● *Otus trichopsis* resident

cially from 1949-1955, from Molino Basin, at the lowest oaks, to the Hitchcock Tree, at about 6000 feet altitude and had found only *asio* at the former place and only *flammeolus* at the latter area, with no members of the genus in the piñon and Arizona cypress stands between. After the conclusion of these studies involving extensive mapping of territories of *flammeolus* (Marshall, Pac. Coast Avif. no. 32, 1957: 42 — Bear Canyon), L. A. Stimson saw a large *Otus* at the Hitchcock Tree in 1957. On *October 26, 1958* a male *trichopsis* was collected there (Bialac, ARIZ); it was in a tussle with a pair of Saw-whet Owls near the tree. Spotted Screech-Owls have been there ever since! The mighty tree, a Douglas fir, has since fallen, to be memorialized by cross-sections at the Palisades Ranger Station and Arizona State Museum.

The Arizona race, *aspersus* (Brewster), is the grayest and largest of all, thus remarkably paralleling the local form of *asio*. In addition to the vocalizations described above, the female commonly utters a short, inflected, whistled query: *kew*.

162. FLAMMULATED SCREECH-OWL; FLAMMULATED OWL; SCOPS OWL *Otus flammeolus* (Kaup)

Abundant summer resident in most of the Transition Zone, particularly where oaks are present. Not rare as a spring transient in lowlands of south, and recorded once at bottom of Grand Canyon (*May 11, 1929* — V. Bailey, US). One or two fall records from non-breeding areas: Burton, southern Navajo County, *September 2, 1949* (A. J. Levine, ARP); and near Desert View, South Rim of Grand Canyon well east of the ponderosa pines, *October 11, 1963* (GCN, *fide* Merrill D. Beal). One winter record, *February 16, 1949*, at Phoenix (Stannard and Margolin, ARP).

The identification, voice, and ecology of this abundant owl are covered above under the genus *Otus*. The single hoot of the male is often preceded by two grace notes at lower pitch. When aroused on his territory he gives a gruff, throaty hoot, whereas ordinarily the call is clear, penetrating, and ventriloquistic. The female is hardly ever seen or heard, especially when the birds are being sought by imitating the territorial call of the male — her call is a high-pitched quavering cry.

The species is not found in cut-over forest nor in pure stands of ponderosa pine; it requires some undergrowth or intermixture of oaks in the forest. Where the oaks or piñons are large and dense, at the lower edge of the pine forest, it gets into these Upper Sonoran Zone associations and occurs along with the Elf Owl. It also is common in fir and spruce forests, at least in the Pinaleno and Santa Catalina Mountains, and in aspens on Pinal Peak.

The incessant hooting of the males can be heard on moonlit nights upon their arrival in spring, if the weather is warm and calm; also they call into the summer, and again in fall, after the molt. In fall, an "Indian summer" of warm, calm weather and bright moon is again required for this resurgence of hooting. All this emphasizes the migratory nature of the owl; for while the male may not answer in the usual cold spring and fall weather, when it doubtless is actually present, yet its failure to hoot on favorable nights immediately on either side of the inclusive dates we give for its stay in Arizona is very good evidence indeed that it is absent from the state in that interval.

Our records on the breeding ground extend from March 26, 1953 and *April 10, 1950* to *October 12, 1951*, all at Bear Canyon, Catalina Mountains, Pima County (Marshall, WF). Further evidence for migration are these nine spring birds from well below any pines, with dates from April 7 to May 15: Paul Spur, west of Douglas, Cochise County (*April 7, 1956* — W. Alexander and E. Ordway, ARP); Tucson (*April 29, 1957* — R. Carpenter, ARIZ); Castle Dome Mountains, Yuma County (April 29, 1959 — Monson); "Tucson" and "Catalina Mts. at 2500 ft." — 2 males probably both from Sabino Canyon (*May 3, 1918* — Kimball, AMNH); San Pedro River east of Huachuca Mountains (*May 5 or 9, 1902*, 2 females — Howard, *fide* Swarth, Pac. Coast Avif. 4, 1904: 9 or *fide* FW files, respectively); Benson, Cochise County (*May 7, 1951* — Hargrave, ARP); and south of Peppersauce Canyon, base of Catalina Mountains (*May 15, 1961* — P. Westcott, ARIZ).

The Flammulated Screech-Owl is so uniformly abundant in ponderosa pine forest that we need not represent its range on a map, which would be coexistent with one of the greatest "empires" of this pine in North America. It is especially abundant in the Rincon Mountains of Pima County and the Flagstaff region, wherever large Gambel's oaks occur within both of these forests; it is also abundant at the lower edge of the pines in the Catalina Mountains, where it likes little clumps of silver-leaf oaks. It is the common owl at Grand Canyon Village. Just beyond the upper end of the road in Madera Canyon, Santa Rita Mountains, its abundance is remarkable because of the few pines there. Apparently this is a "traditional" breeding ground, preserved despite the logging off of the pines — the oaks are especially large there, which gives a forest-like aspect to the place.

Geographic variation in the Flammulated Screech-Owl is minor compared to that in the Common Screech-Owl because the former is so highly migratory, and so likely to mix genes from one population to another through deviations from the migratory path. Were this not so, it would probably show considerable racial variation because its breeding range is broken up into insular populations on mountains. As it is, there is just a gradual and not very spectacular cline of decreasing size and increasing redness from north (Okanagan region, British Columbia) to south (Las Vigas, Veracruz), speaking of the breeding range. The winter range is from Sierra Autlán, Jalisco (WF) to Guatemala (F). In the south, there are numerous very red specimens, but these also turn up as variants in populations to the north, including the Davis, California bird of *October 31, 1935* (Emlen, University of California at Davis). Equally striking, but unfortunately no more consistent in occurrence, are the very black birds with extremely heavy ventral marks taken in the Hualapai Mountains (ARP). However, these latter may represent a well-isolated population, for neither van Rossem nor Marshall found it in the Charleston Mountains, Nevada, just to the west. Until this is investigated, we can only recommend that no races be recognized for the present; otherwise a pretty chaotic idea of migration would emerge from the far-flung occurrence of the extreme black and red variants.

163. GREAT HORNED OWL *Bubo virginianus* (Gmelin)

Fairly common resident statewide, except in densest forests.

Horned Owls are common residents throughout the Sonoran and Transition Zones of Arizona, except in the densest unbroken forests and chaparral, where they are scarce; they are thus scarcest in the high mountains of central southern Arizona. In summer they range up to the Hudsonian Zone (Merriam, N. Am. Fauna 3, 1890: 91), and they are probably resident in the Boreal Zones as well as below; at any rate, they were still present in the lower edge of the Canadian Zone of the White Mountains on *October 26, 1936* (ARP) and Hart Prairie, San Francisco Peaks, *November 24, 1933* (Jenks, MNA).

The most interesting thing about the Great Horned Owl in most of Arizona (except the western deserts) is its late nesting as compared with the birds of the eastern United States. There, they must nest in winter in order that the young may become proficient at hunting by the time food gets scarce the following fall. But in Arizona, with its later snows and more abundant mammal life, this is unnecessary and the birds nest in spring and fledge their young in June. Thus, as Herbert Brandt said, "The species is governed by food for its offspring" — the total environment and its strongest influences, not necessarily photoperiods, latitude, or other mysterious forces.

The Great Horned Owl in Arizona lives almost entirely on rabbits and rodents, also skunks and other mammals. Very few birds are taken, though poorly housed chickens are always a temptation. Despite their beneficial status economically, however, they are much persecuted by ill-informed gunners and trappers. Nevertheless, they have maintained their numbers well, except in the desert around Phoenix.

Horned Owls taken in Arizona have been identified as of three subspecies, but it seems more likely that the variation observable, chiefly the coloration of the feet, depends somewhat on the habitat: birds from wooded areas tend to have darker feet than those from open deserts. Undeniably, there is a tendency to somewhat larger size in the northeast. But we feel that individual variation is so great that no violence to the facts would be done by calling all birds of the southwestern United States *pacificus* Cassin. We know of no good evidence for migration in the state.

164. MOUNTAIN PYGMY-OWL; "PYGMY OWL"; ROCKY MOUNTAIN PYGMY OWL *Glaucidium gnoma* Wagler

Resident, but generally uncommon, of Transition and locally Upper Sonoran Zone woods west to central Arizona: Pajaritos (including Atascosa) and Santa Catalina Mountains, Prescott, south of Ash Fork (Mearns), and once on South Rim of Grand Canyon (nesting—E. D. McKee). Has reached lowest edge of Upper Sonoran

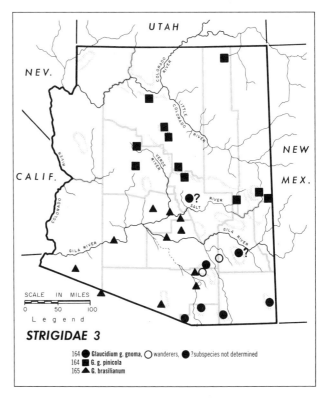

STRIGIDAE 3

164 ● *Glaucidium g. gnoma*, ○ wanderers, ● ?subspecies not determined
164 ■ *G. g. pinicola*
165 ▲ *G. brasilianum*

Zone twice, *January 21* and *April 7, 1950* (Catalina and Galiuro Mountains—Marshall, WJS and WF).

Pygmy-Owls are, next to the Elf Owl, our smallest owls and they are active during the day as well as at twilight. They feed on lizards, mice, small birds, and large insects. They are as apt to be hooting at ten o'clock of a morning as at ten on a moonlit night. The foraging area of each pair is very large in Arizona, where the birds are far less abundant than Screech-Owls and Elf Owls, or even Horned Owls. They differ from other small owls in the sharp dusky streaking below, and the relatively long, expressive tail, which they flick when excited. They have beady little yellow eyes, and a ferocity and strength out of proportion to their size, thus teaching us to keep hands out of hollow trees. There is even a report of a Mountain Pygmy-Owl killing a Gambel's Quail (Kimball, Condor 27, 1925: 209-210)!

The Mountain Pygmy-Owl is found in the wooded or forested parts of eastern Arizona and except as noted above, does not approach the desert range of the Ferruginous Pygmy-Owl. It differs from the latter, in Arizona, in having a dusky tail with narrow white cross-bars and a spotted crown in all but the juvenal plumage. The best distinction, besides the habitat, is the voice, which in the Mountain Pygmy-Owl is mellow, uninflected, and devoid of harsh overtones; also the successive notes are delivered at greater intervals than in the rapid notes of the Ferruginous.

There are two races in Arizona, as shown on the accompanying map. The northern *pinicola* Nelson is a large, gray race, although there is a tendency in Mountain Pygmy-Owls for females to be redder than their mates. This subspecies is found chiefly in coniferous forest. Its voice, like that of the birds of the Pacific Coast, consists of evenly spaced short whistles. Such birds have been heard in the fir forest of the Santa Catalina Mountains (Marshall) and we wonder if they differ racially from the birds at lower elevation there, which have calls like the following race. In any case, the species is unexpectedly rare in the Santa Catalinas.

The southern *gnoma* Wagler is a smaller edition of the above; it favors south-facing slopes of oaks and generally has its notes delivered in two's interspersed with singles.

165. FERRUGINOUS PYGMY-OWL; *Glaucidium brasilianum*
FERRUGINOUS OWL (Gmelin)

Local and generally sparse resident of Lower Sonoran Zone in central-southern and central Arizona, from Saguaro Lake, Superior, and Tucson west rarely to desert ranges of southern Yuma County (Cabeza Prieta Tanks — Monson).

This largely tropical Pygmy-Owl, abundant in South and Central America (as the scientific name implies), is here at its northern limit, where it lives chiefly in saguaros. It also inhabits shady riparian timber, as formerly in the Tucson mesquite forests; it still occurs in cottonwood-mesquite at the mouth of the Verde River. The calls are much more rapidly delivered than in the above species, and usually they are harsh and inflected, like the syllable *poip*. But a male will alternate between these and clear whistles. The female's calls are higher and more aspirate.

At present this bird is most frequently seen at the mouth of the Verde River and eastward. It is now rare and local at Tucson, where it was formerly much more numerous.

The Arizona race is *cactorum* van Rossem, a well-marked pale grayish extreme for the species. The tail of this race is always rufous (until badly worn and faded), with faint broad dusky cross-bars; this distinguishes it from the Mountain Pygmy-Owl. The streaks, instead of spots, on the crown are not noticeable in the field.

166. ELF OWL *Micrathene whitneyi* (J. Cooper)

Common summer resident in southern Arizona lowlands, scarcer north of Gila Valley in central Arizona, ranging up through live oak belt to lowest pines. Scarce to rare in Colorado Valley below Davis Dam, commoner on Big Sandy River. Casual records from Prescott, *June 20, 1892* (Loring, US) lower Oak Creek, *August 29, 1956* (E. Rigby, MNA), and Chloride, Mohave County, *May 27, 1960* (Coppa, ARIZ).

This is the smallest owl in the world; curiously, it was discovered at Fort Mohave by the indefatigable Dr. Cooper — where no one has found it since! Nevertheless it is Arizona's most abundant owl. Though much has been made of its supposed strict association with the saguaro ("elves and giants"), it is actually ubiquitous in any part of southern Arizona furnishing woods with nesting holes, below the heavy pine forest. Elf Owls are especially abundant in Madera Canyon of the Santa Rita Mountains, where they defend their favorite nest sites

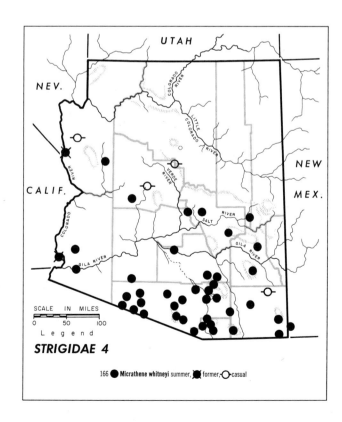

STRIGIDAE 4

166 ● *Micrathene whitneyi* summer, ◼ former, ○ casual

in sycamores from the intrusions of Spotted Screech-Owls.

As seen by flashlight, Elf Owls' ventral streakings are of a blurry, indefinite appearance, and the birds are short-tailed, like a Screech-Owl. For their size, they have an extraordinarily loud voice, which consists of chucklings and yips like a puppy dog. Strictly nocturnal, most of their calling is done at dusk from the entrance to the hole, and again at dawn, following the Cassin's Kingbird chorus. Also they call incessantly during moonlit nights in spring, after which they are hard to detect.

Their occurrence in numbers is from March through September, with extreme dates at Tucson of *February 25, 1940* to *October 10, 1885* (see Phillips, Wilson Bull. 54, 1942: 133). An exceptionally early bird was heard briefly at the foot of the Santa Rita Mountains, February 16, 1945 (Loye Miller and van Rossem); and the latest specimens for the state are two from Paradise, Chiricahua Mountains, *October 13, 1925* (Kimball, MICH).

The Arizona race, *whitneyi*, is larger than *sanfordi* (Ridgway) of northwestern Mexico and usually has more pronounced rufous coloration below. Specimens from the southern Papago Indian Reservation, near the international border, seem to approach *sanfordi*. The winter range of *whitneyi* is still virtually unknown, but will probably prove to be the Rio Balsas Basin of southern Mexico.

167. BURROWING OWL — *Speotyto cunicularia* (Molina)

Rare and local summer resident in Sonoran Zone grass- and farm-lands, except in farm areas about Phoenix and Yuma, where fairly common. Formerly commoner. Also rare fall transient, mainly in northeast and on southwest deserts; one record on Mogollon Plateau (White Mountains, October 9, 1937 — J. O. Stevenson). In winter somewhat commoner in southern and central Arizona, with two northern Arizona records: Springerville, January 8, 1959 (Levy) and Snowflake, *December 22, 1947* (A. J. Levine, MNA).

The Burrowing Owl apparently does not dig its own burrows in the ground, but uses those made by prairie-dogs, banner-tailed kangaroo-rats, and other mammals. In turn these animals require flat, unplowed prairies for their homes. Therefore cultivation and extermination of the prairie-dog villages is responsible for a great decrease in numbers of Burrowing Owls in Arizona, as compared to early times. Yet the bird is marvelously adaptable and mobile, and it finds and utilizes unplowed fields with burrows in unexpected places. This "Billy Owl" or "Ground Owl" was an important member of the cast of characters in old cowboy yarns about the prairie-dog and rattlesnake. It has a picturesque way of bobbing its whole body up and down on its long legs. It flies and takes long glides close to the ground, and has a long wing for its size. The Burrowing Owl does not often molest its mammalian host, but lives principally upon large insects and small vertebrates. At Tuba City, Nelson found that "remains of a cottontail . . . were the only signs of food, though it is scarcely likely that these birds could kill a mammal of this size."

Unlike all other owls, the Burrowing Owl perches conspicuously on mounds and fence-posts, and never seeks concealment except in the ground. It is relatively silent for an owl, in line with the good visibility of its habitat. It feeds in the early morning and late afternoon, but may often be seen sitting quietly during the day, in full sunlight. It is also active on bright nights.

The Burrowing Owl is apparently migratory in northeastern Arizona, where it is found only from the beginning of May through October, save as noted above. While no banding has been done, the populations represented on the accompanying map by both solid and open dots are probably permanently resident.

168. SPOTTED OWL — *Strix occidentalis* (Xantus)

Uncommon resident of the heavily forested mountains and high mesas. Rare in lowlands not far from mountains, perhaps chiefly in winter. Nested near Tucson, *1872* (Bendire, US). Absent from most or all areas inhabited by *Bubo*.

The Spotted Owl is perhaps the hardest of all Arizona owls to understand. It performs no regular migrations, so far as known, and has been taken in every life zone from Lower Sonoran (Tucson) to Canadian (San Francisco Peaks). Yet few indeed are the places where one may confidently go to find a pair of Spotted Owls. These are shaded canyons, usually with dense firs, and often with nearby cliffs, in forested areas that are not inhabited by Horned Owls — which applies especially well to the high mountains of the southeast, each of which supports several pairs. When Phillips began to

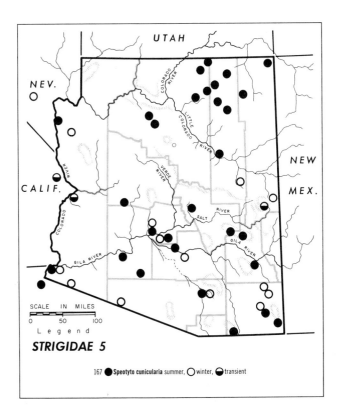

STRIGIDAE 5

167 ● *Speotyto cunicularia* summer, ○ winter, ◐ transient

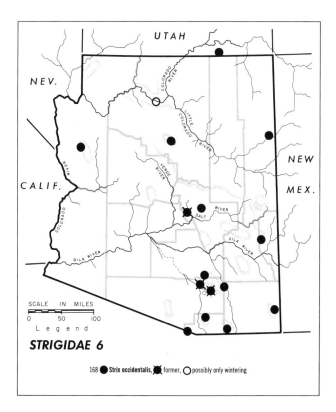

visit the Hualapai Mountains, the Spotted Owl was present each year. But then came a forest fire; jeep roads were bulldozed up from the park headquarters to fight the fire. And apparently this tipped the ecologic balance, for on a visit that fall, Horned Owls were heard for the first time, and Spotted Owls could not be found. None of us has ever seen a Spotted Owl in the lowlands; were it not that Bendire and Willett were collectors and observers of the highest integrity we should be highly skeptical of the records at Roosevelt Lake and all around Tucson. Nesting forsooth! We would point out, however, that there were fine woods of cottonwood and mesquite at Tucson in Bendire's time. Its reappearance in the Tucson Valley of today is unthinkable.

The Spotted Owl's dark eye, mild expression and soft colors give it a benign aspect at variance with its hair-raising screeches, barks, and whistles. The usual song, higher in the female, is of four or five barks which sound like the baying of a hound. The whistle is more terrifying, for it starts low, and ascends in pitch like a siren. The Arizona race, lighter and with more white spotting than that of the Pacific Coast, is *lucida* Nelson.

169. LONG-EARED OWL *Asio otus* (Linnaeus)

Rather rare winter resident generally but of statewide occurrence. Even scarcer in summer, but may breed at almost any point in eastern and central Arizona; has nested as far southwest as Bates Well, Organ Pipe Cactus National Monument, *June 23, 1932* (juvenile — van Rossem, LA). Found breeding commonly, recently, in Tombstone region. Birds taken, freshly dead, from upper Havasu Lake, *July 9, 1948* (Monson, US) and from highway a few miles west in California, *May 7, 1952* (ARP)! Winter flocks may number up to about 50 birds (Sulphur Springs Valley — Jacot, ARIZ).

Supposedly breeding chiefly in southern Canada and the northern United States, the Long-eared Owl would be expected to nest chiefly in the higher mountains of Arizona. Actually there are almost no occurrences in mountains, and no breeding record there. There were in fact hardly any evidences of breeding anywhere in the state before 1932, and these were from the foothills: 10 miles west of Camp Verde (Mearns), Fort Huachuca (Benson), and Santa Rita Mountains (Swarth). It was therefore a great surprise to learn of its nesting far out in the desert at Bates Well. Still more surprising was Stophlet's discovery that these owls were actually nesting in numbers in the foothills near Tombstone (Wilson Bull. 71, 1959: 97-99)! A few of these nests were found well down in the desert by C. M. Palmer, Jr.; and some of the babies that fell out and died are preserved to substantiate these extraordinary data (ARP). Nests were in Emory oaks, with the Lower Sonoran Zone ones in a desert hackberry and a mesquite in the creosote association.

Under these circumstances, the migrations of the bird in Arizona became rather obscure. But it winters as high as the Transition Zone at Flagstaff (Phillips, MNA).

170. SHORT-EARED OWL *Asio flammeus* (Pontoppidan)

Generally rare winter visitant in open grassland, marshes, swales, and fields of southern and western Arizona; also records for extreme desert: one south of Mohawk, Yuma County, November 27, 1942 (Monson); several in Yuma and western Pima Counties, 1959–1960 (Monson). Rare in migration in the north. A report of numbers on the Colorado River in September (Coues) is doubtless erroneous, though Dickerman picked up 4 killed in highway west of Gila Bend, by alfalfa fields, *December 1960* (MIN).

This open-country, semi-diurnal owl is rare in Arizona and it is the only one which is entirely a winter visitor.

Extreme dates in northern Arizona are from the end of September (Ligon, Monson) to April 24, 1953 (Phillips and Dickerman — above McNary). All southern Arizona records are from mid-October to mid-March. The above record by Dickerman and others by Housholder indicate a concentration in the alfalfa fields of the Maricopa-Gila Bend area; and exceptional numbers, up to 9, were in Bermuda grass fields at Yuma at the start of 1959 (E. McGregor, Monson).

171. SAW-WHET OWL *Aegolius acadicus* (Gmelin)

Resident, perhaps fairly common but not often detected, of mountains of eastern, central, and perhaps northwestern Arizona; also possibly winter visitant to some extent in same areas. Rare and irregular in lowlands of

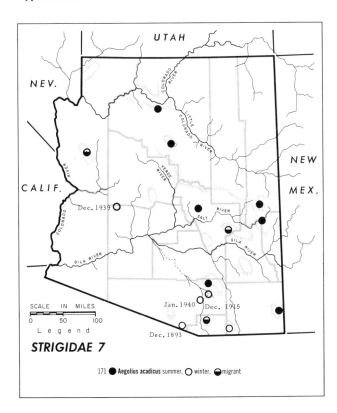

STRIGIDAE 7

171 ● *Aegolius acadicus* summer, ○ winter, ◐ migrant

central-southern and west-central Arizona in winter. Status poorly known.

In contrast to other owls, Saw-whet records are concentrated in the silent months (when owls are not calling much) from September to January. This accounts for our doubts as to its status. Even the specimens from the White Mountains and Blue Range were taken on *October 9* (ARP) and *September 10* (US), respectively; but we can hardly doubt that the birds breed in these extensive fir forests. The only actual nest ever found in Arizona remains that discovered by the redoubtable Dr. Mearns, on the San Francisco Peaks. The only more recent summer records are those of Perry and Art Brown on the south rim of Grand Canyon (1939, young shown to McKee, Behle, and Allen) and of A. H. Miller (Condor 39, 1937: 130) in the Sierra Ancha and Chiricahua Mountains. Dr. Miller is an expert on this owl, and he is somehow able to get them to answer his imitated whistles. Marshall was unable to find the birds at the same spot in the Chiricahuas, but this merely emphasizes what has been found in other states: that the Saw-whet Owl is erratic; it will breed in an area for one or several years, then move on. At the Hitchcock Tree in Bear Canyon of the Santa Catalina Mountains, where night birds have been intensively studied since 1949 (Marshall), a loudly calling pair, in some sort of conflict with a Spotted Screech-Owl, was seen on *October 26, 1958* (Marshall, ARIZ) — never before nor since found anywhere in that range. The fact that it was a pair might mean that the birds had bred there, among these lowest Douglas firs. (Unfortunately the area had not been visited that summer.)

Birds possibly migrating were seen near the summit of the Hualapai Mountains, September 29, 1948 (Phillips); on the Natanes Plateau, southern Gila County, October 17, 1936 (R. T. Peterson); and at Madera Canyon, Santa Rita Mountains, April 29, 1951, but not on May 5–6 (Marshall). Dates of winter visitors to the desert are shown on the accompanying map. Those of *1939–1940* include two specimens taken along the Santa Cruz River south of Tucson on *January 1* and *27* (W. L. Holt, Phillips, ARP) in a year when this owl appeared in unusual numbers in the eastern United States; but the others seem not to be correlated with flights elsewhere.

GOATSUCKERS
CAPRIMULGIDAE

Soft-plumaged night birds, goatsuckers feed on insects caught in the air during continuous flight (nighthawks); also in flight and leaps from a perch or the ground (Whip-poor-will, Poor-will). They are crepuscular, singing and feeding so effectively at dusk and dawn that they can afford to spend the day and rest of the night dozing. However, at full moon, they call all night long. They are easily seen a long way off by their brilliant eye-shine, as reflected from a flashlight or car light. Their wonderfully intricate pattern of delicate black pencillings against silvery, bronzy, or ochraceous hues makes them among the most handsome of all birds, and among them the Poor-will is preeminent in this respect. The intricate coloration of each species renders it invisible against the particular ground or leaf litter upon which it sleeps or nests. Among their peculiarities are a brain the size of a pea, an enormous mouth which opens so wide that the large eyes can be seen through its roof, tiny feet with a comb on the claw of the middle toe (used to scratch the face — see Brauner, Condor 54, 1952: 152-159), and a tendency of some species to hibernate. All forms have very long wings, and they perform sexual displays which utilize their wonderful power of buoyant flight and emphasize striking white patches on the throat, tail, or wings.

There is some confusion concerning the perching habits of our goatsuckers. Our observations show that the much publicized habit of perching lengthwise along a limb, applies only to resting or sleeping Whip-poor-wills and nighthawks. The Poor-will and Preste-me-tucuchillo crouch on the ground when sleeping by day or dozing through the middle of a moonless night. When actively foraging the Whip-poor-will, Preste-me-tucuchillo and Poor-will perch crosswise on twigs, just like flycatchers. Additionally, all three forage from boulders, and the Poor-will (mainly) and Texas Nighthawk (occasionally) feed from a standing position on bare ground or dirt roads.

172. WHIP-POOR-WILL; STEPHENS' WHIP-POOR-WILL
Caprimulgus vociferus
Wilson

Common summer resident of Transition and adjacent Upper Sonoran Zones of southeastern and (in recent

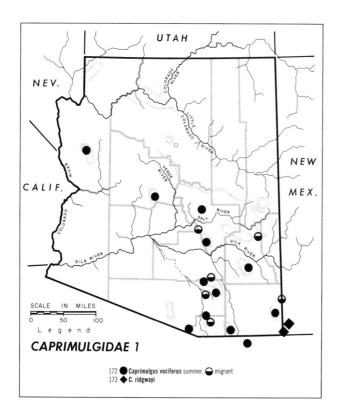

CAPRIMULGIDAE 1
172 ● *Caprimulgus vociferus* summer, ◐ migrant
173 ◆ *C. ridgwayi*

years) central Arizona; ranges west to Pajaritos Mountains, and northwest less commonly (no specimen) to the Hualapai Mountains. Two lowland records; near Roosevelt, *November 4, 1952* (Dickerman, CU), and Tucson, October 6, 1953 (in town — Marshall).

This familiar bird will be instantly recognized by eastern observers, as it differs only slightly from the well-known eastern race. It is, however, limited to the densely wooded mountains, has a rougher voice, and lays plain white eggs. Most arrive in the last few days of April or early May and depart in early October; their breeding stations are represented by solid dots on the accompanying map.

The half-filled circles represent occasional birds found at lower elevations in early March (New Mexico), April, May, October, and November. The meaning of these records is not clear because only one specimen is represented: a *C. v. vociferus* from Roosevelt (as above). All other Arizona specimens are from the mountains and are of the longer-bristled, browner, more coarsely-marked race *arizonae* (Brewster), though some *October* birds are quite blackish.

173. PRESTE-ME-TU-CUCHILLO; RIDGWAY'S WHIP-POOR-WILL; BUFF-COLLARED NIGHTJAR   *Caprimulgus ridgwayi* (Nelson)

One record only, Guadalupe Canyon in extreme southeastern corner of Arizona, *May 12, 1960* (Levy, US). It is believed to breed here and in adjacent New Mexico, where it was first found in *1958* (Johnston and Hardy, KANU).

Although inexperienced observers have reported it in great numbers, Seymour Levy believes that his specimen is the only valid record for Arizona. (The other reports must pertain to the dawn chorus of the Cassin's Kingbird.) The song, indicated by its common name used in Mexico, rises steadily in pitch and usually terminates explosively.

174. POOR-WILL;   *Phalaenoptilus nuttallii* NUTTALL'S POOR-WILL (Audubon)

Common in summer about hills and rocky outcrops of Sonoran Zones (rarely higher) throughout state, less common in southwest. Winters (in small numbers as far as known) in southern and western Arizona.

This is the common goatsucker of desert slopes throughout the state and it ascends locally into the more open parts of the pine and fir forests, as high as Phelps Ranger Station in the White Mountains (*October 26, 1936* — ARP). It is absent from dense forests and level plains.

Much of the supposed evidence for migration of the Poor-will is based upon its polymorphism. Authors have assumed that the darker birds found in the desert were migrants from farther north. There is still no good evidence for this, but it is well established that both dark and light phases occur in the same breeding population anywhere within the range of the species. Amazing proof that some, if not all, of the birds are non-migratory, came from Edmund C. Jaeger's confirmation that the Poor-will hibernates. A few such hibernating birds have been found around Tucson, and surely this concept could apply to all or many of the southern Arizona birds because they appear during warm spells in winter and start singing in full force with the first warm days in March. Captive birds hibernated when food was withheld and they reached body temperatures of 6° centigrade (Marshall, Condor 57, 1955: 129-134). This was with lean birds in spring, and so far no experimenter has duplicated the natural hibernation of a Poor-will in its late fall condition of extreme obesity.

In northern Arizona, hibernation has not been well-proven as yet. There are no records between *October 26* (as above) and April 5, 1933 on the northeast slope of the San Francisco Mountains (Hargrave), except for "March-Nov." in the bottom of the Grand Canyon (Grater).

From the above mention of polymorphism, the reader may deduce our reasons for declining to recognize any race other than the nominate *nuttallii* in Arizona. (Phillips, Jour. Ariz. Acad. Sci. 1, 1959: 28).

175. COMMON NIGHTHAWK; BOOMING NIGHTHAWK; WESTERN   *Chordeiles minor* NIGHTHAWK   (Forster)

Common summer resident, and abundant in migration, in open parts of Upper Sonoran Zone and above, in central and northern Arizona; less common and very local summer resident in grasslands at base of mountains of northwestern, central southern, and southeastern Arizona. Extremely rare transient (no specimens) in Tucson Valley; unknown elsewhere in Lower Sonoran Zone closer than Indian Springs, Nevada.

The Common Nighthawk shares with the Yellow-billed Cuckoo the distinction of being the last of our summer residents to arrive from the south and these are also among our longest migrants. They winter exclusively in South America. It is very exceptional to see a Common Nighthawk in Arizona before the very end of May, and the bulk arrives during early June. From then through August this is one of the most conspicuous and abundant birds in northern Arizona, where it favors level terrain (but may be seen over the tops of the San Francisco Peaks, at least as a migrant). Spectacular numbers moving from communal roosts to feeding areas give the misleading impression that migration is going on all summer. Of interest because of proximity to or overlap with the Texas Nighthawk, are the isolated populations in elevated grasslands in the southeastern part of the state: around Nogales, Sonoita, and the base of the Huachuca Mountains.

There are occasional reports of Common Nighthawks in southern Arizona in "early spring" (Scott) and as early as April 23 (Swarth). These we regard as exceptional if correctly identified. The oft-quoted specimen taken by Mearns, *May 9, 1885* at Picacho Peak (also reported as from Camp Verde) has been reexamined and proves to be a grease-stained Texas Nighthawk (AMNH). A most exceptional occurrence was a flock of about 75 Common and 200 or more Texas Nighthawks near Tucson on May 16, 1942 (Wm. X. and Alma J. Foerster). Coues' report from "April" to "October" at Prescott seems very dubious. Actually the latest fall date in which we have full confidence is *September 15* in the Huachuca Mountains (A. B. Howell, LA) and on the northeast slope of the San Francisco Mountains (Hargrave).

The breeding race of Arizona, unless possibly *howelli* Oberholser breeds in the extreme northeast, is the strongly rufescent, moderately dark *henryi* Cassin, represented by dots. Specimens have been taken from *May 28 (1915 in the Chiricahua Mountains* — van Rossem, LA — where we have not located a breeding population, though there are other specimens in summer, AMNH) to *September 2 (1934 at Springerville* — Stevenson, ARP).

Of the migrant races, the commonest is *hesperis* Grinnell, a more blackish bird. The dates of the various specimens shown by open triangles on the map are chiefly in the first half of *June* and from *mid-August* to *late August* (and *September 13* in southern Nevada). The exceptions are from near Holbrook, *July 2, 1949* (ARP); Winslow, *July 14, 1909* (Birdseye, US); and Mormon Lake, *July 19, 1955* (ARP).

A single specimen from 42 miles north of St. Johns, in eastern Arizona, seems to be still blacker than *hesperis*, and is referred to *minor*. It was found dead a day or two on *September 10, 1956* (Marshall, Bialac, Werner and Phillips, ARP). Unfortunately, Nighthawks move south before molting, and are thus hard to identify in the fall, so that this might be an exceptionally dark *hesperis*.

Only one Arizona specimen seems properly referable to *howelli* Oberholser. This is a young female taken near Prescott on *August 24, 1931* (Jacot, ARIZ). This race is

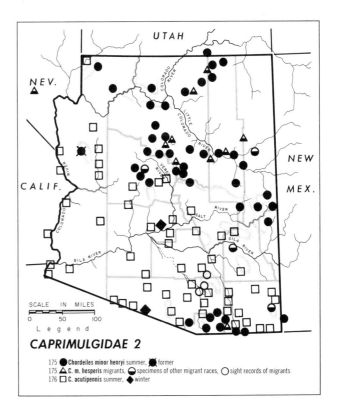

175 ● Chordeiles minor henryi summer, ◉ former
175 ▲ C. m. hesperis migrants, ◐ specimens of other migrant races, ○ sight records of migrants
176 □ C. acutipennis summer, ◆ winter

somewhat paler than *henryi*, and often less rufescent. Certain statements in the literature at variance with this account of the migrant races are considered to refer actually to misidentified specimens of *henryi*.

176. TEXAS NIGHTHAWK; TRILLING NIGHTHAWK; LESSER NIGHTHAWK *Chordeiles acutipennis* (Hermann)

Common to abundant summer resident in Lower Sonoran Zone of southern and western Arizona, except in most of Yuma County away from the Colorado and Gila Valleys. Two winter records: Phoenix, *December 27, 1897* (Breninger, in Bent) and southwest corner of Papago Indian Reservation, January 6, 1940 (Phillips). No record between early January and March. No record for northeastern Arizona, though recently reported in southeastern Utah (Behle *et al.*, Wils. Bull. 75, 1963: 452).

There is some evidence that the Texas Nighthawk can move its eggs, once the nest is disturbed (Levy). The tiny chicks, protectively colored to match the desert pavement which serves for their nest, definitely move after hatching. The ordinary song of the male is a vague purring trill, indefinitely prolonged; and there are interesting sexual displays, accompanied by strange whinnying sounds, as described by A. H. Miller (Condor 39, 1937: 42-43). The trill, lack of a diving *boom,* and white patch nearer the tip than the bend of the wing distinguish the male from the Common Nighthawk. The female's wing patch is buffy and often hard to see, thus distinguishing her also from the Common Nighthawk. Any Nighthawk with no wing patch at all is a female or juvenal Texas Nighthawk. This species is apt to be abroad in numbers through the morning, feeding until around 11 o'clock; also they feed around bright city lights far into the night. Thus they are more diurnal and are active for a longer period than are other goatsuckers.

The season of occurrence of the Texas Nighthawk varies strikingly from place to place in Arizona, though all the birds are summer residents. About Tucson, where our records are most extensive, the bulk appears in early May with a few birds usually detected from the 18th to the 26th of April. They remain common until mid-October, the latest being seen on November 4 (1943, M. Cater, A. H. Anderson, and Wm. X. Foerster). This single bird is the only November record, but there is a surprising number of records of one to three birds between March 13 (1934, A. H. Anderson) and April 7. It is possible that this vanguard is *en route* to the lower valleys farther west and northwest. Thus at Yuma, Herbert Brown saw four on March 3, 1902 and ten the next day, while by March 15 it had "become quite plentiful all through the valley." Additional dates in southwestern Arizona are March 6 (1918 in the corner of the state – Kimball, FW files) and March 8 (1939 at Quitovaquita – Huey). In central Arizona, Yaeger saw one near Chandler on March 1 and later, 1934, an exceptional date for that area, where they do not become common until about the end of April (Simpson and Werner).

In parts of western Arizona Texas Nighthawks remain regularly into November. Some very late dates are two at Gila Bend on November 20 (1960 – Levy) and one at Parker on November 23 (1951 – Pyle and Gould). At Salome, Sheffler sees a few regularly to Christmas week, whereupon they disappear until April! These late dates are all of birds feeding about bright lights in towns.

weather when such food is unavailable. The nests are cemented together with saliva in a protected place (except the tropical *Panyptila*). Some cave-dwelling Malaysian species of *Collocalia* make nests entirely of saliva, which are harvested for bird's-nest soup under strict regulation and conservation practices enforced by the Government. Another feature of some swifts is the bare underside of the hand. They have chipping, twittering calls and very hard feathers, especially those which guard the eye from the terrific rush of air. Curved claws, reversible toes of equal length, and strong shafts of the tail feathers help them cling upon vertical surfaces, where they roost in huddled masses.

[BLACK SWIFT   *Cypseloides niger* (Gmelin)]

[Hypothetical. Status not quite satisfactory. Probably transient in east. One old specimen (AMNH) labelled "Arizona," without further data. Two sight records in fall at Flagstaff (J. Kittredge, Jr. in FW files; Hargrave), one of which is apparently the basis of statement in A.O.U. Check-list that it migrates through Arizona. Also seen twice by Mearns in 1880's (once on May 6, 1887 atop "Squaw Peak, Verde Mountains"); and twice by Gallizioli near Safford, May 30, 1953 (flock of about 35) and August 17, 1954.

The Black Swift is larger than those swifts which have been taken in Arizona, and it is wholly black, including the throat. A fascinating article on the nesting of this bird in nearby Colorado is by Owen Knorr (Wilson Bull. 73, 1961: 155-170).]

SWIFTS   *APODIDAE*

The uninformed call these birds "swallows," from which they are actually quite distinct. Swifts fly by moving the wings only up and down, not backward and forward. In fact, the larger tropical species fly for long periods without beating the wings at all; the smaller species twinkle them rapidly between periods of soaring. The rigidity of the swift's wing, with its muscles and arm bones all telescoped against the shoulder, gives the bird an appearance like a cigar stuck into a bow. Nevertheless, it has the ability, unique among birds, of flexing the primaries slightly at the fingers. All North American swifts are blackish or sooty on the breast and belly, unlike any swallow except the glossy male Purple Martin. Swifts never perch upon trees and wires nor do they alight upon the ground. Their lives are spent in the air; they even mate in mid-air. They roost upon cliffs or the trunks of trees, behind waterfalls, or in chimneys, according to the species. A grounded swift cannot take flight and must starve if it cannot climb. Swifts feed on swarms of insects, often ants, high in the air, and in pursuing these range far and wide. Some species, such as the White-throated Swift, can go into torpor during bad

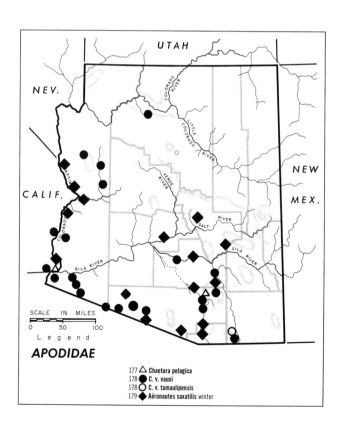

APODIDAE

177 △ Chaetura pelagica
178 ● C. v. vauxi
178 ○ C. v. tamaulipensis
179 ◆ Aëronautes saxatilis winter

177. CHIMNEY SWIFT *Chaetura pelagica*
(Linnaeus)

Only one verifiable Arizona record: non-breeding pair at Tucson, May 30 to *mid-June, 1952* (ARP). Many records of genus not determined to species. A transient reported from California side of Colorado River above Yuma, *May 6, 1930* (Huey, SD).

This well-known species is the only swift found in the eastern part of the continent. It is usually black from soot, but even its normal or washed colors are somewhat darker than *vauxi,* from which it also differs in being larger and with a much louder voice. The one found in Arizona was breaking twigs from a eucalyptus tree (in head-long flight) and roosting inside a brick chimney in the engineering building of the University of Arizona campus, but was not nesting.

178. VAUX'S SWIFT *Chaetura vauxi*
(Townsend)

Fairly common but irregular migrant in central-southern and western Arizona, east at least to Santa Catalina (Scott) and Huachuca Mountains (ARP, CAS, ARIZ).

The Vaux's Swift (pronounced as in hawks), our smallest swift, is brownish rather than sooty black below, with the throat much paler, though there is no white in the plumage. About the size of the Violet-green Swallow, it often flies very low among trees, 'round and 'round in a repeated foraging circuit. In the hand, unless too worn, both Vaux's and Chimney Swifts show needle-like projecting shafts of the tail feathers. The migration periods in southern Arizona are principally from mid-April to mid-May and mid-September to mid-October. The only northern Arizona records are of one or two birds seen on the Anita Road south of the Grand Canyon in fall, 1956 (Hargrave).

The regular migrants belong to the pale northern race, *vauxi*. The race *tamaulipensis* Sutton of Mexico is darker, grayer below, and blacker (glossier) above; it is apparently a casual visitor to Arizona, with one specimen from Fort Huachuca (*May 14, 1950* — ARP).

179. WHITE-THROATED SWIFT *Aëronautes saxatalis*
(Woodhouse)

Common breeding bird at cliffs in the mountains and mesas except in southwest, where it nests only at Parker Dam and perhaps sparingly in the Kofa and Castle Dome Mountains of Yuma County. Found in winter irregularly in southern and western Arizona, but by the hundreds daily during some winters.

The White-throated Swift is the commonest swift in Arizona, the fastest, and the only one with conspicuous white, producing a patterned plumage. It is usually high in the air unless coming in to drink at a pond, where it scoops water in full flight. The voice is extremely shrill, laughing, and metallic. Standing on a ridge, one can often hear these swifts passing like bullets; they are already far away before one can accommodate his eyes to seeing them. They nest far inside crevices of rocks, including those on large cliffs and those on steep hills. Northern Arizona migrations are in late March, early April, and September; extreme dates are March 7 (1935, inside the Grand Canyon — *fide* McKee, Gr. Can N.H. Bull. 4, 1936: 13) to October 9 and 17 (1938, west of Grand Canyon Village — McKee).

We have not seen in Arizona the straightaway steady flights characterizing the migrants up the Pacific Coast. Instead, at least in southern Arizona, the flock at any time of year will suddenly arrive to forage low over a riverbed, then as suddenly they retire again into the sky. There seems to have been a decrease in the numbers of swifts on the San Francisco Peaks since about 1935.

The smaller, typical race is the only one definitely recorded from Arizona. There is a larger form, *sclateri* Rogers, which ranges up through Wyoming and northern Colorado, and which might migrate through eastern Arizona; as yet no specimens have been definitely identified, however.

HUMMINGBIRDS
TROCHILIDAE
by Allan Phillips

Hummingbirds are famous even in the Old World, where they do not occur; the chief writers on the family have been Frenchmen. The hummingbird family includes our smallest birds, some more like a bumblebee or sphinx moth than a bird in appearance; but also some larger than a wren, whose range in the United States is mostly limited to the mountains of southeastern Arizona. Their colors are iridescent, and almost all are "bright green above" as described by many persons who thus ask for their identification! They can fly backwards. They have a long needle-like bill and extrusible tongue with which they eat pollen, nectar, and insects in flowers. They also catch tiny insects in the air. (Most other birds which feed on flying insects are specially equipped with a broad bill or cavernous mouth surrounded with a net of bristles.) The males take no part in nesting; pairing is brief, frequent, and not always too exacting, which leads to an unusual amount of hybridization. Nests are made of plant down bound with cobwebs and adorned with lichens or small leaves — real works of bird art. At least one species, the Black-chinned, continues to build up the nest while incubating the two oval white eggs; these are laid 36 to 48 hours apart in the Black-chinned, and the incubation and nestling periods are surprisingly long for such a small bird. Hummingbirds, when hard-pressed by injury or harsh circumstances, may sink into a state of torpidity at night; but they live at a fast and furious pace by day. The females of most species can only be distinguished from each other as specimens; and since the male plays no role at the nest, the identification of nests and eggs must be based upon collection of the female parent. In fact, hummingbirds are for the expert, who bases his determinations on the exact shape, size, and color of certain tail-feathers! Not

—Continued on page 62

KEY TO WESTERN HUMMINGBIRDS

I Throat (gorget) solidly dark or nearly so, back to a sharp line abruptly limiting the pale chest, or else the whole breast and belly dark green or blackish (adult males). ....*A*

I' Throat pale, sometimes speckled or dark centrally, but not abruptly and solidly contrasted with the pale chest (chiefly females and young). ....*B*

   A Underparts dark green or grayish; gorget blue or green; size medium to large. No reddish in plumage, but bill sometimes red. Mountains and foothills, chiefly. *1*

   A' Underparts (at least chest) whitish, in abrupt contrast to gorget. Bill black; gorget or sides often more or less reddish. Size small to medium. Widespread, including deserts and plains. *2*

      1 Very large, with long black bill; flight and wing-beats relatively slow, the wings fairly visible in flight. ....*a*

      1' Moderate size; bill red at base; flight normal, the wings blurry. ......*b*

         a Wholly dark; the gorget green, in good light much paler than chest
.... *Rivoli's Hummingbird*

         a' Breast gray (not dusky nor greenish), and tail broadly white at corners. Gorget dull blue. ....*Blue-throated Hummingbird*

         b A conspicuous broad, clear white ear-stripe, extending from eye onto sides of neck. Breast and belly pale, dull green mixed with pale buffy or whitish. Front of gorget and of crown purple, rest green (rare).
.................................*White-eared Hummingbird*

         b' Solidly dark nearly everywhere except on crissum. No purple nor extensive head-striping (locally common). *Broad-billed Hummingbird*

      2 Sides of chest, sides and flanks more or less tinged with rufous or tawny buff. Tail often rufous, or with white corners. ....*c*

      2' Sides and flanks greenish or grayish, without warm (buffy or reddish) tones. Tail practically all dark. ....*d*

         c Rufous deep and extensive, covering tail and its coverts, rump, and sides of head. Gorget orange-red. (Common.) ....
...........................*Rufous and Allen's Hummingbirds*

         c' Paler tawny buff, not extending onto head, rump, or tail coverts. Gorget purplish red. (Very rare, except in Big Bend, Texas.) ....*(1)*

            (1) Extremely small (bumblebee-sized); bill shorter than head, straight. Tail-corners broadly white. (Very rare.) ....
.................................*Heloise's Hummingbird*

            (1') Normal hummingbird size and shape. Bill longer than head, decurved. Hardly any white in tail. (Regular in Big Bend only.)
.................................*Lucifer Hummingbird*

         d Gorget conspicuously produced laterally, its posterior margin thus decidedly concave, and bordered above by whitish on sides of head. Size rather small to small. ....*(2)*

         d' Gorget relatively even behind and without white border above. Size rather small to moderate. ....*(3)*

            (2) Gorget and crown solid violet. Size rather small. Sides extensively green, infringing on center of breast. *Costa's Hummingbird*

            (2') Gorget lilac red, mixed with white at chin. Crown green. Small. Sides narrowly pale dull greenish (some buff tinge laterally), the breast all whitish. ....*Calliope Hummingbird*

            (3) Moderately large and dark; not pure white anywhere except tufts of legs and anal region. Crown and gorget crimson. ....
.................................*Anna's Hummingbird*

KEY TO WESTERN HUMMINGBIRDS

(3') Rather small to moderate. Chest (if clean) whitish. Crown greenish ....(a)

(a) Gorget black (a purple fringe posteriorly, seldom visible); widespread. ....Black-chinned Hummingbird

(a') Gorget red .... ∝

∝ Gorget slightly more orange. Wings always silent. (Recorded west to Guadalupe Mountains, Texas, as a migrant only.) ....Ruby-throated Hummingbird

∝' Gorget slightly purer (purpler) red. Wings produce shrill, shrieking rattle in flight, unless molting. (Common in Rocky Mountains and Arizona, migrating across lowlands locally.) ....Broad-tailed Hummingbird

B Underparts snow-white all the way to chin, contrasting with red bill and violet crown; tail solid greenish. ....Violet-crowned Hummingbird

B' Bill largely or wholly dark; underparts at least partly gray or clouded. Tail-corners pale. (Females and young.) 3

3 Sides and flanks distinctly tinged with rufous or cinnamon-buff. Bill all black. (In hand, bases of some tail-feathers rufous.) ....e

3' Sides and flanks greenish or grayish, with little or no buff tinge. Bill sometimes partly reddish. (No rufous in tail.) ....f

e Cinnamon-buff wash spreads evenly across breast. Bill distinctly decurved. (Chiefly Big Bend.) ....female and young Lucifer Hummingbird

e' Mid-line of breast faintly if at all buff-tinged, distinctly paler than sides. Bill straight. (Common.) .... *Female and young of all* SELASPHORUS *and Calliope Hummingbirds*

f Size very large; bill long, black; flight and wing-beats relatively slow *(4)*

f' Size rather small to moderately large; bill sometimes reddish at base. Flight normal. ....*(5)*

(4) Underparts pale gray, slightly mottled, about like the color of the tail-corners. White marks around eye short, not very conspicuous. Tail from above green. *female Rivoli's Hummingbird*

(4') Underparts uniform dark gray, much darker than the extensive white corners of the tail. White lines below, and especially behind, the eye. Tail from above black. ....*female Blue-throated Hummingbird*

(5) A conspicuous white line back of eye, contrasting to darker cheek below. Base of lower mandible pinkish or reddish. ....(b)

(5') White spot back of eye not prolonged into a line, thus head not contrastingly striped. Bill wholly black. ....(c)

(b) Below uniform pale gray. Head-stripes dark gray and dull whitish. (Common locally.) *female Broad-billed Hummingbird*

(b') Below whitish, spotted with green; sides green. Head-stripes blackish and white. (Rare.) .....*White-eared Hummingbird*

(c) Underparts and tail-corners pale grayish; throat spotted, usually with dark spots (often red if in good light); slightly larger. ....*female Anna's Hummingbird*

(c') Below, and tail-corners, whitish. Throat clear or *finely* speckled. Slightly or distinctly smaller. ....*female Ruby-throated, Black-chinned, and Costa's Hummingbirds*

only are hummingbirds' colors variable according to the angle of the light, but discoloration by yellow pollen about the head is frequent. Nonetheless, a key (pp. 60-61) to all hummingbirds recorded in the U.S. west of the plains is offered to help field identifications.

180. LUCIFER HUMMINGBIRD *Calothorax lucifer* (Swainson)

Casual: two old records, Fort Bowie, Cochise County, *August 8, 1874* (Henshaw, US), and "Arizona" (date?, old specimen — Simons, US).

The Lucifer is the only Arizona hummer with a definitely decurved bill.

For further information on the status of this extremely rare species in the United States, see Taylor and Duvall (Condor 53, 1951: 202-203); and Pulich and Pulich (Auk 80, 1963: 370-371). On its life history, see Helmuth O. Wagner, Anales del Inst. Biol. Univ. Mex. 17, 1946: 283-299.

181. BLACK-CHINNED HUMMINGBIRD *Archilochus alexandri* (Bourcier and Mulsant)

Common summer resident in certain deciduous associations of Sonoran Zones, including cities; generally absent from deserts. Migrates across the Mogollon Plateau, at least in fall. Scarce in western Arizona after mid-June.

This is the common hummingbird of southern Arizona towns and rivers. The entire throat of the male appears black in life. The Black-chinned Hummingbird occurs in southern Arizona from the middle of March (rarely earlier) through September; in the latter month, this species is thinning out and the Anna's Hummingbird is arriving to replace it. Apparently most of the adult male Black-chins leave in mid-summer; later chiefly the females and juveniles are about. But the latest record for the state (*October 2, 1947,* at Patagonia — ARP) is an adult male. There is considerable evidence, especially at Phoenix (Hargrave) that this species raises at least two broods in southern Arizona. The same twig may be used for a nest in successive years. Nesting females are found principally among willows, cottonwoods, and sycamores along streams and in olive trees in town. They used to be abundant along the Santa Cruz River near Tucson, but were greatly reduced in the early 1940's through destruction of the batamote thickets, tree-tobacco, and other streamside plants.

In northern Arizona it seems to occur usually from mid-April to mid-September, but data are scanty. Due to the scarcity of deciduous trees in the Upper Sonoran Zone there, it is very local as a breeding species. Spring arrival seems to average earlier in western than in southeastern Arizona. As is generally the case among migratory birds, female Black-chins arrive later than males, usually in late March. Therefore we must doubt the identity of the female "Black-chinned Hummingbird" that was experimented on by Bené (Condor 43, 1941: 237) at Phoenix, February 26 to March 1940. Once here, female Black-chins lose little time in starting their nests; they were building near and in Tucson, March 22 and 29, 1947 (C. T. Vorhies, Alma J. Foerster).

182. COSTA'S HUMMINGBIRD *Archilochus costae* (Bourcier)

Common breeding bird in the deserts of central and western Arizona, but not in Huachuca Mountains region. Disappears almost completely by early July in western Arizona, to reappear in October and spend the winter in western Pima County and in Yuma County from the south side of the Kofa Mountains (GM) southward (east to Rancho Bonito near Quitovaquito — Huey, SD). Young birds remain later on the east side of the Santa Catalina Mountains, to *August 10* (*1884* — Scott, AMNH). Aside from the extreme southwest, it evidently does not normally occur between early August and late January, contrary to the statement (A.O.U. Check-list) that it "winters over most of breeding range," including "Williams River" (*i.e.,* Big Sandy River, February — Kennerly and Möllhausen). Casual males at Phoenix, *December 16, 1959* (G. F. Davidson, MIN) and Clifton, *March 9, 1936* (Jacot, ARP).

This is the dry desert hummingbird *par excellence;* it lives in an entirely different habitat from that of the Black-chinned Hummer, favoring ocotillos, chuparosa, and cacti. Males migrate commonly through the blossoming ocotillos in late February and March; by the end of May they have virtually disappeared from the deserts, and evidence from museum specimens indicates that they have nearly all gone to the Pacific Coast of California and Baja California at that time! Because the remaining females and juveniles are impossible to distinguish in the field from the Black-chinned Hummingbird, we can give only a general outline of the breeding range in Arizona, as noted above.

Swarth's, Willard's, and Brandt's erroneous reports of breeding at higher elevations (upper San Pedro Valley and Huachuca Mountains region) were based on a lichen-decorated type of nest which is one of the variations produced by the Black-chinned and Broad-tailed Hummingbirds as well (RSC).

This hummer, and certain swallows, are the first landbirds to return to Arizona in "spring" migration. There are three records for *January 25* (two near Tucson — Mary Jane Nichols, ARP; one in Bill Williams Delta — Monson), and even one for January 19 (1947, near Parker — Monson)! It was apparently present in some numbers by *February 9, 1854,* in the Big Sandy Valley below Wikieup (Kennerly and Möllhausen). Nesting is correspondingly early. A young male already becoming iridescent was taken at Castle Dome, *April 16* (*1935* — Huey, SD). Other young are still being fed by their mothers in June, and fresh eggs are recorded on May 5 (Sutton and Phillips, Condor 44, 1942: 60; Huey, Trans. San Diego Soc. Nat. Hist. 9, 1942: 366).

Swarth (Pac. Coast Avif. 4, 1904: 17-18) wrote that Costa's Hummingbird "begins to appear in the Huachuca Mountains about the first of July . . . Some adult males also were taken . . . I have seen nests with eggs along the San Pedro River in July. They breed quite commonly all along this valley . . ." These statements seem to be the exception that prove the rule that Swarth was an extremely careful, conscientious worker. The facts are that Costa's Hummingbird has not been proven to nest anywhere in

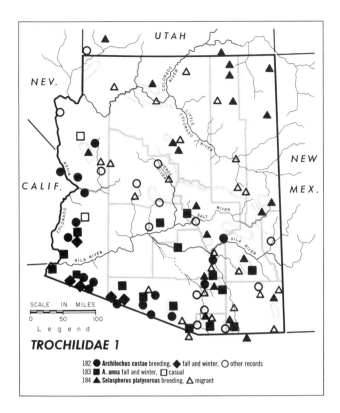

TROCHILIDAE 1
182 ● Archilochus costae breeding, ◆ fall and winter, ○ other records
183 ■ A. anna fall and winter, □ casual
184 ▲ Selasphorus platycercus breeding, △ migrant

those parts of the San Pedro Valley visited by Swarth, even during its actual nesting season; and the only adult male found by Phillips in Swarth's collection (CAS) is dated *May 23, 1896!*

183. ANNA'S HUMMINGBIRD — *Archilochus anna* (Lesson)

Migrates in rather small numbers into southern Arizona in September and early October and winters there until December and rarely to early March. Has not been found east of Huachuca Mountains, Fort Grant (Henshaw, US), and Roosevelt Lake (Willett), nor along Colorado River farther north than Imperial Dam. Recorded casually from Hualapai Mountains, *July 19, 1959* (Wm. Musgrove collection at Kingman).

Anna's is the only regularly wintering hummingbird in Arizona. One wonders whether those wintering in Tucson and other higher parts of southern Arizona ever get back to California for their early spring breeding period there, but most of them must definitely leave by the end of December, if they have not starved to death. On the other hand, they winter quite successfully in the lower Phoenix area, as attested by their regular presence from *October* to late February, and a veritable "flight" in early *March* (at least in *1957* — JSW).

The few that remain in the southern part of the state after December seem to survive best in the mountains; there is not a single definite record after January in the Tucson valley! This peculiar case of a "winter range" deserted in December is discussed by Phillips (Wilson Bull. 59, 1947: 111-113). Young hummingbirds that left their nest in an oleander at Yuma on the extraordinarily early date of March 15, 1962, may have been Anna's (Monson).

184. BROAD-TAILED HUMMINGBIRD — *Selasphorus platycercus* (Swainson)

Common summer resident throughout boreal and Transition Zones, and among deciduous trees along streams in adjacent Upper Sonoran Zone. Migrates uncommonly through lower country between or adjacent to breeding areas, mainly in spring; once heard as far west as the Ajo Mountains, March 23, 1947 (Phillips). We have no fall records in the lowlands of northwestern Arizona, and reports in the adjacent Nevada lowlands would seem to require specimen support. Hybrid with Costa's Hummingbird reported from Rincon Mountains (Huey, Auk 61, 1944: 636-637).

The Broad-tailed is the common nesting hummingbird in the higher mountains. A shrill whistle is made by the wings of the adult male in flight, and a dozen are heard before one is seen. This whistle is louder and more rattling than those of the Rufous and Allen's Hummingbirds. Young or molting males would not make this sound and would thus appear just like the male Ruby-throated Hummingbird, a species which has not yet been taken in Arizona. The deeper red of the male Broad-tailed Hummingbird's throat is different from the brighter, more orange-red of those other Arizona species that have this color confined to the throat. Males display at willow thickets of high mountain meadows, to which the females are attracted for mating, though they nest elsewhere in dense trees from Arizona cypresses and oaks on up into alpine firs. The display is described by Helmuth O. Wagner in Zool. Jahrbüch, Abt. Syst. 77, 1948: 267-278.

This hummer is found in the mountains of southern Arizona from the end of February or early March to late September (possibly the first days of October, but we have seen no October specimen). Its migrations in the Lower Sonoran valleys are hard to understand, for it may be heard or, rarely, seen there at almost any date through the spring and summer, though few or none seem to occur between early June and August 10. We have fall lowland records only as far northwest as Oracle (Scott, AMNH) and Fort Grant (Phillips).

Although found just below the Mogollon Rim on *March 30 (1937,* near Whiteriver — ARP), definite northern Arizona records are from the first half of April to *September 10, (1932,* Flagstaff — Hargrave, MNA) only.

185. RUFOUS HUMMINGBIRD — *Selasphorus rufus* (Gmelin)

Common spring migrant from west slope of Baboquívari Mountains westward (south of the Gila River) almost to Colorado River, mid-February to *April;* very rare in spring farther east, and casual in the north (one record, Flagstaff, about *April 25, 1952* — Wetherill, MNA). Abundant fall migrant in northern and eastern Arizona, occurring in smaller numbers in central and southwestern Arizona. One substantiated winter record: Tucson, December 1950 to *January 14, 1951* (E. D. and V. M. Morton, ARP). There are other recent sight

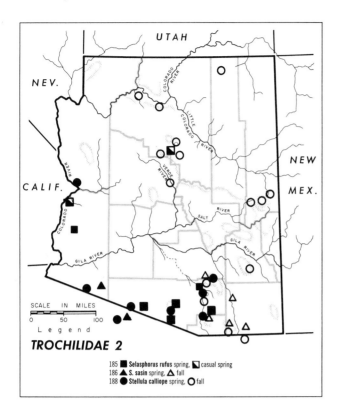

records of this or similar hummingbirds at same place in winter, and specimens from there *November 11, 1938* (adult male, ARP), and near Wickenburg, *November 22, 1959* (Werner, JSW).

This is the abundant hummer of the Arizona mountains during its southward migration in late summer. In the spring it is less common and is virtually restricted to the southwestern deserts; at this season the migration is up the Pacific Coast toward nesting grounds in the northwest. The main fall route is entirely different: down the summit of the Rocky Mountains. The adult males are almost unmistakable, usually being entirely rufous, red, and white. Unfortunately, few of these are seen after August, for they pass through first, being (with other hummers and swallows) the first land birds to start moving south. All the birds seen after August (except as noted above) are in nondescript female and immature plumages.

Northern Arizona records, in fall, are from June 28 and 30 (1938, southwest of Kayenta — Russell) and *July 2 (1936,* near Eagar — Watson, ARP) to *September 25 (1933,* northeast of San Francisco Peaks — Hargrave, MNA), and possibly for a week later (unidentified hummingbirds only).

In southern Arizona, there are occasional February reports from the extreme west, and a male was taken in the Tucson Mountains, *February 21, 1959* (Levy, US). More normal dates are from mid-March to the start of May, and from about mid-July to mid-October. Fall departure at Tucson is clouded by the occasional wintering birds, and by a straggler found (dying) on the University of Arizona campus, *November 3, 1955,* by a student (specimen examined by Phillips — WGG). Late records farther north are *October 11 (1949* in the Aquarius Mountains — ARP) and October 17 (1953, at Phoenix — Phillips).

186. ALLEN'S   *Selasphorus sasin*
    HUMMINGBIRD   (Lesson)

Rather uncommon early fall transient (chiefly in July) in the mountains of central southern Arizona east to Mule Mountains (at Bisbee — Robinette, AMNH) and Benson (LLH). One verified spring record for Heart Tank, Sierra Pinta, Yuma County, *February 18, 1955* (GM) and one southwest of Sonoyta, Sonora, *February 22, 1955* (Marshall, WJS). Owing to confusion with Rufous Hummingbird, may be more common than records indicate.

Allen's Hummingbird has a journey parallel to that of the Rufous, but it is somewhat earlier and much rarer in Arizona, which is at the northeastern limit of its migratory loop. Lack of fall records from northwestern Mexico seems to show that the birds reach the Valley of Mexico non-stop. The breeding area is of course coastal California. The male cannot safely be distinguished in the field from the Rufous Hummingbird, since the color of the back is somewhat variable in both species. It is probable that collecting of exceptionally early fall migrant *Selasphorus* in southern Arizona mountains would show that Allen's is more common than presently supposed.

Allen's Hummingbird has the earliest fall migration of any Arizona bird. Authentic dates are from *July 5 (1901,* in the Huachuca Mountains — Breninger, MCZ) to *August 21 (1890,* Mule Mountains) only. It is true that J. A. Allen, reporting the latter specimen (Bull. AMNH 5, 1893: 36), also stated that Robinette had taken others, including one from Arizona in September; but examination of these by Phillips shows them to be actually Rufous Hummingbirds.

187. BUMBLEBEE HUMMINGBIRD;
    HELOISE'S HUMMINGBIRD;
    MORCOM'S   *Selasphorus heloisa*
    HUMMINGBIRD (Lesson and Delattre)

Accidental: two female specimens, Ramsey Canyon, Huachuca Mountains, *July 2, 1896* (H. G. Rising, US, MVZ).

This hummer appears exactly like a Calliope Hummingbird in the field except that the male's gorget is solidly colored.

The two Arizona specimens are the types of the pale northwest-Mexican race *morcomi* (Ridgway), which breeds in the Sierra Madre Occidental from southern Chihuahua south (see Phillips, Anales del Instituto de Biología, Univ. Méx., 32, 1962: 338-339). It has not otherwise been taken within several hundred miles of the United States border.

188. CALLIOPE   *Stellula calliope*
    HUMMINGBIRD   (Gould)

Uncommon to rare spring migrant in southwestern Arizona, chiefly from Baboquívari Mountains westward. Common fall transient in mountains of northern and eastern Arizona, occasional in lowlands.

The Calliope Hummingbird has a distribution and migration almost identical with that of the Rufous, but it is always less common in Arizona and does not remain so late in the fall. The male is our only hummingbird with a streaked throat. But the female can only be identified by one thoroughly familiar with the calls and appearance of the other female and immature hummingbirds, which would have to be present for direct comparison in the field.

Authentic southern Arizona specimens span only the periods *mid-April* to *April 27* (*1956*, east of Papago Well, southwestern Pima County — ARP), and *early August* to *September* [not "August"] *27* (*1873*, Fort Grant — Henshaw, US). The brevity of these periods is due largely to the difficulty of identifying and collecting this tiny creature. It has been taken in Mexico as early as *March 27* (*1938*, Crater Elegante, west of Sonoyta, Sonora — S. B. Benson and J. E. Simpson, MVZ) and even *June 28* (*1957*, southwest of La Junta, Chihuahua — P. Ogilvie, KANU). There is one May sight record: a male at the mouth of Madera Canyon, Santa Rita Mountains, May 8, 1943 (Vorhies and class).

Northern Arizona records are from *July 14* (*1936*, near Flagstaff — Phillips, MNA) to *September 29* (*1914*, near Springerville — Peters, US).

189. RIVOLI'S HUMMINGBIRD *Eugenes fulgens* (Swainson)

Fairly common summer resident in mixed Upper Sonoran and Transition Zones (males sometimes higher, even to edge of firs), in mountains of southeastern Arizona, north to Grahams and Santa Catalinas; possibly occurs northwest to Sierra Ancha, but no specimens. Casual near Tucson, November 11, 1944 (Alma J. Foerster). A hybrid with *Cynanthus latirostris* taken in Huachuca Mountains (W. W. Brown, AMNH).

This and the next species are large hummingbirds which differ from the smaller species in their slower-beating wings and shrill calls. The male Rivoli's appears all black, while the female has moderate-sized dull whitish corners to the tail. We cannot agree with Swarth that the two sexes inhabit different altitudes in the mountains (though this occurs to some extent in Black-chinned and Broad-tailed species), for both sexes are found at 5000 feet in Madera Canyon, below the pines of the Santa Rita Mountains. It is less restricted to moist canyon bottoms than is the Blue-throated. Quiet, sedate for a hummer, Rivoli's perches erect, showing a rather thin head and neck, with the bill held horizontally. The call-note is a sharp *chip* like that of a Black Phoebe.

Arizona records are from April 5 (1954, in the Santa Rita Mountains — James M. Gates) to *September 24, 1873*, when Henshaw (Amer. Nat. 8, 1874: 241) first discovered it in the United States, on Mount Graham (Pinaleno Mountains).

Separation of the Arizona birds as a distinct race, *aureoviridis* van Rossem (Proc. Biol. Soc. Wash. 52, 1939: 7) appears to be completely unwarranted.

190. BLUE-THROATED HUMMINGBIRD *Lampornis clemenciae* (Lesson)

Uncommon summer resident of moist canyons in the Huachuca and Chiricahua Mountains. Reports from Santa Rita Mountains require verification. Only one specimen record from Santa Catalina Mountains (*May 14, 1884* — Stephens), and one sight record (May 18, 1945 — Brandt). No record whatever at "lower elevations," where said to winter by A.O.U. Check-list; a report for "Tucson" is an error.

This hummingbird is similar to the female Rivoli's, with the same large size and black bill, but it a deeper and purer shade of gray beneath. The call is a loud, long *seep*, given in flight. Both sexes show a snow-white terminal half (as seen from below) of the longer and more rounded tail. The neck is fuller than Rivoli's (like a seal), the posture less rigid, the contours more rounded and graceful in this elegant large hummingbird. Favoring shady streamsides in the mountains, it returns year after year to the same spot to nest, often building on top of the previous nest. This is due to its fussiness about the nest site, which must be sheltered from rain and near water. Helmuth O. Wagner has written of the life history of this bird in Mexico in Veröff. Mus. Natur. Völker Handelsk. Bremen, Reihe A, 2, 1951: 5-44.

The earliest Arizona spring record (Oberholser, Bird-Lore 26, 1924: 248) is April 21 (1912, at "Tucson" — an error; reported by Willard as "Tombstone," but doubtless actually seen in Huachuca Mountains). But Willard also reported that eggs are laid and hatch as early as April 23 and May 10, respectively (FW files); and that young were still in the nest on September 8 (Condor 15, 1913: 41).

Old reports from the Santa Rita Mountains go back to Bendire's statement (*Life Hist. N. A. Birds* 2, 1895: 191) that "it has been taken by Mr. E. W. Nelson" there. But Mrs. Bailey (Pac. Coast Avif. 15, 1923: 25) records no such specimen.

The Arizona race is the pale northwestern *bessophilus* (Oberholser).

191. VIOLET-CROWNED HUMMINGBIRD; "SALVIN'S HUMMINGBIRD" *Amazilia violiceps* (Gould)

Found chiefly in the Guadalupe Mountains of extreme southeast, where it breeds. Has been recorded rarely in summer from Sonoita Creek (ARP), and the Huachuca (F) and especially Chiricahua Mountains (seen nearly every summer recently).

The Violet-crowned is the only Arizona hummingbird that is snow-white from chin to crissum and with a conspicuous livid pink bill. It is also the only one in which the sexes are alike; both these and the immatures have a violet crown. The common call is a chatter, somewhat like that of a Broad-billed Hummer. The Violet-crowned Hummingbird is not likely to be seen except in extreme southeastern Arizona, being primarily Mexican in its distribution. Until 1947 it was known in the United States only from two specimens taken in the Huachuca and Chiricahua Mountains. But it has since extended its range northward, and now can confidently be looked for in riparian trees in Guadalupe Canyon and in Cave Creek of the Chiricahua Mountains, at least from May 22 (Wm. G. George) through August.

Nesting in Guadalupe Canyon was discovered by Seymour Levy (Auk 75, 1958: 350; 77, 1960: 470-471). Arizona

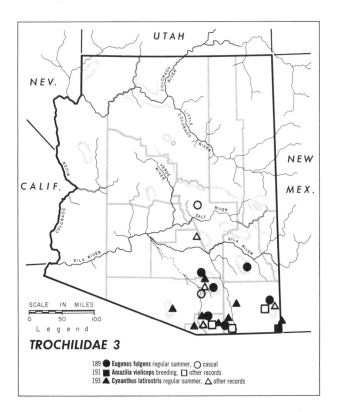

TROCHILIDAE 3

189 ● Eugenes fulgens regular summer, ○ casual
191 ■ Amazilia violiceps breeding, □ other records
193 ▲ Cynanthus latirostris regular summer, △ other records

birds belong to the widespread, greenish-tailed west Mexican race *ellioti* Berlepsch, although long referred to as "*Amazilia salvini* (Brewster)." The type of the latter name has been examined by several ornithologists and proves to be a hybrid with *Cynanthus latirostris;* no such hybrid has yet been taken in Arizona, however. Birds lacking the showy crown, supposed by Griscom (Bull. MCZ, 75, 1934: 378) and Wetmore (Jour. Wash. Acad. Sci. 37, 1947: 103-104) to be the female or immature, are really of another species or well-marked race *viridifrons* (Elliot) of southern Mexico.

192. WHITE-EARED HUMMINGBIRD *Hylocharis leucotis* (Vieillot)

Formerly a rare summer visitant to southeastern mountains, north to Santa Catalinas and Chiricahuas. No authenticated record after *1933* (Campbell, MICH) until *July 4, 1961* (Gould, ARIZ). Supposed winter date listed in A.O.U. Check-list is an error; nor is there any good evidence of its having bred in Arizona.

The White-eared Hummingbird has in the past 40 years been virtually unknown in Arizona, the spate of sight records being probably due to misidentifications, because proper field marks to distinguish it from the Broad-billed have not been made clear. The male is of course unmistakable, appearing all dark except for a prominent white stripe across the side of the head, and flaming red base of the bill. The female, however, is easily confused with other female hummingbirds, particularly the Broad-billed, which also shows a "white ear" (see the key). The note is a metallic clicking like the cricket toy used in changing slides at a lecture.

Helmuth O. Wagner describes its behavior in Zool. Jahrbüch. 86, 1959: 253-302, and A. K. Fisher (Auk 11, 1894: 325-326) writes of its discovery in Arizona. Nearly all of the Arizona records were made from *1894* to *1919*. We owe its recent rediscovery in some numbers, in Cave Creek of the Chiricahua Mountains, to Patrick J. Gould. In Mexico its habitat is pine-oak forest.

Authentic Arizona records all fall in the two-month period from *June 9* to *August 14,* in the Huachuca Mountains (respectively by Fisher, and Swarth, Pac. Coast Avif. 4, 1904: 19-20). We are skeptical of the date "October 1, 1899" on two specimens (AMNH) taken by Lusk, though both are indeed well along in the molt. The race of northwestern Mexico, whence our Arizona birds doubtless originate, is *borealis* Griscom, which is paler in the female and larger than those farther south.

193. BROAD-BILLED HUMMINGBIRD *Cynanthus latirostris* (Swainson)

Common summer resident in mesquite-sycamore associations from the Guadalupe Mountains west along the border, locally, to the Baboquívaris, and north at least to the Santa Catalinas. One record for Chiricahuas, June 6, 1947 (Hargrave). One verified winter record, near Tucson, to December 4, 1960 (Levy). Occurs rarely about Tucson in migration, sometimes remaining for considerable periods in town (Phillips).

The male is the only all-blackish Arizona hummer except for the giant Rivoli's. The female has the entire underparts uniform clear gray, without speckling, even on the throat. This species likes sycamores with mesquites, as where the mountain canyons drop into the desert; but it does not occupy many such areas in the central part of the state and at Portal — which appear suitable. Its call exactly resembles the chatter of the Ruby-crowned Kinglet. The only records in Arizona away from the breeding grounds are out on the desert and around town, where females have been found repeatedly in Phillips' garden at Tucson. This is quite different from the situation in Sonora, where there seems to be a summer movement up the mountains into the oak-pine areas.

Although all specimens come from south of the Gila River, there are three sight records of males in the Pinal Mountains area (one consisting of three males at the Southwestern Arboretum near Superior, April 20, 1947 — Wm. X. and Alma J. Foerster). Arriving in occasional years during the first week of March, it is found more regularly from mid-March to mid-September, with occasional birds lingering to October 1 (1956, in Tucson — Phillips). The exceptional females and young may stay in the town of Tucson for long periods, such as one female that appeared at Olive Road on April 9, 1956 and was present at least until June, when Phillips left. Such birds are of course unmated. The relatively small and green-breasted *magicus* (Mulsant and Verreaux) is the race of Arizona and the northwestern coast region of Mexico.

TROGONS   *TROGONIDAE*

194. COPPERY-TAILED TROGON *Trogon elegans* Gould

Uncommon (irregular?) summer resident of the Huachuca, and in recent years, the Santa Rita and Chiricahua mountains; there are old sight records for the Santa Catalinas (*fide* Scott). One winter record, a young male near Tucson, January 17, 1953 (Marshall *et al*).

The Coppery-tailed Trogon belongs to a pan-tropical family of gorgeous short-billed birds which eat fruit and large insects in foliage. Of jay size, our species is fully representative, possessing further familial traits of broad duplex feathers (with an aftershaft), a most curious tiny foot in which the first and second toes are both turned to the rear, and saw-teeth on the bill. The copper color of the upper surface of the tail is not readily distinguished, but the bright red of the breast and glossy dark green (appearing black in some lights) are certain to be seen and admired in the male. The tail bulges basally and from below seems mostly white; in flight this white is fluttered laterally. The female is less red than the male, with brown upperparts and a light tear-streak on the side of the head. The trogon is found mostly among sycamores in Arizona, where the frog-like call *co-ah* is repeated monotonously; this gives it its Mexican name, Coa. The nests in cavities of sycamores have been much persecuted by photographers. Sedate birds, trogons sit quietly with the tail straight down. For many years they have been protected by law in Arizona, and they occur nowhere else in the United States with any regularity. They are curiously erratic in their occurrence and numbers in successive summers, showing no definite trend in these fluctuations. The visitor must take his chances on seeing the bird, for he may find none where there had been several the summer before.

Trogons have been found in their breeding habitat in the mountains of southern Arizona from *April 3* (*1939*, Santa Rita Mountains – Brooks, MVZ) to September 20 "and subsequently" (*fide* Scott, Auk 3, 1886: 425). Though known from the Huachucas since the 1880's, no actual nest was found in Arizona until 1939 (Allen, Auk 61, 1944: 640-642).

KINGFISHERS *ALCEDINIDAE*

This is a cosmopolitan group, though chiefly tropical. Only one division of the family is of fish-eaters; to this our two species belong. Kingfishers have great heavy bills out of proportion to the rest of the bird, and the head is likewise large, whereas the tail and feet are small. The front toes are curiously united at the base to make a sole of the foot for grasping a slender perch. Kingfishers eat purely animal food and nest in holes. They fly or sit over the water and plunge down headfirst to catch fish in the beak.

195. BELTED KINGFISHER *Megaceryle alcyon* (Linnaeus)

Fairly common to common transient wherever there are permanent fish-inhabited waters, and winters where these do not freeze; also sometimes appears at temporary waterholes on desert. Although there are summer records for central and northern Arizona, there is no good evidence of nesting in the state in the present century.

The Belted Kingfisher is the usual kingfisher of Arizona, of Flicker size, with bushy crest, and a long rattling call. It is colored deep blue and white; only the female possesses the chestnut band across the breast. The past status in Arizona is much clouded in doubt, and even its present role would bear investigating. For early observers reported it as nesting here. Bendire (*Life Hist. N. Am. Bds.* 2, 1895: 35) definitely states that "in southern Arizona . . . I have found Kingfishers breeding in localities where fish must have formed but a small percentage of their daily fare." But Mearns (ms. and FW files) notes them as "first seen" at Camp Verde in September 1884, *July 29, 1885, August 28, 1886,* and September 1, 1887.

Some of the recent mid-summer records are July 3, 1949 at Lakeside (*fide* Louise Levine) and July 4, 1954 at the junction of Sycamore Creek with the Verde River (Wetherill). Late (winter) records from cold places in the north are Black River, in the Transition Zone of the White Mountains, *November 27, 1936* (when it was very cold – ARP) and Springerville, January 7, 1936 (Jacot). The ordinary periods of migration are April to May and August to October. No geographic races of this species are recognized (see Phillips, Anales del Inst. Biol. Univ. Méx. 33, 1963: 336-338).

196. GREEN KINGFISHER; TEXAS KINGFISHER *Chloroceryle americana* (Gmelin)

Rare straggler into Santa Cruz drainage (Tucson and above) and San Pedro Valley (Benson and above) in fall and winter. No authentic record west of Arivaca (*December 23, 1960* – Levy, US); Coues' records for the Colorado River (1865) are very questionable. Though called "casual" in A.O.U. Check-list, there are specimens from Fairbanks (Willard, MVZ), Nogales (Dille, ARP), Arivaca, Patagonia (three seen, two taken, ARP and LLH), and Tucson (ARP), almost certain sight records for the latter two places (A. A. Nichol and W. P. Taylor), and sight records for Benson and St. David. Most of the records are from *October 1* to *February*.

Green Kingfishers are found along small running brooks with shade and roots over the water upon which they habitually perch. They are of sparrow size with a bill larger than a Flicker's. The green upperparts are so dark as to appear black at any distance. They are mostly white below with a dark chest band; the male has a broad area of chestnut on the breast.

The birds at Benson and St. David were both seen the same day – May 21, 1943 (Hargrave and Gordon Pettingill). Coues' sight records lack confirmation and seem impossible. Arizona specimens belong to the race *hachisukai* (Laubmann), characterized by much white spotting above.

WOODPECKERS *PICIDAE*

Woodpeckers, with their zygodactylous feet (two toes backwards as in cuckoos), strong claws, and stiffened tail, cling vertically to the sides of trees while disturbing the peace with their drumming "song" or their excavating for wood-boring insects. Some are exceptional (as the Acorn Woodpecker and Lewis' Woodpecker) in flying out from a perch after passing insects, like any flycatcher; and the Flicker probes ant-hills on the ground. The barbed or gluey tongue can be thrust far out to draw insects from their galleries. They carve out their nests in tree trunks, wherein are laid the eggs — white as in most hole-nesting birds. The young are nearly fully grown before they leave the nest. These young, of both sexes, resemble the adult male more than the female. Adult males differ from female adults usually in having red or more red on the crown or moustache. Most woodpeckers are non-migratory. The principal species in Arizona, ranged by size, are the Flicker (very large), Lewis' (large), Downy and Ladder-backed (small), and all the rest, medium.

197. FLICKER (INCLUDING YELLOW-SHAFTED FLICKER, RED-SHAFTED FLICKER, AND GILDED FLICKER); HIGH-HOLE; YELLOW-HAMMER; *Colaptes auratus* MEARNS' FLICKER (Linnaeus)

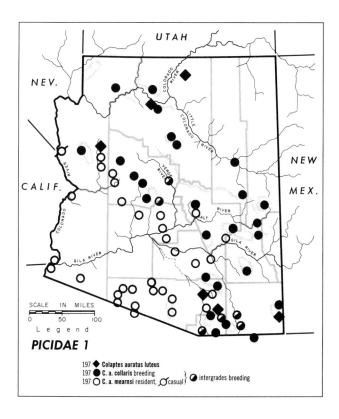

PICIDAE 1

197 ◆ Colaptes auratus luteus
197 ● C. a. collaris breeding
197 ○ C. a. mearnsi resident, ⊘ casual } ⦿ intergrades breeding

Common summer resident of forested mountains; common permanent resident in the wooded Lower Sonoran Zone, including sahuaros, from San Pedro Valley west, scarcer in Yuma County and along the Colorado River, where it ranged 100 years ago up to Fort Mohave. Also winters commonly in areas with trees, below, and uncommonly within, the Transition Zone. Because of extensive intergrading, all Flickers are considered conspecific.

The Flicker is our largest woodpecker in Arizona and is easily recognized. It is brown above and white below, conspicuously barred above and spotted below with black, and with a black crescent across the chest. In flight it flashes a white rump and yellow or red under the wings and tail. This flashing of color, perhaps as much as its characteristic call of *wicka wicka,* accounts for its name.

In summer the distribution of Flickers is much restricted by their need of forests of sizeable trees suitable for the excavation of nesting holes. They eschew the hard oaks and mesquites; but some sizeable stands of cottonwoods are unpopulated at this season, for unknown reasons.

Our taxonomy started with Linné, who believed in special creation. If the coloration of a bird was different from that of its neighbors, then it must have been created that way from the beginning, as a full species. Adjoining populations of flickers, under the old system, would thus be named as yellow-shafted, red-shafted, Mexican, and gilded species. But with the understanding of gradual evolution from a common ancestor this concept has changed. The names have not kept up with this new understanding although the birds themselves, in the case of Flickers, ignore these man-made "species" boundaries. For instance, one winter in Phillips' garden in Tucson, a pair was formed of a Gilded and a Red-shafted Flicker who stayed together until migration separated them in spring.

There is an identity in behavior and voice of all Flickers, and it has long been known that yellow and red forms massively interbreed all down the western plains of Canada and the United States. Test (Univ. Calif. Publ. Zool. 46, 1942: 371-390) showed that the difference between the yellow and red color is a minute final step in the metabolism of the carotenoid pigments involved. If a red-shafted Flicker has to replace feathers lost in winter, these grow in as yellow because of deficiency of animal food at that season (not *vice versa*), suggesting that the attainment of red is due to one or a few genes plus diet (L. L. Short). Grinnell correctly noted the frequency of red variants of *mearnsi* along the lower Colorado River but incorrectly interpreted this as meaning that hybrids reported from other places were similar variants, and that they did not indicate interbreeding between two "species." Phillips (Condor 49, 1947: 121) pointed out that the small red *nanus* Griscom, of yucca desert in northeastern Mexico, has a light cinnamon crown showing that it is most closely related to the small yellow *mearnsi* (and *chrysoïdes*) in deserts of Arizona and northwestern Mexico. Indeed, they are so similar that a fossil from Nuevo León, in the range of *nanus*, was actually identified as *chrysoïdes* by Loye Miller. These interesting forms have converged in small size commensurate with their desert environments, where trees are small.

Over most of their ranges, then, the differently colored forms of Flickers interbreed massively wherever they possibly can. There would be no doubt in anyone's mind that they are all the same species were it not for one peculiar situation in Arizona. Here we find a common, large, red-shafted bird (*collaris*) breeding in coniferous forests of the mountains; below it, separated by several thousand feet of unfavorable habitat, is a small, abundant yellow one (*mearnsi*), breeding in the cottonwood and saguaro stands of the desert. The place to look for intergrades is in cottonwood forests at middle altitudes in this broad hiatus between the two metropoli; these stands are few, far between, and in some of them Flickers are quite rare. The result is that few ornithologists have ever seen specimens from such areas, nor have they heeded the published accounts by Monson (Condor 44, 1942: 223) and Phillips (*loc. cit.*). Actually, perfectly clear examples of intergrades from these cottonwood forests, taken after the early April exodus of wintering *collaris*, are in the collections of Phillips, Monson, Brandt (Cincinnati), and the University of Arizona; they come from the Nogales (collected by Dille), Babocomari, Hereford, and Camp Verde (Marshall and Ambrose) regions, respectively. Later, in 1963, Short collected many additional specimens (AMNH).

The Flickers of the Verde River present an unusually interesting problem. In the 1880's Mearns mentioned only *collaris* and stated that "very few remained in the summer." By 1916 the Biological Survey team of H. H. T. Jackson and W. P. Taylor considered Flickers common there in summer and called them *mearnsi,* though they took only one worn specimen. Mrs. Betty Jackson, at Montezuma Castle National Monument in the 1930's, observed Red-shafted Flickers. Today there is a sizeable population there closest to *collaris,* but with rather cinnamon crowns and an occasional yellow variant. Farther toward the desert, around Mayer and Ash Creek, the birds are closer to *mearnsi* (Jacot, Marshall, Ambrose, ARIZ).

There is striking geographic variation in the Flicker. In general, lowland populations become smaller from north to south, whereas those of the mountains maintain a large size. *Colaptes auratus luteus* Bangs is the large form of the northeastern United States, Canada, and Alaska (*borealis* Ridgway not being satisfactorily separable). It is yellow beneath the wings and tail, has a gray crown, red collar behind this, and brown throat with a black malar patch in the male. This race is a rare transient and winter visitant in Arizona; specimen records only are represented on the map. Midwinter specimens are all from Tucson, and there is a surprising concentration of records in October and April, which would indicate migration were it not for the virtual absence of records farther south. Pure-blooded birds have been taken in southern Arizona from *October 13* (*1949* on the Big Sandy River — ARP) to *April 7* (*1937* near Sonoita — Dille, ARP). The few northern Arizona records are concentrated between *April 17* (*1957*, South Rim of Grand Canyon — Hargrave and W. E. Dilley, GCN) and *May 7* (*1947*, 40 miles north of Cameron, northern Coconino County — K. Warren, MNA — fished out of a well). Intermediates with *collaris* are rare in Arizona compared with the pure-blooded form.

Colaptes auratus collaris Vigors, of the same large size as *luteus,* is red under the wings and tail, has a brown crown and gray throat with a red malar stripe in the male. It almost always lacks the red nuchal crescent. Its breeding distribution in the mountain forests of Arizona is plotted on the map. The birds of the Hualapai Mountains are possibly distinct; two of four molting August and September specimens (ARP) are dark like the northwestern coastal *cafer* and are heavily suffused with pink. The other two appear duller and lighter (though not far into the molt), suggesting that the population is merely an incipient race. In winter, most of our *collaris* quit the mountains to inhabit lowland woods and groves, including shade trees in town, where they are doubtless joined by throngs from the north, as shown by their great numbers. The wintering period in the lowlands extends normally from late September to early April. Extreme dates for southern Arizona (Colorado River — Monson) are from September 16 to April 16; for northern Arizona (Wupatki National Monument — Davey and Corky Jones) September 10 to April 13.

Colaptes auratus mearnsi Ridgway is the small resident desert form, characterized by yellow under the wing and tail, and bright cinnamon crown. Otherwise its color is like that of *collaris*. As seen flying over, the tail is broadly tipped with black — for almost half its length, as contrasted with a narrow tip in *luteus*.

Some variants, especially along the Colorado River, are red under the wing and tail, showing that the yellow wings so stressed in books are not nearly as important a racial trait as are the small size and cinnamon crown.

[PILEATED WOODPECKER   *Dryocopus pileatus* (Linnaeus)]

[Hypothetical. Found once, near Point Imperial on North Rim of Grand Canyon, where one was seen August 30, 1935 (R. K. Grater), and workings photographed. The record, however, was not substantiated, though it is cited in the A.O.U. Check-list.]

[RED-BELLIED WOODPECKER   *Centurus carolinus* (Linnaeus)]

[Erroneously reported on the basis of two undated specimens (B. H. Dutcher, AMNH). These are in a small collection labelled as from "Fort Grant, Arizona," which contains eastern races of birds. They doubtless came from Fort Leavenworth, Kansas, where Dutcher was stationed in the late 1890's.]

198. GILA WOODPECKER   *Centurus uropygialis* Baird

Common resident throughout Lower Sonoran Zone of southern and western Arizona; rather local in extreme southeast. Winters fairly commonly in adjacent Upper

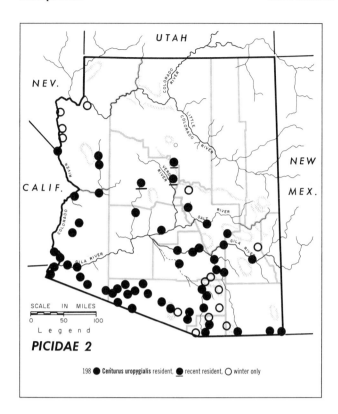

198 ● Centurus uropygialis resident, ● recent resident, ○ winter only

Sonoran Zone, casually reaching the lower edge of Transition Zone (Bear Canyon, Santa Catalina Mountains, *February 7, 1953* — ARP). Occurs occasionally in Prescott region. There are winter sight records north to near Pierce's Ferry, Mohave County (Grater).

This is the commonest and noisiest woodpecker in the saguaro stands of southern Arizona. It is a solidly fawn-colored bird with a black-and-white barred "ladder-back," these bars extending on to the wings and tail. Whiter areas flash in flight from rump and wings. Gila Woodpeckers rear two or three broods a year. After nesting, they spread to areas unoccupied in summer, though never abandoning completely the breeding ground. They are opportunists in their feeding, varying the diet in season with dates, pomegranates, orange juice, saguaro fruit, bits of lizard or tasty nestling birds and eggs — but the main diet is insects. Sliced fruits attract them to the garden. The Gila Woodpecker's activity in opening fruits of the saguaro makes the red pulp available to less well-armed birds. Some frightful concentrations occur during winter in isolated orange groves. In Phoenix they "did quite heavy damage" to the pecan crop in January 1945, and ate black walnuts then and in April of the same year, according to V. H. Housholder, who also "observed a pair drilling for, extracting, and eating angleworms from the lawn."

The Gila Woodpecker is mostly resident; but from the end of August to the first of April, it is more or less common in lower parts of the Upper Sonoran Zone adjacent to its breeding area, at the base of the east slopes of most of the southern Arizona mountain ranges (especially the Baboquívaris, where it is common), and at stations somewhat to the north of the known breeding range — all shown by open circles on the map.

Edgar Mearns collected at Camp Verde from 1884 to 1888 and saw his first Gila Woodpecker there, *February 24, 1888* (US #235018). By 1916 it was fairly common there (Jackson and Taylor, FW files), and is still present and presumably expanding, as shown by diamonds on the map.

The birds of the Colorado Valley have been separated by van Rossem (Condor 44, 1942: 22) as *Centurus uropygialis albescens,* on the very slim characters of slight pallor and greater width of the white markings on the back as compared to the black marks between them. Though a trend towards paleness in that region is well shown in other species, such as the Common Screech-Owl and the Song Sparrow, in the Gila Woodpecker it is scarcely enough to merit recognition. If valid, the eastern race would have to be renamed, since the type locality of *albescens* at Laguna Dam is so close to that of the nominate race at Alamo as to make it a straight synonym.

199. RED-HEADED WOODPECKER
Melanerpes erythrocephalus
(Linnaeus)

Accidental: one taken *about June, 1894* in the Chiricahua Mountains (W. W. Price, location unknown).

200. ACORN WOODPECKER; ANT-EATING WOODPECKER; MEARNS' WOODPECKER; CALIFORNIA WOODPECKER
Melanerpes formicivorus
(Swainson)

Common resident among large oaks in mountains throughout Arizona (except extreme north and in Baboquívaris, where rare and local, or perhaps irregular). Straggles from breeding range at all seasons except early spring (but mainly in fall), in some years even to Colorado River (Monson) and Organ Pipe Cactus National Monument (Phillips).

This white-eyed clown of woodpeckers has a harlequin face; it is clothed in contrasting black and white, with a red crown. It is communistic in its nesting (Ritter, *The California Woodpecker and I,* U. C. Press, 1938), is very social, and has a loud call, "Jacob." Acorn Woodpeckers store acorns in the bark of dead pines — individually, each in a neat hole, or at times piled into cavities. They catch flying insects in late afternoon or following a rain, putting on an impressive show over mountain canyons, for they go much farther and higher for each capture than do the kingbirds and other flycatchers. They drink from tree-holes. In the Hualapai Mountains, in a great hollow limb of a Gambel's oak at the County Park, these common birds used to crowd in to roost. Later the dormitory blew down and the birds were thereafter scarce in this area. In the Flagstaff region, the Acorn Woodpecker is found only where stands of sizeable Gambel's oaks occur, as near Mormon Lake; and it is a very rare wanderer through the

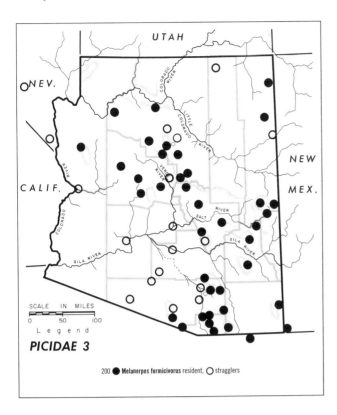

PICIDAE 3

200 ● Melanerpes formicivorus resident, ○ stragglers

rest of the forest where Gambel's oaks are small or absent.

As hinted above, the distribution of Acorn Woodpeckers is governed by the occurrence of suitable cavities in old snags for communal roosts; by smooth, soft trunks of pines in which to store acorns; and of course, by extensive stands of oaks to provide acorns. Whereas most mountain birds that straggle to the deserts do so in winter, the Acorn Woodpecker straggles from May through to the end of the year, beginning before the movements of other woodpeckers. Desert records are of single birds, grouped into minor flight years, coincident with flights of jays and nuthatches, as in the fall of 1934 and of 1950.

The Arizona race is the slender-billed extreme of the species and was appropriately named *aculeata* by Mearns, years ago. Many recent authors, ignoring the meaning of this name, have synonymized it with *formicivorus* of Mexico because its bill is not shorter. All the Arizona specimens examined are *aculeata*, although an example of *bairdi* Ridgway, the heavy-billed coastal race of California, has been taken in the Sheep Mountains, Nevada (van Rossem, LA). Another from the Dead Mountains, Nevada, is unfortunately a headless mummy and cannot be racially identified (M. Sullivan, Lake Mead Nat. Recr. Area collection). A. H. Miller reports a representative of the Arizona race as a vagrant to Eagle Mountain, central Riverside County, California on *October 19, 1945* (Miller, MVZ).

201. LEWIS' WOODPECKER *Asyndesmus lewis* (Gray)

Fairly common summer resident of certain Transition Zone parks in San Francisco Mountains, rarer and local northward and eastward. More or less uncommon (in most years) transient and winter visitant in open Upper Sonoran and low Transition woody areas. Occasionally winters in lowlands, commonly so in flight years.

This handsome black woodpecker, discovered by Captain Meriwether Lewis' expedition, looks and flies like a crow, but shows red underneath. It is fond of soaring out from cliffs and tall dead snags after insects, like its close relative, the Acorn Woodpecker. Therefore it requires broken terrain and burns, and is absent from dense forests. In summer it is strangely local in Arizona, being rare in the White Mountains region, which appears suitable for it. There are no midsummer records there except at Alpine.

It is a great satisfaction to bird lovers to see what seems to be the same birds in successive winters in the same trees. In most winters one to four birds appear at the pecan orchard south of Tucson; they carry a pecan to an "anvil" on a dead cottonwood branch, fitting it into a hollow where it can be whacked open. The species also seems quite regular in winter in open live-oaks farther east.

The great flight of 1884–1885 is graphically described by Brown (Auk 19, 1902: 80-83) and Scott (Auk 3, 1886: 427). Lewis' Woodpeckers were abundant that winter at such diverse points as Tucson, the Mogollon Plateau, and the northeast part of the Hualapai Indian Reservation, but not at Camp Verde (Mearns,

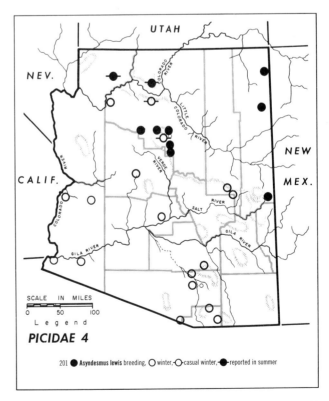

PICIDAE 4

201 ● Asyndesmus lewis breeding, ○ winter, -○- casual winter, -●- reported in summer

ms.) or the upper Santa Catalina Mountains (Scott). Mearns saw a "flock containing hundreds" at Prescott, *November 12, 1887*. Again in the winter of 1946–1947 there was a major flight to western Arizona, where the birds were reported by ranchers as damaging fruit on lower Trout Creek in the Big Sandy Valley.

Migrants, not shown on the map, are uncommon in fall and rare in spring throughout the state wherever there are tall trees; but they are most regular in open pine-oak country. Migration is mostly in April and October. Extreme dates in southern Arizona are September 12 (1885 at Camp Verde — Mearns) to May 10 (1945 at or near Fairbank — Brandt, Hargrave and Oberholser). Northern Arizona dates, away from breeding grounds, are from September 10 at the Grand Canyon (Merriam) to May 7 at Keams Canyon (Monson).

202. YELLOW-BELLIED SAPSUCKER; RED-NAPED SAPSUCKER; RED-BREASTED SAPSUCKER *Sphyrapicus varius* (Linnaeus)

Nests in Canadian Zone of mountains along and north of Mogollon Plateau, and irregularly (?) in the Hualapai Mountains. Very rare except in parts of White Mountains, Blue Range, and farther west at Promontory Butte (Johnson, ARP). Common transient for long periods throughout most of state but very rare in driest open desert areas (GM). Common in winter (except in north where very rare) in Sonoran Zone deciduous trees. A hybrid with *S. thyroideus* reported from the Huachuca Mountains (W. W. Brown, CLM).

This is a mottled black and white woodpecker, giving a dark blurry effect broken by a prominent white stripe down the closed wing. The head varies greatly, according to the subspecies, sex, and age, from all red to mottled brown or patterned black and white. Most Arizona birds have only the crown and throat red, with a trace of this color on the nape. Many show up in dull juvenal plumage, which persists around the head well into winter. This and the next species drill horizontal rings of little holes around trunks of broad-leaved trees such as willows, cottonwoods, aspens, and walnuts, where they work quietly. The Yellow-bellied Sapsucker is abundant in Arizona in October and March and does serious damage to isolated orchards. Hummingbirds, Ruby-crowned Kinglets, Audubon's Warblers and other small birds visit these drillings to feed. There was a remarkable influx of sapsuckers at Cazador Springs, where Correia and Phillips had camped for several snowy weeks in January, 1937, without seeing more than a few sapsuckers in all. But beginning on the 23rd, several new birds appeared daily, indicating that they were already moving north. (Since most of these birds were collected, there can be few duplications in the figures. A total of 13 birds was collected from January 23 to 26!). An interesting chronicle of movements since the ice age, which have led to the present contacts among well-marked races, is presented by T. R. Howell (Condor 54, 1952: 237-282).

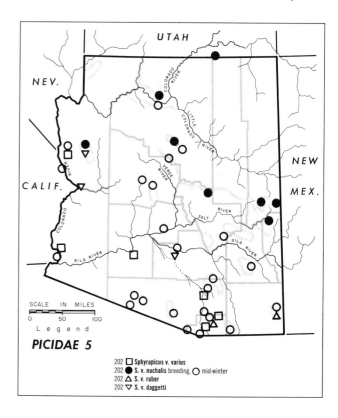

PICIDAE 5

202 ☐ *Sphyrapicus v. varius*
202 ● *S. v. nuchalis* breeding, ○ mid-winter
202 △ *S. v. ruber*
202 ▽ *S. v. daggetti*

Sphyrapicus varius varius, the race which breeds in Canada and the northeastern United States, is a rare and irregular winter resident in southeastern Arizona and is casual northwestward to the Colorado River. Although termed "casual" in the A.O.U. Check-list, specimens have been taken in Arizona in seven different winters, starting in *February 1940*. In *1952–1953* it occurred in some numbers, and several were taken near Tucson (ARP and CU). Specimens from the points shown on the map are in ARP and (near Yuma) GM collections, taken from *November 17 to February 26*. This race is the most extensively white of the species, with minimum red, and least black on the back. Field identification is possible to those familiar with specimens, though the male *varius* is quite similar to many females of *nuchalis*.

The status of *nuchalis* Baird in Arizona is that of the species, given in the opening paragraph above. It is intermediate in characters and distribution between *varius* and the following races, as indicated by its former common name, Red-naped Sapsucker. Both it and *varius* have a black chest patch in the adult.

The extended periods of migration make the winter range of *nuchalis* difficult to determine. One stayed at Flagstaff from November 23 to December 28, 1952 (Eleanor Pugh). But the only January record in this well-known area is a single bird banded at Walnut Canyon National Monument on January 18, 1936 (Louis Caywood).

Migrants are passing through northern Arizona regularly to at least *mid-November*, and one was seen at Flagstaff as late as December 9 (1938 — Kassel, Plateau 13, 1941: 66-67). This, combined with the January influx mentioned above, shows that there are only a few weeks in early winter when migrants are not to be expected. The earliest fall migrants for northern Arizona are September 19 (1889 in the San Francisco Mountains region — Merriam), and possibly August 23 (1917, in the Lukachukai Mountains — Goldman, FW files), if this is not a breeding locality. Spring migration records in the north are from *February 25 (1939,* on the South Rim of Grand Canyon — D. C. Smiley, GCN; approaches *ruber*) to *April.* It must be remembered that the sight records mentioned above may pertain in part to *S. v. varius.*

Southern Arizona records extend from September 8 (1956 — George) in the Chiricahua Mountains, to May 7 (1951 — Junea W. Kelly *et al.*) at Arivaca Junction. It is rare before September 20 and after late April.

The remaining two races, *ruber* (Gmelin) and *daggetti* Grinnell, have a wash of red over the head and chest, virtually concealing the black and white pattern. They are rare in Arizona and apparently have been somewhat confused, though the loss of Gilman's several specimens makes it impossible to verify this. The origin of our Arizona birds is probably at the far northeastern extension of *ruber* across the mountains of British Columbia. The contact between this eastern arm of *ruber* and the quite dissimilar eastern races could produce a wide variety of plumage types, some possibly resembling *daggetti* closely. One example is a female from Portal, Chiricahua Mountains, *December 11, 1955* (Bruce Elliott, ARP). Typical *ruber* is the darkest race of the species, with the least white on the back; and even this is overlaid with buff. Two fairly typical specimens are from the Chiricahua Mountains (*October 27, 1932* — A. Walker, CLM) and Patagonia (*February 2, 1953* — Dickerman, CU). A bird photographed in color at Canelo, west base of the Huachuca Mountains, November 30, 1954, seems typical of this race also (Mr. and Mrs. Erle D. Morton). An occasional intermediate toward *nuchalis* has been taken; one appeared at Globe as early as *October 5-9 (1956* — Betty Jackson, SWAC).

Efforts to confirm early Arizona records of *daggetti* have substantiated only one specimen from Sacaton, *February 9, 1910* (Gilman, Hanna collection). The other specimens have either disappeared or proved to be intermediates. Only Jon Coppa's remarkable summer straggler to the Hualapai Mountains actually represents this race as published from Arizona (Coppa, Condor 62, 1960: 294). At the same time, he also took a breeding pair of *nuchalis,* our only summer record there. An additional specimen of *daggetti* is from the Bill Williams Delta, *January 23, 1953* (GM). *Daggetti* is characterized by its paleness and greater amount of white spotting on the back, as compared with *ruber.*

203. WILLIAMSON'S SAPSUCKER; NATALIE'S SAPSUCKER *Sphyrapicus thyroideus* (Cassin)

Nests in aspens from Mogollon Rim northward, more or less commonly. Winters in Transition and (sparingly) high Upper Sonoran Zones south and west of Mogollon Rim, rarely on Mogollon Plateau, and once at Grand Canyon Village (February 6, 1949 — H. C. Bryant). Very rare winter visitant in the Lower Sonoran Zone, usually near mountains but west casually to Colorado River (two sight records — Cooper, Phillips).

This sapsucker was named for Lieutenant Williamson, who conducted one of the early railroad surveys in California. The sexes are so unlike that they were origi-

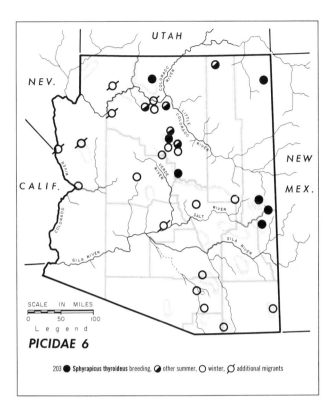

PICIDAE 6

203 ● *Sphyrapicus thyroideus* breeding, ◐ other summer, ○ winter, ◌ additional migrants

nally classed as two species, and were sometimes placed in different genera until the brilliant ornithologist, Henry W. Henshaw, discovered them nesting together in Colorado. The female does indeed resemble a Gila Woodpecker except that the black and white barring above extends down onto the sides and flanks. The male is more like a blackened version of the Yellow-bellied Sapsucker, being entirely black above, except for the same white wing stripe as seen on other sapsuckers. Williamson's Sapsucker winters at much higher elevations than does the Yellow-bellied, and one wonders what it finds to eat in the forests at that cold season.

As shown on the map, the winter and summer ranges of this species are complementary, divided by the Mogollon Rim, except for an occasional straggler to the north in winter, principally in the McNary region. Most of the stragglers to the Lower Sonoran Zone appear during the migrations and soon leave. In the pure Transition Zone around Flagstaff the species is almost entirely a transient. The half open circles show the down-hill wanderers in late summer, and certain other summer records in far northern Arizona which probably do not represent actual breeding localities. It is possible that the winter range extends as far northwest as the Hualapai Indian Reservation; no midwinter visits have been made to this plateau, but the bird was still present on *November 17, 1884* (Mearns, AMNH).

Excepting the summer vagrants, northern Arizona migrations are from September 14 (1922, near Flagstaff — Swarth) to November 6 regularly, and occasionally to *November 20 (1936,* White Mountains region in yellow pines — ARP); and from *February 14 (1934* in Upper Sonoran Zone east of San Francisco Peaks — Hargrave, MNA) and *February 28 (1937,* north of Whiteriver —

ARP) to April 28 (1933, Grandeur Point, South Rim of Grand Canyon — McKee). Southern Arizona records are from August 15 (1961, Tucson Mountains — Levy), *August 30 (1902, Huachuca Mountains — Swarth, CAS)* and *September 21 (1874, Mount Graham — Henshaw, US)* to May 2 (1948 — Brandt) and *May 9 (1880 — Stephens;* present whereabouts unknown), both in the Chiricahua Mountains.

Arizona birds examined all belong to the small-billed Rocky Mountain race *nataliae* (Malherbe), which has been taken west to the Hualapai Mountains (ARP).

204. HAIRY WOODPECKER; WHITE-BREASTED WOODPECKER; CHIHUAHUA WOODPECKER
Dendrocopos villosus (Linnaeus)

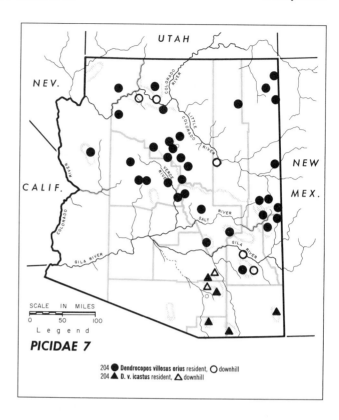

Common to fairly common resident of coniferous forest. Formerly descended uncommonly in winter to adjacent valleys, but no recent record at any distance from conifers.

The Hairy Woodpecker is the common medium-sized woodpecker of Arizona conifers. It is entirely white below and with white stripes on the head and the middle of the back. Elsewhere it is black. It is a species widely distributed from Alaska to Panamá, varying greatly in size and color over this enormous range. In the northwest and extreme south of this range the birds are chocolate color ventrally and have the white outer tail feathers barred with black. The call is a sharp *chink*.

The early explorers found Hairy Woodpeckers fairly common in winter, in riparian timber of the Sonoran Zones. Bendire (*Life Histories* 2, 1895: 53) "shot several near Tucson in winter" in or about 1872; but it has never been found there since by Herbert Brown, C. T. Vorhies, or the authors, who have lived there for long periods. Mearns (Auk 7, 1890: 251-252) states that it descends to the Verde Valley "very rarely . . . and only when the mountain timber is icy or the weather uncommonly fierce." But he took specimens there dated *September 2, 1887,* and *January 20* and *April 11, 1888* (AMNH); saw two on October 23, 1886; and heard one on January 28, 1885. This is a curious discrepancy both as to the weather and to the supposed rarity. Scott (Auk 3, 1886: 425-426) found them "generally . . . rather common" in the lower live-oak belt near Oracle from "early in November . . . until the last of January," taking a male *November 5, 1884* (AMNH). Yet Jacot (*in litt.*), living at Oracle for several years during the 1940's, never saw it there.

Geographic variation in the Hairy Woodpeckers of Arizona involves size and brownness of the underparts; there is no important difference here in amount of white spotting. So the trend is from white, larger birds in the north to browner below and slightly smaller to the south. The northern Arizona bird is *orius* (Oberholser), which extends southward to the Pinaleno Mountains (also called the Graham Mountains and Mount Graham). This bird is nearly pure white beneath, and it is intermediate between the large, pure white bird of the Rocky Mountains, *monticola* (Anthony) and smaller, browner birds to the south and west. (*Monti-*

cola perhaps occurs in the extreme northeast corner of the state, but we have no series from there to establish its presence.) *Leucothorectis* (Oberholser) was described from New Mexico as another such race, slightly smaller than *orius,* but the type series includes several birds not fully matured, and we feel that there is insufficient clinal change to admit two intermediate races.

Dendrocopos villosus icastus (Oberholser) is the slightly smaller, browner bird from the Santa Catalina Mountains southward to northern Jalisco.

205. DOWNY WOODPECKER; BATCHELDER'S WOODPECKER
Dendrocopos pubescens (Linnaeus)

Sparse resident in deciduous trees of Transition and Canadian Zones from White Mountains (where less uncommon), Sierra Ancha, and San Francisco Mountains northward, including (exceptionally) Navajo Indian Reservation. Apparently somewhat less rare in winter, when it reaches Upper Sonoran Zone. Casual in southern Arizona: near Kelvin, Gila County, *April, 1882* (Scott), and at Tucson, *March 13, 1954* (ARP).

A small edition of the Hairy, the Downy Woodpecker's bill is proportionately shorter, by a centimeter. It is much rarer in Arizona, where it is mostly limited to deciduous trees. It is equally widespread as the Hairy in the eastern United States, but here in Arizona it finds its southern limit in the center of the state. There is a remarkable concentration of the few records away from the breeding grounds in October, March, and

April. This might suggest a migration were it not that the species is unknown to the south, in Mexico. In the Pinaleno Mountains, Monson found one Downy Woodpecker on each of these dates: October 20, 1935; May 9 and August 27, 1936.

Arizona birds belong to the large, white Rocky Mountain race *leucurus* (Hartlaub), though one of the specimens from Navajo National Monument (Wetherill, MNA) is sufficiently white-spotted on the wings to suggest *nelsoni* (Oberholser).

206. LADDER-BACKED WOODPECKER; CACTUS WOODPECKER *Dendrocopos scalaris* (Wagler)

Common resident throughout Lower Sonoran Zone, as well as in open parts of Upper Sonoran Zone except in northeast.

The Ladder-backed is the only small desert woodpecker and it is entirely variegated with black and white. Its call is *pick,* much weaker than the similar notes of the Hairy and Arizona Woodpeckers. Though formerly called in Arizona the "Cactus Woodpecker," it is not as dependent on the saguaros as are the Flicker and Gila Woodpecker. For its small size enables it to nest in yucca stalks and branches. Although it scarcely extends north of Arizona, it is very widely distributed over Mexico.

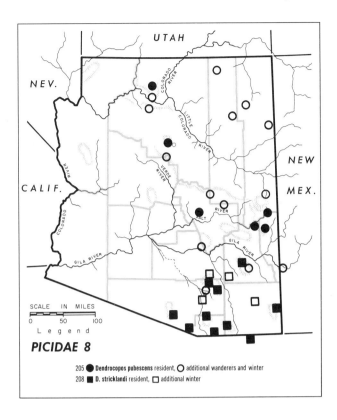

PICIDAE 8

205 ● *Dendrocopos pubescens* resident, ○ additional wanderers and winter
208 ■ *D. stricklandi* resident, □ additional winter

The report of this woodpecker as far northeast as Holbrook (Oberholser, Proc. U. S. Nat. Mus. 41, 1911: 153), as pointed out by Woodbury and Russell (BNav), evidently refers to "a few *Dryobates*" reported by Streator (FW files) as seen at Winslow and Holbrook. This may have referred to *D. pubescens,* as believed by Woodbury and Russell. But the only specimen of any *Dendrocopos* ever taken in that part of the Little Colorado River is Kennerly and Möllhausen's *D. villosus* (US). No member of this genus is recorded there in the present century, and it is quite certain that *D. scalaris* does not cross the Mogollon Plateau at any point.

The Arizona race of the Ladder-backed Woodpecker is the large, rather pale *cactophilus* (Oberholser).

207. NUTTALL'S WOODPECKER *Dendrocopos nuttallii* (Gambel)

Accidental; one taken at Phoenix, *June (or January?) 24, 1901* (Breninger, MCZ).

208. ARIZONA WOODPECKER *Dendrocopos stricklandi* (Malherbe)

Fairly common resident of live oaks of southeastern Arizona, west and north to Baboquívari, Santa Catalina, and Graham Mountains. Rare in winter in adjacent lowlands.

This has a duller, more wooden call than the Hairy's. It is plain brown above, spotted below, and limited to the live-oaks. Henshaw discovered it in Arizona, as he did so many other birds, in his work with the Wheeler Expeditions in 1873–1874. It ranges from here south to Mount Orizaba, but becomes sootier and white-barred, gets into pine forests, and occurs at increasingly higher altitudes — all towards the southeast. In Arizona the nest is usually in a walnut tree, which is easier to drill than an oak.

Our race is large, pale, and solidly brown above. It is known as *D. s. arizonae* (Hargitt), which extends to central Sonora.

209. NORTHERN THREE-TOED WOODPECKER; ALPINE THREE-TOED WOODPECKER *Picoïdes tridactylus* (Linnaeus)

Uncommon resident in boreal zones, rare in Transition Zone, from White to San Francisco Mountains and on the Kaibab Plateau.

This is considered one of the rarer woodpeckers; in our experience it is quiet and easily overlooked. The absence of winter records from the Boreal Zone simply reflects the difficulty of getting there at that time. The pecking made by this woodpecker in its search for grubs inside tree trunks is very loud and resolute, for it is most highly evolved for such activity. It has a heavy bill, large and strong skull, and only three toes. The only good field mark for identification is its barred flank, for the back is nearly as white as in the Hairy Woodpecker. The crown patch of the male is located between the eyes, not on the occiput as in adult male Hairy. At close range it is seen to be yellow — unique among woodpeckers of Arizona.

This is a holarctic bird, here at its southern limit in North America. Arizona specimens are rather small, but are referred nevertheless to the Rocky Mountain race *dorsalis* Baird.

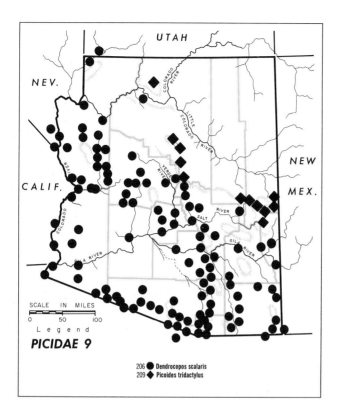

206 ● Dendrocopos scalaris
209 ◆ Picoides tridactylus

PERCHING BIRDS

PASSERIFORMES

Perching birds, or passerines, are considered to be the most highly developed birds, and more than half of our species belong to this order. Most are of small to medium size. They are characterized by a "perching" foot with four toes at the same level, three in front, and the one in the rear with a very large claw. Passerines differ much among themselves in their habits, but usually these are fairly consistent within a family. Therefore it is of extreme importance to learn the family names and groupings in order to understand these birds, let alone to identify them. There are habits ranging from the aquatic Cinclidae, swift-like Hirundinidae, hawk-like Laniidae, crepuscular Turdidae, woodpecker-like Dendrocolaptidae and Certhiidae, and acrobatic Paridae and Sittidae, to such familiar types as flycatchers, wood-warblers, and seed-eating finches. Passerines also include parasitic types both among the cowbirds (Icteridae) and weaver finches (Ploceidae). There are some burrowing flycatchers in South America (*Agriornis* — see Johnson, Goodall and Phillipi, *Las Aves de Chile*, tomo 1, 1946: 139-149), and there are troglodytes such as the Cave Swallow (Hirundinidae) and certain wrens (Troglodytidae). The most strikingly colored and ornamented birds in the world belong to this order; they are the Paradisaeidae. It was the study of one of the passerine groups, the Geospizinae of the Galápagos Islands, which so strongly influenced Charles Darwin in his development of the concept of evolution.

COTINGAS   COTINGIDAE

210. ROSE-THROATED BECARD; XANTUS' BECARD   *Pachyramphus aglaiae* (Lafresnaye)

Local and irregular summer resident along Sonoita Creek, Santa Cruz County; in Guadalupe Mountains in southeastern corner of Arizona; and in Tucson vicinity. A report from Chiricahua Mountains lacks valid basis. Before *1947*, only one record: Huachuca Mountains, *June 20, 1888* (Price, MCZ).

The Rose-throated or Xantus' Becard (to rhyme with checkered) belongs to a large and varied family of tropical American birds, allied to the tyrant fly-catchers. This is the northernmost species, in turn reaching its northern limit in southern Arizona. It is a sluggish, bull-headed, dusky-capped bird, our race being gray above and pale beneath. There is no pattern of prominent markings aside from the dusky cap and, in the male, a red patch on the lower throat. It has a mournful descending whistle. The nest is an immense bushel-basket of strips of cottonwood bark swinging from the tip of a slender branch, normally built entirely by the female, with encouragement from the male. The only Arizona set contained six eggs, near hatching on *June 19, 1948* (Phillips, CM).

The becard seems to be slowly and irregularly pushing northward, but still is not established permanently away from Sonoita Creek. Here it was first found *June 19, 1947* (ARP) but old nest remains indicate that it had been present in earlier years. There were four active nests in 1948 (Phillips, Condor 51, 1949: 137-139). The Arizona Game and Fish Commission accorded it legal protection at that time, whereupon it soon disappeared! Finally it was again reported there in 1955 and has since been seen more or less regularly, though without any spread or increase.

Authentic records elsewhere are few indeed. There are none since *1888* for any of the mountain canyons except Guadalupe Canyon, where an old nest was collected in May 1957 by Levy (US). Here, no bird has been seen. Near Tucson, another old nest was found in the fall of 1958 (George); and a solitary male built a new one in 1959 (George, Marshall) but apparently failed to find a mate.

Arizona records are from May 13 (1950 — Phillips, Marshall) to September 14 (1947 — Phillips); but in northern Sonora, not far south of Nogales, at least three birds were still present *October 12–15, 1954* (ARP).

Twelve
Color
Plates

reproductions of sketches made in the field by

GEORGE
MIKSCH
SUTTON

INDEX TO THE COLOR PLATES

PLATE	Species Name and Reference Number	Text Page
I	Cooper's Hawk *Accipiter cooperii* (65)	20
II	Spotted Screech Owl *Otus trichopsis* (161)	49
III	Spotted Owl *Strix Occidentalis* (168)	53
IV	Arizona Woodpecker *Dendrocopos stricklandi* (208)	75
V	Sulphur-bellied Flycatcher *Myiodynastes luteirentris* (217)	80
VI	Coues' Flycatcher *Contopus musicus* (235)	90
VII	Arizona Jay *Aphelocoma ultramarina* (252)	105
VIII	Curve-billed Thrasher *Toxostoma curvirostre* (283)	123
IX	Red-faced Warbler *Cardellina rubrifrons* (342)	159
X	Painted Redstart *Setophaga picta* (346)	161
XI	Hooded Oriole *Icterus cucullatus* (356)	167
XII	Rufous-winged Sparrow *Aimophila carpalis* (403)	198

PLATE I
Cooper's Hawk

PLATE II
Spotted Screech Owl

PLATE III
Spotted Owl

PLATE IV
Arizona Woodpecker

PLATE V
Sulphur-bellied Flycatcher

PLATE VI
Coues' Flycatcher

PLATE VII
Arizona Jay

PLATE VIII
Curve-billed Thrasher

PLATE IX
Red-faced Warbler

PLATE X
Painted Redstart

PLATE XI
Hooded Oriole

PLATE XII
Rufous-winged Sparrow

TYRANT FLYCATCHERS
TYRANNIDAE
by Allan Phillips

Most Tyrant Flycatchers sit upright on exposed perches where they have a good view of passing insects. A wide bill and long rictal bristles in most species, as well as long wings, aid in the capture of this insect prey in flight. The feet are small and these birds never walk and hardly ever even alight on the ground. The colors are usually dull, running to grays and olives above and white or yellowish below. The sexes are alike. (In coloration, the Vermilion Flycatcher is an exception to all these rules.) Many of the species look very much alike and are best distinguished by their voice and actions. Each has a unique dawn song, surprisingly loud and varied and in some species not even suggestive of the ordinary voice. A very few, such as the Western Wood-Pewee, sing it again at dusk; otherwise it is heard at no other time than at dawn. But the ordinary daytime calls are likewise distinctive, except that a locative *whit* is common to several species.

211. EASTERN KINGBIRD *Tyrannus tyrannus* (Linnaeus)

Rare summer visitant in northern Arizona. There are also records of single birds seen near Parker, Yuma County (Monson), and Marana, Pima County (Burt L. Monroe, Jr.), and one taken at Cottonwood, Verde Valley (ARP), all in early September. Accidental near Tucson, March 19–20, 1943 (Hugh and Margaret Dearing). No Arizona nest yet recorded.

This is the only flycatcher which is dusky above and entirely white below. It is easy to distinguish, for it is usually in the open and has a white tip entirely across the tail.

Should the nest be found in Arizona, it will look like those of our other kingbirds, all of which build open cups in tree-tops or telephone poles, with little attempt at concealment. For protection of the nest, kingbirds rely upon chasing off all predators and big birds in the vicinity.

Summer residence in cottonwoods of northeastern Arizona is still uncertain. According to A. K. Fisher's notes (FW files), W. W. Price saw a pair at Holbrook in August 1894; and Russell and Amadon found one dead on the highway there *June 14, 1938* (MNA). But the species was not seen by Phillips on his visit July 3, 1949. At Kayenta, Hargrave saw one on June 13, 1933, but the bird was not present on his return the next month. Other northern Arizona records seem to be of pure stragglers: Phantom Ranch, bottom of Grand Canyon, May 19, 1929 (Florence Merriam Bailey *fide* McKee, *Prelim. Check-L. Birds Grand Canyon*, 1930: 7); and Flagstaff, *June 19, 1938* (ARP).

212. WESTERN KINGBIRD; ARKANSAS KINGBIRD *Tyrannus verticalis* Say

Common summer resident in open associations below the Transition Zone (local in northern Arizona); and fairly common transient (especially in fall) elsewhere in unforested areas, up to the Transition Zone. No authentic record between October and March.

The Western Kingbird is a paler edition of the Cassin's, with the chest almost white so that it does not contrast with the white chin. The crown is pale gray in definite contrast to the darker eye-line. Unless very worn or molting in the fall, the black tail shows a narrow contrasting white side (white outer web of each outer rectrix). The voice is subdued for a kingbird, an expressionless *kip* being the usual call. This is the common kingbird of the Lower Sonoran Zone. It has some extraordinary migrations which are not fully understood. We have no reason to suppose a post-breeding emigration to the north, the coast, or to the mountains, nor that it nests twice in different regions. Nonetheless, we are far from understanding what really goes on and when, because we can find no characters to divide this species into recognizable geographic races.

Confusion begins in late March when the first birds appear, scattered over the whole area from the Sonora coast to Phoenix and Tucson. The coast birds are certainly migrants, but their destination, whether near (at Yuma), or far, is not known. Whether the Tucson and Phoenix birds will stay to nest there, or move on, is equally uncertain. During April the birds become widespread and common and start nesting in southern Arizona. In May they spread across northern Arizona to southeastern Utah (*April 26, 1892* at Riverview – Rowley, AMNH). Migrants continue to go through Arizona until late May, with occasional birds still present in non-breeding areas to early June (flock of three in Upper Sonoran Zone on North Rim of Grand Canyon, *June 2, 1936* – Grater, GCN; south of breeding range in Sonora, June 1, 1956 and June 7, 1953 – Phillips).

In 1959, kingbirds were already scarce in the Salt River Valley by mid-July (Phillips *et al.*), but James M. Simpson thought this departure was unusually early. At the same season Western Kingbirds appear well south in Sonora and soon become abundant. Also in July, they appear in the Flagstaff region, where none breeds; the earliest dates are July 28 and *29* (1947 and *1939*, respectively – ARP). Meanwhile local pairs at Tucson continue breeding. Grown young were still begging from their parents there August 24 (1961 – Marshall) on a farm already overrun with migrants from the north. After the beginning of August, these kingbirds are virtually restricted, as regards southern and western Arizona, to the Tucson and Benson areas (and southward over Sonora) until a new migration wave comes in September. In the Santa Cruz Valley, hundreds pass southward at dawn about 100 feet up, in loose aggregations; when they alight, every tree and bush has at least one kingbird on it. This great migration continues into the first week of October, after which the numbers rapidly dwindle. Latest records are, for northern Arizona, *October 8 (1937*, near Springerville – Stevenson, ms.), and for southern Arizona, *October 23 (1949* – Marshall, WF).

A straggler at Topock, *October 29, 1953* (GM) is aberrant in that it is a juvenile still molting and was found in western Arizona, where the species normally disappears much earlier.

The Western Kingbird nests primarily in broad-leaved deciduous trees, such as cottonwoods and mesquites. It is not averse to placing its nest on an exposed telephone pole, as long as big trees are in the vicinity. But it shuns forests, dense woods, conifers, and even evergreen oaks. For this reason it is extremely local in northeastern Arizona, where it is limited in the breeding season to a few areas of cottonwoods or shade trees, at lower elevations. At its southern limits, the Western Kingbird has a spotty breeding distribution, being absent from favorable cottonwood habitats at Patagonia and Guadalupe Canyon; yet it nests on the San Pedro River at Hereford (bob-tailed young — Marshall and Phillips), at San Bernardino Ranch, and between there and Apache (nesting — Phillips, Hargrave, and Brandt). It is also absent from most of the southwestern deserts between the Baboquívari Mountains and the Colorado River.

Reports outside of the dates cited above are almost surely misidentifications. Many observers do not realize how much a Say's Phoebe resembles a kingbird, when it sits up on a high wire. Nor do they appreciate how similar to each other the various species of kingbirds appear.

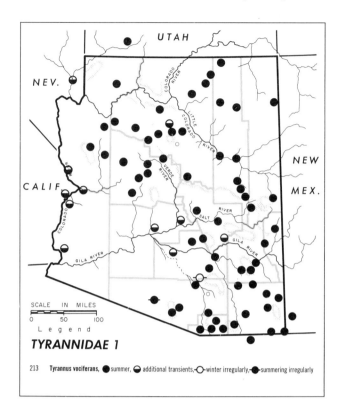

213. CASSIN'S KINGBIRD *Tyrannus vociferans* Swainson

Common summer resident in Upper Sonoran and highest Lower Sonoran Zones, openings in ponderosa pines, and along major streams except Colorado and lower Salt and Gila Rivers, where it is a rare transient. Casual in winter in Tucson (December 1952 — Marshall; 1957–1958 — George), and perhaps the east base of the Santa Rita Mountains (*December 7, 1963* — Levy, US). The supposed January record for Nogales is an error; we have no mid-winter specimen.

Cassin's Kingbird is our only kingbird with a definitely gray chest. The crown from eye to eye is dusky, the tail is black, and the belly is yellow. It is most easily recognized by its ringing *chibéw*. Like other kingbirds, Cassin's, where present in sufficient numbers, gathers to roost in large companies. Dawn songs from such birds have led to overenthusiastic reports of "numbers of Ridgway's Whippoorwills" at Guadalupe Canyon by those unfamiliar with the differences in quality of their songs. Another impressive roost is south of Nogales on the Río Bavasac, Sonora.

Cassin's is our hardiest kingbird, the only one nesting in the piñon and pine zones, and the last to leave Arizona in fall. Like all kingbirds, it prefers tall trees by open spaces; but it nests at altitudes from Transition Zone down to the highest Lower Sonoran Zone — the upper limit for other kingbirds. It overlaps these principally where extensions from its breeding populations push farther down among the Lower Sonoran cottonwoods and mesquites. Here it is usually outnumbered in summer by the Western Kingbird.

Above the pines it is a straggler (Hudsonian Zone of San Francisco Peaks, 1887 — Mearns, Auk 7, 1890: 255). Even in the Transition Zone it is somewhat irregular; at Flagstaff it bred in the 1930's, but not after World War II (except 1949).

Exceptionally early records in southern Arizona are *March 7* and 8 (*1947* south of Tucson — Vorhies; specimen now lost), March 10 (1958 also south of Tucson — Marshall), March 18 (1925 in the Baboquívari Mountains — Bruner, Condor 28, 1926: 233), and March 19 (1940 near Nogales — Dille). In these districts it is regular by the end of March. It has reached Roosevelt Lake and the foot of the Mazatzal Mountains on *March 30* (*1952* — Dickerman, CU), and the Fort Apache area, *April 13* (*1937* — ARP). Northern Arizona spring arrival is in the second half of April; a definite identification is April 27, 1948 near Shumway (Phillips). Migration continues at this time in the south (indicated by a notable concentration at Madera Canyon, Santa Rita Mountains, April 28–30, 1951, which dispersed within a week — Marshall).

The meaning of five records on the Colorado River around Parker and the Bill Williams Delta, June 5 (two seen) to July 23 of three different years (but none after 1955) is not clear. One might suspect wandering from the nearby Bill Williams River; but such wandering cannot account for the bird seen 23 miles above Imperial Dam, June 23 and August 15, 1955 (all Monson)!

The start of fall migration is perhaps indicated by these two records of two birds each appearing where none had been nesting: in Tucson, July 25, 1947 (Phillips), and Flagstaff, August 12, 1954 (Wetherill). The height of passage evidently starts on *September 20*, when there is a grouping of records from the central and western lowlands, where Cassin's is always a rare kingbird: delta of Bill

Williams River (*1949* – GM); Sacaton (Gilman); west of Phoenix (*1953* – Dickerman, CU). In these lowlands there are but three spring records: two at the end of March (Sacaton – Gilman; Parker – Monson), and one on May 17 (1950, Topock – Monson).

Soon after September 20 most have left northern Arizona, where the last record is *October 8* (*1937,* near Springerville – Stevenson, ms.). In the south, migration lasts for a surprising time. Even as far north as the Gila Valley, two were still at San Carlos, *November 8, 1951* (ARP). Cassin's Kingbirds have remained to *November 10* at Tucson (*1893* – Mearns, US) and Benson (1944 – Hargrave), and two were still lingering at Benson on November 22, 1951 (Marshall, Phillips). It thus seems likely that the fat specimen from Patagonia, *December 2, 1939* (ARP), was a belated transient and was not wintering, as first thought by Monson and Phillips (Condor 43, 1941: 109).

Misprints and misidentifications account for the "Jan." (i.e. "Jun.") record from Nogales, and probably the March one from the bottom of the Grand Canyon (Grater, *Check-List of Birds of the Grand Canyon*).

Within our borders, we find no such concentrations of migrating Cassin's as of Western Kingbirds. But just south of Nogales, from Agua Zarca southward in Sonora, Cassin's were abundant on November 1, 1952 (Phillips).

214. THICK-BILLED KINGBIRD *Tyrannus crassirostris* Swainson

Local summer resident in riparian trees of Guadalupe Mountains in extreme southeastern Arizona, beginning in 1958; and of Patagonia beginning in 1962.

This recent arrival, found nowhere else in the United States, is Arizona's loudest bird. Its usual calls are a shrill *cut-a-réep* and *kiterréer*. The breast and belly are pale yellow and the slightly notched tail is solid dark gray. The Thick-billed Kingbird is associated with sycamores. It makes a big show of each food-capturing flight, quivering the wings and keeping the head feathers erected.

When J. T. Wright first explored the Nogales area (Rancho la Arizona) in Sonora in 1929, he found Thick-billed Kingbirds already established there. They were thought to be a remote colony, isolated from its nearest neighbors by the length of the state of Sonora. In 1952-1953, however, Phillips, Yaeger, and Marshall took specimens at several intermediate points; but all searches on the Arizona side of the border, in the Nogales-Patagonia region, were fruitless. Finally the indefatigable Seymour Levy discovered it in Guadalupe Canyon, far to the east, on *June 4, 1958* (US; Auk 76, 1959: 92); here it had not been detected up to 1948 (Mearns, Phillips, Brandt, *et al.*). In 1962, a colony settled on the well-worked Sonoita Creek near Patagonia (Bill Harrison)!

There are no satisfactory data on migrations in Arizona. West and south of Nogales, Sonora, it is recorded at least from *May 7* (*1937* – van Rossem and Hannum, LA ?) to *September 13*–14 (*1952* – Phillips and Dickerman, ARP and CU).

If there really is geographic variation in this kingbird, the northern race would be *pompalis* Bangs and Peters.

215. TROPICAL KINGBIRD; COUCH'S KINGBIRD; WEST MEXICAN KINGBIRD *Tyrannus melancholicus* Vieillot

Nests near Tucson in open cottonwoods; also recently elsewhere in Santa Cruz and possibly San Pedro Valleys and along Salt River east of Phoenix. Wanders rarely in late summer and fall to Colorado River, north to Topock (October 1, 1947 – Monson). One casually at Imperial-Riverside county line, California side of Colorado River, March 22, 1957 (Monson), and one at Pima, near Safford, April 10, 1955 (Gallizioli).

The Tropical Kingbird resembles the Western Kingbird except for its tail, which is dark brown and notched, instead of black-and-white and even. Less noticeable differences are the brighter yellow belly, whiter throat, and longer bill of the Tropical Kingbird. Its call is a long canary-like trill. Its resemblance to the common species of kingbirds and its subdued voice make it easy to overlook, so that its history in Arizona is less certain than that of the preceding.

The Tropical Kingbird was first found in Arizona by H. H. Kimball on *May 12, 1905* (MCZ) near Tucson at Fort Lowell, where the species has never been seen again. Possibly, however, it was present through the years on the other side of Tucson, where Phillips discovered its nesting in 1938 (Auk 57, 1940: 117 and 258). He found about four pairs both in *1938* (ARP) and 1939. Of the three species of kingbirds nesting at this spot, the Tropical Kingbird is the latest to arrive in spring and the first to leave in fall; the extreme dates are *May 10* (*1958* – ARP) to September 19 (1949 – Phillips). To this day, the little colony, located between Midvale Farms and the San Xavier Mission area, has maintained about the same status. It has been about 90 miles removed from the northern limit of the main population in Sonora.

Meanwhile other undoubtedly brand new colonies have sprung up at spots which have been well investigated both before and after establishment: about three pairs were found at Canoa Ranch, and a single bird seen at nearby Kinsley's (Arivaca Jct.) June 30, 1959 (Phillips); and common on the Santa Cruz River east of Nogales during the summer of 1962 (Harrison *et al.*). Also an isolated pair was found building a nest east of Phoenix on *May 19, 1956* (JSW, Simpson and Werner, Condor 60, 1958: 69)! On the desert in the same region, one had been seen May 16, 1951 (Junea Kelly).

Additional records which may have been non-breeding birds come from Tumacacori National Monument, August 20, 1948 (Phillips and Monson) and Pomerene, San Pedro Valley, *September 11, 1953* (ARP). The earliest record for a fall straggler is *August 16* (*1954,* delta of Bill Williams River – GM).

This and the foregoing three species of kingbirds of the same size, behavior, nesting habits, and hunting

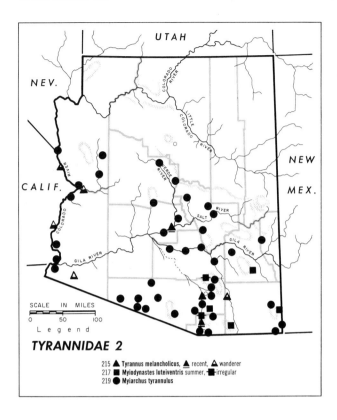

method, all now regularly nest in southern Arizona, with up to three kinds at one place. Ecologists marvel that such similar birds should coexist in the same row of cottonwoods on presumably the same food. Marshall's notes depict aggression by one species against another in various combinations, and conversely their toleration of each other even in the nest tree at other times. Such tidbits of behavior should not obscure the fact that generally these pugnacious birds pay no attention to other kingbirds except those of their own species; the late-arriving Tropical Kingbird experiences no difficulty in establishing itself in cottonwood rows already occupied by two other kinds. But you need not be a theoretical ecologist to admire the marvelous sight of three species of kingbird — Western, Cassin's, and Tropical — all in one bush, as seen by Marvel, Schaefer and Marshall near Nogales in 1962!

Our Tropical Kingbirds are *occidentalis* Hartert and Goodson, a smaller race than *couchii* Baird of Texas.

216. SCISSOR-TAILED FLYCATCHER *Muscivora forficata* (Gmelin)

Rare or casual summer visitant, always singly, chiefly to southeast; specimens from Kayenta, *July 8, 1934* (Russell and Hargrave, MNA), near Willcox, *August 7, 1953* (Gallizioli, ARP), and Pomerene, *May 8, 1957* (LLH). Also records from Saguaro Lake, July 12, 1935 (Hargrave); near Phoenix on Black Canyon Highway, May 18, 1958 (Harry L. Crockett); east of Nogales, August 4, 1962 (Marvel, Schaefer and Marshall), and near Tucson, September 17, 1960 (Marshall). A few others are of the genus, not identified as to species.

The Scissor-tailed Flycatcher is nothing but a Western Kingbird with pinkish belly and long tail streaming behind. Even the call note is similar.

The few Arizona records of this genus are all in 1928 or later, which may indicate that it is becoming more regular; there were four in the 1930's and four also in the 1950's (but none between 1935 and 1952).

217. SULPHUR-BELLIED FLYCATCHER *Myiodynastes luteiventris* Sclater

Fairly common summer resident of sycamore-walnut canyons of Santa Rita, Huachuca, and Chiricahua, rarely Santa Catalina (1907 — Lusk; 1955 — Marshall) and Graham Mountains (1951, 1952 — Marshall). One lowland record reported, Santa Cruz River, 20 miles north of Nogales, *May 27, 1917* (Dawson, ms., specimen presumed destroyed in fire at Santa Barbara Museum in 1962).

The Sulphur-bellied Flycatcher is the only conspicuously streaked flycatcher in the United States, where it is essentially limited to southeastern Arizona. The size of a kingbird, it shares the hole-nesting habit and rufous tail of the next genus, *Myiarchus*. It is the last flycatcher to arrive in the spring, the bulk usually appearing in early June. The birds depart in early September, immediately after nesting.

One wonders at the fate of the newly-hatched young found as late as August 28 (Howard, Bull. Cooper Ornith. Club 1, 1899: 103-104). The earliest eggs are those examined by Brandt on June 17, 1947, in which incubation had begun about ten days before (*Arizona and its Bird Life*, 1951: 420). That year and in 1940 the birds arrived exceptionally early (Sutton, Auk 60, 1943: 346). Nesting throughout the tropical lowlands of Middle America, the Sulphur-bellied Flycatcher winters only in South America — remarkable as one of the few long migrations of tropical birds.

Although the first individuals usually appear in the last of May and the first days of June, four were already present on *May 12, 1945* in Madera Canyon (Foerster and Hargrave, LLH). Exceptionally early individuals have reached the Chiricahua Mountains May 5 (*1956* — ARP) and May 11 (1948 — Hargrave). The latest departure date is September 20 (1913, probably in the Huachuca Mountains — Willard, FW files).

218. GREAT CRESTED-FLYCATCHER; "CRESTED FLYCATCHER" *Myiarchus crinitus* (Linnaeus)

Casual; one taken in Huachuca Mountains, *June 3, 1901* (Breninger, MCZ — of the nominate race, which includes *boreus* Bangs).

219. ARIZONA CRESTED-FLYCATCHER; WIED'S CRESTED FLYCATCHER *Myiarchus tyrannulus* (Müller)

Common summer resident of saguaro, cottonwood, willow, and sycamore associations north to central Arizona, much less common toward Colorado River, where it is found locally north to the southern tip of Nevada; and at Yuma has apparently nested in the city.

The flycatchers of the genus *Myiarchus* are sedate inhabitants of shady trees, whence issues their call, a soft *whit*. The edgings of the primaries are rufous, giving a tinge to the folded wing, and most species have conspicuous rufous on the spread tail. Otherwise they are plain brown above, pale gray on the throat and chest, and pale yellow on the belly. The tail of juveniles looks all rufous, with this color encroaching upon the central pair of rectrices (which are dark brown in adults). These flycatchers build a nest inside a hole, and some species are famous for their use of the cast snake skin invariably to be found wrapped around the nest. For the determination of *Myiarchus* specimens, mouth color, size, wing formula, and the exact pattern on the tail feathers are important, as pointed out by Dickerman and Phillips (Condor 55, 1953: 101-102); Phillips (Anales del Inst. Biol. Méx. 30, 1960: 355-357); and later by Lanyon (Condor 62, 1960: 341-350; 63, 1961: 421-449—containing pictures of rectrix patterns).

The Arizona Crested-Flycatcher is the largest *Myiarchus,* whose thickened bill may be noticed in the field. The somewhat yellower belly than the Ash-throated is useful only when allowance is made for season and age, or if the birds are together and of the same age. For this genus of flycatchers is one of the most difficult to identify among birds — much like *Empidonax*. Fortunately the calls are distinctive; that of the Arizona Crested is a loud *bew,* exactly the same as the second syllable of the Cassin's Kingbird's call. The mouth color, which perhaps is flashed during close encounters to aid in recognition, is grayish flesh color, like the Ash-throated's.

Since it is larger than the Ash-throated Flycatcher, this bird requires larger trees for its nests; therefore it is less widespread on the drier deserts. Saguaros, cottonwoods and sycamores, as well as some shade trees in towns suffice. It is incongruous, perhaps, to find Arizona Crested-Flycatchers both in saguaro deserts and shady riparian timber, but from the bird's standpoint, both habitats are equal in providing woodpecker holes for nesting well above the ground.

This is the *Myiarchus* that is with us for the shortest period, and is perhaps the first of all the summer residents to disappear in the fall at Tucson; here it is usually not seen after the first half of August!

The late records are usually of families and juveniles which must represent late nestings due to failure of the first attempt; examples are juveniles from Canyon on the Agua Fria on *August 22* (*1929* — Jacot, ARIZ) and a family of four at Olive Road in Tucson, September 7 (1952 — Phillips). However none of these rules applies at the mouth of the Verde River, where it was seen on each of three visits in fall by Phillips and Yaeger! The dates were *September 24, 1949* (ARP), September 30, 1950, *October 6, 1951* (ARP).

Ordinarily, spring arrival is the end of April or first of May; exceptional dates are *April 12* and *20* (near Tucson — Kimball, AMNH) and April 20 (east slope of Santa Catalina Mountains — Scott). Therefore we cannot believe that eggs "were laid about the first of May" near Tucson (Brandt, *Arizona and its Bird Life,* 1951: 100-101, 662-3); while the claim of eggs in Arizona on April 4th (Lincoln in Bent's *Life Histories,* U. S. Nat. Mus. Bull. 179, 1942: 127) is absurd. Nesting does quickly follow the actual arrival, for a pair was apparently building on May 11, 1939 at Organ Pipe Cactus National Monument (Huey, Trans. San Diego Soc. Nat. Hist. 9, 1942: 367), and two pairs were carrying nest material at Tumacacori Mission on *May 12* and *13, 1931* (van Rossem, *idem* 8, 1936: 136-137).

The large west Mexican race *magister* Ridgway is the form occurring in Arizona; it is also the largest race in the genus. This species well illustrates size variation in lowland birds—with its impressive increase from south to north.

220. ASH-THROATED FLYCATCHER *Myiarchus cinerascens* (Lawrence)

Common summer resident throughout all but the densely wooded parts of the Sonoran zones, rarely densely wooded parts of the Sonoran zones, rarely straggling higher. Winters in mountains of Yuma County, sparsely along lower Colorado River, east to Phoenix and Casa Grande areas, and rarely to Tucson.

Except in dense oak woods in the southeast, this is our commonest *Myiarchus;* it is the only one to be found in the northeast, or in winter. Characteristically a desert bird, it also inhabits streamside trees and open oak or juniper woods; flycatchers must not be identified by habitat alone. Wintering Ash-throats are apt to linger among the denser mesquite thickets along washes, or along streams, leaving the more barren mesas untenanted at that season. Much like an Arizona Crested Flycatcher in appearance and mouth color, the Ash-throat is best told by its soft voice; its loudest call is a subdued *ka-brick*. In direct comparison, it is paler-bellied than its larger cousin — but so are young of the latter, and this yellow changes with the season and amount of wear of the feathers. Summer Ash-throats, and faded young of both species, are often nearly white-bellied. The Ash-throat's bill, of course, is slighter.

Some Ash-throated Flycatchers attempt to nest in iron pipes out in the desert sun, where the temperature must be outrageous. The outcome of these brave attempts would be interesting to follow.

The Ash-throated Flycatcher is present in southern Arizona from late March (lowlands) or mid-April (mountains) to the middle of August, with individuals lingering into September. Rather extreme dates near Tucson of birds thought not to be wintering are *March 11* and *14* (*1905* — Kimball, Marsden, AMNH), *October 9* (*1947* — ARP), and October 18 (1957 — Marshall). Of about four birds seen at the mouth of the Verde River on October 7, 1951, only one could be found on the following November 1 (Phillips). From non-wintering areas there are records for

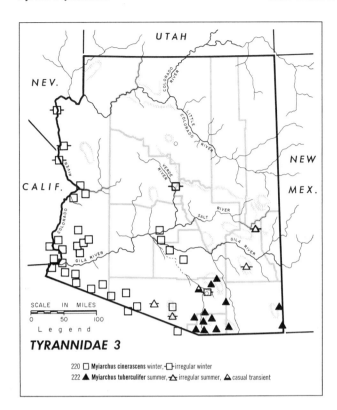

TYRANNIDAE 3

220 ☐ *Myiarchus cinerascens* winter, ⊡ irregular winter
222 ▲ *Myiarchus tuberculifer* summer, △ irregular summer, △ casual transient

the foot of the Mazatzal Mountains, March 30 (1952 – Dickerman) to September 19 (1884 at Camp Verde – Mearns, ms.) and about the last of September in the Santa Catalina and Pinal Mountains (Scott). There are only two occurrences of this species above the Upper Sonoran Zone, both at Flagstaff (Phillips, MNA): *June 10, 1936* and *August 5, 1947;* the latter, a juvenile, was presumably already south-bound.

Northern Arizona records are from April 20 (1935 at the bottom of the Grand Canyon – Mrs. C. M. Shields, fide McKee) and *April 24* (*1892* at Riverview, Utah – C. P. Rowley, AMNH) to *August 31* (*1932* east of Flagstaff – Hargrave, MNA). In northwestern Arizona it was seen near Pierce's Ferry April 16 (1937 – Hargrave).

Arizona birds are of the mainland race *cinerascens* which is distinguished from *pertinax* Baird of peninsular Baja California chiefly by its smaller bill and longer, more pointed wing.

221. NUTTING'S FLYCATCHER; "PALE-THROATED FLYCATCHER" *Myiarchus nuttingi* Ridgway

Casual: one record, near Roosevelt Lake, *January 8, 1952* (Dickerman, ARP).

Nutting's Flycatcher is much like the preceding two in appearance. Colors are essentially like the Arizona Crested, but the size is even slightly smaller than the Ash-throated. It differs from both, as seen in the hand, by a bright orange mouth lining, and in the calls. These are quite staccato, not blurry, and one is a clear rising whistle. Nutting's Flycatcher is a full species in its own right, as any brief acquaintance in its nearest breeding areas near Hermosillo, Sonora, will attest. There its unique voice and preference for arid tropical woods can be contrasted with those of its congeners. There is no foundation for the idea that it hybridizes with the Ash-throated Flycatcher. This erroneous impression of the past was gained from the confusing array of winter and migrant specimens from Mexico, before the migrations and geographic variations in *Myiarchus cinerascens* were properly understood and before the true specific characters were first pointed out (Dickerman and Phillips, Condor 55, 1953: 101-102; Phillips, Anales del Inst. Biol. Méx. 30, 1960: 356-357).

The Arizona specimen belongs to the pale northernmost race, *vanrossemi* Phillips (*op. cit.* p. 357), which is equally large as, but paler than, *inquietus* Salvin and Godman farther south.

222. OLIVACEOUS FLYCATCHER *Myiarchus tuberculifer* (Lafresnaye and D'Orbigny)

Common local summer resident of denser live oaks, and higher Lower Sonoran riparian associations of Santa Cruz River drainage, from Guadalupe Mountains west beyond Nogales; ranges north to the Graham Mountains. There are specimens from Mitchell Peak, south end of the White Mountains, *July 8, 1951* (Marshall, WF); the Baboquívari Mountains, *May 27, 1931* (van Rossem, LA); and even Sells, Papago Indian Reservation, *July 8, 1918* (A. B. Howell, US). It is reported in spring migration below the Huachuca Mountains (Swarth, Pacific Coast Avifauna 4, 1904: 23), and near San Xavier Mission, April 28, 1946 (A. H. Anderson).

The Olivaceous Flycatcher is our only *Myiarchus* with little or no rufous in the tail. Also it is smaller and darker than the others in Arizona. Because of the tail color it has been mistaken for other species such as the Tropical Kingbird and an *Empidonax,* but not by persons already familiar with the characteristic form, posture and behavior of all *Myiarhcus* flycatchers, as seen so well in the Ash-throated Flycatcher. All are slender, bushy-headed, and erect. The Olivaceous Flycatcher has bright orange mouth lining and a distinctive long, mournful, descending whistle, of lower pitch than that of the Say's Phoebe. Like several other tropical birds which reach southern Arizona, there is marked fluctuation in numbers from year to year at the north edge of its range.

In the Santa Catalina and Pinaleno Mountains it was common in 1951, when it even reached Mitchell Peak. But in 1952, 1953, and 1955 it was rare in the Santa Catalina Mountains and it was not detected during brief visits in 1954 and 1962. In the Grahams (Pinaleno Mountains) it was rare in 1952 and not found in 1953 (all observations of Marshall). The old Fort Grant record does not pertain to these mountains; Dr. Palmer's collecting was at the original Fort Grant, on the lower San Pedro River, where this flycatcher does not occur. Swarth was correct in questioning the identification.

The normal span of summer residence in our mountains is from mid-April to mid-August; thereafter they may withdraw to the creeks and watered canyons. Extreme dates are *March 31* (*1931* — Jacot, ARIZ) and April 6 (Swarth; both in the Huachuca Mountains) to October 2 (*1947* at Patagonia — Phillips, G. M. Sutton collection), and *October 6* and 7 (*1945* and 1944, respectively, in the Pajaritos Mountains — van Rossem and L. Miller, LA). It was last seen south of Nogales, in Sonora, October 13 (1954 — Phillips). The race of Olivaceous Flycatcher in Arizona is the pale *olivascens* Ridgway.

223. EASTERN PHOEBE *Sayornis phoebe* (Latham)

Rare fall transient and winter visitor in southern Arizona, chiefly in the southeast but recorded west to the Colorado River. Eleven specimens, north to Salt River (JSW, MNA) and west to Bill Williams Delta (GM).

This is a dark flycatcher above, plain whitish below, distinguished by the dipping and spreading of its rather long tail, exactly as in other phoebes. It is almost always found around ponds and streams and its call note is nearly as sharp as the Black Phoebe's, which also prefers water.

The Eastern Phoebe was unrecorded in Arizona until *February 5, 1902*, when Herbert Brown took one, now re-labeled as from Tucson (ARIZ); but Brown was evidently in Yuma at that date. Only three other individuals were seen in the state (Kimball, LA; Yeager, MNA; see Condor 23, 1921: 57 and 38, 1936: 171) before *1950* (ARP). The winter of *1952–3* produced a sudden flurry of records: delta of Bill Williams River (GM), Patagonia (ARP), and two individuals in the Tucson valley in early March (Mrs. Erle D. Morton, Phillips). Thereafter one or more appeared in southeastern or central Arizona nearly every winter until *December 21, 1958* (Margolin, JSW). Surprisingly, two individuals were seen together once (one collected), at the Southwestern Research station, Chiricahua Mountains (*October 13, 1956* — Cazier, SWRS). Seasonally, Kimball's records for *August 16* and October 6 remain the earliest for Arizona, while March 10, 1953 (Mrs. Morton) is the latest. Again taken in *1963* (Fr. Amadeo M. Rea, MVZ).

224. BLACK PHOEBE *Sayornis nigricans* (Swainson)

Breeds commonly along permanent streams and canals of central and southeastern Arizona; and breeds uncommonly and locally in the bottom of the Grand Canyon, in western Arizona, on the north slope of the White Mountains region, and even up into the Transition Zone. Winters at water throughout Lower Sonoran Zone valleys, including the Colorado Valley, sparingly so east of San Pedro Valley. On migration apt to occur at any water-hole or cattle trough, even in the northeastern part of the state and up into the ponderosa pine zone. Not "resident" as stated in A.O.U. Check-list.

The Black Phoebe is unlike any other flycatcher in color. Were it not for its shape and phoebe-like actions, it might be confused with a junco. It is found near water, and the nearer the better. The bulky nest is plastered over the water on a rock ledge or under a bridge. In the arid climate of Arizona this bird does not

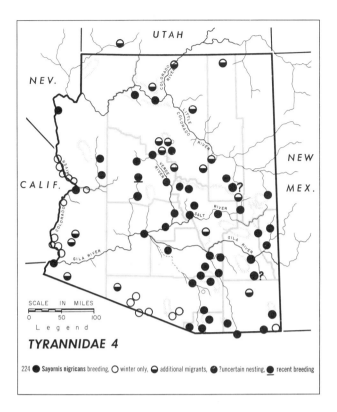

TYRANNIDAE 4

224 ● *Sayornis nigricans* breeding, ○ winter only, ◐ additional migrants, ● ?uncertain nesting, ● recent breeding

build on porches the way it does in coastal California, and as the Eastern Phoebe also does. It utters a sharp *cheep* at every turn and twist in pursuit of a fly, and also each time it alights.

The importance of water in the ecology of this species leads to differences from other Arizona land birds. For example, both the Baboquívari Mountains and the lower Santa Catalinas, where Scott lived, appear quite similar. They are essentially mesquite, live-oak, and sycamore canyons; but the Baboquívaris become dry in the spring, whereas water was available at the east base of the Catalinas, in Scott's day at least. Thus at the same altitude and vegetation, the Black Phoebe was a summer resident there, whereas in the Baboquívaris it is a winter resident. There is certainly no shortage of rocks in either locality!

There have been distributional changes within historic time. In 1864–1865 the Black Phoebe was "not found at Fort Whipple, though detected a very few miles southward" (Coues); this status is similar to those given for the Canyon Wren, Black-tailed Gnatcatcher, Phainopepla, and Abert's Towhee — evidently referring to its occurrence on Date Creek. Yet it was "not uncommon" on Willow Creek just three miles north of that Fort, July 1916 (Jackson and Taylor, FW files); and by the 1920's it bred all along Granite Creek, even into the pines to the south of Prescott (Jacot, ms.).

It is not surprising that the A.O.U. Checklist considers this bird a resident, since its movements are among the most difficult to understand of any Arizona bird. Particularly incomprehensible is the origin and destination of those migrant phoebes that appear in northeastern Arizona, for there are practically no breeding populations to the north or east! Even in areas where Black Phoebes appear to be resident and sedentary, long movements occur; this is proved by a nestling, banded near Bakersfield, California,

April 30, 1948 (McClure) which was collected at Wikieup on the Big Sandy River, Arizona, *September 21, 1948* (ARP — the only banded bird that Phillips has ever collected)! Such an eastward movement may explain the various February, May (to June 4), and August-September records for Nevada, where Black Phoebes breed (if at all) only along the Colorado River (*cf.* Condor 53, 1951: 238, and 61, 1959: 287). But California is a most unlikely source for the birds that appear in northeastern Arizona, judging from the known migrations of more geographically variable birds.

The sparse breeding populations on and north of the Mogollon Plateau consist primarily of single isolated pairs, with the possible exception of those in the bottom of the Grand Canyon. Nesting abundance is therefore insufficient to account for the many wanderers and migrants, which are often far from suitable streams. These migrants are shown on the accompanying map by half-filled circles; they occur principally from mid-July to the end of August. Spring records happen to be restricted to the Flagstaff-Mormon Lake area: March 17 to May 6 (Phillips). The latest northern Arizona record is *September 19* (*1934* — (Stevenson, ARP) at Eagar, where the bird breeds.

Concentration of mid-summer records northeast of any sizeable breeding population suggests a post-breeding emigration from the nesting ground. Evidence of such movement can also be found in southern Arizona. Here the Black Phoebe is recorded from non-breeding sites as early as July 1 (1939 on the University of Arizona campus, Tucson) and July 2 (1934 in the upper Santa Catalina Mountains — both by Phillips). The latest spring records in such places are April 28 (1951 in the lower Santa Catalina Mountains — Marshall) and May 3 (1925 in the Baboquívari Mountains — Bruner, Condor 28, 1926: 334). For non-wintering localities, records extend from March 5 (1947 in Tucson — Phillips and A. J. Foerster) and *March 15* (*1937* near Whiteriver — ARP) to October 14 (1951 at Sullivan's Lake, north of Prescott — Phillips), October 21 (1956 at Globe — Harold A. Marsh), and "till cold weather" (in the lower Santa Catalina Mountains — Scott). Arizona birds are the northern race *semiatra* (Vigors) with wholly white under tail coverts.

225. SAY'S PHOEBE *Sayornis saya* (Bonaparte)

Fairly common breeding bird about cliffs and structures throughout less densely wooded parts of the Sonoran zones, and locally higher. Winters south and west of the Mogollon Rim in Sonoran zones, sparingly north to Springerville, Joseph City, the west side of the Navajo Indian Reservation, and probably inside the Grand Canyon. During fall migration may be found on Kaibab and Mogollon Plateaus away from breeding areas. A post-breeding migration carries most of the birds out of the southwestern part of the state, where virtually absent in July and August.

This is the open-country species of Phoebe, which hovers in the air like a Sparrow Hawk or Mountain Bluebird, when tired of perching temporarily on low weed stalks or wires. Since man has come along, most Say's Phoebes nest in his old wells, bridges, eaves, or wrecked automobiles (which should be of interest to the temperature-minded ecologist). The black tail might lead to confusion at a distance with kingbirds were it not for its regular dipping and fanning. Closer, the uniform head and throat and rusty belly should prevent mistakes — still the bird has frequently passed itself off as a kingbird. The mournful whistle, thin, weak, and descending, is distinctive. The changing history of the Say's Phoebe at Flagstaff is recorded by Phillips (Plateau 28, 1955: 25-28.)

General arrival in northern Arizona lowlands (where not wintering) is from March 5 to 11, but one arrived at Snowflake as early as February 15 (1951 — Albert J. and Louise Levine). The birds leave by early October, but remain to the middle of the month farther west (Prescott), according to Coues.

In the lower deserts of southwestern and central Arizona, and east even to Globe (Harold A. Marsh) and an area near Tucson (Marshall), Say's Phoebes seem to decrease by early June; by the end of the month nearly all are gone (last seen near Tucson July 4, 1959; first on September 21, 1958 — Marshall). Those that remain all summer in Tucson continue nesting, and young were ready to leave one nest late in September (L. W. Arnold). In mid- or late September others start to appear on the deserts; an early date near Yuma is *September 15* (*1956* — Imperial Refuge, GM). In severe winters there may be some withdrawal from the foothills to the desert in December (see Scott, Auk 4, 1887: 18).

The races of Say's Phoebe south of Canada have never been properly worked out. The darker, northern *S. s. yukonensis* Bishop has not been identified from Arizona. Fading in life, and apparently also in the museum, complicates an already difficult taxonomic problem.

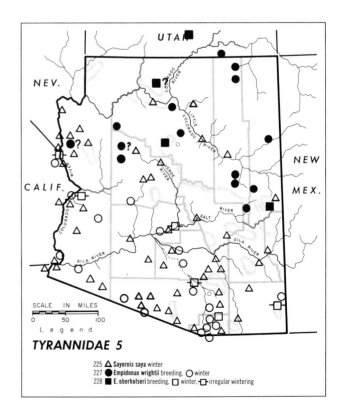

TYRANNIDAE 5

225 △ *Sayornis saya* winter
227 ● *Empidonax wrightii* breeding, ○ winter
228 ■ *E. oberholseri* breeding, □ winter, ⌑ irregular wintering

Genus *EMPIDONAX*

This perplexing genus contains thirteen small flycatchers in North America, of which nine have been recorded in Arizona. All except the Buff-breasted Flycatcher look very much alike, and the beginner should not attempt to identify any more than this and the Gray. *Empidonaces* are characterized by pale eye-ring and wing-bars, small size, a nervous twitch of the tail (in all but the Gray Flycatcher, which dips its tail), and the *whit* call-note. Most kinds remain well within foliage, and are of a protective olive-green tinge. Most also migrate south before the prebasic molt, which means that they look even more alike in fall, due to wear and fading, than in spring. After catching a fly they usually move on to a new perch. An occasional *Empidonax* caught in adverse weather may behave like a vireo, gleaning from twigs; and the Hutton's Vireo has been confused with these flycatchers, both in the field and museum.

The following is a key to those normally occurring in the Southwest, to which may be added for emphasis and summary (though of little use except in the hand) that of the two wide-billed birds, Traill's is brown with a white throat, whereas the Western is green and yellow. The Buff-breasted Flycatcher has ochre color on the chest. The Gray Flycatcher has purer grays and whites, a distinctly white outer web of the tail, and a long bill of two colors. This leaves three similar species which are fiercely difficult; of them the Hammond's is very dark, has a small black bill and clear gray face. Fortunately the dawn song of each species of *Empidonax* is different and very distinctive. The following key is somewhat simplified, so will not work on 100 per cent of the specimens.

A KEY TO SEVEN SOUTHWESTERN SPECIES OF EMPIDONAX

I. Bill ⅝ as wide as long, entirely pale below; tail even, rounded, or slightly double rounded; outer web of outer tail feather not markedly paler than rest of tail.
  A. Throat white; head grayish or olive; back brownish. Sixth primary not sharply cut out on outer web.
            Traill's Flycatcher *Empidonax traillii*
  B. Throat washed with yellowish; head more or less greenish; sixth primary as distinctly cut out as 7th, 8th and 9th.
            Western Flycatcher *Empidonax difficilis*

II. Bill less than ⅝ as wide as long; mandible often dark at least distally. Tail often notched and/or with outer web of outer tail feather often paler. Sixth primary always cut out on outer web.
  A. Small; ochre breast; mandible pale. (Mountains only)
            Buff-breasted Flycatcher *Empidonax fulvifrons*
  B. Larger, breast not ochre.
    1. Outer web of outer rectrix not much paler than rest of tail. Bill and tarsus short; tail notched in center.
      a. Larger. Wing not shorter than 67.4 mm. in males nor 61.8 mm. in females. Bill short; tarsus about 16 mm., rarely up to 17.6 mm. Tenth primary nearly equal to or longer than fifth. Wing exceeds tail by 8 mm.+; tail evenly forked or notched, the notch usually 2 mm. deep or more.
            Hammond's Flycatcher *Empidonax hammondii*
      b. Smaller. Wing not over 68 mm. in the male and 62 mm. in the female, more rounded; tail fork shallower. (Local in the SW.)
            Least Flycatcher *Empidonax pusillus*
    2. Outer web of outer rectrix pale, more or less white; bill and tarsus usually longer; tail variable in shape. Wing rounded; tenth primary shorter than fifth. Tarsus 17 mm. or more (rarely less than 16.5 mm.).
      a. Darker; tenth primary usually shorter than fourth; wing very rounded; wing rarely over 8 mm. longer than tail (usually much less); outer web of outer tail feather whiter than 1 a or 1 b but not pure white.
            Dusky or Wright's Flycatcher *Empidonax oberholseri*
      b. Paler; tenth primary usually between fourth and fifth in length. Like 2 a, but paler and often clearer gray-and-white. Bill long, mandible pale (yellowish and/or pinkish) at base, usually dark distally. Wing bars and edging of primary coverts more whitish. Outer web of outer tail feather white; wing generally exceeds tail by over 6.4 mm.
            Gray Flycatcher *Empidonax wrightii*

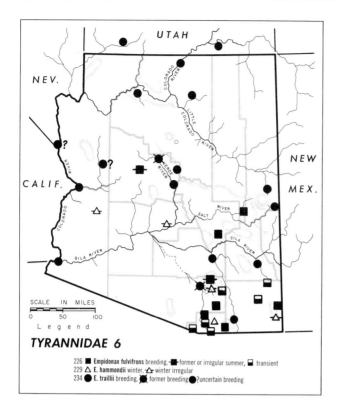

226 ■ Empidonax fulvifrons breeding, ▬ former or irregular summer, ▄ transient
229 △ E. hammondii winter, ⟁ winter irregular
234 ● E. traillii breeding, ◐ former breeding ●?uncertain breeding

TYRANNIDAE 6

226. BUFF-BREASTED FLYCATCHER *Empidonax fulvifrons* (Giraud)

Rare summer resident in Arizona, more common and widespread formerly; at present known regularly only from sycamore groves in the Huachuca Mountains, but probably still occurs in southern parts of White Mountains region. There are old records from as far north as Prescott (*May 9, 1865* — Coues, US) and Fort Apache, also from the Santa Catalina, Rincon (Brown, ARIZ), Santa Rita, Chiricahua, and Patagonia Mountains. In spring migration occurs at foot of mountains, even to Sonoita Creek (*March 30, 1927*—Jacot, ARIZ) and west to Pajaritos Mountains west of Nogales (*April 13, 1947* — ARP).

The Buff-breasted Flycatcher is a tiny *Empidonax* with cinnamon-buff on the breast. In summer this color becomes so diluted by wear and fading as to be useless in field identification. But the bird can still be identified by its choice of habitat and by a dull *pt* for a call note. It is the only *Empidonax* nesting in lower parts of the southeastern mountains. It prefers open stands of pine or riparian groves, generally well below the altitudes sought by breeding Western Flycatchers. However, migrant Western Flycatchers move through these same lower slopes nearly all summer.

In Mexico, where this flycatcher is common, its varied situations of occurrence are alike in being very open, with considerable bare, weedy, or grassy areas among the trees, where the birds forage from the top of a weed stem, or from a low bush. In Arizona the mountain vegetation has become badly choked with brushy junipers, dead branches and young trees, through elimination of the original grass and by adherence to other ingenious programs of misuse. The requirement of openness is not met and perhaps this is responsible for the present rarity of the Buff-breasted Flycatcher in our state; the early ornithologists found it everywhere!

The nest is saddled on the base of a horizontal limb where it cannot be seen from below. Unlike most species in this group, the Buff-breasted Flycatcher molts before it leaves for the south in fall, which means that its elegant new cinnamon plumage could be seen around the last of August. This is its only molt of the year.

The Buff-breasted Flycatcher nests at least as far south as El Salvador. If it is conspecific with *E. atriceps*, then it extends to western Panama, where it is non-migratory.

Arizona records are from *March 29* (*1917* in the Chiricahua Mountains — Smith, AMNH) to *September 12* (or later?, *1876*, at Seven-mile Hill, near Fort Apache — Aiken, ed. Warren, Colorado College Publications Study Series 23, 1937: 62). The latest occurrence below the breeding range is *May 4* (*1903* at the foot of the Huachuca Mountains — Swarth, Pac. Coast Avif. 4, 1904: 27). The Buff-breasted Flycatcher has decreased markedly since 1920.

The supposed record of this flycatcher in the Huachuca Mountains on *October 18, 1893* (Lincoln, in *Bent's Life Histories*, U. S. Nat. Mus. Bull. 179, 1942: 260) is an error; the specimen is *E. difficilis* (Mearns, ms., fide Duvall, in litt.).

227. GRAY FLYCATCHER *Empidonax wrightii* Baird

Common summer resident of piñon-juniper areas from Mogollon Plateau northward; also westward to at least the Juniper Mountains (ARP), and south to Fort Apache, southern Navajo County (ARP, US). Winters sparingly in mesquite associations, usually near water, in southern and central Arizona, north to Topock (Monson), Salome (WJS), and Wickenburg (ARP), and east to the Patagonia region (casually to Chiricahua Mountains — Kimball, MICH). Common transient in more open parts of state, except in southwest, where uncommon.

This species does not jerk the tail, but dips it slowly and evenly downward, like a subdued Phoebe. This in combination with its white wing-bars, pale gray color, and frequent occurrence in open country, makes identification possible in the field. The usual call is a clear *chí-bit*. This is the only *Empidonax* flycatcher likely to be seen in southwestern and central Arizona in winter. Also it is the only one from early June through July on the piñon mesas of the northeast, where it is the earliest of the *Empidonax* group to arrive in April. There is an important article by Russell and Woodbury on its nesting (Auk 58, 1941: 28-37).

Southern Arizona migrations are from April 2 (*1902* in the Huachuca Mountains — Swarth, Pac. Coast Avif. 4, 1904: 26) to *May 21* (*1916* at San Carlos, near Coolidge Dam — Taylor, US); and again from *August 4* (*1939* in the Sulphur Springs Valley near Gleeson — GM) and *August 5* (*1874* at Fort Bowie — Henshaw, Muséum National d'Histoire Naturelle, Paris) to *October 15* (*1956* at Fairbank, San Pedro River — Levy, US). An exceptional winter locality is the live-oaks of the Patagonia Mountains (*February 21, 1948* — ARP).

Northern Arizona records extend from *April 6* (*1937*, eight miles south of Taylor, southern Navajo County — ARP) and *April 17* (*1936* at Flagstaff — Phillips, MNA) to *September 23* (*1934* at Springerville, two young males — Stevenson, ARP).

228. DUSKY FLYCATCHER; WRIGHT'S FLYCATCHER — *Empidonax oberholseri* Phillips

Locally common summer resident of Canadian Zone willows of White and (very locally) San Francisco Mountains, and possibly also the Kaibab Plateau. Winters casually along lower Colorado and Salt rivers and in Chiricahua Mountains, more regularly near Tucson and Patagonia, where not exceptional. Though said by A.O.U. Check-list to winter "casually" in Arizona, nine specimens have been taken near Tucson in the period between November 20 and March 10 of eight different winters; four in two years at Patagonia; four near Phoenix (JSW); one at Parker (GM); one near Needles, California (MVZ); and one or two in the Chiricahua Mountains (Kimball, MICH). An uncommon migrant throughout wooded areas of the state except along lower Salt, Gila, and Colorado rivers, where apparently absent; common in extreme southeast.

This flycatcher breeds in fairly tall, dense, bushy willows with gooseberry undergrowth on the San Francisco Peaks, and along little willow-lined creeks in the White Mountains. For the identification of small flycatchers by their distinctive songs, the next best thing to properly identified sound recordings is Ralph Hoffmann's book, *Birds of the Pacific States*. Unfortunately they do not sing away from the nesting grounds, nor do they choose distinctive habitats while migrating. Most *Empidonaces* (except the Gray Flycatcher) prefer shady, broad-leafed cover at this time, and they winter in brush or trees along streams. But Dusky Flycatchers migrate more commonly in the mountains than in the valleys, at least in spring.

Southern Arizona migrations (birds that seem not to be wintering) are from *April 6* (*1952* near Ventana Ranch, Papago Indian Reservation — ARP) to *May 24* (*1917* in the Chiricahua Mountains — Smith, AMNH) and *May 29* (*1953* in the Sierra de los Ajos, northern Sonora — ARP); and again from *August 6* (*1952*) and *August 8* (*1893*) in the mountains of northern Sonora—Marshall, WF; Mearns, US; and *August 8* (*1908*) to *October 6* (*1932*), both in the Chiricahua Mountains — Kimball, MICH; Walker, CLM; and exceptionally? *October 20* (*1951* at Wikieup — ARP). Migrants occur west to the Mohave or Chemehuevis Mountains (Huey, SD), the Kofa and Castle Dome Mountains (GM), and supposedly at Bard, California (LA; specimen not reexamined by us).

Northern Arizona migrations are from *April 25* (*1948* near Holbrook — ARP) to *May 22* (in the Flagstaff region — MNA), and from *August 9* (*1905* at Laguna, New Mexico — Hollister, US) and *August 11* (*1920* at Pinetop, southern Navajo County — Kimball, MICH) to *September 27* (*1876* in the White Mountains — Aiken, *fide* Warren, Colo. Coll. Pub., Studies Ser. 23, 1937: 62) and *October 14* (*1920* at Taylor, Navajo County — Kimball, MICH).

The winter records of the Dusky Flycatcher illustrate beautifully the misleading results to be obtained from a superficial statistical analysis of biological data. The statistician would conclude that it winters far more commonly at Tucson, where there are more than twice as many records and it is recorded in four times as many different winters. Actually the bird is much more numerous, and probably regular, at Patagonia; here the specimens were all taken in just a few days' visits scattered over the years; while the Tucson records are the results of almost continuous search for a number of years by various competent collectors.

229. HAMMOND'S FLYCATCHER — *Empidonax hammondii* (Xantus)

Common throughout state during migrations. Winters regularly in small numbers on Sonoita Creek near Patagonia, casually near Tucson, once near Phoenix (JSW), in Chiricahua Mountains (ARP), and accidentally at Salome (WJS).

Since migrating *Empidonaces* do not sing, it is practically impossible to identify *hammondii* in the field in Arizona. Nevertheless, the observer must be impressed with the great numbers of small migrant flycatchers in woodlands throughout much of the state, and he may take comfort in the knowledge that most of them are Hammond's. Occasionally Hammond's gives a call nearly as sharp as that of a Pine Nuthatch, thus identifying itself. During some migrations, the woods at middle elevations of the mountains resound with these single *peeps*.

Southern Arizona migrations are from *March 23* (*1887* at Fort Huachuca — Benson, US) and *March 24* (*1940* in the Baboquívari Mountains, several — ARP and GM) to *May 28* (*1917* in the Chiricahua Mountains — Smith, AMNH); and again from *August 11* (*1893* at San José Mountain, northern Sonora — Mearns, US) and *August 26* (*1902*) to *November 4* (*1907*, both in the Huachuca Mountains — Swarth, Pac. Coast Avif. 4, 1904: 25 and Condor 10, 1908: 111). In central and western Arizona, the Salome individual is the only record between *October 24* (*1954*, Kofa Mountains — GM) and *March 31* (twice).

Northern Arizona records, as yet, are all from *August 22* (*1889*, San Francisco Peaks — Merriam, US) to *October 15* (*1920*, on Showlow Creek, southern Navajo County — Kimball, MICH). It will probably be found in spring, however, with further explorations. We know of no authentic June record. A specimen labeled "6–6–1895" (Lusk, PH) was probably mis-dated, either by the collector or in relabeling.

230. LEAST FLYCATCHER — *Empidonax pusillus* (Swainson)

A rare fall migrant in western Arizona: three immature specimens from the Big Sandy Valley, *October 20, 1951* (ARP); one from the Tule Mountains, extreme southwestern Arizona, *September 29, 1956* (GM); also one from Boulder City in Nevada, *September 6, 1950* (Lake Mead Recreational Area collection). Unrecorded farther east.

Perhaps the greatest of the many surprises we have had, in studying birds in Arizona, was the discovery of this strange migration route. The Least Flycatcher is a very familiar eastern bird. Field studies in central and eastern Arizona, and examination of thousands of small flycatchers from Arizona and New Mexico, had given us no reason to suspect its occurrence in the west. Still, on the preceding day or two, Phillips had seen one or two small flycatchers that puzzled him; so when another was found almost in camp early on the 20th of October, he collected it. Upon picking it up, he at once recognized it, and went in search of the others. Crossing the bottomlands to a field on the far side, a round trip of perhaps a half mile in all, three more *Empidonaces* were found, two of which were Least Flycatchers! With two more specimens from western Arizona and adjacent Nevada, the existence of a migration route is well established. As in the Bobolink, the route seems to be used exclusively by young birds; and there are no spring records of either.

231. WESTERN FLYCATCHER *Empidonax difficilis* Baird

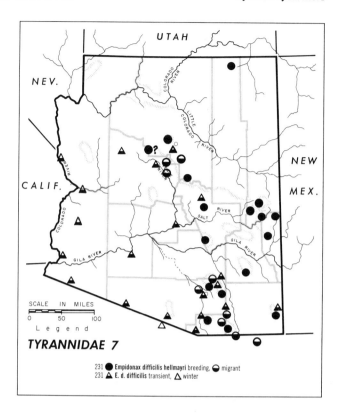

Common summer resident in boreal zones throughout southeastern Arizona, and in the northeast west to the Kayenta region and San Francisco Mountains; breeds down into more shady parts of Transition Zone locally. Absent as a breeder from northwestern and central Arizona and the Grand Canyon. Common transient in southwestern and central Arizona, rare in north (fall records only) and extreme east, and unknown in northeast. Singles found in Bill Williams Delta, *December 13, 1950* and January 5, 1951 (GM, Monson); and there are two winter records for central-northern Sonora near the Arizona line (US, ARP). Records of transients nearly span the summer.

The Western Flycatcher is the only *Empidonax* likely to be seen in Arizona which has yellow and green hues extending forward and covering the head. However, these tinges are barely visible in young birds, worn adults, and certain races. The calls are unusually high-pitched. The summer breeding race of the spruce and fir forests, with a strident *whee-seet* call of definitely two syllables, arrives in mid-May and leaves in September. The Pacific Coast form, whose call is a rising *peeeést,* has an incredibly long period of migration in the lowlands, being absent only around the end of June and the first half of July. The original nesting site was in crannies of banks or hollowed cavities in trees, but in summer resorts the eaves of outhouses have proved an acceptable substitute. Even on migration, Western Flycatchers seek dense shade. They are somewhat irregular, for though normally common in spring around Tucson, none was seen in 1940 or 1952.

Western Flycatchers migrate through southern Arizona from *March 15* (*1939*) and *March 20* (*1948* near Tucson —both ARP) to June 20–21 (1951 in the lower Huachuca Mountains — Marshall), and *June 23* (of two years in Sonora — ARP); and again from *July 13* (*1940* near Tucson — ARP), and *1951* in the delta of the Bill Williams River — GM) and July 28 (1951 in silver-leaf oaks, lower Santa Rita Mountains — Marshall) to October 24 (1947 in the Ajo Mountains — Phillips) and *October 28* (*1883* near Oracle, Santa Catalina Mountains — Scott, AMNH). A report from the Hualapai Mountains in early July (Stephens, Condor 5, 1903: 102) might pertain to either migration, or is casual.

Northern Arizona records are from May 18 (1939 in the White Mountains–Hargrave, and *1937* on West Pueblo Creek, New Mexico, near Blue P. O., Arizona — Mellinger, ARP) to *September 13* (*1876* on Seven Mile Hill near Fort Apache — Aiken, *fide* Warren, Colo. Coll. Publ. Studies Ser. 23, 1937: 62) and September 23 (1950 near Supai — Phillips). Here, too, we have an early July record which may be of a casual stray: Bill Williams Mountain, near Williams (Wetmore, Condor 23, 1921: 62). There are only two northern records of definite migrants (*E. d. difficilis*), both from the Flagstaff region: *August 5, 1934* (Stevenson, MVZ and *August 21, 1952* (ARP).

Four races of the Western Flycatcher may be recognized north of central Mexico; all have occurred in Arizona, but the status of each is different. The breeding race is *hellmayri* Brodkorb, relatively bright green and yellow and rather large. Nearly all the above northern Arizona records are of this race. It arrives on the southern Arizona breeding grounds little earlier than farther north. Southern dates are *May 7* (*1931* in the Santa Rita Mountains — van Rossem, WJS) and *May 13* (*1917* in the Chiricahua Mountains — Smith, AMNH) to *September 9* (*1952* in the upper Santa Catalina Mountains — ARP). *Hellmayri,* as a migrant in southern Arizona off its breeding grounds, is far outnumbered by *difficilis*.

Difficilis is merely a small edition of *hellmayri,* nesting along the Pacific Coast. There are only the two northern Arizona records, from the Flagstaff region in fall, as discussed above. In southern Arizona it far outnumbers *hellmayri* at most times and places, even in early June; this misled Brodkorb (Condor 51, 1949: 38) into considering the breeding birds of Yavapai County and south to be "atypical *difficilis*"; therefore he placed the name *immodulatus* Moore, to which some specimens from the Santa Rita Mountains had been referred, in the synonymy of *difficilis*. Actually most or all of Moore's type series of "*immodulatus*" are *hellmayri*. We have here an excellent illustration of the importance of annotating a specimen's label with its breeding condition, amount of fat (which indicates if it is migrating), and whether it comes from the fir forest or the foothill oaks! The migrations of *difficilis* in southern Arizona are essentially those of the species, given above. The latest definite *difficilis,* an immature somewhat approaching *insulicola* in color, was taken *October 11* (*1932* south of Mammoth, San Pedro River — Jacot, ARIZ).

The early flight through southwestern Arizona in March and early April is composed largely of rather gray birds which appear to be intermediate between *difficilis* and *insulicola* Oberholser. The only typical specimen of the latter, however, is from the delta of the Bill Williams River, *April 12, 1948* (GM). This race is dull colored, darkgrayish above, with whitish wing-bars and a relatively long tail. It breeds on islands off the coast of southern California.

Empidonax difficilis cineritius Brewster breeds only in Baja California. It is strictly accidental in Arizona. As with the preceding race, the record comes from the Bill Williams Delta, being the *December 13* specimen, mentioned in the first paragraph. This race is paler than *insulicola*, with a shorter tail and more rounded wing.

These four races plus the populations of central Mexico show only minor geographic variation, most of which is discussed above; the Western Flycatcher is relatively stable in appearance from Sitka, Alaska to Oaxaca. But suddenly at the Isthmus of Tehuantepec, the birds become much brighter and yellower, so that they have often been considered a distinct species, "*Empidonax flavescens*," from there southward to the limit of their range in Panamá.

232. YELLOW-BELLIED FLYCATCHER *Empidonax flaviventris* (Baird and Baird)

Accidental, one specimen: Tucson, *September 22, 1956* (ARP).

This bird was finally taken after a puzzling call note had been heard during the preceding evenings in Phillips' back yard.

233. ACADIAN FLYCATCHER *Empidonax virescens* (Vieillot)

Accidental, one specimen: Tucson, *May 24, 1886* (Herbert Brown, ARIZ).

This and the preceding species, both greenish, might easily be confused with the common Western Flycatcher.

234. TRAILL'S FLYCATCHER *Empidonax traillii* (Audubon)

Breeds locally in dense willow association and buttonbush swamps of Sonoran Zones (very locally in Transition Zone) throughout the state. Transient throughout state, being especially common in southwest, but possibly only a fall casual in the northeast. Migrants pass through the south chiefly in first half of June and during August and September. Extreme spring dates are *May 15* (*1921* at Bard, California — M. Canfield, LA) to *June 23* (*1932* at Bates Well, Organ Pipe Cactus National Monment — van Rossem, LA).

As a breeding bird, Traill's Flycatcher has decreased in Arizona, though it is hard to be sure just where it used to be. It is now absent from the Santa Cruz River, and whether it still nests at Camp Verde, Charleston (San Pedro River), and the Colorado River is not known. In appearance it resembles the other Arizona species but its song on the breeding ground is as distinctive as its Sonoran Zone nesting habitat — invariably next to water. The song sounds like *fuhreé-beéyer,* often interpreted by other writers as *fitzbew;* it has an explosive, rough, rolling quality. Other names sometimes used for Traill's Flycatcher are Little Flycatcher and Willow Flycatcher.

The breeding race in Arizona is the pale, medium-sized *extimus* Phillips, though some individuals in the east, at Springerville, are darker. This race is the first to arrive and has been taken in southern Arizona from *May 3* (*1933* south of Mammoth, on the San Pedro River — Jacot, ARIZ) to *September 10* (*1884* at Tucson—Brown, ARIZ). Occasional individuals appear away from breeding areas in the first part of *June* and from *July 21* (*1940* near Tucson — ARP) to early *September* (ARP). Northern Arizona data are insufficient to outline the span.

The Great Basin race *adastas* Oberholser is slightly darker than *extimus*. Like *extimus* its main migration route is somewhat east of Arizona. Only two Arizona specimens are typical in both size and color. These are from 25 miles northwest of Kayenta, northern Navajo County *July 28, 1935* (Russell, UT); and Patagonia *August 8, 1940* (GM).

The Pacific Coast race *brewsteri* Oberholser is small, dark, and relatively brown. It is the abundant transient through southern Arizona whose spring migration is given in the first paragraph for the species. Fall dates range from *July 24* (*1932* at Rockville in southwestern Utah — Woodbury, UT) and *August 4* (*1886* at Camp Verde — Mearns, AMNH) to *September 29* (*1949* at Big Sandy River and *1950* at Sullivan's Lake north of Prescott — both ARP). The few northern Arizona records are from *August 3* (*1937* on the Kaiparowitz Plateau, Utah — UT) to *September 2* (*1931* in the Grand Canyon — Vernon Bailey, GCN). It ranges uncommonly up into the Transition Zone (Flagstaff region) and casually into the densely wooded southern mountains (ARP). Normally Traill's Flycatcher, even in migration, shuns wooded or forested areas. Data on specimens labelled "Tombstone 5-16-1893" and "Dragoon, 4-3-1894" (PH) are not believed to be correct; in any case, these have been relabelled.

PEWEES Genus *CONTOPUS*

These dark flycatchers do not move the tail in any way, but they quiver the wings upon alighting. They are powerful, swift fliers with a broad chest, fairly large head, and no pale eye-ring. They hawk flies far out, then return repeatedly to the same high, exposed perch. The things to look for in identifying the species are the color of the lower mandible and of the throat and chest. The three common species can probably be found together at the north boundary of the range of Coues' Flycatcher: Blue Range, Baker's Butte, and Sierra Ancha.

235. COUES' PEWEE; COUES' FLYCATCHER *Contopus musicus* (Swainson)

Common summer resident of Transition Zone of southeastern and central Arizona, northwest to Prescott and sparingly north to Baker's Butte (Mogollon Rim) and south and west slopes of Blue Range and White Mountains. On migration found in adjacent Upper Sonoran Zone; exceptionally in lower Whetstone (Smith, Condor 9, 1907: 197) and Baboquívari Mountains (April 1945 — van Rossem), and even along Salt (*October 8, 1933* — Yaeger, MNA) and San Pedro Rivers (*April 18, 1907* — Smith). Winter records are from Patagonia (*December 3, 1939* — ARP), near Wickenburg (December 27, 1959 — JSW), and even singing in upper Santa Catalina Mountains (December 30, 1950 — Marshall).

Coues' Flycatcher is the only Arizona flycatcher that is completely dull brownish gray underneath, including the throat and belly. This, with its larger-than-bluebird size and wholly orange lower mandible, renders its identification simple. Occasionally it shows a short but pointed crest. Coues' Flycatcher makes up for its dull coloration by its lovely whistled "José María." The ordinary call is a muted *pil-pil* as with the Olive-sided Flycatcher. Oölogists maintain that small birds prefer to nest in the same tree with Coues' Flycatcher because it chases off jays. Like the Buff-breasted Flycatcher, Coues' is widely distributed to the south of us — at least to Nicaragua, if it is not conspecific with South American forms. The nest is out on a horizontal branch of a pine, and the bird is rarely seen, in Arizona, outside of a pine forest or adjacent shady canyon; but in Mexico it winters in wooded lowlands.

Records in our mountains are from March 29 (in the Huachucas — Swarth) to *September 24* (*1874* in the Pinalenos — Henshaw). Migrations are prolonged, for it is recorded away from its breeding grounds as late as May 27 (1951, near Tucson — Marshall) and again from the end of July (1923, lower Santa Rita Mountains — Goldman, FW files) on into fall.

236. OLIVE-SIDED FLYCATCHER *Contopus mesoleucus* (W. Deppe)

Fairly common summer resident in extensive boreal zone forests, less common in adjacent ponderosa pine-Gambel oak association, of northeastern Arizona, west to Kaibab Plateau and south to the entire Mogollon Plateau and (irregularly?) the Sierra Ancha (JSW). Fairly common migrant over entire state, less common toward west side.

The Olive-sided Flycatcher is like a Coues' in size, but is more compact and bull-headed, with a shorter tail. The bird appears sooty, relieved by a white patch covering the throat and extending narrowly down the midline of the breast and belly. The flanks have a dark streaky appearance unusual in flycatchers. In addition to the *pil-pil* call shared with the Coues' Flycatcher, it has a song or territorial call *hip threé beers* (which Phillips was very surprised to hear on pine ridges in Oaxaca and Veracruz in December; it is supposed to winter in South America!). This bird flycatches from the top of the tallest tree, preferably a dead bare one which affords a better view.

Due to its prolonged migrations, the status of the Olive-sided Flycatcher in northern and northwestern Arizona is not altogether clear. Thus, a record for Mount Trumbull, northern Mohave County, August 1, 1937 (Huey, Auk 56, 1939: 323) may represent an early fall transient. It appears to be very local as a breeding bird in northeastern Arizona. On this basis, migrations in the north would be from May 12 near Keams Canyon to June 16 on Bahlakai Mesa (both 1937 — Monson) and from *August 6,* (*1936* at the foot of Navajo Mountain, Utah — Russell, BNav) to *September 15* (*1936* in the White Mountains — ARP).

Southern Arizona records are from April 17 (1942 in the Santa Rita Mountains — Mr. and Mrs. Foerster, Frank A. Scott) to *June 15,* (*1918* at Gila Bend — Howell, US), and from August 5 (1952 in oaks east of Colonia Oaxaca, Sonora — Marshall) and "early August" (1882 in the Pinal Mountains — Scott, Auk 4, 1887: 18) to October 10 (1939 near Canelo, Santa Cruz County — Monson; 1953 in Tucson — Phillips).

237. EASTERN WOOD-PEWEE *Contopus virens* (Linnaeus)

One immature female, near Tucson, *October 7, 1953*, after a hard blow from the east (ARP). (Another from the Chiricahua Mountains, *September 16, 1956* — Bialac, ARP, not typical of either species, is probably a freak *sordidulus*.) Field identification of this species

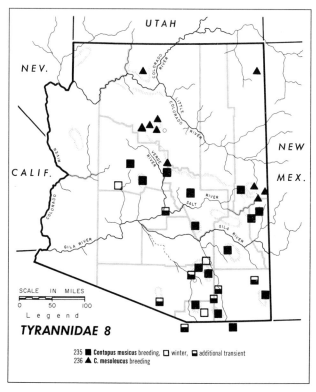

TYRANNIDAE 8

235 ■ Contopus musicus breeding, □ winter, ▣ additional transient
236 ▲ C. mesoleucus breeding

in western North America is not to be condoned, since it does not sing here; the above are the only possible records west of the Great Plains.

With adequate series of both species in the hand, the Eastern Wood-Pewee is slightly paler and less brownish (purer gray, or even olive-tinged) than *sordidulus*, and more often has the chest-band narrow, sometimes even interrupted medially. The concealed feathers along the bend of the wing are paler. But there is marked variation according to the plumage, age and season, worn summer and fall adults being essentially unidentifiable, except by their songs.

238. WESTERN WOOD-PEWEE *Contopus sordidulus* (Sclater)

Common summer resident throughout the Transition Zone, in heavy pinyon stands, in walnut-ash-sycamore associations, and very locally down to cottonwoods of upper part of Lower Sonoran Zone. Breeds west to Baboquívari (locally) and Hualapai Mountains. Common transient throughout state.

The Western Wood-Pewee has a dull dark mandible and chest-band, the latter uninterrupted medially and contrasted with the whitish belly. The upperparts are quite dark. Pointed wings extend almost to the tip of the tail, as in the Olive-sided Flycatcher. It is smaller than this and the Coues', being about sparrow size. The territorial call is a harsh descending *pheer*. The beautiful gray nest, saddled on a horizontal limb, is made of plant down, cobwebs and lichens. Wood-Pewees (except the tropical species *cinereus*) winter in South America, farther south than do other Arizona nesting birds; they are rivaled in this by the Yellow-billed Cuckoo and Common Nighthawk. Coues' Flycatcher is the only *Contopus* in northern Mexico in winter. Reports of Wood-Pewees in the United States and the greater part of Mexico in winter are misidentifications of Eastern Phoebes and *Empidonaces*.

Because Western Wood-Pewees migrate during much of the summer, we are uncertain of their breeding range in the northeast. Those seen in the Hudsonian Zone of the San Francisco Peaks, July 25–26, 1933 (Phillips) would seem, with little doubt, to have been south-bound transients. Local birds continue nesting at this season, for a bob-tailed youngster was seen August 26, 1953 on the South Rim of the Grand Canyon (Phillips). Overall occurrence in the north is from May 3 (1936 at Flagstaff — Hargrave) to October 10 (1936 at Springerville–Phillips).

Southern Arizona spring migration is from April 11 (1950 at Parker Dam — Monson), April 12 (1940 near Tucson — Phillips), and April 17 (Santa Rita Mountains; Colorado River Indian Reservation; and near Mexicali, Baja California) to June 18 (1951 at Parker Dam, California side — Monson) and June 23 (1918 at Gila Bend — Howell, FW files). Fall migration is from July 26 (1951 at Topock — Monson) to October 27 (1942 at Tucson — Mr. and Mrs. Foerster) and October 29 (1907 in the Huachuca Mountains — Swarth, Condor 10, 1908: 111). Monson reports two casual records on the lower Colorado River in early July. It is rarely seen before *April 21,* and the latest record in western Arizona is *October 19* (*1957* above Imperial Dam — GM).

A race described from southern Arizona as "*Myiochanes virens placens* van Rossem" is a synonym of *Contopus sordidulus veliei* Coues, the breeding race of northern mainland Mexico and the United States generally (Phillips and Parkes, Condor 57, 1955: 245; Webster, Pr. Indiana Acad. Sci. 66, 1957: 337-340). Although further study is needed, it appears that *veliei's* status in Arizona is essentially that of the species, as given above, with inclusive dates of *April 28* (*1956* — JSW) to *October 18* (*1932* — Jacot, ARIZ).

The northern *C. s. saturatus* is darker, when specimens are compared with proper segregation by age and sex. It migrates over most of Arizona; concentration of records in the southeast and center of the state probably reflects more collecting there. Dark specimens in southern Arizona are from *April 21* to *June 16,* and from *August 14* (all Tucson — ARP) to *September 26* (regularly) and *October 19,* as above. In the north they are recorded *May 22* (desert east of San Francisco Peaks — Hargrave, MNA) and from *August 13* (Tsegi Canyons near Kayenta — Russell, MNA) to *September 17* (Springerville — Stevenson, ARP).

239. VERMILION FLYCATCHER *Pyrocephalus rubinus* (Boddaert)

Common to abundant summer resident in mesquites, willows, and cottonwoods (always near water in the lower, western valleys), in southern and central Arizona, but rather local along the Salt and Colorado River Valleys. Also nests locally in sycamore-ash-cottonwood associations. Casual in pine-oak-juniper (Natanes Plateau, Gila County, April 15, 1937 — Phillips). Winters in moister valleys of most of breeding range, sparingly or not at all in extreme southeast except at

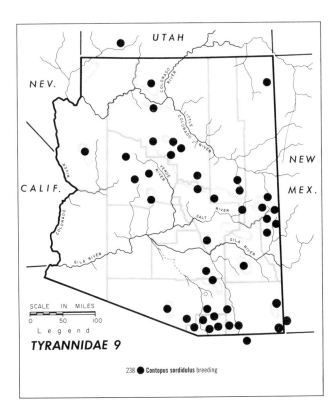

TYRANNIDAE 9

238 ● *Contopus sordidulus* breeding

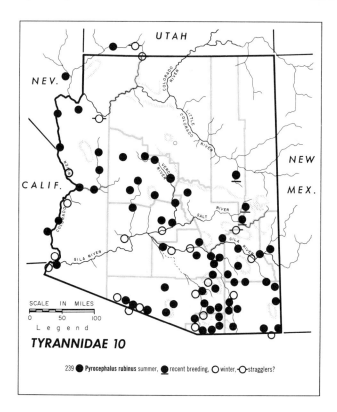

239 ● *Pyrocephalus rubinus* summer, ● recent breeding, ○ winter, -○- stragglers?

San Bernardino Ranch. Uncommon transient at or near water in southwestern deserts.

The Vermilion Flycatcher breaks all the rules of flycatchers. The chest is definitely streaked in the females and young, the sexes are markedly different in color, and the adult male is bright red. A few individuals even go northward to spend the winter. The male prefers to sing in an amazing butterfly-like flutter above the tree-tops, terminating in a dive. W. L. Dawson describes this song as a repeated *tut tut tiddly zing*. This is one of the most spectacular birds of southern Arizona. It often perches on telephone wires or tops of bushes, and is very tame, popular, and easy to see. The call is sharp, though feebler than the similar call of the Black Phoebe, and the bird also has a Phoebe-like dip to its blackish tail, though this starts from a more horizontal position. It is also smaller; its size is slightly less than an English Sparrow's. It probably raises two or more broods. The nest is always in a horizontal fork of a branch, and the three eggs are unique among known flycatcher eggs in being heavily blotched with black.

In a way, the Vermilion Flycatcher is the opposite of a Wood-Pewee. Instead of retiring to South America, individuals may even straggle north in winter; males are among our most brilliantly colored birds; and the winter range is uncertain, rather than the summer range. Midwinter birds are usually seen day after day in the same places, but their numbers vary from year to year. The following two paragraphs illustrate the problem of separating migrants from such sedentary winter or summer residents.

Certainly many Vermilion Flycatchers, unlike Wood-Peewees, arrive in the higher valleys of southeastern Arizona early in the spring migration, in February. Three males were at Canoa Ranch, February 5, 1956 (Phillips *et al.*). Two individuals, additional to those wintering, were seen near Tucson, February 14 (1943 — Hargrave, Jos. Seabury). New arrivals were seen as early as February 17 (1950 near Nogales — Phillips) and daily thereafter (Dille, and *fide* R. J. Hock); February 18 (1936 at San Carlos — Jacot); and along the Big Sandy River by February 27 (1951 — Phillips). It is odd that none has been found wintering even as low on this river as Wikieup, in view of repeated statements that it is a "permanent resident" in southern Nevada. In our own experience, it winters north along the Colorado River regularly to Topock.

In fall, the last are seen in the lower Santa Catalina Mountains, October 1–10 (Scott, Auk 1887: 20). At the mouth of the Verde River, five were seen on the afternoon of October 6, 1951, but only one the next morning (Phillips, Yaeger). At Canoa Ranch south of Tucson, where only one or two winter (Marshall), at least six were seen October 7, 1956 (Phillips, Bialac). The last seen at Wikieup in 1951 were two on October 19 and one the next day (Phillips). In the Yuma region it arrived after an absence since the previous winter on October 1, 1958 (Monson); it is usually quite scarce there in summer.

The Vermilion Flycatcher has recently expanded its range to the Fort Apache area. Here none was found by Henshaw or Aiken, and it was apparently still a casual in the late 1930's: one young male, *August 6, 1936* (Poor and Watson, ARP). But two pairs were seen from the fort to a point 5 kilometers east, April 15 and 17, 1952 (Phillips, Dickerman). Similarly, in the Colorado Valley, it has apparently very rare above the Yuma region 100 years ago (Cooper).

Word of a more surprising expansion — across the Mogollon Plateau into northern Arizona — comes while we are in press. A pair showed up in Snowflake in April 1964 (Albert J. and Louise Levine).

240. BEARDLESS FLYCATCHER *Camptostoma imberbe* (Sclater)

Fairly common summer resident in cottonwood, heavy mesquite, and even sycamore-live oak-mesquite associations, north to Gila River (at mouth of San Pedro River), and from New Mexican border to west side of the Baboquívari Mountains. Winters in the vicinity of Tucson, on the east side of the Baboquívaris, and along the lower San Pedro River.

This tiny nondescript flycatcher, smaller than an *Empidonax,* is best known from its high, thin voice. The common call is *pee-yerp* and the song is a descending series of three to five clear whistles. Each of these notes of the song trails evenly downward in pitch, whereas in the similar vocalization of the Verdin, each note is short and inflected. In summer the Beardless Flycatcher feeds in flycatcher fashion by short flights between which the tail is occasionally flicked nervously. But in winter, much of its food is obtained by gleaning in the twigs like a vireo or kinglet. Its dull colors are much like other flycatchers', but its wing-bars are browner. There is no eye-ring. In the hand, the bill is seen to curve in all aspects, and to be short and vireo-like; the mouth lining is deep orange. The nest is an oval pouch with the opening on one side near the top, placed among spider or caterpillar webs or in mistletoe. Within the United States the Beardless Flycatcher occurs only in southern Arizona and a small area of Texas, though it is extremely wide-spread and common in much of Mexico and occurs south to Costa Rica. There is a bare possibility that it is an *Elainea,* in which case it would be the northernmost representative of a large South American genus. The name "beardless" refers to the suppression of the long rictal bristles, characteristic of most other flycatchers.

Like the Vermilion, the Beardless Flycatcher withdraws in early fall from the higher parts of southeastern Arizona, such as Patagonia. Yet this is the very spot that other small flycatchers, such as Hammond's and Dusky, find most attractive in winter, thus confounding those who would explain migrations in simple terms of food supplies, climate, etc. Here our latest record is September 14 (1947 — Phillips *et al.*); while on visits as late as March 20, it had not yet returned. Nevertheless, it seems likely that spring migration is in March. A small flight in *March 1947* in southeastern Arizona (van Rossem, Phillips) may have been unusual; single individuals were found at Fairbank, San Bernardino Ranch, and Guadalupe Canyon, *March 9, 10, 12,* and *15* (ARP, LA).

The nesting season seems to vary somewhat from year to year. Four nests at Patagonia, May 17–18, 1953, contained young or heavily incubated eggs (S. B. Peyton, Edward M. Hall); whereas the usual nesting is in late May and later, and one was carrying nest material as late as August 4, 1962 at the same place (Marshall). Occurrence in non-wintering situations at Tucson has been from March 30 (1938) to September 20 (1947 — Phillips).

The habits of this diminutive flycatcher have been discussed by Vorhies *et al., Condor* 37, 1935: 246; van Rossem, *Trans. San Diego Soc. Nat. Hist.* 8, 1936: 137-139; Bent; Anderson and Anderson, *Condor* 50, 1948: 163-164; and Brandt, *Arizona and its Bird Life,* 1951: 151, 249-254, and 666-667.

LARKS *ALAUDIDAE*
by Allan Phillips

241. HORNED LARK — *Eremophila alpestris* (Linnaeus)

Nests in open grasslands throughout the state, and possibly in some farmlands; thus absent during the breeding season in open areas without grass. Winters commonly in same areas, also in fields, parks on the plateaux, and sometimes on barren shores of rivers and lakes. Absent from all brushy or wooded areas. Casual above timberline in White Mountains (Poor, ARP).

The Horned Lark is the characteristic bird of the open plains of eastern and central Arizona, but is much scarcer toward the southwest, where grasses are less prominent. Their high *chleep* and *chee-lee,* and the contrast of mostly-black tail with tan (above) and white body, readily identifies them, even in winter when the plains swarm with sparrows and longspurs. Even the boldly white-spotted juveniles are distinctive. Horned Larks are gregarious among themselves, but flock with no other species save McCown's Longspur. They are spaced out on individual territories only from March to June, in Arizona. While on the ground they are almost invisible, so closely do their colors match the soil and the dead grasses. When curious, alarmed, or singing they often mount a rock or fence-post, but they shun bushes and trees.

The Horned Lark has achieved fame for its many geographic variations, which enable it to match the soil of whatever region it inhabits. These color races are much less pronounced than those of the Dark-eyed Junco, Song and Fox Sparrows, however. Thus the determination of the several races known from Arizona is difficult; none is recognizable in the field.

The study of Horned Larks is also complicated by their peculiar breeding seasons and movements. As van Rossem showed (Condor 49, 1947: 39), the many *occidentalis* found as far southwest as Aguila, Maricopa County, in *mid-June* were already away from home, since the race nesting there is *leucansiptila.* Thus the determination of breeding ranges for the various Arizona races must rest on specimens taken from late March through May! Out-of-season nestings are not yet known from Arizona, but south of Sonoyta, Sonora, young out of the nest and a bird in unworn first basic plumage were taken on *March 25, 1947* (van Rossem, LA). The latter must have been hatched in winter! At higher altitudes farther north and east, we do not find young Horned Larks out of the nest before May.

Study of good series in almost all the major collections gives this somewhat revised picture of our races:

A. Dark-backed northern and eastern races. Some of these may reach extreme northern Arizona (where virtually no winter collecting has been done), but the only report

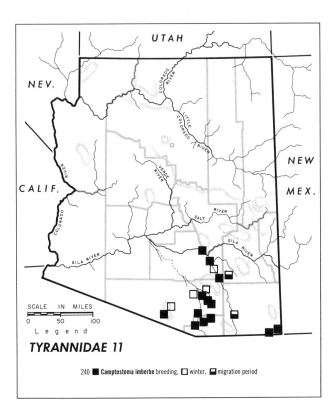

240 ■ *Camptostoma imberbe* breeding, □ winter, ▨ migration period

TYRANNIDAE 11

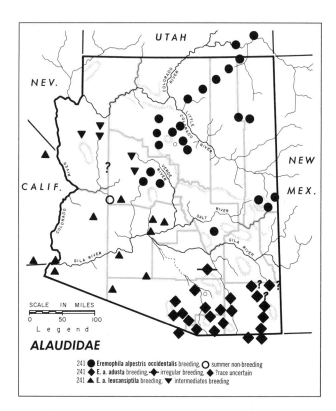

ALAUDIDAE

241 ● *Eremophila alpestris occidentalis* breeding, ○ summer non-breeding
241 ◆ *E. a. adusta* breeding, ◆ irregular breeding, ◆ ?race uncertain
241 ▲ *E. a. leucansiptila* breeding, ▼ intermediates breeding

so far is for Camp Verde. Here the eastern race *praticola* (Henshaw) was recorded by Oberholser, but the identification was questioned by Swarth. Reexamination of the specimen (AMNH) confirms Oberholser's determination, but the bird was actually taken in Minnesota and later given a label with "Fort Verde, Arizona" printed on it! Thus none of the really dark races has a valid claim to appear on the Arizona list.

The darkest race yet found in Arizona is the intermediate *lamprochroma* (Oberholser) of eastern Oregon, known here only from Tucson (*December 12, 1937* and *February 19, 1938*) and from the Aquarius Mountains, southeastern Mohave County (*March 1, 1951,* number 2353; all females —ARP, identifications of first two confirmed by van Rossem).

B. Deep cinnamon-naped California races. These are relatively sedentary; only the nearest, *ammophila* (Oberholser) has been recently identified from Arizona. It lives in eastern California and is the palest-backed form there. A male was collected at Tucson, *February 2, 1917* (Thomas K. Marshall — ARP). Three more dark birds found in grassland at the south end of the Aquarius Mountains (*March 1, 1951* — ARP, see above) seem closest to *ammophila*. If they represent the breeding population, then they are most curiously surrounded by an entirely different, pale race (*leucansiptila*). *Actia* (Oberholser), a deeper red on the back, may be expected in southwestern Arizona for it has been taken on the California side of the Colorado River at Fort Yuma (*January 29, 1913* — A. B. Howell, LA 9824). We cannot find the *actia* reported from Fort Huachuca by A. K. Fisher (Condor 6, 1904: 80).

C. Pale interior races. All of our Horned Larks belong to this group except the few specimens discussed above. A cline runs from dull gray in the north to pinker or redder in the south.

The dullest and grayest race, with little pink or yellow, is *enthymia* (Oberholser) of the northern Great Plains and apparently also northern Utah (where the heads are a trifle yellower). *Enthymia* is fairly common in the northeast and uncommon to rare elsewhere; it is the only race of Horned Lark that comes as a regular winter resident without breeding anywhere in the state. Records in the San Francisco Mountains region are from *September 16* (*1940,* Mormon Lake — Phillips, GM) to *February 28* (*1932,* east slope— Hargrave, MNA). Southern Arizona records are from *August 27* and *September 15* (*1931,* near Prescott — Jacot, ARIZ) to *March 11* (*1915* in the Chiricahua Mountains — van Rossem, LA). Recorded south of the Gila River from late *November* (New Mexico) onward.

The breeding Horned Larks of central and northern Arizona, despite considerable local variations, seem best referred to *occidentalis* (McCall), which includes *leucolaema* (Coues). Those in northeastern Arizona are rather gray and pale; those in the upper White Mountains are rather dark; and most of the populations are moderately reddish-backed. But local grayer populations occur in the Puerco Valley and south, and at (but not around) Springerville. *Occidentalis* is an abundant resident over most of Arizona, and is irregularly common in the rest of the state in winter. (It should be noted, however, that the large series— CAS—of early fall *"occidentalis"* reported by Swarth from Patagonia are mostly dull examples of *adusta*.) The southwestward movement in June, already noted, appears to reach Yuma occasionally (faded male, *August 18, 1902* — H. Brown, ARIZ). Grayish birds have been taken as early as *August 27* (*1931,* near Prescott — Jacot, ARIZ). From the Salt River Valley south, however, no wintering northern races of Horned Larks occur between mid-March and early October. This is the more noteworthy because the race *adusta*, breeding so near Tucson, has been taken well south of its nesting grounds in northern Sonora by mid-*August* (ARP). From Phoenix and Tucson south, actual specimens of *occidentalis* are all from mid- or late *October* to the start of *March,* with a published record of *"leucolaema"* in southeastern California as late as *March 21* (Dickey and van Rossem, Condor 24, 1922: 94).

E. a. adusta (Dwight) is the race of southeastern Arizona, distinguished by its redder, more "scorched," back, which in many individuals hardly contrasts with the nape. This race is largely resident on these plains, where it possibly reached a peak of abundance during the initial overgrazing in 1880–1905, before brush became tall and prominent. At its northwestern limits, below the Santa Catalina Mountains, *adusta* is irregular, breeding in some years, absent in others (Monson and Phillips, Condor 43, 1941: 110; see also Law, idem 31: 219). The colony at Ventana Ranch (Sutton and Phillips, idem 44: 61) proves, with better material, to be nearer *leucansiptila*. True *adusta* shows considerable individual variation, which probably accounts for conflicting reports by various authors. Its eastern limit is the New Mexico border, since the birds just east in the Animas Valley seem nearer to *occidentalis*.

Though migrating regularly well south into Sonora (ARP), *adusta* is known in Arizona only from its breeding grounds and from the Tucson Valley, where it occurs from *December* (Jacot, ARIZ) to *February 24* (*1917* — Thomas K. Marshall, in Ivan Peters collection). The use of this race to delimit a "faunal area" by various writers (*cf.* Mearns, Swarth, Law, and van Rossem) is misleading. It occupies all of its grassland habitat, which is but one of our various biotic communities. The fact that it does not extend into the brushy deserts (which Behle, Univ. Calif. Publ. Zool. 46, 1942: 283, curiously describes as constituting its habitat) is of no more significance than the failure of any bird to exceed its required habitat, in whatever direction.

E. a. leucansiptila (Oberholser) is the pale, pinkish race of western Arizona. Long thought a bird of extreme desert conditions, it appears in recent summaries as restricted to the "Colorado Desert region" of California and the narrower edges of adjacent states (Behle, op. cit.: 277; A. H. Miller, Pac. Coast Avif. 33, 1957: 104), though van Rossem informed us that Behle's "Nord Ranch" is south of Salome in *eastern* Yuma County. In 1947 van Rossem (*loc. cit.*) extended its range east to near Congress Junction, near the foot of the mountains in much higher Lower Sonoran Zone. Subsequent studies showed, to our great surprise, that *leucansiptila* influences even the birds of Upper Sonoran Zone plains in the Juniper Mountains region! West of these mountains, to be sure, almost no material is available; but some specimens from Williamson Valley, farther east, are nearly typical *leucansiptila*. About an equal number of Williamson Valley birds are nearer true, reddish *occidentalis*, which occupies the valleys immediately east and north in typical form. Breeding *leucansiptila* is irregular (found only in wet years) in southern Yuma County and above Gila Bend in the Gila and Salt River valleys. But a juvenile taken at the unexpected locality of Pima, northwest of Safford, *July 8, 1953* (Gallizioli, ARP) seems to be good *leucansiptila!* Otherwise, it is apparently non-migratory, moving short distances only.

SWALLOWS *HIRUNDINIDAE*

by Allan Phillips

From ancient times men have marveled at swallows' regular comings and goings, though tales of their arriving on the same day each year, leap-years included, are amiable myths. They are also famous because of their diurnal migrations, at low altitudes, which make migration a highly visible affair; the steady swarms following coasts and river valleys are indeed impressive. As we shall see, however, they also migrate and feed at night.

Many people confuse swallows and swifts. Swallows bend their wings in normal small-bird fashion, and take more abrupt turns than do swifts; their voices are throatier and they sing at times; they alight on trees, telephone wires, or even the ground; and in general they are more familiar birds, particularly in western North America, where Chimney Swifts are so rare. Except for the adult male Purple Martin, swallows have a white or pale belly. Like swifts, they subsist entirely on flying insects, caught with amazing dexterity in midair. Swallows are among our earliest migrants to arrive in both "fall" and "spring," appearing regularly in July and February.

242. VIOLET-GREEN SWALLOW *Tachycineta thalassina* (Swainson)

Common summer resident in most of Transition and Canadian Zones, and locally in Upper Sonoran Zone cliffs near water in northeast. There are isolated Lower Sonoran colonies in Havasupai Canyon, near Camp Verde, and along the Colorado River (nests on California side) at Parker Dam. Common throughout state on migration, but not in extreme west in fall. The early return of some birds in late January and February gives an erroneous impression of wintering; as a matter of fact, there are no December specimens, and only a few valid sight records of one or two individuals each. The only November specimen is an immature taken from a cat in Tucson *"about the middle of November"* 1944 (Ted McKee, MNA). Thus the statement (A.O.U. Check-list) that it winters up the Colorado River to Needles, California, is erroneous.

This is the common swallow of our mountain forests, where it nests in holes in tall, dead trees. It is also our whitest swallow; in adults, pure white covers the entire underparts and extends up around the eye (in the male) and unto the sides of the rump, giving the impression of a white rump unless the bird is seen from above. A high-pitched double call-note often calls attention to Violet-green Swallows high overhead, where they join White-throated Swifts over steep canyons, feeding higher than most swallows. But with other swallows they skim the alfalfa fields in spring migration. Their flocks and colonies add up to a total that exceeds any other swallow in Arizona, numerically, though the less gregarious Rough-winged Swallow is more widely distributed in summer. In fall, in most of southern Arizona and even to the Big Sandy River, there is often a spectacular rush of Violet-green Swallows, which pass in numbers for two or three days and then are completely gone. This happens just after October 15, and a few of the birds

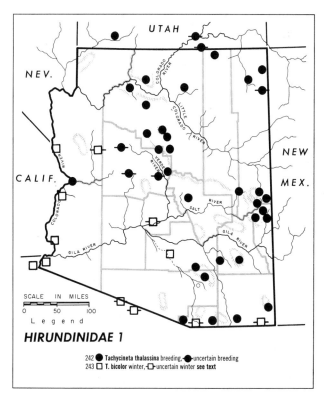

HIRUNDINIDAE 1

242 ● *Tachycineta thalassina* breeding, ◐ uncertain breeding
243 ☐ *T. bicolor* winter, ⬚ uncertain winter **see text**

may reach the Colorado River, where otherwise they are rare as fall transients, in marked contrast to their numbers in spring.

Normal occurrence is from early or mid-February to about October 20 in southern Arizona, and from late March to mid-September in the north. Extreme migration dates, in the south, are from January 27 (1950, eleven in Bill Williams delta — Monson) to May 30 (1940 at Feldman, lower San Pedro River — Monson and Phillips; 1955 on Santa Cruz River, Sonora — Marshall and Phillips); and from June 30 (1951 at mouth of Verde River — Phillips and Yaeger; 1959 at Canoa Ranch, near Continental — Phillips) to October 26 (1953, near Parker — Dean Amadon), November 11 (1953, at Hereford — Phillips), and November 24 (1947, Bill Williams Delta — Monson). Numbers at Binghampton Pond, Tucson, June 6 and 23, 1934 (Anderson) were perhaps nesting in the mountains not far away. In northern Arizona the extent of migration is uncertain. Extreme dates are *March 3* (*1936* at Flagstaff — Hargrave and Phillips, MNA) to October 2 (1926 at Mormon Lake — Griscom).

On *July 31, 1950,* Phillips was trying to get a July specimen of Bank Swallow from the swarm of Violet-greens about the Flagstaff city reservoirs. Dusk was falling when a small-looking swallow flew over and was taken as the Bank Swallow; but in the hand it proved to be a full-grown young female Violet-green, of such small size that it seems necessary to regard it as a stray of the small race *brachyptera* Brewster, which nests in the giant cacti of the coasts of northwestern Mexico. The wings (chord) measure 95.2 and 96, the tail 37.3 millimeters. Otherwise, we refer all Arizona birds to the medium-sized *lepida* Mearns, with greenish back and purple upper tail-coverts. It should be noted, however, that of three specimens from the Colorado River colony (GM), the only adult appears intermediate toward *brachyptera*.

243. TREE SWALLOW *Tachycineta bicolor* (Vieillot)

Winters commonly along the lower Colorado River, and occasionally eastward as far as Picacho Reservoir in Pinal County; probably winters also in extreme southeastern Arizona and the Big Sandy Valley. Generally distributed during migrations, usually along streams or at lakes and ponds. It has been recorded throughout June at Topock, but is usually absent from Arizona between *mid-May* and *early July*.

Except for its lower-pitched, harsher call, the Tree Swallow resembles the Violet-green, and the two are easily confused. Besides, the abundance of the latter and the supposed scarcity of the Tree Swallow in Arizona have led most observers to call any white-throated swallow a "Violet-green" until proven otherwise. Actually, after October 20 the Tree Swallow is by far our commonest swallow. It differs from other swallows in having a pure white throat abruptly contrasted to dark sides of the head, but with pale gray (immature) or white chest. (Young Violet-greens have the throat as dirty white as the chest, and this color passes more gradually into dark brown of the crown). Birds mashed in the highway by cars are often too distorted or bloody to judge colors, and are best identified by the larger feet of the Tree Swallow, particularly its longer hind toe and claw, and certain differences of the outer primaries and the adjacent feathers along the bend of the wing. The white of the underparts does seem to pass up onto the sides of the rump in some young Tree Swallows, which makes this area of little or no use in field identification in autumn. Migrants (particularly the earliest ones in winter) are often glimpsed too briefly to be sure of the species, and such records are perforce ignored in the following accounts.

Another difficulty is imposed by the irregular wintering and early return of the Tree Swallow. It is certain that the Tree Swallow does not normally winter at or above Tucson, yet Dille saw "considerable of a bunch" on the Santa Cruz River near Nogales, November 28, 1936, and a very few swallows, probably of this species, were seen the same winter at Tucson (Wm. X. and Alma J. Foerster). Other records indicative of at least sporadic wintering are from San Bernardino Ranch, east of Douglas, November 22, 1951 (Phillips), and Sonoyta, Sonora, on or before January 25, 1894 (Mearns). The meaning of birds seen at Quitovaquita, west of Sonoyta, in late January and/or early February 1894 (Mearns) and at Wikieup, Big Sandy River, January 28 and 31, 1951 (Phillips) is uncertain. None of these points, except the Santa Cruz Valley, has been studied through the entire winter.

Excluding these and the above-mentioned Topock records, southern Arizona migrations are from January 29 (1956 at Canoa Ranch near Continental — Tucson Bird Club) to May 22 or 23 (1948 at San Bernardino Ranch — Hargrave), and from July 2 (1959, Martinez Lake above Yuma — Monson) and *July 7* (*1955* at Tucson — ARP) to October 7 and 19–20 (1949 in Big Sandy Valley above and near Wikieup, respectively — Phillips); casually to *November 16* (*1957* near Phoenix — Johnson and Werner, from flock of 60; JSW).

Aside from *July 7* (*1936* at Tuba City, east of the Grand Canyon — Phillips, MNA), all northern Arizona records are in the brief periods *April 6 to May 6 and July 15 to September 21,* as yet.

244. BANK SWALLOW *Riparia riparia* (Linnaeus)

Fairly common to rare transient at lakes, ponds, irrigated fields, etc., throughout Arizona.

Known as "Sand Martin" in Europe, the Bank Swallow is almost a cosmopolite. It is a small swallow with rapid, butterfly-like wingbeats. The chest-band is narrow, sharp, and of the same dusky brown color as the upperparts, which separates it easily from the young Tree Swallow, to which its colors are otherwise similar. Such a pattern occurs in no other swallow except perhaps an occasional young Cliff Swallow, with much white on the throat, which could still be separated by its more irregular pattern below and pale rump and forehead. In spite of this ease of identification, the Bank Swallow has been confused with the Rough-winged Swallow (completely unpatterned below!); all supposed breeding records actually pertain to the latter species.

Bank Swallows may migrate partly at night. At Binghampton Pond in Tucson, about 60 apparently roosted on *May 17, 1940,* for they gathered into a flock late in the evening and swooped about low over the cattails, which were extensive there during those years, amounting to perhaps half an acre (a fifth of a hectare). But on our return early the next morning they had completely gone (Phillips, Sutton).

Most Bank Swallows pass through Arizona in relatively short periods; late April to mid-May and mid-July to mid-September. Extreme dates of valid records, in southern Arizona, are April 5 (1929 at Tucson — Ellen C. Rogers, FW files) and April 10 (1943 at Imperial Dam California side — Monson) to May 23 (1951 at Topock and 1954 at Parker — Monson); and about this date (1948, at San Bernardino Ranch — Brandt, Hargrave, and Phillips); and from July 2 (1952) to October 7 (1957, both at Martinez Lake above Yuma — Monson) and October 8 (1933 at the mouth of the Verde River. — Hargrave).

The Bank Swallow is recorded in northern Arizona only in late April (Holbrook — Phillips) and from July 7 (1936 at Tuba City, east of the Grand Canyon — Phillips) to September 16 (1940 at Mormon Lake — Phillips) and September 30 (1938 at Snake Butte, Navajo County, southwest of Keams Canyon—Monson, Condor 41, 1939: 168).

245. ROUGH-WINGED SWALLOW — *Riparia ruficollis* (Vieillot)

Common summer resident in dirt banks of streams throughout Sonoran zones of state, irregular at Flagstaff. Rather common transient at and along waters. Winters rarely along Colorado River, where its spring arrival is in *January;* one was seen at Picacho Reservoir, November 19, 1958 (Levy).

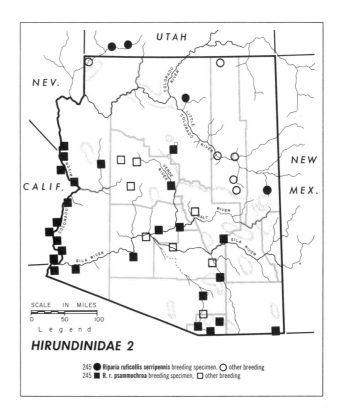

The Rough-wing is our plainest and, in June, most widespread swallow. Its grating call is devoid of any pretense of music. It is our only swallow that nests in dirt banks; this, with its plain brown color, has led to misidentification by many ornithologists as a "Bank Swallow." It is, however, a heavier bird than the Bank Swallow, and does not nest in close colonies. The throat, definitely tinged with pale brown or even pinkish buff in young birds, is diagnostic; it shades without any contrast into the white belly. The leading edge of the outer primary is serrated in the adult male, giving the species its name. The flight is more floating and buoyant than in other small swallows.

The Transition Zone at Flagstaff is marginal habitat. A pair or two were always to be seen between town and the Museum of Northern Arizona along Rio de Flag in the 1930's, but after the war it was absent for a decade. Finally, on *June 21, 1957,* a pair was again found and the female collected to determine the race.

The Rough-winged Swallows of the United States and northern Mexico comprise two races: a northern *serripennis* (Audubon) and a southern race now known as *psammochroa* Griscom which is paler, particularly on crown and rump. The slight racial variation is far outshadowed, however, by phenomenal color changes due to season and foxing. It is impossible to identify a recent specimen without freshly taken specimens of both races in comparable plumage for comparison.

R. r. serripennis, in typical form, nests only in extreme northern Arizona: one specimen, Tuba City, east of the Grand Canyon (Phillips, MNA). Birds of the White Mountains region, judging by very few specimens, are intermediate. *Serripennis* also migrates over the state, though specimens are not numerous. It can hardly be doubted that this is the usual race in northern Arizona on migration. Records here extend from *April 6 (1937* at Snowflake, Navajo County, ARP) to June 2 (1949) and from August 12 (1947; both at Flagstaff — Phillips) to September 15 (1931, north side of San Francisco Peaks — Hargrave; 1934, North Rim of Grand Canyon — Borell *fide* McKee).

Serripennis has been taken in southern Arizona from *March 5 (1928* on the Colorado River at latitude 32°15′ in Baja California — Lamb, MVZ), *March 28 (1887* at Camp Verde — Mearns, AMNH), and *April* to early or mid-*May;* and from *August 1 (1892* in extreme southeastern Arizona — Mearns and Holzner, US) to *September 20 (1886* at Camp Verde — Mearns, AMNH), and casually *November 10 (1952* on the California side of Havasu Lake, two taken from flock of five — GM!).

R. r. psammochroa is known in northern Arizona from the Flagstaff female only. It occupies all of southern Arizona north to Prescott and tip of Nevada. It arrives much earlier than *serripennis,* and stays a trifle later in fall. Southern Arizona occurrence is usually from mid- or late February to late September or early October. Extreme dates are *January 23 (1953* in Bill Williams delta — GM) to *October 9 (1952* at Wikieup — ARP) and exceptionally to *October 22 (1938* at Salome — WJS). A wintering bird was taken at Blankenship Bend above Havasu Lake, *January 10, 1954* (GM). Most reports of "wintering," however, pertain to misidentification or (Evenden, Condor 54, 1952: 36) to migrants returning in February and March.

246. BARN SWALLOW — *Hirundo rustica* Linnaeus

Local summer resident in eastern Arizona towns and ranches from Sonoita, San Pedro, and Sulphur Springs valleys north to Show Low, Snowflake, and Holbrook and east (also breeds in Nogales, Sonora); irregularly in Santa Cruz Valley and at Mayer, Yavapai County (*fide* Phillips). Common in migration at and along waters and over fields throughout state. Has been found in winter in small numbers at Picacho Reservoir, Pinal County (ARP, LLH); and probably wintered once in extreme lower Colorado Valley ("common" from Yuma south, March 7 and later, 1894; Mearns).

Known in Europe as "the Swallow," and subject of a small book by Hosking and Newberry (1946), this species is familiar to most people. The expression "swallow-tailed" comes from the long, slender, deeply cleft tail of adults, which in Arizona are usually pale pinkish on the breast and belly. The extremely graceful flight, streaming tail, and shrill alarm call make adults easy to recognize at a distance. Their semi-metallic twittering song is pleasant, though interrupted by dry stuttering spells. Young birds (until their prealternate molt in late winter) are more difficult to recognize, as their tails resemble those of other swallows except for a hardly visible white spot at the outside edge. The dark throat in abrupt contrast to whitish or pinkish breast separates them from most others, while the dark rump eliminates the remaining Cliff Swallow.

Barn Swallows, as the name tells us, nest familiarly about houses and barns in most of the northern hemisphere. Nesting spots include hanging electric-light-bulb shields in the Hotel Gadsden garage in Douglas and (from 1938 to 1942) on the University campus in Tucson, and the railroad station at Patagonia. But nesting is erratic even within the main eastern Arizona breeding range: in 1922–1924 none was seen at Douglas, while in 1933–1934 they nested commonly at San Bernardino Ranch (Harry L. and Ruth Crockett); but in 1947–1948 they bred at Douglas but not at the ranch (Brandt, Hargrave, Phillips)! At Benson they nested in the 1930's, and to 1942 (Crocketts, Hargrave), but not thereafter. Brood after brood is raised in quick succession, often in the same nest, and even in northern Arizona, nesting goes on into September (Stevenson, Wilson Bull. 63, 1951: 339-340). In this, as in the Cliff Swallow, nesting may still be in progress when lack of food or, more likely, the instinct to migrate drives the parents south, leaving the last young to die in the nest (Cooper, Proc. U.S. Nat. Mus. 2, 1880: 243-244, 246; Bard, *fide* Pettingill, Wilson Bull. 58, 1946: 53). Such desertions have not yet been reported in Arizona, however.

The Barn Swallow was the first bird recognizably reported as nesting in Arizona. At San Xavier Mission (where most or all of the Tucson swallows nested in the 19th Century) on June 16, 1858, Way (in Eaton, SW'n. Hist. Quarterly 36, 1933: 187) found "the birds are its only occupants and they sing praises here from morning until night – they build their nests on the heads of the saints and warble their notes of joy while perched on their fingers – they do not respect the sacred image of Christ for a noisy swallow has built her nest in the crown of thorns that circles his brow and at this moment is perched on his bleeding hand scolding loudly at my near approach."

In the clear skies of Arizona, the hazards that beset night-migrating birds elsewhere are non-existent. They are not forced low to collide with television towers, buildings, etc., nor are they attracted to powerful "ceilometer" searchlights. The principal danger is to waterbirds, which often can find no water on which to alight. It is indeed strange that the only accident to a migrating land-bird that we know of occurred to the supposedly diurnal Barn Swallow – a bird that even the indefatigable Stoddard rarely found migrating at night (Tall Timbers Research Sta., Bull. 1, 1962)! This bird was fluttering with a broken wing under a tall pine snag at about 9,000 feet (or 2,700 meters) altitude on the southwest slope of the San Francisco Peaks in mid-morning, May 5, 1934 (Phillips). It had obviously flown into the mountain while migrating in the dark of night.

To one familiar with the great southward migration along the South Carolina beaches in mid-summer, Barn Swallows in Arizona start to migrate remarkably late. In southern Arizona we usually see them from late April to late May, and from the first days of September to mid-October. There are, however, several records in late March and early April, chiefly in the western half of the state, and a few June records, even to June 29 (1961 at Martinez Lake above Yuma – Monson). Extreme fall dates are August 18 to November 4 (1959 at Imperial Dam and 1960 near Yuma, respectively – Monson).

HIRUNDINIDAE 3

246 ▲ Hirundo rustica breeding, ▲ irregular breeding, △ winter

Northern Arizona migrations are from April 25 (1953 at Springerville — Phillips) to May 28 (1938 in the San Francisco Mountains — Hargrave), and again from August 22 (1938 near Tuba City — Woodbury and Russell, BNav) to October 13 (1938 near Snake Butte, southwest of Keams Canyon — Monson).

Records indicative of sporadic breeding at outlying points are: Camp Verde (bred regularly 1884–1892 — Mearns, A. K. Fisher); Sells, Papago Indian Reservation (pair seen July 10–11, 1918 — A. B. Howell, FW files); and Hoover Dam region, (1938 — Grater). It seems probable that Barn Swallows nest every year in the Springerville-Snowflake-Holbrook region, but only sporadically elsewhere. The northernmost locality of probable breeding is Window Rock, Apache County, where several were seen in early August, 1936 (Hargrave).

247. CLIFF SWALLOW *Hirundo pyrrhonota* (Vieillot)

Nesting colonies are found almost throughout the state in vicinity of water, mainly on cliffs, dams, bridges, and culverts, but occasionally on buildings. Common transient at rivers, lakes, fields, etc., statewide, the southward migration apparently starting in late June. Occasionally seen in late February and early March along lower Colorado River, and one found at Tucson, *February 10, 1948* (ARP), presumably the "swallow" seen at same spot the preceding December (Erle D. and Virginia M. Morton).

Cliff Swallows are moderately chunky swallows with a light rump contrasting to otherwise blackish upperparts. Adults have a dark throat, likewise contrasting to white breast and belly; young often have the throat more or less extensively mixed with white, until the first prebasic molt (which usually occurs in South America). Cliff Swallows are highly gregarious. Their crowding is advertised by the rows of mud retorts plastered against rock or concrete under a protecting overhang. A study of nesting was made by Emlen (Auk 71, 1954: 16-35), and Willard (Condor 25, 1923: 138-139) discussed some colonies in southeastern Arizona.

Cliff Swallows usually arrive in central, eastern, and even northern Arizona in late March or early April; they stay to September (and into October in central and southern Arizona — even to October 19, 1948, west of Arivaca — Amadon and Phillips). Generalizations are difficult because of year-to-year variations, pronounced differences according to locality, and the attraction of migrants to local breeding colonies, there to mix confusingly with the inhabitants before continuing their trip. For example, a flock of 23 was flying in and out under a bridge near Granite Reef Dam, Salt River Valley, on *June 9, 1957*, presumably entering nests; but a male taken (testes large, but moderately fat) proved, even at this date, to be a migrant *H. p. pyrrhonota*, and when the site was revisited five weeks later all the birds were gone (Simpson and Werner)! Cliff Swallow colonies are strangely absent from apparently suitable mud-and-cliff habitats in the entire Tucson and San Fran-

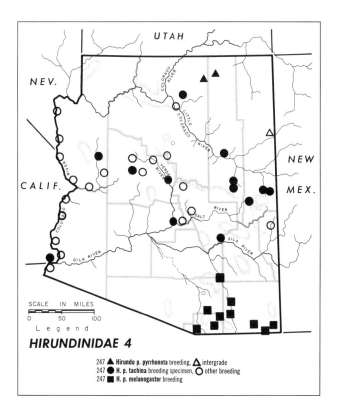

HIRUNDINIDAE 4

247 ▲ *Hirundo p. pyrrhonota* breeding, △ intergrade
247 ● *H. p. tachina* breeding specimen, ○ other breeding
247 ■ *H. p. melanogaster* breeding

cisco Mountains regions, and probably much of the Gila and upper Colorado valleys.

Hargrave made a special study of the Cliff Swallows of the San Pedro Valley, particularly the colony at St. David. His earliest arrival dates (for the local breeding birds, an easily distinguishable race) were *April 21 (1945)* and April 28 or 29 (1947). Yet there are many records farther west and north, even in Upper Sonoran Zone localities, much earlier! Some examples are: Yuma, March 8 (1961, six — Monson) and March 9 (in 3 years — Mearns, H. Brown, and Monson); near Parker, February 27 (1948) February 20 (1952) and Topock, March 3 (1949 — all Monson); mouth of Verde River, March 14 (1931 — Harry L. and Ruth Crockett); Camp Verde, March 27 (1886) and March 31 (1888 — Mearns); Williamson Valley, near Prescott, *April 2 (1929* — Jacot, ARIZ); and in the north: Snowflake, southern Navajo County, "late March" (Albert J. and Louise Levine); and Lupton, central Apache County, April 10 (Wetherill)! They were "abundant" by March 24 (1933 at Granite Reef Dam, Salt River — Yaeger), and at Yuma Herbert Brown estimated "probably 400 feet" (122 meters) of them perched side by side on wires, March 23, 1900.

Even after this very late arrival, the local birds at St. David were not long in sole possession of their bridge. Hargrave took migrants of northern races there on the same arrival date *(April 21, 1945)* and again on *July 8, 1944* (these both *pyrrhonota*, the last "from migrating flock") and on *July 17, 1945* (*hypopolia*); the last two were doubtless southbound, having begun their prebasic molt. Evidence elsewhere indicates that Cliff Swallows are migrating north in southern Arizona commonly to late May, if not to *June 9* (as above; *cf.* also Jeter, Condor 61, 1959: 434, whose assumption that the birds with pale foreheads would remain to nest at Fairbank seems dubious).

Records of probable southbound transients are Tucson, June 14, 1951, June 16, 1939, June 23, 1951 (about thirteen), and July 5, 1938, and farther south near Sahuarita and at Canoa Ranch, June 29 and 30, 1959; and in northern Arizona, Mormon Lake, June 21, 1938 (all Phillips). Two birds seen near Flagstaff, June 10, 1953, but not there in late June (Dickerman) may have been going either way!

We have mentioned that, contrary to newspaper fairytales about these swallows in California, Cliff Swallows vary from year to year in the date of their arrival in spring. The most conspicuous example was the extremely early arrival in 1948. That year they arrived in central and central-southern Arizona at dates when, normally, they are just reaching the Colorado River. Thus, four were seen at Agua Caliente, west of Gila Bend, February 28 (Erle D. and Virginia M. Morton); two near Tucson, March 4, and about three at Kinsley's Arivaca Junction the next day (Phillips, Margaret Nice, Wm. Foerster); and others regularly near Tucson later, including two specimens *March 15* (ARP). In more normal years, arrival at Tucson is sometime from March 20 to early April.

Though Cliff Swallows show marked racial variation, we plead guilty to not having tried to unravel their fall migrations. At that time, most of them are in ragged plumage and hard to identify. Juveniles are excessively variable, at least in color. Spring adults are usually easily identified, however. The four races we recognize all occur in Arizona. The scarcest here is *H. p. hypopolia* (Oberholser), the largest race (chord of wing about 112 millimeters or more), and with the forehead white. There is one fall record (*July 17*, as above), and a very few in spring from *April 15* at Tucson (ARP; and if verifiable, Kimball, Condor 23, 1921: 57) to *May 4* (*1894* in southeastern California at Laguna Station, San Diego County — Mearns, US), including one from northern Arizona (Holbrook, *April 26, 1948*—ARP).

H. p. pyrrhonota is similarly white on the forehead, but its wing measures 107-110 millimeters, rarely 105.5. It is apparently the breeding race in the extreme northeast (see map), and is known as a fall transient farther south near Safford (*August 18, 1937* — S. L. Green, ARP) and at St. David (*July 8*, as above, and *July 13, 1944* — LLH). At other seasons, its occurrence is better documented. The one winter record pertains to this race, and there are a dozen southern Arizona specimens from *April 15* (Tucson and Fort Apache — ARP) to *May 17* (*1940*, Tucson — Sutton, Auk 60, 1943: 347). Additionally there are specimens that seem nearest this, but approach the next race, as early as *April 5* (*1918*, Tucson — Kimball, LA and AMNH) and as late as *August 24* (*1931*, Lonesome Valley near Prescott — Jacot, ARIZ). Limits of its stay in northern Arizona are unknown, but it was present on the San Juan River in southeastern Utah by *April 28* (*1892* at Riverview — C. P. Rowley, AMNH).

H. p. tachina (Oberholser) is the most widespread nesting race in Arizona. It is small like the next race (wing 99.2 — 107, rarely 108.3 millimeters), and the forehead is variably intermediate in color — usually buffy, but varying from dull white to dull brown. The browner foreheads indicate intergradation toward the next race, although some ornithologists have suggested that the two are distinct species. Such birds range north to Fort Apache (ARP) and the Salt River Valley, and the single breeding bird from Wikieup on the Big Sandy River (ARP) is of this type. This is apparently the earliest race to migrate in spring; the many spring arrivals listed above probably all belong here. The five typical migrants from Tucson are from *March 15* (*1948*, two — ARP) to *April 19* (*1918* — Kimball, LA). Typical fall specimens are all from known or probable breeding areas. Presumably they stay into September, for young were still in the nests at Lupton, on the Rio Puerco, northeastern Arizona, August 18, 1946 (Wetherill).

The most distinct race is *melanogaster* (Swainson) of southeastern Arizona and Mexico, with its deep chestnut forehead. There is a cline of decreasing wing length toward the south, exactly the reverse of that claimed by the supporters of "*minima* van Rossem and Hachisuka." A male and two juveniles from southeastern Arizona (ARP) measure 105-107, while five specimens from Zacatecas and Nayarit (ARP) measure 101-104.8, and six from Guanajuato and Hidalgo (Instituto de Biología, Universidad Nacional de México) measure 99-104.1.

The late arrival at St. David seems to be a general rule for *melanogaster* in southeastern Arizona; but the date of the only Tucson record, given as April 10 (Condor 23, 1921: 57), we read on the label as *April 15, 1918* (Kimball, AMNH). Farther north, at Coolidge Dam, Hargrave saw two Cliff Swallows on March 26, and twenty to forty on April 3, 1956. All those well seen on the latter date appeared to have chocolate foreheads, though both the date and the locality are outside of the known range of *melanogaster* in Arizona; no specimens were taken, however, and those from nearby San Carlos, *May 1916* (US) are *tachina*. Likewise the status of *melanogaster* in the relatively well-known Santa Cruz Valley remains in doubt. It certainly has not nested in Nogales in recent years (since Dille moved there about 1936); yet Holzner took 3 typical and 1 atypical *melanogaster* there on *May 20, 1893* (US), and no other race — a record that hardly indicates migration. To this date, there is but a single other positive record for this valley or west (the Tucson one noted above)! The bird from Continental, *August 16, 1918* (A. B. Howell, US), reported as *melanogaster* by Mrs. Bailey (Pac. Coast Avifauna 15, 1923: 46), is too faded for positive identification as between this race and *tachina*.

Cliff Swallow colonies in Arizona, localized by the need of mud, are relatively permanent. We do not know how many broods are raised in a season, nor whether our birds leave their last young to starve, a strange custom of certain birds in California, discovered by the indefatigable Dr. Cooper (Proc. U.S. Nat. Mus. 2, 1880: 243-244, 246).

248. PURPLE MARTIN *Progne subis* (Linnaeus)

Breeds in Transition Zone of open parts of the entire Mogollon Plateau region, even to such areas as Williams, Mount Trumbull (Huey), the Natanes Plateau, the Sierra Ancha, and the Prescott region; also in the Chiricahua Mountains, but absent from the other mountains of southern Arizona, the Grand Canyon, and the northeast. Also breeds in saguaro associations of south-central Arizona west to the Ajo Mountains and north to near Picacho (Pinal County), Florence, Roosevelt Lake, and the lower San Pedro Valley; nesting away from saguaros in this region only at one point (Arivaca — van Rossem). Very rare outside of breeding ranges, but strays occasionally to the lower Colorado River on migration (Monson).

Martins are large swallows, about the size of starlings. They attract attention by their flocking habits and, when invisible overhead, their loud, gurgling musical calls. Most swallows are on the wing at dawn, but

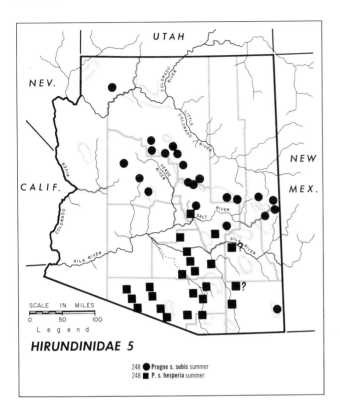

HIRUNDINIDAE 5

248 ● *Progne s. subis* summer
248 ■ *P. s. hesperia* summer

the martin beats them all, and may be heard flying over the desert in the wee hours of the morning, long before daylight. In one experiment in Michigan, two females taken from their nests and released in the evening and night 185 and 234 miles away were back home the next morning (Southern, Wilson Bull. 71, 1959: 256-259)!

Hardly purple, martins are steely black above and, in old males, below. Females and young are white below (pale gray on the throat, without contrast to the breast); the pale color extends up in a collar behind the dark cheeks. (One-year-old males usually have a sprinkling of black feathers beneath.) Martins nest in holes in trees, and in the eastern states they use multiple, apartment-house bird-boxes. An interesting study of their nesting was made by R. W. Allen and Nice (Amer. Midland Naturalist 47, 1952: 606-665).

Something like the larger swifts of the tropics, martins glide effortlessly through the sky, usually feeding higher in the air than the smaller swallows. In fact, their extreme scarcity as migrants away from breeding colonies suggests that they migrate at altitudes beyond human vision from the ground. Special studies by Hargrave, Cater, and Wm. X. Foerster at the famous Tucson roost (Cater, Condor 46, 1944: 15-18; Anderson and Anderson, Condor 48, 1946: 140-141) showed, however, that these migrants descend to join roosts en route to their winter home in South America. For a *September 26* series contained both races. Likewise, these and other martins join the roosts of *P. chalybea* in Chiapas on migration (Phillips).

The Purple Martin presents some truly extraordinary parallels to the Cliff Swallow. Both are highly gregarious birds whose local status may easily be confused by the attraction of migrants or strays to colonies of another race. Both are strangely absent from some apparently suitable regions in Arizona. Both are extreme long-distance migrants, normally wintering entirely in South America (unlike any of our other swallows; see Phillips, Revista Soc. Mex. Hist. Nat. 22, 1962: 309). Both vary geographically in size and in the color of the forehead. And in both, amazingly, the desert populations of central southern Arizona arrive in spring long after those of the cold plateaux of the north!

But there are also striking differences. Purple Martins need no mud nor overhanging cliff or building, but instead depend on not-too-closed "forests" with plenty of large holes for gregarious nesters. They range over far greater areas each day. They present, as adults, our only example of conspicuous sexual dimorphism in the family, and the only case of a dark-bellied swallow. And they differ conspicuously from all other swallows in their extreme scarcity away from breeding areas, and their nocturnal activity.

Over-all occurrence of martins in Arizona is usually from early April to early October; extreme records thus far in the north are *April 12* (*1937*, near McNary, White Mountains region — formerly ARP) to September 21 (1950 Flagstaff — Phillips). Adult males have left Arizona by about September 20, and the females and young dwindle sharply at the end of that month; the latest record is October 7 (1938), when one was seen passing over Tucson and others were still fairly common at Binghampton Pond in the evening (Phillips).

Like the Flicker, the martin is a beautiful example of ecologic races. Both species, together with the Sparrow Hawk and Violet-green Swallow, are widespread nesters in holes of trees over much of North America, and all of these as well as the Screech and Elf Owls are of decidedly smaller size in the cardón-saguaro deserts around the Gulf of California. Most of these species, however, show this reduction in size only in Mexico (and a hint in the lower Colorado Valley region), being of the usual large size (or absent, in the case of the swallow) in the saguaros of the Tucson region. Thus only the Flicker and Purple Martin races are clearly ecological, extending as far as known to the limits of their respective lowland distributions. Both are necessarily locally distributed, since wide areas of open desert, grasslands, and live oaks are uninhabitable. In the case of the martin, the two habitats (and distributions) impinge on each other most closely in the Roosevelt and Coolidge Lake areas where the saguaros approach the pines. Here further collecting is most urgently needed.

Particularly critical is the situation in the Sierra Ancha, where the large, widespread montane race *subis* was taken on Aztec Peak (A. H. Miller, MVZ). The species is apparently not regular there (Johnson, Simpson, Werner), at least in the higher parts. Presumably the large race has been seen commonly at Payson (Phillips). The small desert race (discussed below) has been taken at Globe (LLH) and on the highway thence north to Young at a point two miles south of the Salt River (Werner and Simpson, JSW). This is presumably the race, found around Roosevelt itself, that bred in saguaros near Coolidge Dam in 1934 (Yaeger). Identification by altitude, however, is a hazardous business not to be recommended, as the reader will soon discover.

The small race is currently known as *hesperia* Brewster, described from southern Baja California. The Arizona desert bird has been separated as *oberholseri* Brandt (*Arizona and its Bird Life,* 1951: 669-670); but we are not in a position to judge the validity of this separation. Certainly Arizona highland and lowland birds do not differ in color (though both are paler on the forehead, in females and young, than true *subis* of the eastern states). Our use of the name *hesperia* is thus tentative.

Though a great deal has been written about the supposed restriction of certain owls, woodpeckers, and flycatchers to the saguaros in Arizona, it is actually this race of martin (strangely overlooked by these authors) that really does depend on the saguaro forests for its nesting holes. The lone exception to this rule is that "they were apparently nesting in certain buildings" at Arivaca, more than 30 kilometers from any saguaros, in *June 1932* (van Rossem, Trans. San Diego Soc. Nat. Hist. 8, 1936: 140; LA). This does not mean that they remain all summer, or even all day, among the saguaros. The great roosts, to which all but the incubating females repair each evening, are in groves of cottonwoods in the valleys; and it is presumably these desert martins which, attracted by a forest fire, were once seen high over the pine forests of the supposedly martinless Rincon Mountains (Hendrickson, Condor 51, 1949: 230). Stranger yet is a first-year male, apparently out to see the world, that joined a local flock of *subis* and was among the three specimens taken at one shot near Flagstaff, *August 2, 1939* (ARP; wing chord 137.4 and 138, tail 66.7 millimeters)! Besides these wanderings, many individuals must move many miles every day to and from the roosts. The location of these roosts makes a tremendous difference in the apparent status of martins locally. The Tucson roost was established at least by 1909 (Visher), and persisted to 1947. During this period martins were among the most conspicuous birds flying over town; but since then they are hardly ever seen except on the desert among the saguaros, and even there, of course, in much smaller numbers. For their gregariousness at night is much less evident by day, and it is unusual for more than one to three pairs to nest in the same or neighboring saguaros (Cater); the largest concentration we know of is of about six pairs that chose one particular saguaro south-southwest of Tuscon in 1943 (Erle Morton). Another intermittent roost was at Benson: in its last year (1947) the first male appeared on May 5, and on the next two evenings about thirty and about a hundred (respectively) came in — 98% adult males (Hargrave). Specimens taken during the incubation period in 1945 had been almost all males, plus one female that was not nesting (LLH). This roost broke up in 1931 and again in June 1947 (*fide* Hargrave).

In Arizona, as elsewhere, *P. s. subis* does not congregate in these spectacular roosts during the breeding season. Because *hesperia* enters the United States only in central Arizona, and martins there are seldom seen before the last days of April or first week of May (starting to enter roosts April 26 to May 5), it is probable that all records before then, and all from other parts of the state pertain to *subis*. This would place migrating *subis* in southern Arizona, but none as yet in the north. A difficulty here is that we do not know how far *hesperia* may go to drink or feed; specimens of *hesperia* were taken from a "loose flock" of nine or ten martins feeding over the Gila River at Gila Bend, *June 12, 1918* (A. B. Howell, US) — many miles from any known nesting area. It is thus impossible to state the race of the "several" martins at Sacaton, Gila River, May 13, 1914 Gilman, ms.), the three at Oracle, May 20, 1934 (Phillips), or the several (?) once seen drinking south of the Winchester Mountains, northwestern Cochise County (Gallizioli). Birds seen not far from this last point, at Hooker Hot Springs, May 17, 1948 (Brandt, Hargrave, Phillips) were presumably *hesperia*.

At points more remote from the range of *hesperia*, migrant martins are seen, at best, about every other year, and then usually only single birds. (There are still no records for the upper Gila, upper San Pedro, upper Santa Cruz, or Big Sandy valleys). The records are widely spaced in spring, as to dates, from March 19 (1948 near New River, north of Phoenix — Pulich) and April 2 (1888 at Camp Verde — Mearns) to May 28 (1953 at Peoria, north of Phoenix — R. Roy Johnson). Fall records include "several" at Camp Verde, September 3, 1887 (Mearns); extreme dates are August 27 (1953, Bill Williams delta) to September 30 and October 8 (1955, Martinez Lake above Yuma — all Monson). The report of one "occasionally" over the highest parts of the Huachuca Mountains in late June, 1945 (Brandt, *Arizona and its Bird Life,* 1951: 525-526) would seem to require substantiation; and the statement that martins were "flying at all elevations," including the desert adjacent, in the Chiricahua Mountains in summer (Tanner and Hardy, Am. Mus. Novitates 1866, 1958: 6) is directly contrary to everyone else's experience there, where it is largely limited to Barfoot Park and remains within the pine forests (see also Kimball, Condor 23, 1921: 58).

In marked contrast to this apparent scarcity of migrant *subis* in Arizona, specimens taken at one shot into the edge of the Tucson roost, *September 26, 1943* (LLH), and found there just before and after, proved to be no less than 40-45% *subis*. Mr. Coy, the proprietor, noted the first arriving martins at the roost on April 26 (1944). Presumably, therefore, scattered records of one to four individuals near Tucson from March 22 (1942, four — Alma J. Foerster) to April 18 represent migrant *subis;* there are such records for 1939, 1942, 1950, and 1956, and in *1940* an adult male *subis* was actually taken, *May 17* (Sutton, Auk 60, 1943: 347), in company with a female *hesperia*.

CROWS AND JAYS

CORVIDAE

The corvids are another cosmopolitan family, which includes the largest perching birds and those thought by some to be the most intelligent of birds. (Parrots, titmice and chickadees have also been accused of this.) The general colors of the Arizona forms are black, gray, white, and blue. Most are permanently resident, but several are subject to periodic irruptions, with northern or montane species appearing far out on the deserts. They all build big stick nests on a tree or cliff, and are omnivorous. Besides the Arizona types, this family includes such familiar birds as the Rook and Jackdaw.

249. GRAY JAY; CANADA JAY; ROCKY MOUNTAIN JAY
Perisoreus canadensis (Linnaeus)

Common resident in fir and spruce of the White Mountains.

The Gray Jay is also called the Whiskey-jack. These are quiet birds, of very fluffy soft gray plumage, and are tame. Their movements are restricted to short distances, with no post-breeding wanderings or irruptions. The Gray Jay is slightly larger than a Robin, and slightly smaller than most jays. The area of Boreal Zone forest occupied in Arizona includes the Blue Range (Hargrave).

250. STELLER'S JAY; LONG-CRESTED JAY
Cyanocitta stelleri (Gmelin)

Common resident of pine, fir, and spruce forests throughout state. Winter range extends into oaks of southern Arizona. Large flights occur in some years into southern and western lowlands, when the birds may be found over the entire desert, even to near Yuma.

As its former name implies, this is the only normally occurring jay of Arizona which has a crest. It is blue-bodied and black-headed, with handsome white little marks or "diadems" on the forehead and eyelids. This is the common jay of the mountain forests. In southern Arizona it descends regularly in winter to the oak-clad foothills of some of the higher mountains, such as the Huachucas and Chiricahuas, where it may stay into June. Its common call is of much lower pitch than other jays. Picnic grounds in the mountains and South Rim of the Grand Canyon are frequented; there the birds become tame and learn to accept food from the hand. The young in juvenal plumage are entirely sooty. The Steller's Jay is a western representative of (but a different species from) the Blue Jay of the eastern United States, which has not been taken west of western New Mexico.

V. H. Housholder writes that he found "two killing a gopher snake 26 inches long on the highway south of Show Low. From marks on the snake it would appear it had been hit by a car previously. I opened the snake and found three young jay birds inside."

The principal winters of irruption onto the desert were 1893–94, 1910–11, 1934–35, 1943–44, 1950–51, and 1960–61. These generalities may be made: (1) the movement usually takes place in the same winter when Scrub Jays, Piñon Jays and Clark's Nutcrackers irrupt; (2) Steller's Jays arrive about a month later than do the Scrub Jays; (3) there is no massive exodus from pine forest of the desert mountains, and the birds are too numerous on the desert to be accounted for by the local population — therefore they must come from farther north, and this has been substantiated by some racial determinations; (4) two races sometimes occur in the same flock; (5) almost all specimens taken prove

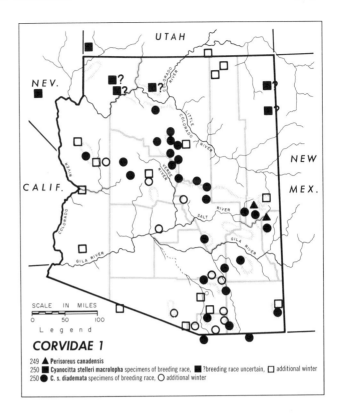

CORVIDAE 1
249 ▲ *Perisoreus canadensis*
250 ■ *Cyanocitta stelleri macrolopha* specimens of breeding race, ■? breeding race uncertain, ☐ additional winter
250 ● *C. s. diademata* specimens of breeding race, ○ additional winter

to be immatures (distinguished by thin transparent cranium early in the winter; dull, unbarred alula and primary coverts; and by worn, narrow, pointed, and light colored primaries and rectrices); (6) the flocks are actually on the move, progressing southward in late October and early November across the desert or through the city; (7) they arrive in suitable wooded terrain, usually of oaks, and remain there through the winter and into the spring, after which there is a northward movement in brushy desert or miscellaneous habitats; (8) for a year or two following a big flight year, birds again may spend the winter. Chronologically, the records away from normal areas are as follows:

1873: Upper Gila River, Greenlee County (Henshaw).
1884: Two males taken in "oak bushes near Agua Fria" between Prescott and Camp Verde, *April 25* (Mearns).
1886: Two seen at Camp Verde, March 4, after a storm (Mearns, ms.).
1887: Several seen in Sonoran zones near Camp Verde, October 11 and 13 (Mearns, ms.).
1889: Lower edge of pygmy forest, northeast side of San Francisco Mountains, September (Merriam).
1893–94: Common at cienega at head of Babocomari River, near Huachuca Mountains, *October 18-19, 1893;* common in Bear Valley near Warsaw Mills, *December 3–5, 1893;* taken *January 12, 1894* at Sonoyta, Sonora, where "dead birds, eaten by hawks, were also seen"; and feathers found as far west as Quitovaquita (all Mearns, US and ms.).
1894–95: "Common" in Dragoon Mountains and seen south of Willcox, Sulphur Springs Valley (Osgood).
1901: Two taken at Redington, San Pedro River, *March 11* (Lusk, AMNH).
1907: Whetstone Mountains, *September 29* and a bit later, "a few" seen (Austin Smith).

1910–11: Sacaton, maximum of seven seen, October 17 to March 1 (Gilman, Condor 13, 1911: 35, and ms.).

1925: Twenty in Deadman and Indian Flats, northeast side (?) of San Francisco Mountains, October 30 (Taylor, US).

1934–35: Common on Papago Indian Reservation, September to November (Monson, Condor 38, 1936: 176); four flying across Whitlock Valley in southeastern Graham County, October 4 (Taylor, US); Phoenix, singles in December and January and a "large flock" on January 13, 1935 (Mr. and Mrs. Crockett); Camp Creek northeast of Phoenix, *December 2* (an adult – Dille, ARP); near Tucson, November 5 (Anders Anderson) and March 23 (H. K. Gordon, ms.); "fully fifty," at Blythe, California, *February 23* (Reis, Condor 40, 1938: 44); Castle Dome, near Yuma, *April 18* (S. G. Harter, *fide* Huey, Condor 37, 1935: 257); also seen about this time in mountains northwest of Nogales (*fide* Phillips) and in the bottom of the Grand Canyon (Grater).

1943–44: Near Tucson, eight records of one to two birds, October 25 to March 30; also one collected (LLH); Phoenix, *October 27* (Housholder, US).

1944–45: Pajaritos Mountains, October 12 (van Rossem); near Tucson, March 24 (A. J. Foerster, S. Monson) and also seen once in Tucson (McKee, ms.); Patagonia Mountains, October (Loye Miller, where thought not to breed according to Marshall, Peter Westcott).

1948: Patagonia *January 17* and *March 5* (ARP); still in Lower Sonoran Zone hills east of Douglas between May 22 and 24 (Brandt, Hargrave, Phillips).

1950–51: Gila River at Geronimo, September 20 (Phillips); common at Salome (W. J. Sheffler); one on Christmas Census at Phoenix (Blue Point – Phillips); about 40-50 thought present at Wickenburg, and photographed (Wm. G. Bass); Baboquívari Canyon *December 3*, and San Xavier Reservation March 3 (one each place – Marshall, WF); Imperial Dam (Halloran *et al.*); specimens from Bill Williams Delta (Monson), Tucson, Ute or Black Mountains, and five from Big Sandy River above Wikieup (all ARP); flock of 6 or 7 near Benson, June 7 (Phillips).

1957: One near Yuma on November 21 (Monson).

1960–61: Thomas Canyon, Baboquívari Mountains, flock of more than 20, five specimens, *December 18* (Schaldach and Crossin, RSC); north end Coyote Mountains, one, September 25 (Levy); Kofa Mountains, one September 28 (Monson); Tucson, flock of 25 or more, passing south over exactly same route as flock of *coerulescens* a month before (Marshall); numerous other reports; two singles seen during following winter at northwest side of town (Peter Westcott).

Inland races of the Steller's Jay, differing in their white diadems from the blue-fronted birds of the coast, were revised prematurely by Phillips (Condor 52, 1950: 252-254), who at the time had not found the good series of fall *diademata* from Durango and Zacatecas (US). Should these be compared with his fall *browni*, described as an intermediate race of Arizona between the dark form to the north and the pale form to the south, they might prove identical, in which case *browni* would fall as a synonym of *diademata*. Until further studies are made, the two races of Arizona may be called *macrolopha* and *diademata*. There is no significant variation in size.

Cyanocitta stelleri macrolopha Baird. This darker bird is the breeding form of the Great Basin and Rocky Mountains. Some of its populations have been called *cottami* Oberholser and *percontatrix* van Rossem, which we consider synonyms. Probably *macrolopha* is the race of the birds in the Chuska and Lukachukai Mountains, but no fresh-plumaged specimens are available to prove this. However, it definitely was the predominant wintering bird of south-central Arizona in the flights of 1893 and 1950. The Warsaw Mills and Sonoyta specimens of the 1893–94 flight apply to this race (Mearns and Holzner, US). Of thirteen specimens taken west of the Santa Cruz Valley in this remarkable flight, only one is not *macrolopha*. From the 1950–51 flight, specimens mentioned above from Bill Williams Delta, Tucson, Ute or Black Mountains, and Big Sandy River are true *macrolopha*. (*Diademata* was collected from the same Big Sandy flock – ARP.) We also include under this race the Castle Dome bird of *1935*, which Huey (*loc. cit.*) recorded as *percontatrix*.

Cyanocitta stelleri diademata (Bonaparte). This lighter race nests in all the coniferous forests of the state except the extreme northeast, where probably it is replaced by *macrolopha*. In the northern part of the state, where there is so much forest, its movements are not as noticeable as on the southern mountains, where it moves down into oaks and cypresses (as early as September 2, Santa Catalinas – Marshall) of the Upper Sonoran Zone in winter, and may linger into summer (June 11, Huachucas – Marshall). Most of the specimens from the winter desert flights have been examined by Phillips and found to be of this race, as shown on the accompanying map.

251. SCRUB JAY; WOODHOUSE'S JAY   *"Aphelocoma" coerulescens* (Bosc)

Common resident of dense Upper Sonoran brush and woodlands, west and south to Hualapai and Baboquívari Mountains. In most winters a few descend to Lower Sonoran brush, orchards, and trees, mostly along streams of central Arizona, and to brush at Harquahala and Kofa Mountains; and in occasional years it is quite common in such lowland areas throughout the state. Migrates across the Mogollon Plateau at Flagstaff in some autumns.

The various races of this species were formerly known as the Florida, Texas, Woodhouse, California, etc. Jays. Adults are recognized by the contrasting pattern of the head and throat: the white throat becomes streaked with bluish toward the chest; the cheeks are blackish; and there is a thin white superciliary. But young in their first summer lack these markings, and may be distinguished from Arizona Jays by their smaller size, their calls, and floppy flight with the longer tail bobbing up and down. The calls are a very distinctive rising *shreeep* and a chatter. This is the only jay that regularly migrates into the desert, though this movement is much more pronounced in some years than in others. The really major flights are usually accompanied by smaller numbers of Steller's and are the only occasions when Scrub Jays are likely to be seen in more than family-sized flocks. For they are well spaced out in their normal home of chaparral. Even so, they attain a considerable density in the chaparral of the Hualapai and Pinal Mountains, with loose flocks up to 25 in winter (Westcott). Such abundance falls off immediately to the south of our state.

Big flights to the lowlands have occurred in the same years as important flights to the Steller's Jay: in 1934–35 (Dille's notes say it was unusually common in the Salt River Valley, according to Carlos Stannard, from *January* to April, *1935;* also taken near Tucson, *October 16* — Jacot, ARP); in 1950–1951 in flocks all over the state; and in 1960–1961. In the latter winter, they were found from October through December in the Cabeza Prieta Game Range area between Ajo and Yuma, and one lingered in the Kofa Mountains to June 28 (J. L. Phillips). In the residential part of Tucson, a flock of 17 or more worked southward on September 26, followed by others on subsequent days, to eight on October 3, 1960 (Marshall, Phillips). A month later, a similar flock of Steller's Jays came over the same route, deployed in the same manner over a block-wide front, and utilized the same trees! In these major flight winters, Scrub Jays are actually moving across the desert in September, October, and again from late April to early June, presumably en route, respectively, to and from winter quarters in canyons, river-bottoms, and similar densely brush-lined washes, where they remain stationary.

As in the Steller's Jay, the only important geographic variation shown among birds that enter Arizona is a trend of paler coloration from northeast to south and west. Two races can be distinguished in fall and winter specimens, but variable wear and fading obscure the differences by spring.

The darker northeastern form is unnamed. It nests in Colorado, and perhaps also in northeastern Arizona — whence we lack specimens from non-flight years. The flight of 1950–1951 was made up mostly of this race (Salt River Valley and Tucson — ARP). The westernmost records during such irruptions are Kanab, Utah (*December 30, 1946* — Greenhalgh, UT #8372) and Bill Williams Delta (*December 7, 1946* — ARP); the southernmost is Cananea, Sonora (*September 23, 1961* — Westcott, ARIZ).

Aphelocoma coerulescens woodhouseii (Baird) is the paler form which nests in suitable chaparral and junipers over most of Arizona. Springerville birds are somewhat darker. The only lowland specimen of this race taken during the big jay flight of 1950–1951 was on the Colorado River near Palo Verde, California (*January 4, 1951* — Pulich, ARP). The type specimen of *woodhouseii* may have come from Fort Webster; it is actually a pale bird. Therefore the name *woodhouseii* should supplant *nevadae* Pitelka and should be used for the southern and western population — not an eastern population as in the A.O.U. Check-list.

252. ARIZONA JAY;
MEXICAN JAY *"Aphelocoma" ultramarina* (Bonaparte)

Common resident of live oaks in southeastern and central Arizona, west to Baboquívari and Santa Catalina Mountains, and north sparingly to a number of points below the Mogollon Rim as far as the northwest corner of Gila County. There are a very few records outside the Upper Sonoran Zone, mainly from the adjacent pine forests.

The Arizona Jay is the abundant jay of the live

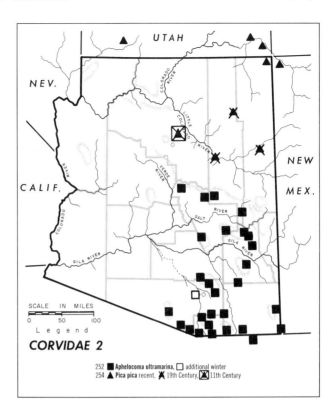

CORVIDAE 2

252 ■ *Aphelocoma ultramarina*, □ additional winter
254 ▲ *Pica pica* recent, ✖ 19th Century, ⊠ 11th Century

oak woodlands near the international border. It is gregarious even in nesting. It is slightly larger than other jays, with a plain head. Its best distinction is the call, a series of ringing *week*'s. Arizona Jays are tame and can be hand-fed in picnic grounds. Although sometimes ascending into pine forest, they practically never descend from the oaks to lower zones. Most reports to the contrary result from confusion with the Scrub Jay, which surprisingly few Arizonans seem to know! Even the redoubtable Major Bendire apparently confused the two (see Auk 7, 1890: 29 and *Life Histories,* 1895: 380).

References to habits are Gross (Condor 51, 1949: 241-249), Scott's "On the breeding habits of some Arizona birds" (Auk 3, 1886: 81-86), Wagner (Veröff. Uberseemus. Bremen, Reihe A, 2, 1955: 327-328), Jerram Brown (Condor 65, 1963: 126-153), and Westcott (Master's thesis, Univ. Ariz., 1962).

There are only two "good" records of birds at any distance from normal habitat: Tucson, a flock of four, September 30, 1956 (Phillips *et al.*) and one, April 18, 1934 (Phillips). Arizona specimens belong to the pale northwestern race *arizonae* (Ridgway).

253. SAN BLAS JAY;
BLACK AND BLUE JAY *"Cissilopha" san-blasiana* (Lafresnaye)

Accidental. Two taken from flock of about eight near Tucson, *December 19, 1937;* one in same locality, *December 19, 1938;* and one *January 15, 1939* (Jacot, Phillips, ARP, LLH).

This is a medium-sized slender jay with entirely black head and underparts. The upperparts except for the head and neck are bright blue. Young birds show a slender crest in front, rising from the forehead. It is native from San Blas, Nayarit, south along the coast to Guerrero.

The birds were seen by Mr. and Mrs. Foerster, Jacot, and Hudspeth, in addition to Phillips. Repeated efforts have produced no evidence that they were introduced or accidentally transplanted by man; and the fact that the first two were immature birds in fresh plumage makes this very unlikely. (See Phillips, Condor 52, 1950: 86). The specimens were identified by Friedmann and Duvall as the northern race, *nelsoni* Bangs and Penard.

Amadon (Amer. Mus. Novit. 1251, 1944) has shown that the supposed genera of American jays are highly unsatisfactory; this species gives additional evidence by the great difference between the pronounced crest, like a bushy *Lophortyx,* extending over the bill of the immatures and the very slight evidence of a medial crest on the crown of adults. This casts doubt on the crest as a generic trait, separating such supposed groups as *Cyanocitta* and *Aphelocoma.* These would be left with only color as a criterion, which is not enough; genera should be based on morphological differences. Eventually, anatomical studies such as that of Sutton and Gilbert on *Psilorhinus* (Condor 44, 1942: 160-165) and Hardy's on their biology (Univ. Kansas Sci. Bull. 42, 1961: 13-149) may prove the existence of valid genera in these jays.

254. BLACK-BILLED MAGPIE *Pica pica* (Linnaeus)

Only two specimens: near Winslow, *December 8, 1853* (Kennerly and Möllhausen), and nestling at Rio Puerco at Navajo Springs, Apache County, *June 27, 1873* (Newberry). (Also a fresh skin found at Tuba City, *1938.*) Prior to about 1885 said to be common in parts of northeastern Arizona (*fide* Mearns, Amer. Anthropologist 9, 1896: 399). Now found only in San Juan River drainage of extreme northeast, where one was seen at Tees-Nos-Pas, September 19, 1936 (Monson). No verified record for Mogollon Plateau or southward.

The Magpie, a well-known holarctic species, is, as the name implies, pied black and white. This with the elongated tail renders it unmistakable. It was here, at its southern limits, in the 1880's. The matter of its decline is discussed by Woodbury and Russell (BNav). The skin referred to above was found in the street at Tuba City by H. C. Lockett (MNA), where presumably it had been dropped by a Navajo who had secured it in that general region.

255. COMMON RAVEN; AMERICAN RAVEN *Corvus corax* Linnaeus

Fairly common resident almost throughout open parts of Arizona wherever nesting cliffs are available. Rare in Flagstaff area, lower Colorado Valley below Ehrenberg, and as a summer resident near Phoenix. Large congregations occur in northern and eastern Arizona at times.

This largest of perching birds prefers to nest on a cliff. It also builds in tall trees and on power poles. Notable concentrations of Common Ravens occur in summer on the Kaibab Plateau, and there seems to be a large roost west of Ash Fork. In southern Arizona, no large flocks are known, but this may be due to the difficulty of distinguishing this species from the White-necked Raven.

In the Salt River Valley, this raven was formerly a resident all year, and a pair raised a brood ¼ mile west of Gillespie Dam in 1920 (Housholder). Now it is found chiefly in winter and eastward. The Common Raven is the only resident corvid of the Organ Pipe Cactus National Monument.

256. WHITE-NECKED RAVEN *Corvus cryptoleucus* Couch

Common summer resident of yucca-mesquite-grassland association of southeastern Arizona north to Safford area, ranging uncommonly into lower, less grassy brush; limits of nesting range uncertain, but may extend to west edge of Papago Indian Reservation. Winters over most of breeding range, although large numbers appear to leave in conspicuous migrations in mid-November.

There is a myth that this species can be identified merely by seeing the white bases of the neck feathers. Not only is the reflection of the sun likely to mislead, but also the Common Raven has at least light gray bases to these feathers, to say nothing of the fact that these white bases are perfectly concealed in life and the birds generally face the wind so that they are not likely to be ruffled up. Actually the White-necked Raven is impossible to identify in life unless seen right beside the Common Raven, when its smaller size can be discerned. On one occasion Phillips saw 100 *Corvus,* obviously of two sizes, in a field south of Sahuarita; but whether the few smaller birds were Crows and the bulk White-necked Ravens, or whether these were respectively White-necked and Common Ravens, he still does not know. Even birds in zoos have proved, upon having their feathers forcibly ruffled up, to be other than as labelled. As a general rule it appears that the White-necked Raven occurs in larger flocks in southeastern Arizona than any other *Corvus,* and that it is the dominant species in yucca-grassland from the Huachucas east. Willcox is a good place to observe it. The call is somewhat like a prolonged loud caw of a crow, but the voice of the Common Raven varies considerably and would have to be studied intensively before a means of distinguishing its relative by sound could be worked out. A raven nest in a yucca or low mesquite would almost certainly be a White-necked. The white at the bases of the neck feathers is made conspicuous in aggressive display, but so far this has been seen only in captive birds (Johnston, Auk 75, 1958: 350-351). The white flare can be dramatic, as seen once in a strong wind, when it blossomed out like a spot-light (Marshall).

Due to the difficulty of distinguishing this species afield, the exact limits of breeding and winter range are still uncertain. The northernmost specimen in Arizona is from Winkelman, *November 16, 1920* (Kimball, AMNH); the westernmost — just feathers picked up — 1½ miles west of Ventana Ranch, Papago Reservation, *May 23, 1947* (Pulich, ARP). There is no evidence of any contraction of

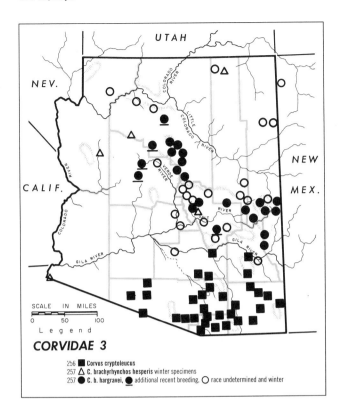

CORVIDAE 3

256 ■ *Corvus cryptoleucus*
257 △ *C. brachyrhynchos hesperis* winter specimens
257 ● *C. b. hargravei*, ⬤ additional recent breeding, ○ race undetermined and winter

range in historic time in Arizona, other than as governed by local conditions (see Vorhies, Condor 36, 1934: 118-119).

257. COMMON CROW
Corvus brachyrhynchos Brehm

Locally common resident of open parts of entire Mogollon Plateau, and down Salt River into Lower Sonoran Zone; also in the Chuska Mountains and the Defiance Plateau of the Navajo Indian Reservation, and perhaps elsewhere in the north. During winter may be seen in adjacent areas, rarely as far from breeding range as the Big Sandy or even lower Colorado valleys. No specimens as yet south of the Gila Valley, which is apparently its limit of occurrence in any numbers.

The Crow, so familiar in the eastern United States, is local and usually uncommon in Arizona. The White-necked Raven, nearly the same size, is a much more common species, and one which accounts for the many sightings of "crows" in southern Arizona. Both occur at Safford in winter, where about fifty, of one species or the other, were watched feeding on pecans by V. H. Housholder in April, 1956. Just before dark they took off southward in one large flock, toward a cottonwood tree roost that was said to be located ten miles away. They never stopped flapping, and they looked just like Crows going to roost. It is possible that an occasional Crow straggles farther south than this, but there are no specimens.

Contrasted to the ravens, the Crow has a small, slender bill and a less graduated tail, but these are not too satisfactory field characters. The Crow's flight is with much deeper and more frequent wingbeats, and its caw is less resonant and shorter. In the hand, Crows can immediately be distinguished from ravens by their normally rounded (rather than pointed) feathers of the throat. Since crows and ravens are not protected, it is best to attempt to secure a specimen, so that occurrence out of the normal range can be verified.

There is considerable racial variation in size within the Common Crow. The population of the northeastern United States is about the size of the White-necked Raven. In the West, the Crow is smaller, with more slender bill and tarsus. This is the subspecies *Corvus brachyrhynchos hesperis* Ridgway, which has been recorded several times in winter at the places shown on the map (CU, MNA, ARP, US, and Musgrove collection at Kingman).

Corvus brachyrhynchos hargravei Phillips is like *hesperis,* but with longer wing and tail, about as long as the typical, eastern race. It breeds principally along the Mogollon Plateau, winters there and in the valleys just under the Mogollon Rim, and has been taken in winter as far south as the Pinal Mountains (Scott, AMNH). Crows apparently bred at Glen Oaks (near Prescott) and at Seneca (north of Globe) in 1949 (Phillips), but nests were not found. Crows were widespread north of Prescott in the winter of 1953-54 (Wetherill); and a flock of 8 was seen there on July 15, 1961 (Peter Westcott).

Crows have been seen sparingly at all times of year in the northeast part of the state; and in winter north of the Mogollon Plateau in the east (Jacot, *et al.*), but in the absence of specimens, the race is unknown. Hoffman (Amer. Naturalist 10, 1876: 239) mentions "numbers of crows and ravens" seen "frequently" above the Colorado River in the "Grand Canyon," which may be the present Hualapai Indian Reservation, but Phillips has not traced the itinerary nor dates of this party.

258. PIÑON JAY
Gymnorhinus cyanocephalus Wied

Common resident of juniper-piñon regions in northern and central Arizona (south possibly to Prescott area) eastward to Natanes Plateau north of San Carlos, and west to at least the Hualapai Indian Reservation and the Mount Trumbull area. Sometimes invades adjacent forests. Wanders erratically in fall and spring (rarer in winter), in some years even to the Mexican border and the lower Colorado River.

The Piñon Jay looks like a small blue Crow, and like it, walks rather than hops. It travels in immense flocks that feed upon the ground, the birds in the rear constantly flying up and passing over the rest, to alight in front. The throat is vaguely streaked with whitish and females are duller and grayer than males, but essentially the bird appears all blue. The shape is that of a Crow or Starling, not long-tailed like the other jays. The querulous, thin *caaaar* is also somewhat suggestive of a Crow's voice. Piñon Jays nest in colonies, and while the females are on the nest in March and April, the roving flocks are composed wholly of males seeking food for their mates. Though a characteristic bird of the junipers and piñons of northeastern Arizona, in some years it may be seen commonly in the pines around

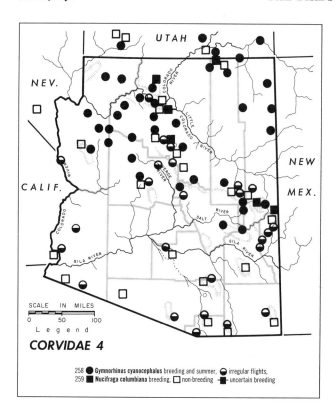

258 ● *Gymnorhinus cyanocephalus* breeding and summer, ◐ irregular flights,
259 ■ *Nucifraga columbiana* breeding, □ non-breeding ■ uncertain breeding

Flagstaff. Unlike the true jays, the flights of Piñon Jays into southern Arizona are usually of short duration.

In flight years, the birds are underway in late August; curiously, they are in southern localities in spring (including a pair, one banded, at Oracle, May 28 to June 19, 1961 — Westcott), just when nesting is underway in the breeding range. A heavy flight to the Huachucas in the winter of 1950–51 (C. Wallmo et al.) was hardly detected elsewhere. These birds were unusual in that they stayed all winter and were seen until the first part of May. A flight of 1955–56 was more widespread, reaching southwestern Arizona (Monson et al.), the Verde Valley, and Prescott. Fifteen miles west of Prescott, the first ten or twelve jays were seen on August 28, 1955. They increased to over one thousand in a flock by December (S. Gallizioli). A 1960 flight also reached Yuma County (Monson), as well as the Baboquívari Mountains in December (W. J. Schaldach, Jr., and R. S. Crossin) and the Santa Rita foothills in October and November (Westcott). In 1961, there were thousands from Oracle Junction to northern Sonora in the period August 20 (one, Tucson Mountains — Marshall) to December (Westcott).

259. CLARK'S NUTCRACKER *Nucifraga columbiana* (Wilson)

Common resident in boreal zones to timberline in the San Francisco and White Mountains, perhaps also on the Kaibab Plateau; has bred casually on South Rim of Grand Canyon (1943 — Amy M. Bryant). During fall and winter there are occasional large-scale invasions of other mountains, when it ranges also into lower country, even to the lower Colorado River (1955 — Monson); following these the birds may linger well into the summer in mountains, and even at Boulder City, Nevada (July and August, 1951 — Gullion et al., Condor 61, 1959: 288-289).

The Nutcracker has almost exactly the shape of the Piñon Jay but is pale gray, with black wings and tail prominently pied with white. The vocal long *kerr* has a very peculiar and long-remembered nasal quality which is impossible to describe. The best place to see Nutcrackers is the end of the San Francisco Peaks road, near timberline. This bird along with Lewis' Woodpecker was discovered by the Lewis and Clark Expedition, sent out by President Jefferson. Mewaldt (Condor 58, 1956: 3-23) in Montana has found that Nutcrackers nest in February in the snow-laden trees.

In Arizona, we have fully-grown juveniles by late April, indicating that here also the birds nest while the snow is deep. After this, they spread out over the adjacent parts of the pine plateaux, and possibly even migrate at times; the "fully fledged young" seen in the Santa Catalina Mountains, May 25, 1904 (Willard, Condor 18, 1916: 159-160) doubtless had not been raised there. Other records possibly in this category are the Huachuca Mountains "during *April, 1895*" (Kimball, *fide* Swarth) and June 21, 1902 (Swarth, Pacific Coast Avifauna no. 4, 1904: 33); Chiricahua Mountains, summer of 1934, seen "commonly" (Frank Hands, *fide* Hudspeth, ms.); west of Kayenta, *June 19–20, 1934* (LLH) and June 25, 1938 (Russell; see BNav).

The most notable southward invasions have been in the winters of 1935–36, 1950–51, 1955–56, and 1961–62. A chronicle of these "irruptions" follows:

1864–65: "Abundant at irregular intervals" near Prescott until March (Coues).

1865: Near Prescott, October 17 (Coues, AMNH).

1910: Sacaton, October 17 (Gilman, Condor 13, 1911: 35).

1919–20: South Rim of Grand Canyon, "noted" November 8 and/or 9 (Taylor, US) and "some . . . could be seen about . . . buildings at any time," December 18–20 (Swarth and Swarth, Condor 22, 1920: 79); Huachuca Mountains, series (W. W. Brown, AMNH).

1924–25: Santa Rita Mountains, seen (Townsend, Bird-Lore 27, 1925: 311).

1935–36: Kayenta *August 4*; Tsegi Canyon, August 6–8 and elsewhere in the Navajo Country until May 4 (Wetherill and Phillips, Condor 51, 1949: 101); "Extraordinarily great influx . . . during September" on both rims, beginning September 1, and still abundant on November 5 (*fide* McKee, Grand Canyon Nat. Hist. Bull. 4, 1936: 14-15); common in September at Navajo National Monument (Wetherill, *fide* Woodbury and Russell, BNav) and at Flagstaff (Kiessling, ms.); common in fall in Pinaleno Mountains and still present on April 26 and June 28 (Monson, Condor 39, 1937: 254); near Phoenix one each on October 5 and 20 (Mr. and Mrs. Crockett); "several" seen in Hualapai Mountains October 21 (Paul Russell, *fide* Stevenson, Condor 38, 1936: 245); one seen at Bates Well south of Ajo, October 22, and one between Cabeza Prieta and Tinajas Altas in southernmost Yuma County, October 23 (Taylor and Vorhies, Condor 38, 1936: 42); found by Jacot (ms.) "tolerably common" at Show Low and McNary in January (but not the previous August), rare to the east in the White Mountains, and absent in pines at Lakeside and off the forested Mogollon Plateau to the north and south; so the distribution was spotty. Huachuca Mountains, "small numbers" May 7 to 18 (Brandt, Auk 54,

1937: 64); Santa Rita Mountains, June 27 (Flock, ms.); south rim of Grand Canyon to at least June 5 (H. K. Gordon, ms.); north side of Navajo Mountain, Utah, to August 11 (Woodbury and Russell, BNav); also found about this time in bottoms of Grand and Zion Canyons, etc. This invasion thus covered the entire state.

1943: Santa Catalina Mountains, one seen November 11 (Cater, ms.).

1945: Huachuca Mountains, six seen on June 20 (Brandt).

1947-48: Santa Catalina Mountains, two, October 25 (Mr. and Mrs. Foerster, V. M. Morton *et al.*); Chiracahua Mountains, one, August 3 (Arthur Aronoff).

1950–51: The start of this flight was witnessed by Myron Sutton at the South Rim of the Grand Canyon, September 3, when a flock of 25 circled, landed, then flew on "obviously migrating" — more arrived later that month, but they tapered off in October (H. C. Bryant *et al.*); ca. 25 miles north of Williams, in pine and juniper, seven on October 28 (V. H. Housholder); seen on Coconino Plains December 16 (Phillips); Santa Catalina Mountains, September 3 (R. Jenks) to May 19 (Marhsall); Huachuca Mountains, common all winter (C. Wallmo), and in flocks up to 30 through *May* (Marshall, WF) and June, with last individual on July 16 atop Miller Peak (Mr. and Mrs. George Olin). Also Boulder City as above.

1955: Prescott, first seen on September 12, and from then through December (S. Gallizioli); first noted in the Santa Catalina Mountains, September 25 (W. E. Lanyon).

1958: Santa Catalina Mountains, several *October 18* (W. G. George, ARIZ).

1960: Santa Catalina Mountains, *November 5* (Wm. J. Schaldach, Jr., ARIZ).

1961–62: Santa Catalina Mountains, *September 29* (ARIZ) through May; Santa Rita Mountains from October 2 (in oak savannah) to February; White Mountains, abnormally high numbers in November (all observations of Peter Westcott). One at tank in Growler Valley, western Pima County, October 26 (Monson).

TITMICE, BUSHTITS AND VERDINS *PARIDAE*

The "tits" hardly differ structurally from jays and crows but are very much smaller. The largest American species are smaller than sparrows. Arizona forms are largely gray, black, and white. Most are sedentary and go in flocks. The Verdin, Plain Titmouse, and Mexican Chickadee form family-sized groups only.

260. BLACK-CAPPED CHICKADEE   *Parus atricapillus* Linnaeus

The only Arizona specimen is from Betatakin Ruin, Navajo National Monument, *October 23, 1936* (Wetherill, MNA), where up to ten at a time were seen throughout October 1935. Most other reports probably refer to *gambeli,* which could have its white eyebrow stripe obliterated by summer, due to abrasion. But in severe winters *atricapillus* may possibly reach the Mogollon Plateau: Milton Wetherill, who took the only Arizona specimen, believes he saw two at Walnut Canyon National Monument, near Flagstaff, February 3–16, 1937 (*fide* Hargrave).

261. MEXICAN CHICKADEE   *Parus sclateri* Kleinschmidt

Common resident in pine and spruce-fir forests of the Chiricahua Mountains. There is a winter sight record for adjacent, lower Swisshelm Mountains (January 20, 1924 — Mr. and Mrs. H. L. Crockett). A specimen labelled "Huachuca Mts." (W. W. Brown, AMNH) was doubtless actually taken in Chihuahua.

The Mexican Chickadee is a somber-looking chickadee with completely black cap and grayish sides. The voice is husky and buzzy, unlike our other Arizona chickadees. It barely enters the United States in the Chiricahua Mountains and the Animas Mountains of adjacent New Mexico. In August and other months outside the nesting period the Mexican Chickadee can be seen in the Chiricahuas, not only in coniferous forest, but also at lower altitudes, in pine-oak woods and in the groves of Arizona cypress.

Nesting is described in Herbert Brandt's book, *Arizona and its Bird Life.* Arizona specimens (AMNH) are the perfectly valid, pale northwestern race *eidos* (Peters).

262. MOUNTAIN CHICKADEE   *Parus gambeli* Ridgway

Common resident in pine and spruce-fir forests, locally into piñon-juniper in northeast (Woodbury and Russell, (BNav), throughout mountains except Hualapais and Mexican border ranges. In winter ranges uncommonly into Upper and rarely Lower Sonoran areas adjacent to its breeding range in northern Arizona, casually to Big Sandy River (February, 1880 — Stephens, ms.).

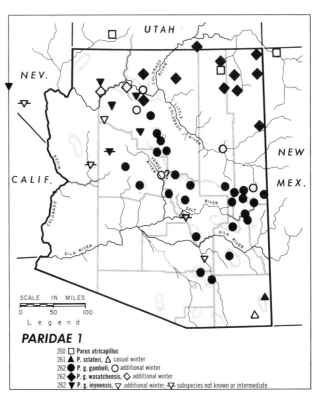

PARIDAE 1
260 □ Parus atricapillus
261 ▲ P. sclateri, △ casual winter
262 ● P. g. gambeli, ○ additional winter
262 ◆ P. g. wasatchensis, ◇ additional winter
262 ▼ P. g. inyoensis, ▽ additional winter, ▽ subspecies not known or intermediate

Except for the juncos, this is the most conspicuous bird in the winter forests of northern Arizona. The wandering flock forms a nucleus to which other small species are attracted. The white forehead line extends back above the eyes and is completely surrounded by black. Chickadees are arboreal acrobats that can hang upside-down as they explore the twigs and tree trunks for insect eggs. Worn midsummer birds may wear off most or all of the white forehead line and thus resemble other species. The voice is clear, sweet, and thin. The chickadees and titmice clean out or excavate their own nest holes in soft wood. Notice (on the accompanying map) that although the Mexican Chickadee and Mountain Chickadee divide between them the coniferous forests of the state, without overlapping, yet they leave unoccupied three important mountains supporting adequate fir forest: the Hualapai, Santa Rita, and Huachuca Mountains. Also barren of chickadees are all the mountains covered with ponderosa pine to the south of these, in Sonora. The southernmost population of *gambeli* in Arizona is on the Rincon Mountains, where it is common (Marshall, Condor 58, 1956: 93).

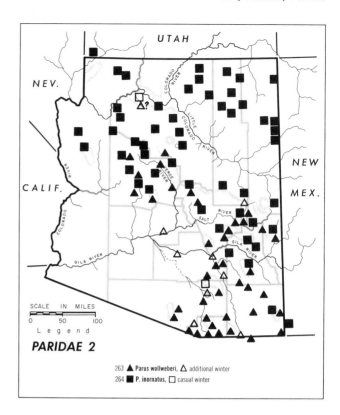

Geographic variation among the Mountain Chickadees of Arizona involves a darkening in color toward the southeast. The typical dark race, *gambeli,* occupies the central and southern portions of the state, and uncommonly descends from coniferous forest in winter as shown by open circles on the accompanying map. The Holbrook record is of four specimens, *November 15, 1947* (Phillips, ARP). In the southern half of Arizona, this race is not to be expected below coniferous forest in winter; for the only lowland records south of the Salt River prove not to be of the local race.

P. g. inyoensis (Grinnell), a very pale form, occupies the Williams area and westward (Hualapai Indian Reservation — ARP), intergrading in the Juniper Mountains with *gambeli* of the Prescott area. Two stragglers to low country are from mesquites at Mammoth (*December 22, 1961* — RSC) and junipers near Hyde Park (*November 20, 1947* — ARP). The latter is small and may represent *abbreviatus* (Grinnell).

A northern race, of the eastern part of the Great Basin, is *wasatchensis* Behle (Condor 52, 1950: 273-274); it is intermediate in color, and this characterizes the preponderance of birds (mostly Hargrave, GCN) from the rims of the Grand Canyon, and an immature male from near Supai, *September 23, 1950* (Phillips, GCN). A few additional specimens in the same series (Hargrave, GCN) qualify as *gambeli* (Grand View Fire Lookout), and as *inyoensis* (one of two taken five miles west of Grand Canyon Village, *December 20, 1956*). Thus we can regard the Grand Canyon district as being a meeting place of the three races, or more likely, as supporting a mixed population tending toward *wasatchensis*. The identification of northeastern Arizona birds is uncertain, but the only fresh-plumaged examples, from Kayenta (Hargrave, MNA), seem to be *wasatchensis*.

263. BRIDLED TITMOUSE *Parus wollweberi* (Bonaparte)

Common resident of Upper Sonoran woodlands of southeastern and central Arizona, north to Mogollon Rim (as at Oak Creek Canyon), and west to Juniper, Weaver, Pinal, and Baboquívari Mountains. Nests (irregularly?) in Lower Sonoran Zone cottonwood-willow-mesquite locally, at least in Camp Verde region (FW files; Mrs. Jackson) and at mouth of Aravaipa Creek (Monson). Regular winter visitant to willow-cottonwood association along larger streams in southern and central Arizona, west to Tempe and Sacaton.

A handsome little chickadee with a gray and black crest, this species is fond of live oaks, but it descends regularly into the cottonwoods of desert streams in winter. The calls and actions are very chickadee-like. Juveniles have the throat less black than adults, which they otherwise resemble. The bird is tame and confiding. Although widely distributed in Arizona, it is found elsewhere in the United States only along the adjacent border of New Mexico.

Most of the Bridled Titmouse population stays in the oaks all year, as is indicated by the status of "common resident" above. Those families or small groups that do move well below the breeding range are seen there from about late September to early April. Examples are September 20, 1940 at Tucson (Monson), September 28, 1938 near Colossal Cave, and April 2, 1939 in mesquite-grassland near Arivaca (both by Phillips). Wintering birds, seen in cottonwood forests of the Verde, Salt, Gila, San Pedro and Santa Cruz rivers are usually within the local large flock of small woodpeckers, White-breasted Nuthatches, Brown Creepers, Ruby-crowned Kinglets, Hutton's and other vireos, Orange-crowned Warblers, Yellow-rumped Warblers, and often Black-throated Gray Warblers. In the giant cottonwoods at "Indian Dam," on the Santa Cruz River ten miles south of Tucson, we have not yet found the nest or seen young, although Marshall's ornithology class has found a stationary pair or singing male in April or May of every year from 1957 to 1961 except in 1959, when, however, the bird was seen by W. G. George in late March and on July 1st. Prior to 1957 it was only a winter resident there, and indeed it was unknown in the Tucson Valley in Herbert Brown's day, the earliest record being in 1915–1916 (Howell, Condor 18, 1916: 214).

264. PLAIN TITMOUSE; *Parus inornatus* GRAY TITMOUSE Gambel

Fairly common resident in Upper Sonoran Zone of northwestern, northern, central, and locally southeastern Arizona; west to Mount Trumbull and the Cerbat, Hualapai, Bradshaw, Graham, and Chiricahua Mountains. Casual at foot of Santa Catalina Mountains, *November 28, 1928* (L. Miller, *et al.*, Condor 31, 1929: 77), and in bottom of Grand Canyon (Supai), *November 18, 1912* (Nelson, US).

This is a plain gray bird with a crest; it is slightly larger than the other parids of Arizona, and it has a wide variety of calls. We have never seen the Plain Titmouse in flocks in Arizona. A life history study was made by Dixon (Condor 5, 1949: 110-136). Primarily a bird of piñon-juniper woodlands, at least in the Great Basin region whence the form occupying Arizona is derived, it becomes extremely rare and local as the required habitat is pinched out by encinal in the southeast. After Marshall had written (Pac. Coast Avif. no. 32, 1957: 13) that piñon-juniper is totally lacking there, he noticed the typical Great Basin pygmy conifer forest of *Pinus cembroides* and *Juniperus monosperma* dominating the landscape for several square miles along the road between Portal and Paradise, in the northern Chiricahua foothills. Here he found some Plain Titmice in *1957* (WF) along with Scrub Jays and even Bridled Titmice (limited to the oaks along the wash passing through the pygmy conifer area). Monson has encountered the bird once at the south end of the range (GM), and Wm. G. George once at the western part (WGG); these are almost the only birds any of us have found in southern Arizona!

The purely gray, interior Plain Titmice show a mosaic of non-clinal geographic variations in size and darkness of the gray tones (Phillips, Jour. Ariz. Acad. Sci. 1, 1959: 28). Therefore no true races can be defined, and we apply the subspecies name *ridgwayi* Richmond to the whole group, including all the populations of Arizona.

265. BUSHTIT *Psaltriparus minimus* (Townsend)

Rather common resident of Upper Sonoran woodlands, and even chaparral and scrub oaks, throughout Arizona, including Harquahala Mountains. Wandering flocks are sometimes found in other zones, including fir forest, July to March. Has been taken in Mohave Mountains (*April 27, 1938,* Huey, SD) and Ute Mountains (ARP) just east of Colorado River, but for the entire Colorado Valley there are just the two sightings (Pulich) shown on the map. It is notable that the last three records were all from the fall and winter of 1947.

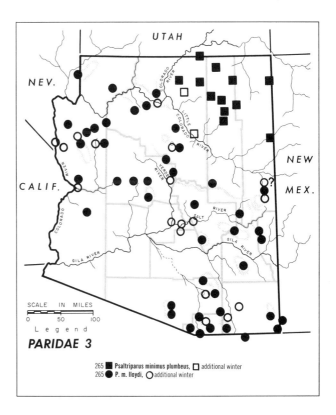

PARIDAE 3

265 ■ *Psaltriparus minimus plumbeus*, □ additional winter
265 ● *P. m. lloydi,* ○ additional winter

Much of the year, Bushtits roam in large flocks. They are tiny long-tailed gray birds without any noticeable pattern. Some Arizona juveniles, however, have more or less black on the checks. Males and nestlings have dark eyes; females have whitish eyes. A remarkable, long hanging nest is built with the entrance high on one side. In this a second clutch is often laid before the first young have left, and these subsequently help feed their younger brothers and sisters. Later in summer, some families wander out on the desert, or more regularly up into the firs.

The literature is full of misconceptions concerning these interesting little birds because of difficulties that well-meaning zoologists have experienced in determining the age and sex of individuals. Various authors have disagreed as to whether the iris color is the same or different in the two sexes; black-eared juveniles have been classed as adults; adult males have been thought to be in great excess and to assist the pair in feeding nestlings in the southern part of the range; and two full species have been assigned to the mountains of the Big Bend region of Texas, where they were thought to be ecologically separated by altitude. Several of these concepts have been discussed and corrected by Phillips (Anales del Instituto de Biología 29, 1959: 355-360). Marshall has studied the birds in the Big Bend and in the Sierra Madre Occidental of Chihuahua and Sonora, has examined all the specimens (including the types) in the American Museum of Natural History and United States National Museum, and has found a way of determining the age of specimens. Let us review all these findings in order to understand the species as a whole, and the geographically variable polymorphism it exhibits — for this has been the major stumbling block.

Polymorphism is the occurrence of two or more alternate "all-or-none" traits among birds of one population of a species. The traits are hereditary. Examples within a single family of Bushtits are cream-colored eyes in some, dark brown eyes in others (but none of intermediate hue); black sides of the head in some, non-black (plain light brown) in others; and pink sides in some, gray flanks in others. Because early ornithologists were looking for new, different species to describe to science, and because most species are distinguished from each other by outstanding differences in plumage color, striking polymorphism was misinterpreted, in the case of the Bushtit, Flicker, hawks, herons and others as indicating that two full species were existing in the same area. It remained to show that the different color phases freely interbreed, and that they occur in the same family as do the red and gray phases of the Screech-Owl, in order to prove that only one species is involved.

Along the Continental Divide from the New Mexico boundary southward into Chihuahua, Bushtit pairs consist of a black-eared male with a plain female, or of two plain-eared birds. Families of either may have some black-eared juvenile males, in varying frequencies. The black-eared trait increases toward the south: in the San Luís Mountains only a few of the juveniles have black ears. Farther south, at the upper reaches of the Rio Gavilán, many of the adult males, most of the juvenile males, and some juvenile females have black ears (WF, WJS). Black-and non-black-eared families occur within the same continuous woodland habitat at equal elevation and have identical voice, behavior and nests.

In the Chisos Mountains of Texas, the piñon pine and oak woods, to which Bushtits are there limited, are not of sufficient extent to allow for a subdivision of the population into two altitudinal or ecologic segments. The type of *Psaltriparus "melanotis" lloydi* Sennett (AMNH) from the nearby Davis Mountains of Texas, a black-eared juvenile male, has the collector's original age designation (of "young") crossed out and changed to "adult" in pencil; others of the topotypical series similarly have the sex and age changed in pencil. Evidently Sennett was loath to believe his collector's evidence that only the juvenile males of this population have black ears. Adults and fall birds past the prebasic molt all are plain! Marshall determined the age in all these Texas specimens (AMNH, US) by the shape and size of the outermost (tenth) primary. That of the juvenile is large, broad, and rounded; in the adult it is short, narrow, and dagger-like. All except one or two birds which had lost this primary in the molt were successfully aged. The first prebasic molt, like succeeding prebasic molts of the adults, is complete, and there remains only one small point to settle — whether a rare inidvidual, right in the middle of the molt, might gain the sharp outer primary *before* losing the black side of the head. The conclusion (which is not new), based on this aging method, is that only juvenile males in the Davis and Chisos Mountains, and only some of the adult males in the Chisos Mountains, have black ears; therefore *"melanotis"* is the same species as *minimus* and the name *lloydi* applies to a non-black adult population, thus superseding *"santaritae* Ridgway" and part of the race *plumbeus* (Baird).

The Bushtit enjoys an enormous geographic range from southwestern British Columbia to Guatemala. Throughout that area, in summary, the following traits apply: newly hatched young and all males have dark eyes; juvenile and adult females have a light-cream-colored iris; freshly molted males in fall show a beautiful suffusion of pink or vinaceous on the flanks, and these are plain in the females; juveniles in summer, before their first prebasic molt, show the large outer primary as mentioned above. As in most passerines, the juvenal feathers of summer can be distinguished from those of the worn adult at that season by their flimsy, lacy texture, and unworn tips; also, young in their first few months of life have the top of the skull very thin and transparent. (This condition should be noted on the collector's label.) With time the first-year bird's skull becomes mature and indistinguishable from that of the adult. For those who shun collecting, the nesting female can be sexed, not only by her cream iris, but by her brood patch — she can be caught, and the feathers of the belly can be blown aside to reveal the completely bare, wrinkled, thickened skin which gives the whole belly the appearance of a water-blister. Juveniles also have the belly bare, but they do not show this watery, blistered, richly vascular condition, which is an adaptation in the female for transferring heat to the eggs.

Also within the range of the Bushtit, the frequency of the black-eared phase increases steadily from north to south, and it involves gradually more of the sex and age groups as we proceed toward Guatemala. Right near the International Boundary, from extreme southeastern Arizona on east to the Big Bend of Texas, is the beginning of black-eared types — mostly in juvenile males. From there southward, progressively more of the juvenile males are black, then the adult males, and last of all, the juvenile females (some of which are partly black farther north). In Guatemala, all the young of both sexes, and all the adult males are black. If you should suppose that the black trait belongs only to adult males, you would naturally conclude that there was an overwhelming preponderance of males in the population, and that these "excess adult males," having nothing better to do, must help feed the young of other parents. Actually, of course, these are merely the young males and females of the previous brood.

This clinal variation in frequency of the polymorphic types is only part of the picture of geographic variation in the Bushtit. Pacific Coast races (none of which is black) have the top of the head brown; they differ in the tone of this brown and in the depth of gray or brownish-gray on the back. Interior United States races, including those of Arizona, are pure gray, with the head and back practically concolor. There is some color variation here, but as in the Plain Titmouse, it seems not to be clinal. However, birds of northeastern Arizona, of the race *plumbeus* (Baird), are larger, especially in tail length, than those over the rest of the state, which should therefore be called *lloydi* Sennett.

266. VERDIN *Auriparus flaviceps* (Sundevall)

Fairly common resident of whole Lower Sonoran Zone except bottom of Grand Canyon; also among mesquites in country otherwise mainly Upper Sonoran. Has evidently increased with the spread of mesquite.

The Verdin is another tiny gray bird, with a tail not as long as a Bushtit's. Juveniles are colored almost exactly like a Bushtit, but may be distinguished by the yellowish base of the mandible. Adults have the head yellow. The bill, unlike that of the Bushtit, is straight-edged and sharply pointed. The Verdin has a variety of staccato calls and pipings, loud out of proportion to its size. It builds large ovals of thorny twigs, strong and compact; at one end of the bottom is a vertical passage into the snug, covered nest, used both summer and winter. A Verdin that Phillips watched starting a nest in cholla had first carefully clipped each spine from the branches where it was to build. Verdins do not form flocks; nevertheless, the solitary Verdin is always the first bird to see a hawk and sound the warning, sometimes when the *Accipiter* is at an incredible distance.

The warning, recognized by other species, is a churring *gee-gee-gee*.

The Verdin is one of the few birds which feeds in creosote-bush (*Larrea tridentata*), another being the Black-tailed Gnatcatcher. Wallace G. Heath observed a Verdin at Tucson which would reach out and grasp a creosote-bush twig in its bill, bend it and transfer it to the foot, and would then peck along the twig, releasing it after a thorough probing, apparently for minute objects on the bark.

Essentially resident wherever found, the Verdin was thought by Mearns (ms.) to be found chiefly in summer at Camp Verde in the 1880's. However, it is still rare there, and none was found in May 1962 (Marshall, Ambrose).

Geographic variation in the Verdins of Arizona involves darkness versus lightness of general coloration. Darker birds, of the race *ornatus* (Lawrence) to the east, extend westward across the state apparently to the Black Mountains (ARP); the pale subspecies, *acaciarum* Grinnell, is narrowly confined to the Colorado River and the extreme southwest. In between are intermediate specimens, represented by open symbols on the accompanying map. (Some specimens of *acaciarum* from railroad towns of Topock, Needles, and Yuma are sooted – ARIZ, MVZ.)

NUTHATCHES *SITTIDAE*

These are small, short-tailed, dark-capped, gray and white birds which walk up and down trees with abandon. The tail is practically hidden beneath the wing tips. Their large feet and long hind claw permit them to hop along the underside of a branch. The powerful beak is used to excavate their own nest holes in dead wood. Though similar in appearance, the voice of each species is distinctive. They flash black-and-white patterns of the wings and tail in flight.

267. WHITE-BREASTED NUTHATCH; ROCKY MOUNTAIN NUTHATCH *Sitta carolinensis* Latham

Rather common resident throughout Transition and lower Canadian Zones, also locally among larger trees of the Upper Sonoran Zone (though no nesting record for Baboquívari Mountains); also locally in riparian Lower Sonoran Zone (Santa Cruz Valley). Fairly regular, August to early April, in nearby Upper Sonoran Zone and cottonwoods of Lower Sonoran Zone, in major flight years even to lower Colorado River (Bill Williams Delta, *November 10, 1950, tenuissima*–GM).

The White-breasted is our largest nuthatch, though still smaller than a sparrow. The black cap does not reach the eye, for the whole side of the head and the underparts are white. The call is slight and nasal. This is the most migratory of the three species. Phillips once took one from the stomach of a black-tailed rattlesnake in the Hualapai Mountains; presumably it had descended to the ground to feed, as nuthatches seldom do. They probably pair early in life, for by early November they are always in pairs on the breeding grounds, and

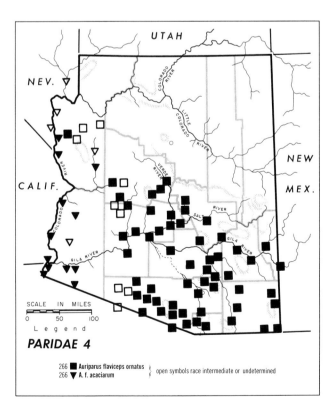

PARIDAE 4

266 ■ *Auriparus flaviceps ornatus*
266 ▼ *A. f. acaciarum*

open symbols race intermediate or undetermined

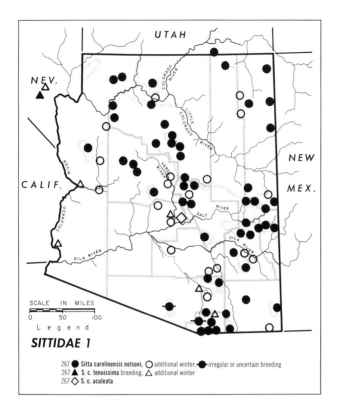

even winter residents in the lowlands are usually paired. In other words, the White-breasted Nuthatch is not a flocking bird, although it usually is a member of the mixed flocks of tree-foraging birds that one finds in winter. There was a big flight in the winter of 1950–1951 which reached the Big Sandy, Gila, and Salt Rivers. It included all three races (ARP).

Sitta carolinensis nelsoni Mearns is the resident Arizona race, characterized by long wing and stubby bill. The above account applies to this form, except as noted. Extreme lowland dates are from *July 31 (1952* at Tucson — ARP) and *August 27 (1937* near Safford — S. Leamann Green, ARP) to *March 22 (1894* near Tucson — L. H. Miller, MVZ). Sight records range from July 31 (Pulich) to April 9 (Brandt), and in the north July 21 (1950 at Snowflake — Louise Levine).

S. c. tenuissima Grinnell, which breeds in the mountains of California and the Great Basin, has a longer and more slender bill and is perhaps a trifle paler. It comes into western Arizona with fair regularity during flight winters.

S. c. aculeata Cassin, breeding on the northwest coast, is small and brown-tinged below. The only record away from its breeding areas is a bird in male plumage from the mouth of the Verde River, *September 30, 1950* (Yaeger, ARP — wing chord 83.3 mm., exposed culmen 17 mm.).

268. RED-BREASTED NUTHATCH *Sitta canadensis* Linnaeus

Resident in all, or nearly all, of boreal zones. During the fall sometimes found in Transition and Sonoran Zones, usually in large trees but casually in desert shrubs, ranging west to the Colorado River in flight years; most regular as a fall transient in mountains and mesas of the north, west to Hualapai Mountains. One casual summer record from Tuba City, *July 2, 1936* (Phillips, MNA).

This is a small nuthatch whose black cap and black eye-stripe are separated by a white superciliary; the underparts are rufous. Its call is an insistent nasal *aaanh,* given four or five times in leisurely succession. It is easily imitated to attract the bird. Like the White-breasted Nuthatch it does not form flocks of its own species. It has the interesting habit of smearing pitch around the entrance of its nest hole. The Red-breasted Nuthatch is quite common in the more extensive Boreal Zone forests of northern Arizona, where it prefers spruce and fir. The southern end of the species' breeding range (except for Guadalupe Island) is reached in all the southeastern Arizona border ranges which have firs — except the Rincon Mountains (Marshall). It is common in the Hualapai Mountains in fall, but we have not been there in summer to see if it nests.

In the mountains south of the Gila River, the Red-breasted Nuthatch rarely, if ever, makes a winter descent into the foothills. Exceptions are for the Catalina Mountain foothills (Scott) and Chiricahuas (Kimball), both taken in *October.* In this region the species is a rare transient and winter visitant to desert riparian cottonwoods, having been seen from September 19 (1947 at Tumacacori National Monument — Earl and Betty Jackson) to April 15 (1956 at Tucson — Tucson Bird Club).

The fall migration period is principally from late September to early November (but has begun as early as *September 5, 1950,* Bill Williams Delta — GM). Major flight years, when the birds spent the winter in lowlands, were 1947–1948, 1950–1951, and 1953–1954. Careful comparison of recent fresh fall skins from Maine and Arizona fails to reveal racial differences, in agreement with Oberholser's findings. *Clariterga* Burleigh must be a synonym.

269. PINE NUTHATCH; PYGMY NUTHATCH; BLACK-EARED NUTHATCH *Sitta pusilla* Latham

Abundant resident in ponderosa pines throughout Arizona, and to some extent in adjacent heavy piñon-juniper. Wanders to timberline, and once to bottom of Grand Canyon (sight records). One specimen from Lower Sonoran Zone: Yuma, *September 30, 1902* (now lost — Brown, formerly ARIZ).

Although Robert Norris (Univ. Calif. Publ. Zool. 56, 1958: 119-300) has adduced many interesting differences between two populations of these nuthatches in Georgia and coastal California, and has concluded that they represent two species, we feel that the great overall similarity, particularly as compared with the other two good species of *Sitta* in North America, indicates that they are races only. Their ecology, behavior, voices and appearance are very similar.

This noisy, gregarious little nuthatch is almost entirely restricted to ponderosa pines in Arizona. It is at once distinguished from the Red-breasted Nuthatch by its brown, not black, cap and lack of a superciliary stripe. Its call, a loud *peep,* is given in a variety of

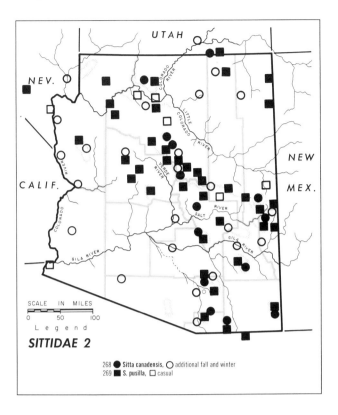

SITTIDAE 2

268 ● Sitta canadensis, ○ additional fall and winter
269 ■ S. pusilla, □ casual

pitches, expressions, and sequences, and usually in quantity. Flocks wander some miles, as across the Grand Canyon (McKee, Plateau 15, 1942: 13), but generally they remain in the pines. In winter this is the most abundant bird of the pine forest.

All the western and Mexican races of the Pine Nuthatch are weakly differentiated; but the somewhat paler birds, with a blacker auricular region and whiter nape spot, constitute the race *melanotis* van Rossem. It has a vast distribution in the interior of the continent, including Arizona.

CREEPERS *CERTHIIDAE*

270. BROWN CREEPER; ROCKY MOUNTAIN CREEPER; MEXICAN CREEPER *Certhia familiaris* Linnaeus

Rather common summer resident of boreal zones and, in the south, Transition Zone, throughout Arizona, wintering uncommonly through Transition Zone and Upper Sonoran woodlands. Rare winter visitant in large trees along rivers, particularly in Camp Verde and Tucson areas (where not uncommon in some winters), and west rarely to the lower Colorado River.

The Brown Creeper has been aptly compared to an animated bit of bark. He spirals up the trunk of one tree, flies to the bottom of another, then starts up again. He is streaked dark brown and white, with a slender decurved bill. The tail is much like a woodpecker's but softer, and is similarly used as a prop. The call is a very high-pitched *seee*, long drawn-out. The nest is tucked behind a slab of bark. This is a holarctic species, which extends in the New World south to Nicaragua.

In northern Arizona, the Brown Creeper apparently leaves boreal and high mountain forests after the nesting season. It appears in the pines around Flagstaff in late July and August and again in March to mid-April, though some winter there (Eleanor Pugh). It is found far below and beyond the breeding grounds from October to early April.

In contrast, creepers of the southern Arizona mountains do not seem to migrate at all. Those appearing in the lower riparian cottonwoods of the south come from distant points; they are found from late October to late March. It is a surprise occasionally to see one of these birds foraging in mesquites. Migrants may be seen in mountains where they do not breed from *October 1* (*1948* in the Hualapais — ARP) to *April 11* (*1935* in the Gila Mountains, Graham County — Yaeger, ARP).

Certhia familiaris montana Ridgway is the race to which the above account essentially applies. It is brown and white striped above, tawny on the rump, pure white beneath and on the superciliary, and has a long bill. It breeds in those mountain areas of Arizona north of the Rincon and Chiricahua Mountains. The intergrades in these two ranges are peculiar in that they encompass various motley combinations of characters of extreme *montana* and extreme *albescens* (Marshall, Condor 58, 1956: 93-94, WJS for the Rincons; and Max Minor Peet collection, MICH for the Chiricahuas). *Leucosticta* van Rossem of southern Nevada is considered a synonym (LA).

Starting with *1948–1949*, Phillips took *montana* in the cottonwood forest south of Tucson in five of nine winters; but this race never occurred in the numbers that *americana* did in the flight of 1952–1953 (specimens each year — ARP).

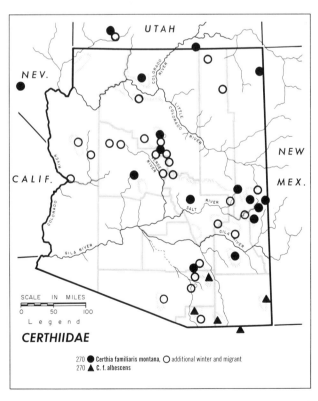

CERTHIIDAE

270 ● Certhia familiaris montana, ○ additional winter and migrant
270 ▲ C. f. albescens

Certhia familiaris albescens Berlepsch breeds and remains in the forests and uppermost heavy oak stands of the Santa Rita and Huachuca Mountains, and from there down into Mexico. Anyone who has seen specimens of this beautiful, soft-colored little bird will agree that it is a real taxonomic treat to contemplate such distinctness and such stability of characters over its wide range — contrasting abruptly with the equally stable *montana* immediately to the north. As noted earlier, the birds from the Rincon and Chiricahua Mountains offer rather bizarre amalgamations of the contrasting traits — an instance of intergradation notable for its narrow restriction geographically.

Albescens is a dark little fellow, the most conspicuous feature of which is its sooty-gray underparts, contrasting with the pure white throat. Above, it is blackish-brown, narrowly streaked with white; the rump is deep chestnut; the bill is short.

C. f. americana Bonaparte is like *montana* but the light marks on the head, and especially the superciliary, are buffy; its bill is short. Breeding in the northern part of North America (southward in the west to northern British Columbia) it is our commonest wintering race in the Arizona lowlands from Tucson eastward. This does not imply that any creepers are common in the lowlands — except that there was a great flight of *americana* in the winter of *1952–1953* (ARP), and probably also *1936–1937* (one specimen — Correia, ARP, from Sabino Canyon). Lesser flights occurred in the same cottonwood forest south of Tucson in *1948–1949* and *1953–1954* (ARP). This made three winters of occurrence there as opposed to five for *montana* during the same nine-year period. The relative numbers of these two races in the lowlands vary from year to year. Although no *americana* are known from north of the Gila River valley, this eastern race and *montana* are about equally frequent in the south and west, even in the Colorado Valley. Arizona records of *americana* are all from Lower Sonoran Zone valleys and they extend from *November 4* to *March 20* (ARP).

C. f. zelotes Osgood of California is blacker above than *montana,* but has a white belly and tawny rump instead of the chestnut rump of *albescens*. It has been taken at the mouth of the Verde River (*February 26, 1949* — Yaeger, ARP) and near Tucson (*January 16, 1948* — ARP). *Occidentalis* Ridgway of the northwest coast is unrecorded in Arizona, but the reddish intermediate, *caurina* Aldrich from the interior northwest, is known from Cazador Spring, north of San Carlos (*January 8, 1937* — Correia and Phillips, ARP).

OUZELS CINCLIDAE

271. WATER OUZEL; DIPPER
Cinclus mexicanus Swainson

Fairly common resident along the few clear, swift, permanent mountain streams of Arizona, mostly along the southern rim of the Mogollon Plateau from Oak Creek east to Black River and Eagle Creek in the White Mountains region, and in the bottom of the Grand Canyon. It is also found sparingly to rarely in the rougher parts of the Santa Catalina, Graham, and Chuska Mountains, and the Sierra Ancha. Rarely wanders at almost any time of year to other mountain streams, as in eastern Mohave County near Fort Rock, November 7, 1959 (Musgrove). Two records from

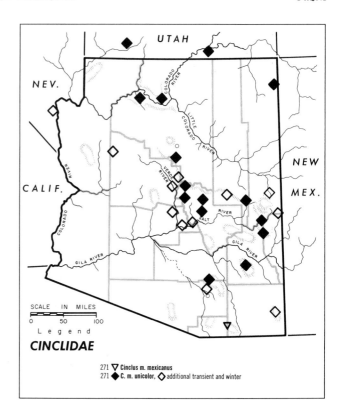

CINCLIDAE

271 ▽ *Cinclus m. mexicanus*
271 ◆ *C. m. unicolor,* ◇ additional transient and winter

Huachuca Mountains, August 1902 (Swarth) and *May 28, 1903* (Breninger, F); also more frequently from Chiricahua Mountains, *March 20, 1881* (Stevens, MCZ), late February 1917 (Austin Paul Smith), and *recently* (Cazier, ARP).

The Water Ouzel is a passerine bird turned into a waterfowl. It perches, occasionally bobbing up and down, on rocks in the roaring torrent, and dives underneath to feed. It is plain slaty gray all over, with a slightly browner head and a very short tail. Underwater the silvery nictitating membrane covers the eye and seems to shine. The bird is the size of a Spotted Sandpiper but chunkier. The nest is a globe of moss placed above the creek; a favorite site is a pothole on the underside of Tonto Natural Bridge. Bakus (Condor 61, 1959: 410-425) studied Ouzels and their movements in Montana.

The Ouzel has not nested in the Santa Catalina Mountains for many years, although Sabino Creek is still a fairly permanent stream with some good waterfalls for nest sites. Every few winters, though, one or two of the birds are seen along lower Sabino Creek.

C. m. unicolor Bonaparte is the gray race found throughout the presumed breeding range outlined above. Southward into Mexico, Ouzels become browner on the head and darker. Blake (Auk 59, 1942: 578-579) has identified the Breninger Huachuca Mountains specimen as a casual of the nominate race *mexicanus*. The closest breeding locality to Arizona would have been the Rio Gavilán, in the Sierra Madre Occidental west of Casa Grandes, Chihuahua (Alden H. Miller, 1948). Marshall did not find it there by 1951, when the water had become polluted with sawdust.

WRENS *TROGLODYTIDAE*

Wrens, as far as Arizona is concerned, are small (except for the Cactus Wren) brown birds with more or less black barring across the wings and tail. They are almost strictly a South and Middle American family, although the Winter Wren has reached Alaska and the Old World. Most species have musical songs contrasted with very harsh calls. Duetting is indulged in by the tropical species, which also build large nests. They all have slender, long bills, slightly decurved. It is possible that a correct classification of wrens can be reached by consideration of the various distinctive kinds of nests.

272. WINTER WREN *Troglodytes troglodytes* (Linnaeus)

Local and rare winter resident, generally in the densest brush of the more permanent streams, of Transition and adjacent zones in various parts of the state. A very few high Lower Sonoran Zone records. One record from the lower Colorado Valley, Parker, *November 1, 1953* (GM). The number of records, particularly on Oak Creek, where the most haphazard visits have already produced two specimens, justifies a status superior to "casual." Other specimens are from "Grand Canyon," and White, Huachuca, and Santa Catalina Mountains.

The Winter Wren is a dark, stubby-tailed bird whose upperparts are sooty brown and whose underparts are not much paler. The call is a very wooden *chimp, chimp* much like that of Pileolated Warbler, but given in two's instead of singly. It is very rare in Arizona and other wrens which have lost their tails should not be identified as the Winter Wren.

Dates in Arizona extend from *October 28* (Wetmore, US) to *April 2* (Swarth, CAS). Our birds belong to the very dark western mainland race, *pacificus* Baird. Since foxing is so pronounced in museum specimens of this species, it is possible that *salebrosus* Burleigh, of Idaho, is merely the unfoxed version of *pacificus*.

273. HOUSE WREN; PARKMAN'S WREN; APACHE WREN; BROWN-THROATED WREN *Troglodytes aëdon* Vieillot

Common summer resident in dense brush and fallen trees from Transition Zone to timberline in all mountains possessing forests. Winters commonly in better-vegetated areas of Lower Sonoran Zone of southwestern Arizona and the lower Colorado Valley east to the Phoenix region, Tucson, and Patagonia, casually farther east (even in upper Chiricahua Mountains — ARP). In migration common in southwestern Arizona, uncommon northeastward and to tops of border mountains; in northeast chiefly restricted to major rivers.

The House Wren is a plain wren, brown above and pale grayish below, occasionally with a light superciliary which becomes common from extreme southern Arizona southward. In this same region the chest becomes definitely tinged with tan. The tail is moderately long, about equal to the body without the head and neck. Although the relative tail length is hard to express here it is important in identification, and the observer can quickly learn to judge it. It is often held at a jaunty angle above the back. The House Wren nests in holes in trees, which it fills with sticks about to the level of the opening before placing its grass and feather cup. When feeding, this wren loves to clamber about and under roots and fallen logs. Important life history studies in the east have been made by Baldwin and Kendeigh (summarized and cited in Bent's *Life Histories*). A local Arizona-Sonora study dealing with the species question is by Marshall (Condor 58, 1956: 93-96).

Normal occurrence in southern Arizona lowlands is from the last of August to early May. Extreme dates are August 16 (1954 in Bill Williams Delta — Monson) to May 24 (1934 at Tucson — Phillips). At somewhat higher altitudes, where only migrants occur, there are late records to October 18 (1951 at Valentine, eastern Mohave County) and to May 25 (1957 at the Southwest Research Station, Chiricahua Mountains — both Phillips). Residence on the summit of the Santa Catalina Mountains (Marshall, WJS) and other southern breeding grounds is about as long as at Flagstaff, where Mrs. Eleanor Pugh has noted it from April 20 to October 8. In many recent summers it has not bred around the Museum of Northern Arizona there. The latest acceptable northern Arizona record is *October 18* (*1936* in the White Mountains — ARP).

Geographic variation in color of the House Wren is so spectacular that the United States and Mexican forms were long considered to be different species. Intergrades had

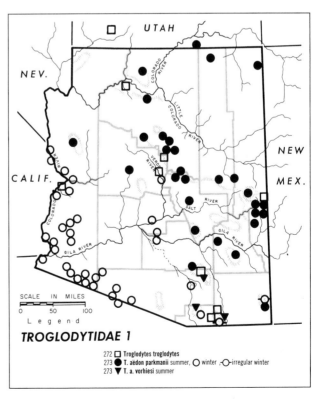

TROGLODYTIDAE 1

Legend
272 □ Troglodytes troglodytes
273 ● T. aëdon parkmanii summer, ○ winter, -○- irregular winter
273 ▼ T. a. vorhiesi summer

been collected from the border ranges but they had remained unnoticed in collections (US, MVZ) until Brandt discovered such birds in the Huachuca Mountains in the summer of *1945* (Univ. Cincinnati collection). The trend from north to south is from a gray-brown back to reddish-brown back; and from entirely gray underparts to tan (first on the throat and chest, farther south all over the underparts), with buff superciliary and blackish bars on the flanks. The remarkable thing about this "cline" is that the transition from one extreme to the other takes place in so narrow a zone of intergradation — confined to some scattered, isolated peaks between the Santa Catalina Mountains and the north end of the Sierra Madre Occidental of northeastern Sonora. Marshall (*op. cit.*) connected the two forms, previously designated as full species, and this was adopted by Paynter (in Peters, *Check-list of Birds of the World* 9, 1960: 422-423).

Troglodytes aëdon parkmanii Audubon is the gray-breasted western United States race, which breeds in Arizona as far south as the White and Santa Catalina Mountains. All the foregoing account of status, migration, and wintering applies to this race.

Troglodytes aëdon vorhiesi Brandt is the name applied to the intergrades from the scattered peaks along the border. Just to prove that taxonomy is not a dead pursuit, and that there are lively issues to be argued, we can say that we disagree among ourselves as to the advisability of recognizing this race. The birds are a motley crew, ranging from extreme *parkmanii* to passable *cahooni* in color, and divided about fifty-fifty in the Santa Rita and Huachuca Mountains (Marshall, WJS and ARIZ). With the type series and further specimens to add to the figure in Marshall's paper (*op. cit.*: 95), it is barely possible to squeeze out a 70% recognition of *vorhiesi* from the Santa Ritas, Huachucas, and adjacent Sonora mountains — if the tiniest trace of buff on the throat is counted. Thirty per cent of the birds are indistinguishable from *parkmanii*. For purely utilitarian reasons Phillips prefers to recognize the race, because the few extreme examples of it can satisfactorily be recognized on wintering grounds far to the south: Sierra Madre, 12 miles west of Milpillas, west of Fresnillo, Zacatecas, *September 26, 1955* (ARP); Omilteme, Guerrero, *October 19, 1947* (MICH #117452).

All such birds can be distinguished from *cahooni* by their white bellies. On the other hand, Marshall prefers to consider these variably intermediate populations to be merely intergrades (followed also by Paynter, *loc. cit.*). In this case *Troglodytes aëdon cahooni* Brewster can be considered to reach Arizona in somewhat diluted form, in the Santa Rita and Huachuca Mountains. Birds from the Rincon (WJS), Chiricahua (ARP), and Pinaleno Mountains (US) and adjacent Mount Turnbull have increasing preponderance of the *parkmanii* phenotype, in that order. At the time he wrote the above-mentioned paper, Marshall was in such despair over the whole concept of subspecies — that they should be thought of as little species with definite boundaries done in zipatone on a map — (see below, under Lincoln's Sparrow) that he refused to use subspecies names in the paper. Fortunately, Alden H. Miller, the editor, put names in the legend of the figure!

Two further remarks on *cahooni* are (1) the extraordinary conformity of the type series from the remote Sierra de Oposura in north-central Sonora with the main body of the population far to the southeast in the Sierra Madre; and (2) the statement in the A.O.U. Check-list (under *vorhiesi*) that the bird is a resident in southern Arizona. All the United States populations of the House Wren are entirely migratory, as are at least some in northern Mexico

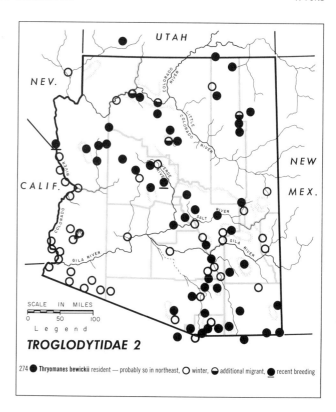

TROGLODYTIDAE 2

274 ● *Thryomanes bewickii* resident — probably so in northeast, ○ winter, ◐ additional migrant, ● recent breeding

(Phillips, Anales del Inst. Biol., Univ. Méx. 30, 1960: 363-364).

274. BEWICK'S WREN; BAIRD'S WREN *Thryomanes bewickii* (Audubon)

Common resident in Upper Sonoran brush and woodland south of the Salt and west of the Verde Valley; summer resident locally and generally uncommonly in piñon-juniper zone over the rest of the state, and in mesquite-willow-cottonwood association along parts of Lower Sonoran Zone rivers. Migrates across the Mogollon Plateau. Winters throughout that part of its breeding range that lies south and west of the Mogollon Plateau; also among dense weeds and brush of the Lower Sonoran Zone, as well as on the Hualapai Indian Reservation, the west side of the Navajo Indian Reservation, and even at Springerville. Recently has become rare in extreme western Arizona, where up to 1950 it was a common winter resident; more common again in fall of 1960 (all by Monson).

This is a prettily patterned little wren, solidly brown above and clear grayish-white below, with a well-defined white superciliary. The long tail appears gray from above but when spread in flight it shows much black. It is often waved expressively as the bird clambers around tree-trunks. The song is of fine clear whistles and musical trills.

Bewick's Wren is another bird whose status at Camp Verde has changed. In the 1880's Mearns found it a winter resident only, though common. But by 1916 it had become a breeding species (FW files) and Mrs. Jackson found it all year at nearby Montezuma Castle National Monument by the 1930's. Another change is its absence during the 1950's as a winter resident in the lower Colorado Valley, mentioned earlier.

Bewick's Wrens are recorded at southern Arizona points where they do not (or did not then) breed from September 15 (1946 near Parker — Monson) and September 25 (1910 at Sacaton — Gilman) to *March 28* (*1888* at Camp Verde–Mearns, AMNH). The interesting migration across the Mogollon Plateau refers to birds near Flagstaff on *March 13* and *April 17, 1936* (Phillips, MNA); fall migrants in the north are from August 28 (1937 at Keams Canyon — Monson) to *September 29* (*1922* on the north slope of the San Francisco Mountains — Swarth, MVZ). Arizona birds belong to the large, pale grayish race *eremophilus* Oberholser. The smaller darker-brown *T. b. charienturus* Oberholser, of coastal southern California and Baja California (with synonymous *carbonarius* and *correctus* Grinnell, apparently foxed) has been taken on the opposite side of the lower Colorado River, and should be sought in southwestern Arizona.

275. CACTUS WREN
Campylorhynchus brunneicapillus
(Lafresnaye)

Common resident almost throughout the Lower Sonoran Zone, especially in cholla cactus, but also in open mesquite and shade trees in towns; lacking only at bottom of the Grand Canyon and in the Verde Valley, where no confirmed report.

The Cactus Wren, our state bird, is the largest United States wren — almost the size of a Starling. It is much variegated above and adults are heavily black-spotted below. It builds a conspicuous, neat, retort-shaped, grass nest, whose size is sure to attract attention. (Some English Sparrow nests are built in trees and might be mistaken. They are of a sloppy construction and have just a hole in the side, for an entrance, instead of the tubular top-floor vestibule of the Cactus Wren). The song is a prolonged *churring* destitute of music. The tail, though expressive, and fanned out in certain displays, is not cocked over the back as in other wrens. Much of the feeding is done on the ground, and the Cactus Wren is fond of dust-bathing. We have never seen one drink. The Cactus Wren has learned to pry insects from car radiators.

Anders H. and Anne Anderson have made an exhaustive study of the species, based on over 20 years' work with color-banded birds at Tucson (Condor, 1957–1963). Arizona specimens belong to the large pale race *couesi* Sharpe, which is further characterized by having the ventral black spots concentrated into a definite chest patch.

276. LONG-BILLED MARSH WREN; WESTERN MARSH WREN
Cistothorus palustris
(Wilson)

Local resident along the lower Colorado River and in marshes of the lower Salt River. Common winter resident and migrant at reed-grown ponds and canals, except possibly in the northeast; less common in winter at frozen marshes on and near Mogollon Plateau.

Resembling the House Wren in size and form, the Long-billed Marsh Wren differs in its black-and-white striped back and sharp white superciliary. It is sometimes hard-put to find a marsh in Arizona and then

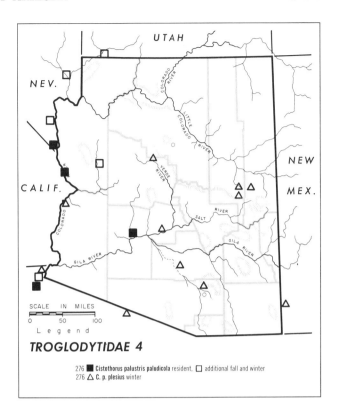

TROGLODYTIDAE 4

276 ■ *Cistothorus palustris paludicola* resident, ☐ additional fall and winter
276 △ *C. p. plesius* winter

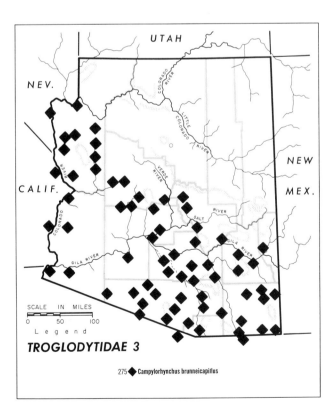

TROGLODYTIDAE 3

275 ◆ *Campylorhynchus brunneicapillus*

occurs in dense weeds along ditches and ponds. Distinguishing it on the basis of habitat alone is risky; at Mormon Lake Phillips once rescued a House Wren which had attempted to pass between two bulrushes and had become stuck across the neck. The harsh calls and scratchy song may be heard all winter. Highly curious, it comes up to inspect people who stand quietly in the marsh.

There is no question that there are geographic differences in color among populations of the Long-billed Marsh Wren in the western United States. But they seem to be of a helter-skelter sort, complicated by foxing of old museum skins. The broad outlines can be resolved, however, as comprising only two races (partly in consideration of Swarth's own series in CAS):

Cistothorus palustris paludicola Baird is the darker race, resident along the coast and inland to the lower Colorado Valley and middle Gila River, Arizona. It extends up the Colorado only to the southern tip of Nevada; claims for its residence farther north into Utah seem to be the result of comparing fresh Utah birds with old, foxed museum skins. Some winter records evidently represent a post-breeding dispersal: east of Searchlight, Nevada, *January 13, 1934* (Linsdale, Pac. Coast Avif. 23, 1936: 93); Overton, Nevada, *October 30, 1917* (Austin Paul Smith, AMNH); a dark male 3 miles south of St. George, Utah on *December 18, 1939* (Behle, UT); and near Wikieup on *August 25, 1948* and *October 19, 1951* (ARP; the first, a juvenile in heavy first prebasic molt).

C. p. plesius Oberholser is the paler, common migrant and winter bird throughout Arizona. Southern Arizona records are from *September 18* (*1938* Tucson — ARP) to *May 7* (*1888* at Ft. Yuma — Stephens, CAS). There are sight records for the species from August 22 (1954, Sullivan's Lake, north of Prescott — Phillips).

Northern Arizona specimens are from *August 28* (*1934* at Springerville — Stevenson, ARP) to *January* only, but marsh wrens have been seen to April 25 (1953 at Springerville — Dickerman).

277. CAÑON WREN *Catherpes mexicanus* (Swainson)

Rather common resident about cliffs, hills, and adjacent buildings and even high dirt river-banks in Sonoran zones throughout Arizona, but quite uncommon and variable over the years in southwest and along lower Colorado River. Ranges exceptionally into low Transition Zone during late summer, but also breeds there locally, south of Williams (Wetmore).

This is a deep rusty-brown wren with a gleaming white throat. It has a remarkably long bill and flattened head for poking into crevices of rocks. Its clear, descending whistles, echoing from canyon walls, supply some of Arizona's finest bird music. It is especially common in the barren parts of the Navajo Country, where there is nothing but cliffs, mesas and rocks (with a few surviving sage bushes); of course the Grand Canyon was made to order for it. However, individual Cañon Wrens range widely during the day and they apparently hold large territories, which causes them to be widely spaced and never very numerous. Though individuals roam far out along steep-banked desert washes (especially after heavy mountain snows) and up mountain gullies in the non-breeding season, there seems to be no true migration. This is a characteristic bird of house-tops and church spires above the narrow streets of Mexican cities, but in Arizona it rarely enters towns.

The small, pale, northern race *conspersus* Ridgway occupies Arizona.

278. ROCK WREN *Salpinctes obsoletus* (Say)

Common summer resident in open rocky situations from timberline and above down to Upper Sonoran Zone, though locally scarce recently (*e.g.*, Hualapai Mountains, San Francisco Peaks); less common in Lower Sonoran Zone (but in numbers some summers to the low desert mountain ranges of Yuma County and in the Colorado Valley as far south as Parker Dam). Winters commonly in open, broken areas of the Sonoran zones of southern Arizona, but mostly absent then in northeastern Arizona east of Tuba City area and north of Holbrook.

The Rock Wren shares the tastes of the Cañon Wren, but has less predilection for sheer cliffs. It is a paler bird, the palest of our wrens, being gray-brown above and whitish below. It is our only bird that shows buff corners on the spread tail in flight. Both it and the Cañon Wren bob as they perch on a rock. The clear song, broken into paired phrases, suggests a miniature Mockingbird. The nest is under a rock, and the bird paves a path to it with small flat stones as if to announce its presence. This is a highly migratory wren. Single birds will spend the entire winter in and around deserted adobe dwellings on the San Xavier Indian Reservation far from nesting habitat.

The Rock Wren is of rather general occurrence in the Lower Sonoran Zone from mid-September to early April. Extreme dates for localities where it probably does not breed are from September 3 (1939 near Tucson) to May 14 (1939 south of Wickenburg, on the Hassayampa River — both by Phillips). It is recorded in the Transition Zone of the Mogollon Plateau from *March 18* (*1936* near Flagstaff — Phillips, MNA) to *October 25* (*1936* in the White Mountains above Greer — ARP).

There is virtually no geographic variation in this species on the mainland north of Guatemala; at least all the United States and northern Mexico birds belong to the typical race.

MOCKINGBIRDS AND THRASHERS *MIMIDAE*

This is a small American family which reaches its center of abundance in Arizona and Sonora. Renowned as songsters, mimids are of Robin to jay size and long-tailed, with slender bills which are elongated and decurved in some species. They are all plainly colored in grays, browns, or whites, and they build bulky nests in bushes. Most species have yellow or whitish eyes. In the genus *Toxostoma* the eggs and calls are more distinctive than is the appearance of the bird.

The ecology and peculiar adaptations of Arizona thrashers have been discussed by Miller (Ornith. als Biol. Wissenschaft 1949: 84-88), Engels (Univ. Calif. Publ. Zool. 42, 1940: 341-400), and Ambrose (Master's thesis, Univ. of Ariz., 1963).

279. MOCKINGBIRD *Mimus polyglottos* (Linnaeus)

Rather common summer resident in the less densely wooded parts of the Sonoran zones that afford thornbrush; less common except in wettest years west of the Baboquívari Mountains, the mouth of the Salt River, and the Kingman region. Common resident of Maricopa County cities and Yuma. Winters commonly in most of southwestern Arizona west of Altar Valley and south from Roosevelt Lake, Wickenburg, and Davis Dam, and in upper Gila Valley; less commonly in and near Tucson, and exceptionally in southeastern Arizona and near Supai (Mearns). Migrant and wanderer on Mogollon Plateau. Only a straggler in open pine-grass (Transition Zone) in Arizona, though common there locally in northern Sonora. There are peculiar seasonal fluctuations in the numbers of Mockingbirds in southern Arizona not yet fully understood.

This famous songster of much of North America (except the northern parts) has gray upperparts and white underparts without spotting or patterning when perched. Flying, it shows a prominent white-and-dusky pattern of wings and tail, like a longer-tailed shrike. The white, however, is nearer the tip of the wing than with the shrike. The bill is short in proportion and the eye is yellow. Juveniles are spotted beneath. The Mockingbird is famous for raising or "flashing" the wings, the object of which has been much discussed. Marshall has seen a Mockingbird spreading its wings wide and pronating them so as to direct the pattern forward as it runs and leaps around a concrete cistern, scaring insects off the wall and gobbling them up. But this is quite different from the precise, rigid type of wing-flashing, which looks like "setting-up exercises" or the manual-of-arms. Its tropical representative, which lacks the white, does the same thing.

Mockingbirds are fond of singing on moonlit nights. They establish feeding territories in winter from which they chase other birds as vigorously as they do in the breeding season. In the latter period, in towns, they also sometimes chase people, dogs, and cats. The call is a harsh, snapping, *chack*. This bird has done damage on occasion to Arizona vineyards (V. H. Housholder).

At Tucson, the Mockingbird is found all year, but in changing roles of migrant, summer resident, and winter resident. Only a good banding study there could reveal the actual times of bulk migration, arrival and departure. Meanwhile Marshall's observations on the San Xavier Indian Reservation show an influx of new birds, singing in new spots, in latter March; eggs and young develop in May, and most or all of the birds are gone by the middle of August (except for an occasional late family of young still following their parents). Wanderers can be seen in late August and in September; then winter residents arrive and establish their feeding territories in late September.

In southeastern Arizona, some birds may winter about Nogales, but most arrive in May (even in June of 1948 at Patagonia – Phillips) and leave in August, or in mid-September on the Sonoita Plains (where there are positive observations before that and negative ones afterwards, by Phillips and Marshall, in several years).

After the big rains and snows of fall-winter in 1951–1952 and again in 1957–1958, Monson witnessed a sudden and spectacular flourishing of ephemeral plants over much of the southwestern desert, including the Havasu Lake region and the base of the Sierra Pinta. Here Mockingbirds, Western Meadowlarks, and other species appeared "from nowhere" and nested abundantly.

In northern Arizona the breeding population arrives in latter April and mostly leaves by mid-August. Migrants were taken at Flagstaff, *April 18* (*1936* – Hargrave and Phillips, MNA) and in the lower pines at the north base of the San Francisco Peaks, *August 13* (*1931*); others were later taken and banded farther east, in the Upper Sonoran Zone, through *September 8* and *24* (all Hargrave, MNA).

We are unable to recognize a western race of the Mockingbird, as the darker coloration of the eastern population seems due to more soot and less fading.

280. CATBIRD *Dumetella carolinensis* (Linnaeus)

Locally a fairly common summer resident in dense willow-brush association (Upper Sonoran Zone) in the Springerville region, and probably in Oak Creek Canyon recently; possibly on the west side of the Chuska Mountains. There are records from Cow Springs, Coconino County, April 1934 (Wetherill, *fide* BNav), and Lee's Ferry, August 25, 1909 (Nelson, *idem.*). Casual in the

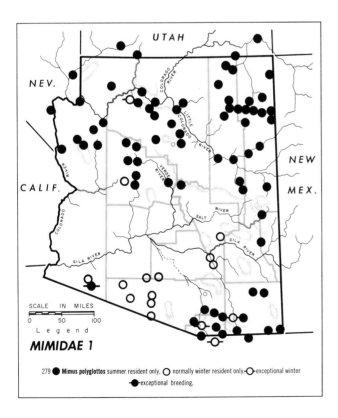

MIMIDAE 1

279 ● *Mimus polyglottos* summer resident only, ○ normally winter resident only, ◐ exceptional winter ● exceptional breeding

Chiricahua Mountains, *September 16, 1956* (immature — Peter Marshall, SWRS).

The Catbird is our smallest thrasher, solid gray with a black cap, and usually chestnut under the blackish tail. In the White Mountains region it is chiefly limited to the Eagar area, but a nest with young was found at Show Low, July 1937 (Poor and Sanborn), and a straggler as high as Greer (R. Roy Johnson). The first of the Oak Creek reports was for June 1955 (Bialac).

All Arizona records refer to the western race, *ruficrissa* Aldrich (Proc. Biol. Soc. Wash. 59, 1946: 132), which seems a perfectly good subspecies when recent specimens are studied. In particular, a freshly-molted bird from Eagar (*August 12, 1950* — ARP) has a lighter crissum than any recent eastern skins seen by Phillips; there are still very few fall western examples for comparison, as the birds leave so early — in August! Because of sooting, abrasion, and differential fading in life, as well as post-mortem fading in the museum, Rand and Traylor's study of variation (Auk 66, 1949; 25-28) cannot be digested, divorced as it is from any mention of the crissum or date and year of collection. They make the astonishing announcements (p. 28) that (1) "It is only by studying adequate samples of migrating or wintering populations that one can determine what their breeding area was," and (2) the western population of Catbirds "presumably winters in the southwestern states and western Mexico." To date, the Catbirds have most ungraciously refused this kind offer of a new winter home, for they migrate far east, then south, as do the Veery, Red-eyed Vireo, and other eastern birds which have penetrated westward, apparently in Recent times.

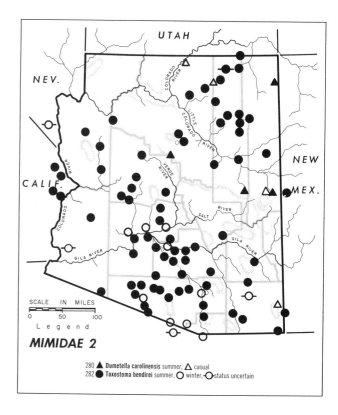

281. BROWN THRASHER   *Toxostoma rufum* (Linnaeus)

Rare fall and winter visitor to Sonoran zones of southeastern Arizona, chiefly in recent years, with records extending west and northwest to Prescott, Phoenix, and Tucson (including four specimens *October* to *January* — CAS, ARP, SWRS; but none from northwest of Tucson). Also taken in Zion National Park, Utah (Grantham, Condor 38, 1936: 85).

Our only richly colored thrasher, this one is rufous-brown above and black-spotted below. The seasonal span of the Brown Thrasher extends from August 20 (1959, at Tucson — Bruce Cole) to April 20 (1949, Phoenix — Mrs. Charles Stevens), with most records from mid-October to March.

All Arizona specimens are *longicauda* (Baird), the western race, which is lighter and larger than that farther east, except one. This is an immature male from near Portal, *October 17, 1956* (J. G. Anderson, SWRS) which is small and dark, yet of a gray cast, and is tentatively referred to *rufum*.

282. BENDIRE'S THRASHER   *Toxostoma bendirei* (Coues)

Rather common to local summer resident of the open parts of the Sonoran valleys almost statewide, except for Sulphur Springs Valley (where breeding is unproven), the plains north of Williams, and the immediate Colorado Valley. Found especially where stretches of open ground meet tall dense bushes and/or cholla cacti. Rare, and possibly only a migrant, in the Sulphur Springs Valley. Winters sparingly from the lower Salt River south to Tucson, but not farther west than the Phoenix region and possibly the Ajo Mountains, nor farther east than Tucson; October to December records are all from a narrow strip from central Arizona southeast to the Tucson region. Migrants apparently return to Tucson in late January.

Bendire's Thrasher is extremely similar in appearance to the abundant Curve-billed Thrasher. Indeed, when the redoubtable Major Bendire sent Fort Lowell specimens to Washington, they were identified as the female of the latter species. It was only upon the Major's insistence that the nests and eggs of the two are entirely different that Elliott Coues finally described the species as new, dedicating it appropriately to its discoverer. It differs from the Curve-billed Thrasher in its smaller bill, with the mandible quite straight. The eye is clear yellow, not orange-yellow. The spots on the chest are triangular, with their narrow points upward, toward the chin. These distinct spots stand out against the background more than do the larger and more diffuse spots of the Curve-billed Thrasher. There is no difference in back color of fresh fall birds; both are rich gray-brown which fades to sandy by spring and summer. There is no white in the wing of Bendire's Thrasher, but the tail corners, when unworn, show a white spot. James Ambrose discovered that the most useful field character is the pale ramus of the lower mandible of Bendire's; this is blackish in the Curve-billed Thrasher. In the hand, he finds that the gonydeal angle made by the rami is rounded in Bendire's, acutely angled in the Curve-billed. These traits serve to identify all ages, even before the bill and feather-coat are fully grown. Bendire's is more partial to open farmlands. It often cocks its tail somewhat over the level of the back, when running along the ground. Calls are not often heard. Though thought by Herbert Brown and also by Herbert Brandt to be very reticent to sing, it can be heard incessantly, and its song well merits the praise it has been given, being a continuous sweet warble with a double quality, as if the bird were singing two songs at once.

Bendire's Thrasher differs from all other thrashers of our southern Arizona lowlands in its pronounced migrations. And it is remarkable that a bird that is gone from the northern half of the state by late August should return to Tucson in late January. Although the range covers most of Arizona, it hardly extends into adjacent states, where it is rare except in Sonora, being common down to Hermosillo. This and the Rufous-winged Sparrow were almost the last genuine full species of United States birds to be made known to science — both discovered at Old Fort Lowell, Tucson, by Major Bendire.

Bendire's Thrasher is divided into two Arizona populations by the forests of the Mogollon Plateau. These are in turn broken into spotty local colonies by the bird's avoidance of all heavy vegetation, continuous brushy cover, and grassland. Replacement of grass by scattered bushes may account for its appearance in the Sulphur Springs Valley. It is favored by man's clearing and agricultural activities, and is probably much commoner now than originally. Nevertheless, there are some serious ups and downs in numbers. Usually a common bird around the Indian farms on the San Xavier Reservation, just south of Tucson, only one was seen there (on Marshall's towhee study area) between May 5, 1959 and February 1, 1961 during regular visits several times a week in most months. Meanwhile it has been found commonly, and nesting, along the Santa Cruz floodplain on the northwest side of Tucson, and on rolling desert land both on the western and eastern outskirts (Crossin, Ambrose).

Migrations in southern Arizona are difficult to decipher owing to the winter residents. Apparently most leave in September and return in early February. Presumed arrivals at Tucson, of birds making themselves conspicuous in areas where none had been seen in the preceding interval, are January 27 (*1938* – ARP), January 29 and February 6 (*1943* – Hargrave), February 1 (*1961* – Marshall), February 9 (Brown, Auk 18, 1901: 23), and February 9 (*1962* – Ambrose). Crossin found a nest with two dead nestlings there on *March 9, 1962*. A juvenile was on the wing on *April 6* (*1952* – ARP) west of Santa Rosa, on the Papago Indian Reservation. Evidently this thrasher arrives a month later farther east, where the earliest dates are *February 21* (*1936* at San Carlos – Jacot, ARP), *February 24* (*1915* at Willcox – van Rossem, LA), and *February 27* (*1956* in Arizona opposite Rodeo, New Mexico – Cazier and Ordway, ARP). The same is true in western Arizona, where the earliest records are for the first few days in March (Monson).

Northern Arizona records are from "March" (1931 east of Flagstaff — Hargrave) and April 17 (1937 at Copper Mine, Navajo Indian Reservation) to August 25 (1937 south of Polacca, Navajo County — both Monson, *fide* BNav). There are several interesting early accounts of this bird that were skipped over in Bent's *Life Histories;* for instance, Herbert Brown's (Forest and Stream 24, 1885: 367 and Auk 5, 1888: 116-118).

283. CURVE-BILLED THRASHER; PALMER'S THRASHER *Toxostoma curvirostre* (Swainson)

Very common resident of cholla cactus association, rather common in other dense thorny brush and even in towns, of the Lower Sonoran Zone west to the Growler and Kofa Mountains and the Big Sandy River; occasional west of Growler Mountains to Cabeza Prieta Mountains. Local in extreme southeast. Rare straggler to the Colorado River (Huey, Condor 22, 1920: 73), Camp Verde (Mearns, US), Fort Apache (Phillips; Pitelka, MVZ), southeastern Nevada (Gullion), and other peripheral localities shown on the accompanying map.

In the preceding account of Bendire's Thrasher we have listed identifying traits of these two similar species. The Curve-billed Thrasher is the most abundant thrasher on the deserts of central and south-central Arizona. It is a very plain bird, brownish gray above and paler beneath with faint, round chest spots. White corners to the tail and white wing bars adorn those birds from the Patagonia Mountains eastward along the international boundary; occasional variants of this type crop up elsewhere, especially around Tucson (Ambrose, ARIZ). An explosive two-whistled call gives the bird the Mexican name "Cuitlacoche." The song is the usual thrasher series of repeated phrases, some of sweet quality, others guttural. This thrasher is not averse to a city life and it occurs in Tucson and Phoenix during most of the year, withdrawing to cholla areas to nest if no cholla is available in town. It is particularly fond of these "cholla meadows." In the open farmlands near Phoenix (Bendire's Thrasher country), it was not found by R. Roy Johnson, who noted its limitation to better-wooded areas such as citrus orchards. It has definitely increased its range to the northwest, for instance into the Big Sandy Valley, where Frank Stephens, who was very interested in thrashers, did not find it in 1880.

Important accounts of this thrasher's nesting are by Brown (Forest and Stream 27, 1887: 464; Auk 5, 1888: 116-118; and 17, 1900: 34; Zoe 3, 1892: 243-248), Gilman (Condor 11, 1909: 50-54), and Willard (Condor 14, 1912: 54-56; 15, 1913: 41; 25, 1923: 122); and of development and behavior of young by Rand (Bull. Amer. Mus. Nat. Hist. 78, 1941: 213-242); see also Stafford, Auk 29, 1912: 363-368, and Hartranft, Oologist 25, 1908: 86.

Geographic variation of the Curve-billed Thrasher in Arizona consists of the gradual increase southeastward in white on the wing coverts and tail corners as mentioned

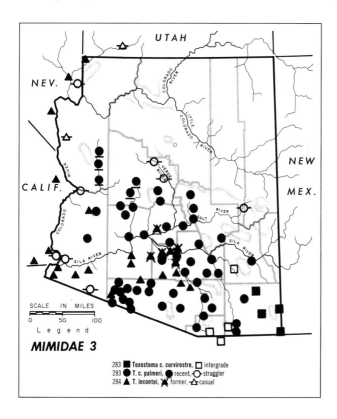

MIMIDAE 3

283 ■ Toxostoma c. curvirostre, □ intergrade
283 ● T. c. palmeri, ◉ recent, ○ straggler
284 ▲ T. lecontei, ✕ former, △ casual

above. Also, towards the southeast, the chest spots become larger and more conspicuous, owing to their standing out against a lighter background. These are attributes of the typical race *curvirostre,* which is better developed farther east and southeast. Our only populations which satisfactorily show these traits are those at the east base of the Chiricahua Mountains and in Guadalupe Canyon. From there west to the Patagonia Mountains the birds are a variable lot, and we indicate a suspected interdigitation of atypical *curvirostre* at higher altitudes (open squares on the map) with *palmeri* (Coues) in the valleys (solid circles). *Palmeri,* with no white and more blurred spotting, resides through the rest of the species' range in Arizona. We consider *celsum* Moore to be a synonym of *curvirostre*.

284. LE CONTE'S THRASHER
Toxostoma lecontei Lawrence

Uncommon, usually very local, summer resident in the open creosote-bush deserts of extreme western and southwestern Arizona, east (at least formerly) in the Gila Valley to the Phoenix-Florence-Picacho Peak region and to east of Ventana Ranch in northern Papago Indian Reservation. Permanent resident in Yuma County south of Gila River and east of Gila Mountains.

Le Conte's Thrasher bears the name of two brothers who were well-known Georgia entomologists. They made a trip from California to the Pima Villages in about 1849, when one of them secured the type specimen.

This thrasher is an almost-white desert wraith or will-o'-the-wisp which runs very rapidly on the smooth desert pavement, and flies off behind a bush almost before one can see it. Except for our white waterbirds, this is the palest of North American birds, and it is limited to the driest desert regions. Intolerant of man and his activities, it has retreated from the newly farmed areas of central Arizona. Its movements are little understood, and its former seasonal status in central Arizona will probably never be known. Except for the Catbird, this is our only brown-eyed thrasher (*crissale* is pale gray). Coues' animated account of his chase of this bird is to be found in his *Birds of the Colorado Valley* (U. S. Geol. Survey Misc. Publ. no. 11, 1878: 72). This was in Union Pass (where Phillips has not found the species) in September, when it might have been migrating. From October through December it is not definitely recorded in Arizona north of the Gila River, nor east of Yuma County. Presumably it returns in late January; one was taken west of Havasu Landing, Havasu Lake, on *January 22 (1947 — GM)*.

There is no basis whatever for the statement that it is abundant in the Santa Cruz Valley region (Engels, U. C. Publ. Zool. 42, 1940: 343); it is unrecorded there, as well as in many other parts of Arizona erroneously included in its range by the same author (map, p. 390). Equally erroneous are his ideas of iris color. Engels' collections of *lecontei* (MVZ) are skeletons from Coolidge and Florence only. The southern limit of Le Conte's Thrasher in that area was long ago determined to be near Picacho Peak (see the classic accounts by Mearns, Auk 3, 1886: 300-307, and Merriam, Auk 12, 1895: 54-60). West of there, it was found in numbers on *March 28, 1963* (bob-tailed juvenile, SHL; John P. O'Neil, in press). The habitat there is sandy drainageways with occasional chollas and mesquites (the latter for nests) within the bleak creosote desert; Black-tailed Gnatcatchers were very common.

285. CRISSAL THRASHER
Toxostoma crissale Henry

Common resident in dense, tall brush along rivers and larger washes of Lower Sonoran Zone, locally in dense broad-leafed Upper Sonoran chaparral, south and west of the Mogollon and Kaibab Plateaus; but absent west of Growler Mountains and Papago Well in western Pima County. Uncommon fall migrant in northern Arizona: 2 taken *September 22 (1932 — Yaeger, MNA)* from south of Grand Canyon Village; also several records from the lower Little Colorado Valley (*September 11* and *22, 1933 — Hargrave, MNA*), and taken at Shumway, *September 3 (1953 — Phillips, ARP)*! These and sight records from the northeast, especially south of Concho, Apache County on or about July 2, 1876 (Aiken, Colorado College Publications, Studies Series 23, 1937: 21 — if they really pertain to this species) should spur someone to seek a breeding colony in that district.

The Crissal Thrasher is the longest and slimmest of the thrashers, with a long and dark tail, the bill greatly decurved, and the iris grayish straw-color. The most secretive of Arizona thrashers, it seldom ventures out of dense mesquite bosque (its lowland home) or chaparral (its habitat in the mountains). In southern Arizona the Crissal Thrashers of manzanita chaparral,

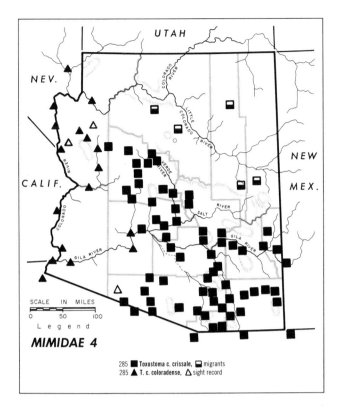

285 ■ *Toxostoma c. crissale*, ▢ migrants
285 ▲ *T. c. coloradense*, △ sight record

MIMIDAE 4

at around 6000 feet in the Catalina, Rincon, and Huachuca Mountains, are isolated by considerable altitude and unsuitable vegetation from the main lowland population. The Crissal Thrasher may be rather numerous without being suspected. At dawn and dusk, however, it voices a *toit-toit* churring call, and reveals its true abundance. The song is like a Curve-billed Thrasher's but more varied and eloquent and much less often heard. A patch of deep rusty under the tail (on the crissum) gives it the name of Crissal Thrasher. The eggs are blue.

If the change from *Toxostoma crissale* to "*Toxostoma dorsale*" is condoned, then you have zoology being written by printers' devils instead of by zoologists! In the original description, the first sheets printed gave identical names and type localities for this thrasher and the Red-backed Junco, an obvious lapse on the part of the typesetter. The name *dorsale*, quite appropriate for the junco but ridiculous for the thrasher, was promptly refuted by Henry, who had the page canceled and reissued, and by Baird in his appendix (Pac. RR. Rep. 9, 1858: 923). The fact that Baird makes the correction in an appendix indicates that he took the name *crissale*, employed in the body of the work, from Henry's manuscript. It is thus obvious that errors were made in the original printing (see also Bailey, *Birds of New Mexico*, 1928: 740); and under Article 19 of the International Code they need not be perpetuated. We cannot condone the exhuming of ancient misprints for the sole purpose of changing long-standing names, nor the interpretation (Oberholser, U. S. Nat. Hist. Mus. Proc. 84, 1937: 330) which destroys the value of Article 19 and raises every typographical error on an unknown, suppressed sheet to the dignity of a new name. The author of this exhumation from the wastebasket admits that the inappropriate *dorsalis* was published "apparently by mistake" (Sci. Publ. Cleveland Mus. Nat. Hist. 1, 1930: 97). This is not nomenclatural stability. We are in full accord with Delacour and Mayr (Wilson Bull. 57, 1945: 5) and Friedmann (Bull. U. S. Nat. Mus. 50, 1941: 62, rejecting Oberholser's proposal) and therefore continue to use the familiar and appropriate name of this bird, rather than "*Toxostoma dorsale*."

Birds along the Colorado and Big Sandy Rivers are paler than those of the typical race over the rest of the state. The former are called *Toxostoma crissale coloradense* van Rossem, to which may belong the birds of the Gila River up to Arlington. Pale variants around Prescott (ARIZ) and the northeast, further add to the mystery concerning this interesting bird, although some *coloradense* might wander directly up the Colorado and Little Colorado Rivers.

Among the interesting early accounts of the Crissal Thrasher which were skipped over in Bent's *Life Histories* is one by Major Bendire (Forest and Stream, 7, 1876: 148).

286. SAGE THRASHER
Oreoscoptes montanus (Townsend)

Fairly common summer resident of sagebrush areas of northeastern and possibly northwestern Arizona, south possibly to Springerville (*July 5, 1936* — F. G. Watson, ARP). Fairly common migrant on open plains, not so common in extreme southwest. Rather common winter resident in sparse brush of Sonoran Zone plains of southern Arizona east to at least the San Pedro Valley, but irregular and less common along lower Colorado River; is also found in winter in the lower Little Colo-

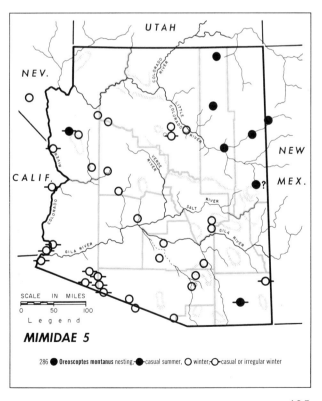

286 ● *Oreoscoptes montanus* nesting, ◐ casual summer, ○ winter, ⊖ casual or irregular winter

MIMIDAE 5

rado Valley. Movements poorly understood. Rare migrant across higher plateaus. Casual (juvenile) specimen from near Gleeson, Cochise County, *June 4, 1940* (GM); and one seen at Flagstaff, December 11, 1938 (Kassel, Plateau 13, 1941: 66-67).

The Sage Thrasher is smaller than the Robin, which it resembles in its shape, stance, and mannerisms when running over the ground. The chest spots are dark, and the tail corners are white. The bill is short and Robin-like, the iris yellow. This is our only thrasher with a fairly definite facial pattern. It is rather erratic in its occurrence in southern Arizona, but is abundant in the north, for instance in the region of Wupatki National Monument, in early fall.

It is an occasional migrant in pine areas, as at Grand Canyon Village, the Kaibab Plateau (Nelson), and Flagstaff. Migration in northern Arizona is in August to October and March to May, principally.

There is a peak in abundance of this thrasher in the northeast in September, which seems to represent a migration. If so, it is uncertain whither the birds are bound, because they rarely show up in the Tucson district until December! There, the bird is most common from January to March, and the earliest arrival is October 4 (1961 at Tucson — Marshall and Ambrose). Westward, singles were seen in the Bill Williams Delta as early as September 1 (1954) and southeast of Quartzsite September 16 (1958 — both by Monson).

In the 1870's to 1890's the early ornithologists Henshaw, Mearns, McGee, and Anthony (in New Mexico) had records and specimens (US, AMNH) suggesting that the bird was common in southeastern and central Arizona below the Rim in September and October, appearing as early as *August 29 (1884* at Camp Verde—Mearns, AMNH). Possibly this was a big migration of birds moving southeastward and southward down the eastern border of Arizona, which has escaped observation in recent years.

THRUSHES TURDIDAE

Thrushes are a diverse, almost world-wide family; the type genus *Turdus,* which includes our American Robin, is one of the largest among birds. Most thrushes are superb and renowned singers, including the European Blackbird and Nightingale. The solitaires of Central America and the West Indies are undoubtedly among the best singers in the world. All thrushes have spotted young, and in the hand they show a "booted" tarsus, whose smooth surface is unbroken by scale borders. A water hole among the junipers of northern Arizona is a wonderful place to see such thrushes as robins, bluebirds and solitaires in fall and winter.

287. AMERICAN ROBIN *Turdus migratorius* Linnaeus

Common summer resident of openings in Transition and boreal zones, and locally in moist Upper Sonoran woodland along the main canyons, throughout Arizona except Hualapai Mountains. Reported recently in summer in Cerbat Mountains of Mohave County (Mus-

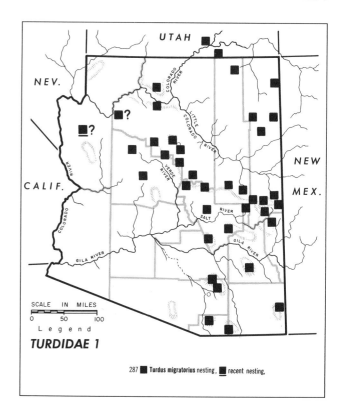

TURDIDAE 1

287 ■ *Turdus migratorius* nesting, ■ recent nesting,

grove). Casual summer Lower Sonoran records: Parker, August 16, 1953 (Monson), and near Kingman, *June 12, 1959* (Coppa, ARIZ). Winters in Sonoran zones, especially in berry-bearing Upper Sonoran woodland and in large towns, somewhat irregularly in mistletoe-mesquite association, and uncommonly in Transition Zone.

This bird is so familiar on lawns in most parts of the United States as to need no description. In Arizona, however, you see it on lawns only during its somewhat irregular but spectacular flight years, or in mountain towns such as Flagstaff. Some of the low-altitude breeding stations, near the edge of the desert, are in mountain canyons that broaden just before emerging from the oaks, such as Madera and Florida Canyons in the Santa Rita Mountains, Fort Huachuca (at least since 1946 — Brandt), and Portal Ranger Station in the Chiricahua Mountains.

In northern Arizona, Robins start to move into juniper areas adjacent to the pines after the middle of July. Although a few linger in the high Transition-Canadian Zone country to at least November, there are records well away from breeding areas from latter August and September to early May. Numbers at Flagstaff increase in March.

Southern Arizona wintering is from just after October 15 to early May; even as far as Yuma, the species was recorded from November 1 to May 2 by Herbert Brown. The race of Robin in Arizona and the west generally is the large, pale *propinquus* Ridgway, which (unlike Robins in the eastern United States) shows little or no white in the tail-corners as it flies up.

288. RUFOUS-BACKED ROBIN
Turdus rufo-palliatus Lafresnaye

Casual straggler from Mexico: one near Nogales, *December 16 to 18, 1960* (Bill Harrison, ARIZ).

The Rufous-backed Robin is a bird of deciduous tropical woods along the Pacific slope of Mexico. Its northern limit there is the mouth of the Rio Yaqui, Sonora (CU, ARP). Superficially it closely resembles the American Robin ventrally, though it is paler. Both are of about the same size and habits. The rufous back is the most obvious difference. The calls are considerably weaker, but the song is clearer, more liquid, not at all wheezy, and altogether beautiful. The sexes are closely similar. Unlike some other Mexican thrushes, this robin is not commonly seen in captivity; and Harrison's specimen has perfect wing-tips and the little dart-like points to the rectrices, so characteristic of all the thrushes when in fresh plumage. This shows that it had not recently been in captivity. The bird was in the company of American Robins.

The characters adduced by van Rossem for his supposed northern race, *grisior*, are of no value; but the race may yet prove valid on the basis of slight differences in the color of the crown in fresh fall plumage.

289. VARIED THRUSH
Hesperocichla naevia (Gmelin)

Two specimens near Tucson: female, January 25 (Phillips, Lanyon) and *February 5, 1956* (Phillips, Virginia M. Morton and Guy Emerson, ARP), identified by A. H. Miller as the pale, gray interior race *meruloides* (Swainson); male (males not identifiable to race), *March 22, 1958* (Phillips and Dickerman, MIN).

This is the only prominently patterned thrush that has ever been found in Arizona.

An *1858* female (MICH) was taken by the Ives Expedition, supposedly near Fort Yuma. It seems more probable that this, like the Red-breasted Nuthatch and Heermann's Black-tailed Gnatcatcher specimen, was really taken farther west. As to the generic name *Ixoreus*, which has been misapplied to this bird, it is purely the result of mistaken identity and should not stand. Bonaparte based his genus on a South American flycatcher which he incorrectly thought was *Turdus naevius* Gmelin. (Coincidentally, while writing this account we are looking at and listening to Robert Dickerman's caged Aztec Thrush, *"Ridgwayia" pinicola*. These birds must be congeneric!)

Genus *CATHARUS*
by Allan Phillips

290. HERMIT THRUSH
Catharus guttatus (Pallas)

Common summer resident of boreal zones and down locally into shady canyons in the Transition Zone. Rather common winter resident in Sonoran zones of southern and central Arizona, especially in dense Upper Sonoran woods and brush and in shady parts of major river valleys. During migrations found even in the most arid desert mountain ranges, where also irregular in winter. Transient, but rare in most non-breeding areas, in northern Arizona, from early September to early November, and again in late April.

The Hermit Thrush is our finest singer. Only the Swainson's Thrush, Scott's Oriole and Western Meadowlark begin to compare. It is the only common spotted-breasted thrush in Arizona (except in May, when Swainson's migrates). The tail is decidedly redder than the back and is often slowly raised while the wings are somewhat spread and lowered, in an attitude of attention. One common call is a soft *chuck*.

The observant reader will already have noticed that there are several groups of birds whose status and movements in Arizona can only be understood with the aid of shotgun, mist-net, or trap. Among such are the hawks, sandpipers, gulls, terns, swifts, hummingbirds, flycatchers, and ravens. The difficulty normally arises from the extreme similarity of a number of different species in external appearance. In the Hermit Thrush, however (like the Horned Lark, Cliff Swallow, and Yellow Warbler), we may identify the species in the field and make counts, but still not know what is going on around us. If we did no collecting, we would surely suppose that our Hermit Thrushes moved downslope in October and back up in late April and early May, in a classic altitudinal migration; the extent of their migrations, as well as the pronounced eastward trend of some populations in fall, would certainly escape our knowledge. These complex fall movements of different races have already been sketched by Phillips (Wilson Bull. 63, 1951; 131). Actually, migrant Hermit Thrushes from other western states reach Arizona in early September and are rather common by the middle of the month, but this is hardly noticeable at first when nearly all migrate along mountains which are already occupied by our local nesting populations! An almost complete turnover occurs in mid-to late October with the birds that nested at lower latitudes being largely replaced by others that come to us from Canada and Alaska. These winter here, and the reverse process occurs chiefly from late March to early May.

We have extolled certain species for showing great geographic variability, but in most cases they are sedentary — which indeed promotes isolation and distinctness among their scattered populations. But the utility of their races in the study of migration is thereby limited. The thrushes of the genus *Catharus*, on the other hand, not only show consistent, marked racial variations, but those species breeding in the United States and northward are completely migratory — most of them going long distances. Preeminent in this variation is the Hermit Thrush. Who can fail to be impressed by the different sizes and shades of these birds, migrating through the Santa Rita Mountains, for instance? What a contrast between the tiny Alaskan birds wintering in our state and the large Rocky Mountain birds spending the winter on lawns beneath the pepper trees of the University of Mexico campus! And color is even more important than size.

Color, however, has been neglected and confused, owing to ornithologists' prevailing reluctance to separate fresh from worn, and recent from foxed skins. Birds of a given population turn gray to a varying extent in life through wear, and later turn back to reddish in the museum with age. Identification of fall migrants and early wintering birds must be made by comparison with reasonably contemporary birds on known breeding grounds actually in or just finishing the molt. Fortunately most thrushes molt before migrating in the fall. To the objection that such birds still on their breeding grounds are shy and hard to obtain, and that they might be mixed with early migrants, we can only reply that it is the business of the ornithologist to get there at the right time and to learn the situation from actual field experience with the birds and careful attention to molt. Linsdale has done this for Nevada populations (MVZ), so has Phillips for Arizona (ARP), and Swarth did so for parts of British Columbia (MVZ).

Fall and early winter specimens, if properly housed and not too recently taken, can be identified according to the latest revision (Phillips, Anales del Inst. Biol. Univ. Méx. 32, 1962: 347-361) covering all known races, which follow:

A) Medium-sized northern races with a brownish wash over the flanks in fresh plumage:

1. *Catharus guttatus nanus* (Audubon). The richest colored race, reddish brown above, with flanks brown. Winters mostly in the eastern United States; accidentally to Arizona (one young female, Parker, Colorado River, *November 29, 1953* – GM).

2. *C. g. munroi* Phillips. Paler brown and slightly smaller; flanks paler and less brown. Breeds in the interior of central and northern British Columbia and possibly southwestern Yukon. Migrates down the western Great Plains to eastern Texas and northeastern Mexico; not yet identified in the United States west of the Pecos River or the Rocky Mountains.

B) Small Pacific Coast races with gray flanks (little or no trace of brown there). Coloration varies elsewhere, especially on upperparts. The wing (chord) rarely exceeds 92 millimeters in the male (87 in the female):

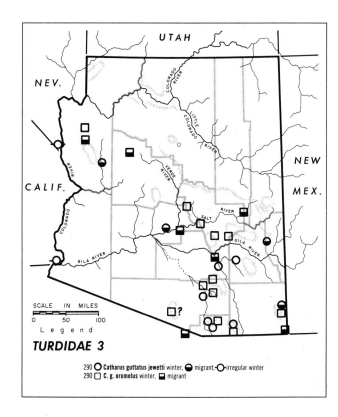

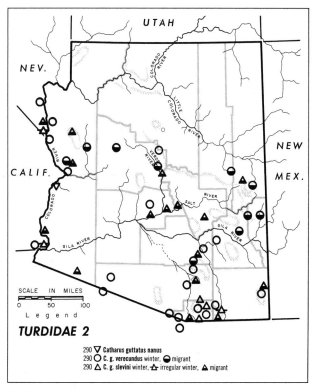

3. *C. g. guttatus* (Pallas). Dark grayish brown above, the tail deep reddish brown. Chest buff-tinged, with rather small blackish spots. Breeds in most of Alaska and apparently along the west slope of the Cascade Mountains of Washington. Winters along the Pacific coast of the United States and east to western Texas, including northern Mexico.

4. *C. g. verecundus* (Osgood). Browner above, less gray than *guttatus,* but just as dark or even darker. Breeds on the islands of British Columbia and in southeastern Alaska. Winters in the same regions as *guttatus* but does not extend eastward as far as the Rio Grande.

5. *C. g. jewetti* Phillips. A brown race like *verecundus,* but paler. Known to breed only in the Olympic Mountains of coastal Washington. Winters principally in southern Arizona, Sonora, and Sinaloa.

6. *C. g. oromelus* (Oberholser). Grayer than any of the previous three races above, less brown; and paler than any except *jewetti*. (Intermediate between *guttatus* and *slevini*). Breeds from the Rainbow Mountains, British Columbia, south along the east slope of the Cascade Mountains to the Warner Mountains of southern Oregon. Winters mainly from central Arizona to northern Mexico (Coahuila and Nuevo León).

7. *C. g. slevini* (Grinnell). Gray like *oromelus,* but still paler above (sometimes brownish, but very pale). Breast-spotting grayish, not black, and the spots often are small, on a white background. Some *slevini* are smaller than the other races.

Slevini breeds along the coast of central California and north to Oregon and possibly to Mt. Adams, Washington (along the west slope of the Cascade Mountains?). It winters below the pines, from extreme southern Arizona south over all of Sonora and Sinaloa, and in southern Baja California (more sparingly southeast to Michoacán and Hidalgo); on migration it reaches western Texas.

C) Larger, pale grayish races of the interior of the western United States. Flanks gray.

8. *C. g. sequoiensis* (Belding). Exactly intermediate between the last and the next. Wing (chord) 92.5 to 98.2 millimeters in male, 88 to 92.3 in female. Breeds in southern and interior California and apparently in southeastern British Columbia and adjacent parts of the United States. Winters in northeastern Mexico, west sparingly to the west slope of the Sierra Madre Occidental in northeastern Sinaloa.

9. *C. g. auduboni* (Baird). The largest race: wing (chord) 98 to 106 millimeters (94 to 101.7 in female). Above pale and gray like *slevini* (but usually sootier in Hualapai Mountains). Chest-spots black and larger than in any other race, the background usually white. Breeds in Rocky Mountain-Great Basin region from extreme southeastern Washington and possibly Montana south. Winters in pine-forested highlands of the mainland of Mexico and Guatemala. *Polionota* (Grinnell) is a synonym.

C. g. guttatus is our commonest race in Arizona from mid-November to early March, particularly in the mountains. Normal occurrence is from October 26 to the start of May, with extreme dates in southern Arizona *October 10* (*1959*, Mitchell Peak, Greenlee County — Peter Marshall and Wm. George, ARIZ) to *May 16* (*1956*, Kofa Mountains — GM). The only definite dates in or near northern Arizona are *October 9* (*1939* on the South Rim of Grand Canyon — Louis Schellbach, GCN) and *October 19* (*1906*, high in the Mogollon Mountains, New Mexico — Vernon Bailey, US).

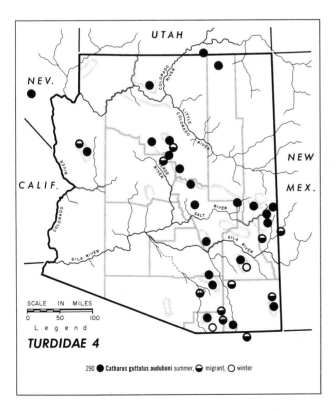

TURDIDAE 4

290 ● Catharus guttatus auduboni summer, ◐ migrant, ○ winter

C. g. verecundus is almost equally common except north of the Gila Valley in the east, where (east of the Verde Valley) it is known only as an uncommon transient. It occurs mainly from October 16 to mid-March; extreme dates in southern Arizona are *October 1* (*1924*, "Paradise??," — Chiricahua Mountains? — Kimball, MICH) to *April 11* (*1935* on Turtle Peak, Gila Mountains, Graham County — Yaeger, ARP). Northern Arizona records are *October 18* (*1936* in the White Mountains — ARP) and *October 27* (*1920* on Showlow Creek, Navajo County — Kimball, MICH).

C. g. jewetti is a fairly common October transient in southeastern Arizona, and it winters in relatively small numbers (chiefly in the lowlands) north and east as far as Aravaipa Creek, Graham County (WJS) and west (casually?) to the Colorado Valley at Bard, California (MVZ 81415), with one specimen (somewhat intermediate toward *slevini*) from as far north as the southern tip of Nevada (MVZ 64192). The latest record, and the only certain spring transient, is *March 17* (*1940* at Tempe — LLH). The fact that records virtually come to an end in the first half of February may well mean that the warm brown tones that mark this race are of short duration, spring birds in Arizona becoming gray like *oromelus*. This is probably true of many *verecundus* as well, since from mid-March to the end of April *verecundus* forms only about 10% of the small Hermit Thrushes of southern Arizona, while in October these two brown races constitute about 50%. The alternative view to fading and loss of the warm brown tones in the Arizona sun would be an earlier spring departure to the northwest coast, with its less rigorous winters than the snow-bound interior. Either viewpoint is perfectly logical. Fall records of *jewetti* are concentrated in the middle third of October. At Tucson, there are very few records between *November 11* and *March 7* (both ARP); these outside dates may represent migrations.

C. g. oromelus is a fairly common winter resident over the entire Arizona winter range of the species, except that there are no typical specimens from the Colorado River Valley. It winters north to near Kingman (Coppa, ARIZ), and near the Mazatzal Mountains (Dickerman, CU), in lowlands and foothills. Arizona records are from *October 2* (*1948* in the Hualapai Mountains — ARP) to *May 2* (*1925* in the Chiricahua Mountains, one or two males — Kimball, MICH). It is peculiar that although *oromelus* is about equal in numbers to *verecundus* in winter, it is far outnumbered by that race in fall and far outnumbers it in spring. This means that identifications of late winter, spring, and summer Hermit Thrushes are to be taken with at least one grain of salt! This warning applies even more strongly to reports of gray races east of the Great Plains.

C. g. slevini winters rather commonly in the lowlands and at the foot of the mountains, chiefly in the Patagonia area of extreme southern Arizona. Marginal winter stations are "California Lakes" (extreme southeastern California) and the southern tip of Nevada (MVZ 81417 and 64193, both approaching *oromelus* slightly); and Fort Huachuca (Cahoon, MCZ). Southern Arizona migrations are from *September 14* (*1949* in the Hualapai Mountains; no record farther south before *October 2* at Tucson — all ARP) to *October 19* (*1956* at Globe — LLH); and from *February 27, 1908,* and *March 3, 1920,* both near Tucson (Kimball, LA and MICH) to *May 12* (*1956*, Colorado River opposite Picacho — GM). Casual in northern Arizona: one male, Showlow Creek, southern Navajo County, *October 27, 1920* (Kimball, MICH).

C. g. sequoiensis is a fairly common transient in southeastern Arizona and northwest to the Hualapai Mountains, sticking closely to the higher mountains in fall but migrating in the lower parts of the mountains and the foothills in spring. It is rare in winter, when known only from the Chiricahua Mountains; specimens that now seem referable to *sequoiensis* were taken there *March 6, 1917,* and *February 5, 1919* (Austin Paul Smith; AMNH 376412 and 376414) and an adult male on *January 15, 1919* (Kimball, MICH). Elsewhere, southern Arizona records extend from *September 9* (*1952* in the Santa Catalina Mountains — ARP) to *October 10* (*1935* at Granville, Greenlee County — Jacot, ARP), and again from *March 27* (*1933* in the Huachuca Mountains — Jacot, ARIZ) to *May 6* (*1939* at the foot of the Santa Rita Mountains — ARP) and probably *May 16* (*1907* in the Whetstone Mountains near Benson — Austin Paul Smith; AMNH 376421). The only records definitely in northern Arizona are *September 1, 1950,* on the San Francisco Peaks (ARP) and *April 25, 1948,* at Holbrook (ARP).

C. g. auduboni, besides being our breeding race, is also a rather common transient through the mountains, rarely appearing in the valleys (Tucson, *September 15, 1948,* and Hooker Hot Springs, west-northwest of Willcox, *May 17, 1948* — both ARP). It is probably only casual in winter; Patagonia, *March 10, 1940* (LLH), and Graham or Pinaleno Mountains, *November 26, 1949* (Marshall, WF). There are two specimens from Flagstaff, *August 1,* (ARP) and *September 3* (Hargrave, MNA), which probably represent the post-fledging dispersal of young. Otherwise, all northern Arizona records are from breeding areas, from early or mid-*May* to *October 9* and *26, 1936,* in the White Mountains (ARP), and *October 28, 1920,* at nearby Pinetop (Kimball, MICH). Southern Arizona migrations are from *September 15* (*1948,* as above) to *October 12* (*1922* in the Huachuca Mountains — Jacot collection); and from *April 20* (*1935* at Granville, Greenlee County — Jacot, ARP) to *May 23* (*1917* in the lower Chiricahua Mountains — Austin Paul Smith, AMNH). But spring arrival is probably somewhat earlier; in the upper Santa Catalina Mountains, a female with "eggs large" was taken *April 22* (*1885* — Scott, AMNH), while another had a nest about one-third built on April 24 (1954 — Sheffler).

291. SWAINSON'S THRUSH; RUSSET-BACKED THRUSH; OLIVE-BACKED THRUSH *Catharus ustulatus* (Nuttall)

Rare summer resident of the cork-bark fir forest of the San Francisco Peaks. Rather common spring migrant through southern and western Arizona from Huachuca Mountains west, north to Kingman region, uncommon farther east and in central Arizona; only three May records in the north. Rare and more restricted in the state in fall, when limited mostly to the southern border, chiefly in the mountains, east to Reserve, New Mexico (Kimball, MICH).

This is like a subdued Hermit Thrush with a liquid whistle, *whoit,* and little or no contrast in color between the tail and the back. Except in southern Arizona in May, it is quite rare. Marshall has tried in vain to elicit answers to the territorial and migration call, *wheee,* overhead at night in May and September. The species is

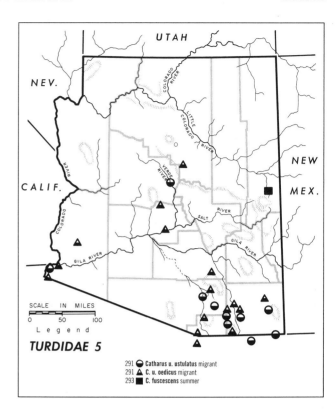

TURDIDAE 5

291 ● *Catharus u. ustulatus* migrant
291 ▲ *C. u. oedicus* migrant
293 ■ *C. fuscescens* summer

either too rare a migrant in Arizona to give us the spectacular night calling from the sky that is to be heard in most parts of the United States, or else it flies higher here. The song of lovely soft whistles has a doubled quality and rises steadily in pitch. We have heard this sweet music in Arizona only among the lofty spires of cork-bark firs on the San Francisco Peaks, in late July.

Elsewhere in northern Arizona there are only three records for the entire species, all in the period *May 6* to *28.* Southern Arizona dates are usually from the last days of April to the first days of June and from early September to early October, with stragglers into November but no authentic winter record. Extreme dates are April 16 (1886 at Camp Verde — Mearns), *April 20* (*1917* in the Chiricahua Mountains — Austin Paul Smith, AMNH), and *April 23* (*1894* at Gardner's Laguna in Baja California — Mearns, US) to June 7 (1946 at Devil's Elbow, Colorado River, California side — Monson); and again from *September 4* (*1892* in extreme southeastern Arizona — Mearns and Holzner, US) and *September 5* (*1956* in the Castle Dome Mountains, Yuma County — GM), to *October 24* (in northern Sonora — Phillips and Amadon, Condor 54, 1952: 166), *November 7* (*1892* in the Huachuca Mountains — Kimball, LA), and *November 9* (*1953* on the San Pedro River at Hereford — ARP).

There are color differences between two groups of races. They must be studied in comparable specimens with cognizance of the great seasonal change and of foxing — which is pronounced in thrushes even under the best museum curating practice. Slight but definite reddening in the tail and its upper coverts, browner sides and flanks, and usually smaller and less sharply defined dark spots on the chest and especially the sides of the breast characterize

Pacific Coast birds; in fresh plumage their ground color of the chest is deeper (richer, more ochraceous). Birds of the Great Basin, Rocky Mountains, and north and east are more contrasted below and completely uniform above, usually grayish or olive brown. We call these latter birds *swainsoni* (Tschudi).

Swainsoni, our breeding race, is represented by a single female, still mostly in juvenal plumage, *August 29, 1952* (Dickerman, MIN). Since the breeding range of this race includes the whole area north of us, and northwest to the eastern parts of the Pacific Coast states and most of British Columbia, one would naturally suppose that *swainsoni* was our common race on migration. But, as usual in *Catharus* thrushes, migration actually moves sharply eastward, and there is but a handful of Arizona records; Camp Verde, *May 11, 1887* (Mearns, AMNH 53878); Tucson, *May 7, 1887* (Herbert Brown, ARIZ 1606); Huachuca Mountains, *May 27, 1901* (Breninger, F 18067; a very pale, gray bird); Chiricahua Mountains, *May 27, 1917* (Austin Paul Smith, AMNH 376730); and still more casually *November 9,* as above. The main migration of *swainsoni* lies east of the 100th Meridian.

The Pacific Coast group comprises two rather well-defined races: a darker, reddish north coast *ustulatus;* and a paler, usually grayer-brown California race, *oedicus* (Oberholser). *Oedicus,* even in spring, is apparently rather rare in Arizona and none winters. It is, however, just as widespread as the more numerous *ustulatus,* the lone northern Arizona specimen (Quaking Asp Settlement, between Mahan and Hutch Mountains, near Mormon Lake, *May 28, 1887* — Mearns, AMNH) being *oedicus.* Other *oedicus* are from *April 20* and *23,* as above, to *May 24* (*1946* in the Huachuca Mountains — Hargrave, University of Cincinnati). The only fall specimen seen is from the Castle Dome Mountains (as above). Normally, *oedicus* does not stop in fall north of southern Sonora.

True *ustulatus* is the commoner race in southern Arizona at all times except at the start of spring migration. Specimens have been taken from *April 27* (*1956* east of Papago Well in extreme southwestern Pima County—ARP) and *May 1* (*1938* near Tucson — ARP) to *June 4,* (*1892, 1899,* and *1946,* all in Huachuca Mountains — US, MCZ, and Univ. of Cincinnati); and again from *September 4,* as above, to *October 13* (*1924,* Chiricahua Mountain region — Kimball, MICH) and *October 24* and *November 7* as above.

292. GRAY-CHEEKED THRUSH *Catharus minimus* (Lafresnaye)

One specimen: Cave Creek, Chiricahua Mountains, *September 11, 1932* (Alex Walker, CLM).

The wing of this male in first basic plumage measures 100.2 mm., so it is of the northern race, *minimus.* Color differences among the races have not been fully worked out, owing to lack of fresh fall specimens from the breeding grounds, and to foxing. For a model life history study of this thrush see Wallace (Proc. Boston Soc. Nat. Hist. 41, 1939: 211-402).

293. VEERY; WILLOW THRUSH *Catharus fuscescens* (Stephens)

Doubtless breeds in the willow association of the Upper Sonoran Zone southwest of Springerville, where taken *July 3* and *4, 1936* (Poor, Watson, Jenks, ARP). Otherwise no authenticated record.

The Veery, known only from one spot in Arizona, is almost identical with the nominate race of Swainson's Thrush, differing only in having the flanks pale gray in contrast to the tawny upperparts, as well as in slightly reduced spotting on the chest. This western race, *salicicola* (Ridgway), is the most heavily spotted and least reddish form of the species, and persons acquainted only with eastern Veeries will be amazed at its resemblance to the northwest coast Swainson's Thrushes. The voice is of course very different from Swainson's. The most remarkable point is that the Veeries breeding in Arizona and elsewhere in the Rocky Mountains must migrate first due east and then make a long, non-stop flight, since the species is virtually unknown anywhere in Mexico or the southwestern United States! The wintering ground is in South America.

BLUEBIRDS Genus *SIALIA*

Bluebirds are chunky, short-billed, long-winged thrushes with notched tails. Unlike other thrushes they perch high and swoop down to catch insects in the grass. The legs are short and not fitted for running. Both sexes show blue on the wings and tail, but females and immature males are quite dull-bodied. They nest in holes in trees. There are only three species in this North American genus, and all occur in Arizona. Here the Eastern Bluebird is the scarcest.

294. EASTERN BLUEBIRD; AZURE BLUEBIRD *Sialia sialis* (Linnaeus)

Rare and local resident in the live oaks and nearby pines from the Huachuca Mountains west to beyond Nogales; a recent nesting in Chiricahua Mountains also. Rare or casual in winter in the Tucson vicinity, where specimens taken *February 12, 1915* (Ivan Peters collection) and where seen January 19 to 21, 1952 (Phillips); and at Patagonia, *November 24, 1950* (Marshall, WJS).

The Eastern Bluebird has a white belly, and the brick-red of the breast extends up the sides of the neck and forward onto the throat, contrasting with a rather dark submalar streak. The back is uniform blue in the male. On a flat, sparsely grown with small pines across Cave Creek from the Southwestern Research Station, Chiricahua Mountains, Guy Emerson discovered a pair of Eastern Bluebirds while he, George and Marshall were looking for Buff-breasted Flycatchers (which were found) on May 30, 1960. The Bluebirds' nest was later observed there by Dr. and Mrs. Leon A. Wiard.

Only two races appear to be recognizable in the United States and northern Mexico. These are a darker, more purplish blue, eastern race *sialis,* which is also darker reddish on the breast; and a western, paler race *fulva* Brewster.

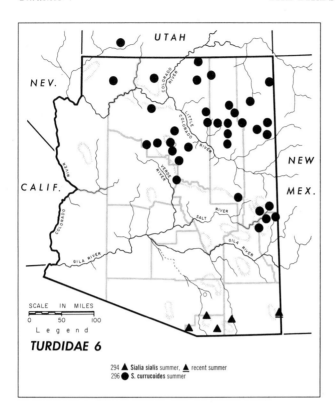

TURDIDAE 6

Legend
294 ▲ *Sialia sialis* summer, △ recent summer
296 ● *S. currucoides* summer

The former has been taken in Arizona only in the lowlands (the specimens detailed above). On the other hand, *fulva* appears to be non-migratory, having been taken in its mountain home in *December* (ARP) and *February* (Jacot, ARIZ).

295. WESTERN BLUEBIRD; CHESTNUT-BACKED BLUEBIRD; MEXICAN BLUEBIRD
Sialia mexicana Swainson

Common summer resident in open Transition and lower Canadian Zones (and in northeastern Arizona, among the larger trees of Upper Sonoran woodlands) west to the Santa Catalina, Bradshaw, and Hualapai Mountains, and Mount Trumbull; now scarcer in the southern mountains. Winters abundantly in berry-bearing Upper Sonoran woodland, uncommonly in open Transition Zone woods, and somewhat irregularly in farmlands and on the desert wherever mistletoe occurs.

In this species the head and neck are uniform, without any red; they are deep blue in the male and more gray in the female. The male often shows much chestnut on the back, but the neck and throat are solidly blue. The belly is gray. Both sexes show definite reddish below and are darker than Mountain Bluebirds. In winter these bluebirds congregate in areas affording mistletoe, juniper or other berries.

When the Western Bluebird invades the lowlands, it is usually found from late October to mid-March. Extreme dates are from October 11 (1887 east of Camp Verde — Mearns) and October 14 (1888 at Tucson — Brown) to *April 13* and *29* (*1884* at Camp Verde — Mearns, AMNH). It is usually uncommon in the pines at Flagstaff from December 11 (1938 — Kassel) to March 3 (1955 — Wetherill).

All color varieties of Western Bluebirds are apt to turn up in any breeding population in the United States and northern Mexico. Also you can find any type within a series varying as to season and wear. Interior birds certainly average more extensively chestnut above and larger than birds from the northwest coast. But the enormous individual variation renders hazardous any guess as to where a given wintering bird has originated. Therefore we prefer to call all Arizona birds *occidentalis* Townsend. This again emphasizes the utilitarian aspect of subspecies, which we discussed above under the House Wren. If subspecies cannot teach you something about evolution or migration, they are not worth having.

296. MOUNTAIN BLUEBIRD
Sialia currucoides (Bechstein)

Common summer resident of the open parts of northern Arizona from piñon-juniper woodland up to timberline, south to the entire Mogollon Plateau region and west to Ash Fork and beyond Mount Trumbull. Winters abundantly in more open berry-bearing parts of Upper Sonoran Zone, uncommonly in Transition Zone openings, and commonly in most years south of its breeding range in farmlands, grasslands, and open berry-bearing woods and brush.

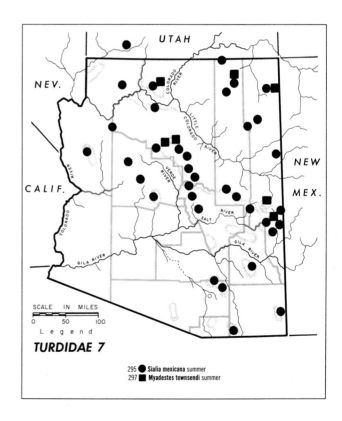

TURDIDAE 7

Legend
295 ● *Sialia mexicana* summer
297 ■ *Myadestes townsendi* summer

The Mountain Bluebird is to thrushes what the Say's Phoebe is to flycatchers, and the Sparrow Hawk is to hawks. It is the open country species which hovers, poised in mid-air, ready to pounce upon its prey. The male is wholly pale sky-blue; the female is dull, but without any rufous tinge. The less musical voice is more burry than in other bluebirds.

Southern Arizona migrations are like those of the Western Bluebird but usually in different winters. Records are from October 20 (1886) to *April 8* and 9 (*1884*, all at Camp Verde — Mearns, AMNH), and April 8, (1946 at Picacho Reservoir near Coolidge — Mr. and Mrs. Wm. X. Foerster; and 1956 at Heart Tank, Sierra Pinta, Yuma County — Monson). An early arrival in numbers — at least 80 — was from near San Carlos to Cazador Spring on November 8 (1951 — Phillips). The birds were still present at Tuba City, in farmlands below the juniper belt, on March 25 (1937 — Monson); but because of the wide breeding range there are no other satisfactory data on northern Arizona migrations.

297. TOWNSEND'S SOLITAIRE *Myadestes townsendi* (Audubon)

Rather common summer resident in high Transition and boreal zones of northern Arizona, west to the Kaibab Plateau and San Francisco Mountains and south to White Mountains. Migrant and post-breeding wanderer in other parts of Transition Zone. Winters commonly in berry-producing woods of Upper Sonoran Zone, rarely in the better-wooded Lower Sonoran valleys and canyons, and even at times in the desert mountains of extreme southwestern Arizona, remaining occasionally to early June. One summer sight record in Chiricahua Mountains (July 11, 1956 — Tanner and Hardy).

Townsend's Solitaire was named by Audubon in honor of John Kirk Townsend, the ornithologist who explored the Columbia River in 1835 and discovered the species. The Solitaire never joins into flocks the way robins and bluebirds do. It is the northernmost representative of a genus of mountain-dwelling birds widely spread to the south of the United States, most of which are famous singers (especially *M. unicolor* of Middle America). Townsend's song is loud and spirited, but it is also rough and with a lot of clicking sounds — quite a disappointment for this group of birds. The call is a creaking high whistle, *eek*. The bird itself is gray with a whitish eye-ring. In flight it shows traces of white at the sides of the tail and a prominent buff wing stripe. It has a long tail but almost no bill and looks all-in-all somewhat like a dwarfed, dull Mockingbird. It perches almost vertically upright, habitually atop a small tree or bush, or high in forest trees. It is peculiar that the arboreal solitaires should descend to the ground to nest — in dirt banks beneath roots or fallen logs. The nest is a small cup of grasses. Solitaires may be seen in impressive numbers in winter at water holes in the juniper woods.

There are records too numerous to list, pertaining to the long wintering period in southern Arizona. The birds show up in the mountains in early to mid-September, and out in such desert localities as Yuma and Wikieup in early October. They stay through May. Extreme dates are September 5, (1942 — A. J. Foerster, south of Prescott) to *May 27* and *June 2* (*1939* near Tucson — A. A. Allen, Phillips, Monson, ARP).

At Flagstaff, birds regularly appear in the first half of August; doubtless they have merely descended from the nearby San Francisco Peaks. The main fall migration in northern Arizona is from *September 3* (*1953* in junipers near Shumway — Louise Levine and Phillips, ARP) and September 24 (1956, already common south of Grand Canyon Village — Hargrave) to October 12 (1936, in the White Mountains–Phillips). Spring migration dates extend from March 23 (1936 near Flagstaff — Phillips) to *May 23* (*1933* at the northeast slope of the San Francisco Mountains region — Hargrave, MNA).

There is a prodigious gap in the breeding distribution of the Townsend's Solitaire. The northern race, *townsendi*, characterized by a paler and duller buff wing-patch, nests as far south as the San Jacinto Mountains of southern California, and the White Mountains and Blue Range of east-central Arizona. Then at the Sierra Huachinera of eastern Sonora and adjacent north end of the Sierra Madre Occidental in Chihuahua is the northern limit of the Mexican race *calophonus* Moore, with its deeper buff wing-patch. In between there are many scattered peaks, some with extensive fir and spruce forests, such as Mount Graham, but with no breeding solitaires.

OLD-WORLD WARBLERS, GNATCATCHERS, AND KINGLETS *SYLVIIDAE*

Sylviids are small to very small birds either with whitish wing-bars or bluish-gray upperparts. They comprise a big family in the Old World; here they are represented only by these three minor groups: Gnatcatchers — tiny blue-gray and white birds the shape of a Bushtit, with a long black-and-white tail; Kinglets — tiniest of all passerines in the United States, olive-green and buffy; and a single warbler-like species, the Olive Warbler, which George (Amer. Mus. Novitates no. 2103) avers is not a parulid. All have very slender, pointed, black bills. They catch insects both in foliage and in the air.

298. BLUE-GRAY GNATCATCHER; WESTERN GNATCATCHER *Polioptila caerulea* (Linnaeus)

Rather common summer resident in open woodlands and chaparral of the Upper Sonoran Zone, west to the Ajo, Castle Dome, Kofa, and Hualapai or even Mohave Mountains, and locally in higher parts of adjacent mesquite associations, at least in the west. Common winter resident of wooded river valleys of Lower Sonoran Zone from the Tucson area and the Salt and Big Sandy Rivers southward, and west to the Colorado River. Rare fall transient in higher forests (a few records, mid-July through August); common transient in western deserts north of Gila River.

The outer tail feathers of the Blue-gray Gnatcatcher are wholly and conspicuously white. The black of the male's head in summer is restricted to a narrow crescent across the forehead and behind the eye. Gnatcatchers are active birds, feeding through the trees and bushes, with the tail waved airily over the back. The Blue-gray frequently gives its call, a plaintive and nasal *peeee*. Sight identification of any but summer males requires a close look at the underside of the tail, which fortunately is possible because the birds are tame. The nest is a work of art — a tall cup of plant down, bound with cobwebs. An interesting case of the nest being removed, presumably to make way for a new one, in the San Francisco Mountains, is recorded by Hargrave (Wilson Bull. 45, 1933: 30-31). The observer may be thankful that there is only one species of gnatcatcher with entirely white outer tail feathers in the United States, for in Mexico there are four, all very much alike; two of them are in Sonora.

As the accompanying distribution map shows, the Blue-gray Gnatcatcher is found in summer in appropriate woodlands and chaparral of the Upper Sonoran Zone over the entire state. The period in northern Arizona is from the latter half of April to September, but the birds are paired and singing in the southern mountains by the first week of April.

At the west, in the arid mountains along the Colorado River, the Upper Sonoran Zone is limited; there the gnatcatcher breeds well down among scrub oaks of the main canyons. The time of common occurrence in the Lower Sonoran wintering area includes the migration period and it extends from the middle of September to at least the *20th* of *April* (Big Sandy Valley — ARP). There are a few *August* specimens from Tucson (Phillips, Jacot, ARP) and San Bernardino Ranch (Mearns and Holzner, US).

299. BLACK-TAILED GNATCATCHER; PLUMBEOUS GNATCATCHER — *Polioptila melanura* Lawrence

Fairly common resident of open brush in the lower Sonoran Zone, commoner westward and rather local east of the San Pedro Valley; not found in Verde Valley or the bottom of the Grand Canyon, nor of course anywhere in the northeast.

This is much like the Blue-gray except that the underside of the tail shows much less white. From late February or March to August the male has a black cap, acquired by a partial prealternate molt. But in winter there is the merest hint of black above the eye of the male. Females have no black whatever on the head; they are a trifle browner than the female Blue-gray Gnatcatcher. The call is usually double, harsher, and more wren-like (less plaintive and smooth) than that of the Blue-gray. Black-tailed Gnatcatchers are non-migratory and are found almost throughout the year in pairs in the more barren desert brush. They inhabit washes running through creosote brush; they never come into town; and only rarely do they visit irrigated or riparian areas.

The Black-tailed Gnatcatcher is a strictly Lower Sonoran Zone bird. Except for the upper Verde Valley and perhaps parts of northwestern Arizona it ranges throughout that zone. The Blue-gray enters its range only during migration and in winter, except very locally in the Ajo Mountains and northwest.

The Arizona subspecies, pale and long-tailed, is *P. m. lucida* van Rossem.

300. GOLDEN-CROWNED KINGLET — *Regulus satrapa* Lichtenstein

Uncommon resident in boreal zones from Santa Catalina and Chiricahua Mountains north and northwestward to Kaibab Plateau. Fairly common fall and winter resident in some of the higher forests of northern Arizona, especially White Mountains, reaching lower Transition Zone. Occasional winter visitant in Lower Sonoran woodlands, with records mostly from Tucson vicinity, but casually west to near Phoenix, Wickenburg (flock of 20!), and even to the Colorado River (23 miles north of Yuma, November 21, 1957 — Monson), and to the Sierra Pinta of Yuma County (*December 27, 1957* — GM)!

The Golden-crowned Kinglet is a handsomely marked little bird whose olive-green colors are relieved by a black-bordered white superciliary, and by pale wing bars which are also set off by dusky. The golden center of the

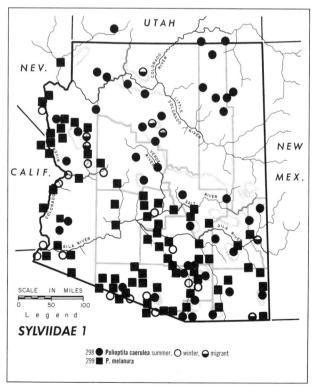

SYLVIIDAE 1

298 ● *Polioptila caerulea* summer, ○ winter, ◐ migrant
299 ■ *P. melanura*

crown is seldom seen, as the birds are usually high in the tops of the firs. The call is a very high-pitched seeping, *tsee-see-see*. Specimens taken in the desert are of a race other than that of the adjacent mountains, thus disproving the idea of vertical migration. The habitat sought by these wintering birds is usually of the largest available trees — giant cottonwoods and willows along streams in the desert — where they join the local mixed flock of warblers, nuthatches, creepers, and Ruby-crowned Kinglets.

In order to understand the movements of Golden-crowned Kinglets in Arizona, we have to deal separately with the two races, a composite account of which was given in the opening paragraph. *Regulus s. satrapa* is the one which is the occasional winter visitant in Lower Sonoran woodlands. It is a grayer-backed bird and it breeds in the northeastern states. It has a shorter bill, more white on the tertials, and other differences in length of superciliary stripe and in wing-bars, as compared with the bird which nests in Arizona. There are ten or so *satrapa* specimens from Arizona, and most are in the months of November and December. James Simpson and James Werner found the flock near Wickenburg (JSW) in the major flight year, 1959. The latest of all the lowland *satrapa* is March 7 (*1953* at Tucson — ARP). One perplexing specimen, if it belongs to this race, is the only one from the mountains; it was taken in the Hualapai Mountains *November 29, 1947* (ARP).

The race *Regulus s. apache* Jenks is the one which nests in the highest mountains of Arizona and of the west generally, except the humid northwest. It reaches its southern limit in the Chiricahua Mountains, beyond which there is a big gap in the distribution of the species. This bird has the longer bill and greener back, less gray — a description based on autumn birds from the White Mountains. (The race *"amoenus* van Rossem," from farther west, is apparently just the fresh plumage of *apache;* its yellower tinge of green is what would be expected, since yellows are fugitive colors in this group of birds, and would be lost by spring.) *Apache* generally stays in the boreal forests of spruce and fir all year, with the following exceptions. One was down at the level of Upper Sonoran Zone oaks of the Santa Catalina Mountains, *February 5, 1950* (Marshall, WJS). Others have appeared in forests outside the breeding area in the vicinity of Prescott, *December 15 and 31, 1926* and *January 29, 1931* (Jacot, ARIZ); the Hualapai Mountains (including *October 1, 1948* at the summit — ARP); the Charleston Mountains, Nevada (series, ARP); and Douglas fir forest on the Santa Rita Mountains (*March 11, 1957* — Marshall, WJS, race not checked). For further details, see the accompanying map.

301. RUBY-CROWNED KINGLET — *Regulus calendula* (Linnaeus)

Common summer resident of the more extensive boreal zone forests of northern and locally southeastern Arizona, west and south to the Kaibab and entire Mogollon Plateaux and the Graham, Santa Catalina (very locally, 1959–1961), and Chiricahua Mountains. Common transient throughout state. Winters commonly in Lower Sonoran Zone, more sparingly in Upper Sonoran Zone and forests south and west of the Mogollon Plateau, and rarely upon or north of, that plateau.

In appearance this is very much like the Golden-crown except for a plain head, varied only by a white eye-ring (which is interrupted at the top). The call is a long chatter, but the song is of remarkable beauty and variety, commencing with three notes like the call of the Golden-crowned Kinglet. It is an abundant bird in Arizona, and in most parts is the only kinglet likely to be seen. It must, however, be carefully distinguished from the Hutton's Vireo, of identical plumage coloration but with a proportionately thicker, light colored bill. This vireo shares the kinglets' habit of flicking the wings (but less frequently).

As a wintering bird, the Ruby-crowned Kinglet is present abundantly in southern Arizona from October through April. The winter records from the northeast are *January 20 and 23 (1936* at Lakeside — Jacot, ARP); and it remained to December 19 (1936 at Tuba City — Monson), but was not seen there later. Some December records at Flagstaff may be of late migrants, for perhaps this species has a "frost flight," as the Dutch term a mid-winter migration in response to cold weather.

There are numerous records from the mountains in September and May; probably these are migrants of the breeding race (which must prefer to migrate at high altitudes), for they are fairly numerous on the tops of the Pinaleno, Catalina, and Chiricahua Mountains where the species breeds only sparsely; and they are on the heights of other mountains where there is no summer population: Hualapai, Santa Rita, and Huachuca. Northern Arizona migrations are from *mid-March* to mid-May

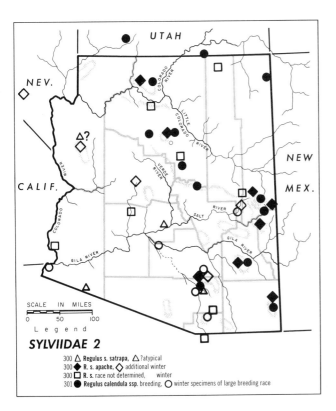

SYLVIIDAE 2
Legend
300 △ *Regulus s. satrapa*, △?atypical
300 ◆ *R. s. apache,* ◇ additional winter
300 ☐ *R. s.* race not determined, winter
301 ● *Regulus calendula* ssp. breeding, ○ winter specimens of large breeding race

and from *latter August* to *late November,* all near Flagstaff (Phillips, Hargrave, MNA).

As a breeding bird, the Ruby-crowned Kinglet occupies forests of alpine fir (also called corkbark fir) and spruce, high in the mountains. On the north slope of the summit of the Santa Catalina Mountains is a forest of corkbark fir in which this kinglet has been sought for years. (Lusk's and Willard's alleged breeding records are doubted because they do not mention the much more common Golden-crowned Kinglet). Finally on *May 29* and June 1, *1959,* W. G. George discovered at least four loudly singing males in the corkbark fir grove and in Douglas firs he collected one (WGG) which had enlarged testes. The following year, he found them singing again on April 20, and considered this the arrival date. Then at the same spot in 1961, Marshall recorded the song of one on tape — of interest because the bird had the order of phrases backward from the usual pattern! Of course wintering birds and migrants display and sing in spring, so that mere song does not prove breeding, and we still lack the evidence of a nest in the Santa Catalinas.

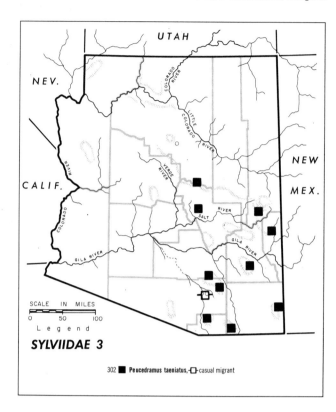

SYLVIIDAE 3

302 ■ Peucedramus taeniatus, □ casual migrant

The Ruby-crowned Kinglet shows geographic variation in size and in color, but an understanding of the latter is in great confusion owing to the profound fading that makes summer birds look wholly different from fall specimens of the same population. In fresh plumage, all these kinglets are exactly alike except for the distinctive race on Guadalupe Island and for the small, dark, richly-colored race *grinnelli* Palmer of the humid northwest coast. *Grinnelli* is not considered a bird of Arizona, though periodically we line up and compare the darkest of all our winter specimens and wonder if *grinnelli* could fade that much during a month or two spent at Yuma. Such a specimen of *November 24, 1902* (Brown, ARIZ no. 3075) is not quite rich enough below nor small enough (wing 55.5 mm) for a female *grinnelli;* there are several others. The two races which are well substantiated as Arizona birds are of lighter color, which is no different in the two; they differ only in size. One still lacks a name!

Regulus calendula calendula is small (though larger than *grinnelli*), with the wing of males around 59-60.5 mm. It breeds across the north of the continent and down the Pacific Coast states in the mountains. It is the abundant winter bird in the lowlands of Arizona. Dates of specimens are from *September 9 (1952,* Santa Catalina Mountains — ARP) and *September 10 (1940,* Flagstaff — Phillips, MNA) to *April 29 (1931,* Huachucas — Jacot, ARIZ. and *May 10 (1933* at Wupatki National Monument — Hargrave, MNA). This race is not shown on the map, for it is ubiquitous, except for the northeast.

Regulus calendula ssp. is a larger bird, wing over 61 mm. in breeding males, which nests in the Rocky Mountains and the highest mountains of Arizona. Breeding locations and the few wintering records based on large specimens are shown on the accompanying map. It is recorded in the nesting area in the White Mountains from *May 15* (LLH) to *October 21 (1936* — Phillips and Correia, ARP). These large Rocky Mountains birds winter west to Yuma, and north sparingly to the Gila Valley and apparently the Whiteriver area (10 miles north, *December 9, 1936* and *March 2-20, 1937* — Phillips and Correia, ARP). This race is recorded off its breeding area from *August 19 (1936)* and *August 26 (1954,* both Flagstaff — Phillips, MNA and ARP) to *April 12–19 (1935* at Granville — Jacot, ARP), and exceptionally *May 10 (1953,* Colorado River at the southern tip of Nevada — ARP).

The type of *Regulus calendula "cineraceus* Grinnell" was collected *May 9, 1896* (Grinnell, MVZ) at Mount Wilson, California, which is some miles from any breeding habitat of this kinglet on the summit of the San Gabriels. It is very worn and gray, as any kinglet would be after this much wear and fading. A male, its wing measures only 60 mm, which is below the minimum of 60.5 for breeding Arizona males. Therefore it cannot be a member of the large Rocky Mountain race, which has heretofore borne its name, and it should be considered a synonym of *calendula.*

302. OLIVE WARBLER   *Peucedramus taeniatus* (Du Bus)

Fairly common summer resident, scarcer northward, in Transition and (locally) Canadian Zones of southeastern Arizona, north to the south edge of the Mogollon Plateau and west to Santa Rita and Santa Catalina Mountains. Uncommon winter resident in most, if not all, of its breeding range.

This is a gray and white bird with white wing bars and white outer tail feathers. The female's head and lower throat is yellow, and the first year male is the same, in the Arizona race (*arizonae* Miller and Griscom). Here the first year male breeds in this plumage; he and the female thus bear considerable resemblance to the Hermit Warbler of the family Parulidae, but the dusky wedge back of the eye is heavier in the Olive. Adult males are easily distinguished by their golden-tawny head. The call is a descending whistle, *kew;* the song is of repeated phrases like that of a titmouse. The nest is made of rootlets and white fibers of the silver-leaf oak, decorated on the outside with pine-needle bracts and lichens. The eggs are dark.

The bird has been studied by George (Amer. Mus. Novit. 2103, 1962: 1-41). The line of reasoning by which the Olive Warbler is considered migratory is its rarity on the mountains in winter, and the presence of the species at Tepic, Nayarit in winter, where none occurs in summer. In the Santa Catalina Mountains it usually cannot be found in its favorite groves in winter, although there are several December records of solitary birds (including specimens — George, *op. cit.*: 36). A pair was in Bear Canyon on February 7 (1953 — Marshall). By March, groups wander through the pines: a flock of 17 escaped snow on the summits on March 29, 1961, by moving downhill to the last pines at Bear Canyon, at about 6000 feet (Marshall). Favorite breeding areas in the Catalinas are ponderosa pines of the summit ridges; these trees are low owing to control by the wind, and in them the birds and their nests can be more easily seen than in tall white firs, which are preferred at least in the Rincon Mountains (Marshall, Condor 58, 1956: 92).

As shown on the accompanying map, the breeding range of the Olive Warbler extends northwest to Baker Butte (Mearns, Auk 7, 1890: 261), and includes the Sierra Ancha (R. Roy Johnson). It must be extremely rare in the Pinaleno Mountains, where found only by Henshaw (*September 25, 1874*). It is absent from forests of the Bradshaw and Pinal mountains, as well as of the San Francisco Peaks. Bailey (Pacific Coast Avif. no. 15, 1923: 49) claims an occasional descent by the Olive Warbler to the Upper Sonoran Zone, but this must be very exceptional, for Marshall's only such record below the lowest pines is of a bird in Emory oaks at the east base of the Santa Rita Mountains on January 29, 1955. A single lowland report is of a male at Indian Dam, on the Santa Cruz River, ten miles south of Tucson, May 25, 1947, seen by Erle D. and Virginia M. Morton.

WAGTAILS AND PIPITS
MOTACILLIDAE

303. WATER PIPIT; AMERICAN PIPIT *Anthus spinoletta* (Linnaeus)

Breeds above timberline in the San Francisco and White Mountains. Generally distributed about water, and even occurs on open plains in Sonoran and Transition Zones on migration. Common winter resident at streams, ponds, irrigated fields, etc., throughout the Lower Sonoran Zone; rare to uncommon in winter in open Upper Sonoran Zone.

The Water Pipit belongs to another Old World family, represented in the New World, south of Alaska and Greenland, by only a handful of *Anthus* (with several species in South America too). These are sparrow-sized birds of open country, with thin, pointed bills and larkspurs (long hind claw). Their behavior and calls are lark-like. Only the Water Pipit is likely to be seen in Arizona by the casual observer. It is a dark-backed, pale-breasted bird with white outer tail feathers. Its color varies conspicuously with the season, for there is a prealternate body molt (involving also a replacement of the long tertials and central tail-feathers which protect the flight feathers from wearing against the grass). Winter birds are streaked across the breast and down the flanks, but in summer, Arizona breeders are almost unstreaked. Colors are decidedly richest, with pinkish-buff below, right after molting; then they fade rapidly. The call is a shrill *pipit,* of two or three syllables at the same pitch. The birds walk along the ground, bobbing the head back and forth and wagging the tail, while they pick up their food from the surface. They may perch on rocks, but never in leafy trees. Quite gregarious in winter, large flocks can be seen in irrigated fields; yet scattered individuals will be found along the edges of streams and lakes. This is the only exclusively Arctic-Alpine Zone bird which breeds on our Arizona peaks.

The wintering period of this pipit in southern Arizona is principally from October to April. Extreme dates are from September 20 (1949 in Bill Williams Delta — Monson) and September 22 (1951 at Wikieup — Phillips) to May 3 (1887 at Camp Verde — Mearns) and May 9 (1947 on California side of Havasu Lake — Monson). Northern Arizona migrations are from March 21 to May 13 (both 1936 near Flagstaff — Phillips) and September 16 (1940 at Mormon Lake — Phillips) to November 30 (1956 at Valle on the Coconino Plains south of the Grand Canyon and south of Holbrook). At this last period of November 1956 (LLH) the birds were easy to see because they clustered along the edge of the highways through the grasslands, in order to get out of the snow. They fed under the Russian thistles there, far from any water. The 30th was the last day they were seen, for there were none December 2 or later.

Study of geographic variation in the Water Pipit, in which foxing and fading occurs to an outrageous degree, is further complicated by definite linkage between rich coloration and paucity of ventral streaking, two traits of the southern race. In other words, if a northern variant is rich, it will also have reduced streaking to make it look like the southern bird, from which it can be distinguished only by size. Other hazards are that most western mountains are unrepresented by recent, fresh-plumaged specimens. A single *September* bird from the Olympic Mountains of Washington (Jewett, US) is as dark-backed as eastern birds. If it represents a new race, then there is no telling whether our few dark Arizona winter birds come from the Olympics or the northeast!

Size increases to the south, and color becomes richer and darker to the east. The breeding bird of Arizona and the southern Rocky Mountains is *A. s. alticola* Todd. This is a large, pale bird, richly-colored beneath, which usually has a short hind claw. (The other races are all equally small and long-clawed). In fall plumage it is distinguished by large size and fainter streaking on the flanks; adult males are paler and grayer than females. *Alticola* has been taken on top of the White Mountains from *July* to *October 9* (*1936* — Poor, Watson, Phillips, ARP). Aside from three specimens at Springerville (*September 29, 1934* and *1937*, ARP) it is almost unknown in winter or migration. This race apparently migrates in a long flight to the Mexican Plateau, and doesn't touch down in the lowlands. The next station of known occurrence to the south of us is at El Salto, Durango, *September 30 (1955*–ARP).

The other three northern races are of equally small size, long hind claw, and (usually) heavy streaking below. *A. s. rubescens* (Tunstall) of the northeast is dark above. An occasional bird from Arizona may belong here: Yuma, *December 22, 1902* (Brown, ARIZ); Stone's Lake north of Ash Fork, *November 7, 1884* (Mearns, AMNH #53583) and La Noria, Sonora (just south of Lochiel, Santa Cruz County), *November 6, 1892* (Holzner, US #127325).

The northwestern races are pale like *alticola*. *A. s. pacificus* Todd breeds in the northwest except for Alaska and is the grayest race. Infrequent in Arizona, it occurs here from *October 22 (1951* at Wikieup – ARP) to *April 30 (1949* at Aguirre Lake near Sasabe – ARP).

Anthus spinoletta geophilus "Oberholser," insofar as it can be distinguished from *pacificus,* appears to be the abundant transient and winter resident in Arizona. It breeds in Alaska and is characterized by being browner above than *pacificus* (Lea and Edwards, Condor 52, 1950: 267). Migration and wintering periods given above, for the species as a whole, probably apply almost entirely to this form, though the actual specimen records are from a somewhat shorter span.

304. SPRAGUE'S PIPIT *Anthus spragueii* (Audubon)

Rare in fall and early spring in fields and grasslands south and west of the Mogollon Plateau; only seven records (all of single birds), extending from Sulphur Springs Valley (Robinette, MCZ) to the Colorado River (Monson), the northernmost near Wikieup in the Big Sandy Valley (ARP). One winter specimen: Tule Desert, Yuma County, *December 30, 1958* (GM).

Sprague's Pipit is named for Isaac Sprague, companion of Audubon on his trip up the Missouri River. Its piercing call is even higher-pitched and thinner than that of the Water Pipit. If seen on the ground, which is unlikely because of its concealing coloration and fondness for tall grass, the back shows pale streaks. In the hand, the hind claw is seen to be very long, from about 12 to 15 millimeters.

Like the longspurs, Baird's, Clay-colored, and Cassin's Sparrows, Lark Buntings and Dickcissels, this plains bird migrates from the northern Great Plains to the southern Great Plains, spreading westward across central New Mexico to southeastern Arizona. It is, however, much scarcer than any of the above-mentioned fringillids in Arizona, and we would call it casual but for the discovery of small numbers south of Sásabe, Sonora (Phillips and Amadon, Condor, 54, 1952: 166). Arizona records are from September 27 (1949, Topock – Monson) to *April 4 (1905* near Tucson – Kimball, MCZ). The remaining locality of record is Fort Huachuca (Cahoon, MCZ).

WAXWINGS AND SILKY FLYCATCHERS
BOMBYCILLIDAE

This is a small family of sleek, soft-plumaged, crested birds, larger than sparrows, with moderate to rather long tails.

305. BOHEMIAN WAXWING *Bombycilla garrulus* (Linnaeus)

A rare but fairly regular winter visitant to northern Arizona; extremely rare south of the northern border, but to be looked for almost anywhere in years of great flights.

The Bohemian Waxwing is solidly brownish-gray below with the chestnut crissum (under tail coverts) abruptly marked off. Otherwise it is very like the much more abundant Cedar Waxwing except for a little white in the wings, slightly larger size, and slight difference in the head-pattern. The call is a trifle rougher, or less sibilant. It is Holarctic in distribution, though rare in eastern North America and western Europe.

In Arizona it has been found most regularly at Grand Canyon Village, to May 9 (1922 – Kennard, Condor 26, 1924: 77). Hargrave saw one at Benson, December 5, 1948. In the great flight of 1931–1932, this species reached the Mexican border area (Baboquívari Mountains – Phillips, Condor 35, 1933: 229-230). The late spring dates of some Arizona records seem remarkable.

The only specimens, remarkably enough, are all from the Colorado River: Fort Mohave, *January 10, 1861* (not 1871 – Cooper, MVZ); Willow Beach below Hoover Dam, *April 30, 1938* (Grater, apparently now lost); and Davis Dam, *March 6, 1959* (two specimens – Musgrove collection at Kingman Union High School).

306. CEDAR WAXWING *Bombycilla cedrorum* Vieillot

An erratic winter visitant in Sonoran Zones south and west of Mogollon Plateau, often abundant in May in valleys and canyons, with occasional birds or flocks remaining even into June. Elsewhere in the state, occurs as a migrant but not known to remain all winter.

The Cedar Waxwing is a smooth gray-brown bird with a little black around the face and a yellow band across the tip of the tail. The belly shades to pale yellow, which extends onto the crissum. Waxwings derive their name from extraordinary shiny red tabs stuck to the secondaries of some individuals, regardless of sex, and possibly also irrespective of age. The Cedar Waxing calls a prolonged, thin, drawling *seeeee* which rises to a piercing chorus just before the flock flies off. These birds are eminently gregarious and have been seen to indulge in a curious habit of passing a berry up and down a line of birds.

Southern Arizona records begin with August 20 (1936 near Fort Apache – Poor and Watson) and *September 11 (1873,* thirty miles south of Fort Apache – Henshaw, US). A notoriously late spring migrant, the Cedar Waxwing's occurrences at that period have extended to *June 4 (1899,* Yuma – Brown, ARIZ); during the period of June 5–15 (1932, seen regularly in cherry trees near Prescott by H. L. and Ruth M. Crockett); and the "first third of June" (in flocks, southeastern Arizona – Brandt). One bird, really lost, was seen at Parker by Monson on June 27, 1953, and apparently the same individual on the 12th of July!

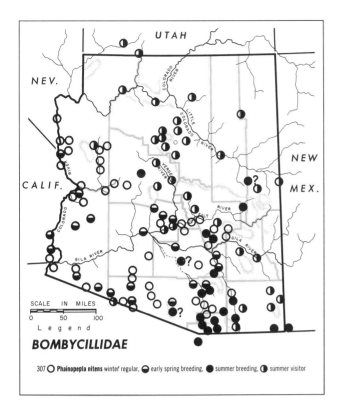

Northern Arizona dates extend from September 4 (1953, White Mountains – Phillips) to June 7 (1937 at Keams Canyon – Monson), June 10, 1943, and June 14 and 15 (1950, all at Grand Canyon Village – H. C. Bryant). There are no records between the end of November and the last of February, as yet.

Genus *PHAINOPEPLA*

by Allan Phillips

307. PHAINOPEPLA *Phainopepla nitens* (Swainson)

Present in different areas in varying roles, seasonally; probably few, if any, individuals are sedentary or non-migratory. It is also entirely possible for two populations, of different seasonal status, to be at the same place. Common winter resident in Lower Sonoran Zone of southern Arizona (except Verde Valley and most of the southeast); most common in southwest. Nests almost throughout Lower (and locally Upper) Sonoran Zone in brush containing mistletoe; later it absents itself from nearly all of western Arizona (east to include Big Sandy Valley and Papago Indian Reservation) from June until late August, or, most usually, early October. Abundant summer resident (and breeds) in parts of southeast, and bred in 1876 near Fort Apache in Upper Sonoran Zone (C. E. H. Aiken). There is a sprinkling of summer records from most of northern Arizona, and it is said to breed locally in the Snowflake area (A. J. and Louise Levine), and perhaps near Mount Trumbull in the northwest. In occasional winters remains far north and east of usual winter range: Fort Huachuca, *1886–1887* (Cahoon, MCZ); and in *1955–1956* to Sedona (Mr. and Mrs. Douglas Rigby) and the Chiricahua Mountains (Cazier, SWRS).

The slim male Phainopepla, with his red eye, is a handsome glossy black, hence the name, which means "shiny cloak." In flight, the wings show a white ellipse near the tip. The female is gray, with the wings nearly uniform. While the migrations are less erratic than the waxwings', they are nevertheless very extraordinary, as detailed below. Phainopeplas perch nearly upright atop mesquite and palo verde trees. The call is *lerp*. There has been much argument about the song, freely uttered in the dense nesting population which comes to the Tucson mesquite forest in spring. It is a sweet gargling, suggestive of the shrike in quality. Elsewhere, where less crowded, Phainopeplas rarely sing. In winter, in western Arizona, they feed largely on the sticky berries of mistletoe; the favorite perches are easily spotted by the pyramids of these seeds glued to the branches below.

Phainopeplas have no respect for the rules. Books tell us that a bird nests in only one place, which is in the coldest part of its habitat. Phainopeplas, however, do not read books, and certainly violate the second part of this, if not the whole rule. They have no love for the hot, dry summers of western Arizona, and they move out *en masse* as soon as the young are independent, in May and probably early June. They are virtually absent thereafter until the end of August, when an occasional bird may appear, but they do not become common in the Big Sandy and Colorado Valleys until October!

Conclusions on molts and ecology, based on the assumption that Phainopeplas are sedentary, must be reexamined. Hensley (Wilson Bull. 71, 1959: 90-91) failed to make regular counts on the desert in summer, we feel sure, unless indeed 1948 and 1949 were exceptional years on the Organ Pipe Cactus National Monument. Virtually all had left the Ajo Mountains region, on and adjacent to the Monument, by early June (1940–Sutton and Phillips, Condor 44, 1942: 62) and by May 21 (1947 – Phillips and Pulich, Condor 50, 1948: 272); later in 1947, Superintendent Supernaugh saw them only at Quitovaquita (except once in June) before their general return starting about October 7. Thus we can hardly believe they were "common throughout" July in 1948 or 1949. (In late July 1961, when organ-pipe cactus fruit was ripe, some males appeared just northwest of Quitovaquita in the Agua Dulce Mountains – Monson).

A second population of Phainopeplas swarms into the mesquite and riparian trees of the river valleys of the southeast (Nogales – Patagonia region, south of Tucson, etc.), spreading less numerously to central Arizona, in April or May. In the mesquite-elderberry *bosque* of the San Xavier Indian Reservation, arrival seems to vary greatly, but the males come first; a few were seen on April 15 (1956 – Tucson Bird Club) and many by April 16

(1958 — Marshall); but in the abnormally wet spring of 1952, very few were in evidence as late as May 4 (Phillips). On Salt River at Tempe, new arrivals were common April 21 (1940 — Hargrave Yaeger). At Benson the first was seen April 20 (1943) and the bulk came the next month (Hargrave). Nesting starts as soon as the females arrive in this whole region, and may continue to mid-July (near Fort Apache — Aiken). Yet a flock of eight flying south over Benson, July 6 (1944 — Hargrave) may have been migrating already; for the last was seen on the San Xavier Reservation bosque on August 5 in 1959, and in mid- to late August in later years, with one record of two birds on September 30 (1956 — Marshall *et al.*). Likewise they leave the Patagonia region from late July to about the middle of September (Phillips). In 1956 very few were seen near Globe between July 23 and October 12, nor between March 28 and May 13 (Harold A. Marsh). It was thus unexpected to find one still at Prescott, September 8, (1949) and at least six in the junipers 23 kilometers north, September 13 (1951 — Phillips)!

In the northeast the Phainopepla is a local and uncommon visitor in summer, not known to breed except possibly at Snowflake (Mr. and Mrs. A. J. Levine). Extreme dates for the scattered occurrences — all in wooded parts of the Upper Sonoran Zone — are *June 14* (*1938* north of Williams — Hargrave, MNA) to September 3 (1953, at least four seen by Mrs. Louise Levine and Phillips near Shumway).

The result of all these movements is that Phainopeplas are almost ubiquitous in the deserts and river valleys in early May, whereas by mid-September they can be found only locally. This enigma might be solved by wholesale netting and marking. One is tempted to guess that the May exodus from the deserts goes largely to the coast, as is certainly the case with the males of Costa's Hummingbird. The peculiarities in life history (see Scott, Auk 2, 1885: 242-246; Rand and Rand, Auk 60, 1943: 333-341) and the unusual migrations make this one of the most interesting of Arizona birds, and one on which a great deal of detailed information is needed. Here is a field for research!

In his *Notes of a Military Reconnaissance,* 1848: 83, Emory states that on November 11, 1846, between the Gila River at a point 8 to 9 miles east of the Pima Villages and the "Casa Montezuma" ruin to the north "we secured . . . our long sought bird, the inhabitant of the mesquite, indigo plumage, with top knot and long tail. Its wings, when spread, showing a white ellipse." Obviously referring to the Phainopepla, this is filed under "Magpie" (FW files) and is doubtless responsible for some of the misconceptions concerning the Magpie's former distribution in Arizona.

We shall leave this difficult and interesting bird for the present, with an attempt at a summary. The Phainopepla exists in three roles in Arizona: (1) as a winter resident in the southwest, where it breeds in spring and leaves the deserts in summer; (2) as a summer resident in the south-central and southeastern parts of the state; and (3) as a summer visitor to the northeast — probably mostly post-breeding birds. Numbers (1) and (2)

above are entirely different populations which differ in migration, ecology, and nesting time. Unfortunately no racial morphologic traits have yet been discovered by which these populations could be identified and traced, but we at least know that their overall breeding distribution overlaps through parts of south-central Arizona, as at Globe and Tucson. At Tucson, the "winter to spring" population lives principally in the open mesquite desert along the Rillito Valley north of town, where there is plenty of mistletoe. The "summer" population is a dense one inhabiting the thick mesquite bosque, wherein grow many elders, preferred for nest sites and feeding.

SHRIKES *LANIIDAE*

Shrikes are a large Old World family, barely reaching the New, where there are two closely similar species, both found in Arizona. They are known as butcher birds because they hang meat on thorns. Ours are mostly gray, white, and black in a prominent pattern.

308. NORTHERN SHRIKE *Lanius excubitor* Linnaeus

Only four verifiable records: Prescott, *February 6, 1865* (Coues, US); near Flagstaff, *January 7, 1936* (Kiessling, MNA); extreme northwestern Coconino County, six miles south of Johnson, Utah, *November 27, 1937* (Arnold Ross, UT); and near Show Low, *December 21, 1959* (Levy, US). Probably fairly regular in north, but almost no work done there in winter, and very hard to identify.

This species might not be as rare in Arizona as the few records indicate, were it easier to identify. As it is, brown-washed individuals with distinct fine barring below may be confidently identified as young Northern Shrikes. But the adults are almost indistinguishable from the Loggerhead Shrike. In the hand, they would be identified as stated in Zimmerman's well-illustrated article on field and hand identification (Wilson Bull. 67, 1955: 200-208). The few shrikes occurring on the Mogollon Plateau in winter merit a careful scrutiny, for it is uncertain whether some or all might be of the Northern species.

Presumably the Arizona specimens all belong to the pale race of northwestern North America, *invictus* Grinnell.

309. LOGGERHEAD SHRIKE; WHITE-RUMPED SHRIKE *Lanius ludovicianus* Linnaeus

More or less common summer resident throughout open parts of state (except brushless grassland) below Transition Zone, rather uncommon (at least in mid-summer) along Mexican border west of Baboquívari Mountains.

Arizona
Birds
In Color
reproductions of photographs made in the field by
ELIOT
PORTER

INDEX TO THE COLOR PLATES

Species Name and Reference Number ( ) and Text Page Number	Illustration Page
White-winged Dove *Zenaida asiatica* (151) 41	one
Roadrunner *Geococcyx californianus* (157) 45	two
Common Screech-Owl *Otus asio* (160) 47	two
Texas Nighthawk *Chordeiles acutipennis* (176) 57	two
Black-chinned Hummingbird *Archilochus alexandri* (181) 62	three
Rufous Hummingbird *Selasphorus rufus* (185) 63	three
Rivoli's Hummingbird *Eugenes fulgens* (189) 65	four
Blue-throated Hummingbird *Lampornis clemenciae* (190) 65	four
Violet-crowned Hummingbird *Amazilia violiceps* (191) 65	five
Broad-billed Hummingbird *Cynanthus latirostris* (193) 66	five
Gilded Flicker *Colaptes auratus* (197) 68	six
Gila Woodpecker *Centurus uropygialis* (198) 69	seven
Hairy Woodpecker *Dendrocopos villosus* (204) 74	eight
Arizona Woodpecker *Dendrocopos stricklandi* (208) 75	eight
Ladder-backed Woodpecker *Dendrocopos scalaris* (206) 75	nine
Ash-throated Flycatcher *Myiarchus cinerascens* (220) 81	ten
Dusky Flycatcher *Empidonax oberholseri* (228) 87	ten
Vermilion Flycatcher *Pyrocephalus rubinus* (239) 91	eleven
Violet-green Swallow *Tachycineta thalassina* (242) 95	twelve
Arizona Jay *Aphelocoma ultramarina* (252) 105	thirteen
Bridled Titmouse *Parus wollweberi* (263) 110	fourteen
Verdin *Auriparus flaviceps* (266) 113	fourteen
Cactus Wren *Campylorhynchus brunneicapillus* (275) 119	fifteen
Rock Wren *Salpinctes obsoletus* (278) 120	fifteen
Curve-billed Thrasher *Toxostoma curvirostre* (283) 123	sixteen
Black-tailed Gnatcatcher *Polioptila melanura* (299) 134	seventeen
Olive Warbler *Peucedramus taeniatus* (302) 136	seventeen
Phainopepla *Phainopepla nitens* (307) 139	eighteen
Bell's Vireo *Vireo bellii* (312) 142	nineteen
Plumbeous (Solitary) Vireo *Vireo solitarius* (315) 144	twenty
Orange-crowned Warbler *Helminthophila celata* (323) 146	twenty one
Lucy's Warbler *Helminthophila luciae* (325) 148	twenty two
Gray-headed (Virginia's) Warbler *Helminthophila ruficapilla* (324) 148	twenty two
Grace's Warbler *Dendroica graciae* (336) 155	twenty three

INDEX TO THE COLOR PLATES

Species Name and Reference Number ( ) and Text Page Number	Illustration Page
Red-faced Warbler *Cardellina rubrifrons* (342) 159	twenty three
Painted Redstart *Setophaga picta* (346) 161	twenty four
Hooded Oriole *Icterus cucullatus* (356) 167	twenty five
Scott's Oriole *Icterus parisorum* (357) 168	twenty six
Western Tanager *Piranga ludoviciana* (365) 174	twenty seven
Cardinal *Cardinalis cardinalis* (369) 176	twenty eight
Pyrrhuloxia *Cardinalis sinuatus* (370) 177	twenty nine
Common (Lazuli) Bunting *Passerina cyanea* (373) 179	thirty
House Finch *Carpodacus mexicanus* (381) 185	thirty
Rufous-winged Sparrow *Aimophila carpalis* (403) 198	thirty one
Black-throated Sparrow *Aimophila bilineata* (409) 202	thirty one
Yellow-eyed Junco *Junco phaeonotus* (411) 206	thirty two

White-winged Dove
Zenaida asiatica

Roadrunner
Geococcyx californianus

Common
Screech-Owl
Otus asio

Texas Nighthawk
Chordeiles acutipennis

PAGE TWO

Rufous Hummingbird
Selasphorus rufus

Black-chinned Hummingbird
Archilochus alexandri

Rivoli's Hummingbird
Eugenes fulgens

Blue-throated Hummingbird
Lampornis clemenciae

Violet-crowned Hummingbird
Amazilia violiceps

Broad-billed Hummingbird
Cynanthus latirostris

Gilded Flickers
Colaptes auratus

Gila Woodpecker
Centurus uropygialis

Hairy Woodpecker
Dendrocopos villosus

Arizona Woodpecker
Dendrocopos stricklandi

Ladder-backed Woodpecker
Dendrocopos scalaris

Ash-throated Flycatcher
Myiarchus cinerascens

Dusky Flycatcher
Empidonax oberholseri

Vermilion Flycatcher
Pyrocephalus rubinus

Violet-green Swallow
Tachycineta thalassina

Arizona Jays
Aphelocoma ultramarina

Bridled Titmouse
Parus wollweberi

Verdin
Auriparus flaviceps

Cactus Wren
Campylorhynchus brunneicapillus

Rock Wren
Salpinctes obsoletus

Curve-billed Thrasher
Toxostoma curvirostre

Black-tailed Gnatcatcher
Polioptila melanura

Olive Warbler
Peucedramus taeniatus

Painted Redstart
Setophaga picta

Hooded Oriole
Icterus cucullatus

Scott's Oriole
Icterus parisorum

Western Tanager
Piranga ludoviciana

Cardinal (female)
Cardinalis cardinalis

male

Pyrrhuloxia
Cardinalis sinuatus

female

Common (Lazuli) Bunting
Passerina cyanea

House Finch
Carpodacus mexicanus

Rufous-winged Sparrow
Aimophila carpalis

Black-throated Sparrow
Aimophila bilineata

Yellow-eyed Junco
Junco phaeonotus

Uncommon transient in Transition Zone. Winters commonly in Lower Sonoran Zone, less commonly in open Upper Sonoran Zone.

The Loggerhead Shrike somewhat resembles a Mockingbird in its general coloration, love of exposed perches, and song. It is, however, blunt-headed and it has a purposeful, direct flight with a sweep upward at the end. The wing-beats are very rapid. A black mask through the eye, set off by whitish above and below, unites with the stout black bill to form a strong head-pattern, unlike a Mockingbird. The folded wing also shows contrasting black. The call is long and harsh. Loggerhead Shrikes are among the earliest passerines after the thrashers to nest on the deserts of western and central Arizona. Young out of the nest were taken by van Rossem and Phillips in Sonora south of Ajo on *March 25* (*1947* – LA, ARP); and at Tucson, a nearly fully grown juvenile was seen on April 1 (1956 – John J. Stophlet and Phillips). For a thorough study of shrikes, see Miller's *Systematic revision and natural history of the American shrikes* . . . (Univ. Calif. Publ. Zool. 38, 1931: 11-242). Miller regretted never having seen a shrike carry prey in the feet, of which there had been a number of recorded instances. (A recent account of a Horned Lark being carried fails to mention whether it was in the bill or the feet – Wiggins, Condor 64, 1962: 78-79!) Doubtless the bird's instinct to impale the prey, in order to secure it while tearing it up with the bill, is a compensation for weak feet as compared with those of hawks. But to carry a bird as large as a sparrow in the bill would certainly make the shrike front-heavy and its flight difficult. The following observation by Marshall, then, is of interest in showing the capacity of the feet of the Loggerhead Shrike for carrying large prey. This bird flew across a street of Tucson with something that seemed nearly as big as itself, carried aligned beneath the belly as an airplane carries a torpedo. The shrike was flying a yard or two above the ground and it continued for 75 yards across a field, without any apparent difficulty. But it did come to the ground beneath a little bush, leaving the prey there when we ran up to see what it was. An adult male House Finch, which was still warm and which apparently had been robust, was the dead prey. We retired while the shrike returned to the bush, hopped around, and finally flew off from the ground with the House Finch again in the feet, head-first. This flight terminated in the usual ascent, to about 20 feet in a tree, but it was not learned if the finch had been dropped just at the moment of climbing.

We recognize two races of the Loggerhead Shrike in western North America, north of Mexico: *excubitorides* Swainson of the interior, pale above and on the chest; and *gambeli* Ridgway with a more barred chest and darker upperparts, sex for sex. *Lanius ludovicianus excubitorides* is the breeding bird of Arizona. There are several types of status (MNA, ARP): (1) Resident, but more common in winter, in southwestern Arizona. A post-breeding emigration in May and early June is suspected. (2) About equally common at all seasons, in Lower Sonoran valleys of southeastern, central, and northwestern Arizona. (3) Mostly a common summer resident, in northeastern Arizona (with a few winter records). (4) Transient on the Mogollon Plateau, principally in *April* and *August*.

Miller (*op. cit.*) has subdivided *excubitorides* into several races which certainly show considerable average differences in measurements and amount of white on the tail. But individual variation is so great that migrants cannot be safely assigned, and these races cannot teach us anything. He had belts of intergradation across the Mogollon Plateau and near Mazatlán, Sinaloa, but no shrikes breed in either area.

L. l. gambeli (MNA, ARP) is probably state-wide in winter; it has been taken in northern Arizona from the last of *August* to *May 3* and it outnumbers *excubitorides* there in winter. In southern Arizona, where the two races are then about equally common, typical *gambeli* have been taken from *October 2* to *March*. Migrants across the Mogollon Plateau are known from late *August* to *September* and again in *April*.

STARLINGS *STURNIDAE*

310. STARLING *Sturnus vulgaris* Linnaeus

Since its first known occurrence in the state in 1946 (near Lupton – Wetherill), the Starling has become established as a common breeding resident in the Phoenix area, and to a lesser extent at Kingman, Tucson, and Yuma. It occurs in winter in any settled location in the Sonoran zones, chiefly in farmlands, and abundantly near Kingman and Holbrook. Numbers seen in southern Utah as early as January 1941 (Grater), but a report from southern Nevada in August 1938 (*fide* Cottam, Condor 43, 1941: 293-294) seems doubtful.

The Starling, about the size and shape of a Meadowlark, is an all-black bird in summer, but is heavily speckled with white in winter. Juveniles are plain gray. The bill turns yellow in February or March and stays that color through the breeding season. Starlings walk on strong legs and do not hop. No other black bird in Arizona has so short a tail. Their recent increase in Arizona bodes ill for our native woodpeckers and other hole-nesters such as the Purple Martin, small owls and *Myiarchus* flycatchers. Thus far Starlings have confined their nesting in Arizona to towns, irrigated farmlands, and adjacent saguaro desert. Perhaps they will not extend far out into the saguaros, but at any rate it is disgusting to see the Martins arriving in May to inspect saguaro holes already full of the abominable Starling families – sometimes two families in a single saguaro! These birds should have been left in Europe, where Starlings belong. There, some very important migrational studies have been conducted, both by observation (including the use of radar) and experiment.

Roosts and winter flocks run into the thousands of individuals. A flock of 100,000 was seen at the Litchfield Park cattle pens, end of January, 1958 (Housholder, Simpson, and Werner). The first known Arizona nest was near Glendale in the spring of 1954 (Bernard Roer, Forrest L. Stroup). The second-hand report from southern Nevada in June 1947 (Johnson and Richardson, Condor 54, 1952: 358) seems just as unlikely as the earlier one (above). As a winter visitant, however, the species was widespread by 1947 (Foersters et al., January, at Patagonia). But it remained almost static for some years thereafter, and none was again seen south of the Tucson Valley (where it was first discovered on the University of Arizona campus on December 30, 1949, by Raymond J. Hock) until 1959 (one, July 23, at Nogales – Phillips).

VIREOS VIREONIDAE

The size of this Pan-American family depends on whether you choose to include the extremely similar *Smaragdolanius, Vireolanius,* and *Cyclarhis* from southern Mexico on south. In any case it is not a numerous family, and our species are all smaller than a sparrow. They are plain birds whose name means "I am green," though most Arizona forms are gray. The head is rather large, with a somewhat thickened bill, distinguishing vireos from the sharp-billed and more active kinglets and wood-warblers. The nest is a handsome gray cup suspended from the fork of a twig. Vireos sing on the nest.

[WHITE-EYED VIREO *Vireo griseus* (Boddaert)]

[Erroneously reported from "Cyanthanis" (= Grantham's), Sulphur Springs Valley; the specimen has been examined and proves to be a Gray Flycatcher, *Empidonax wrightii* Baird.

This vireo was credited to Arizona on the basis of an *Empidonax* identified by Harry C. Oberholser at the Fish and Wildlife Service and marked on the reverse of the label as "*griseus* HCO." Unfortunately the other side of the label said "*Vireo*," and the bird was accordingly filed as *Vireo griseus* by a secretary!]

311. HUTTON'S VIREO; STEPHENS' VIREO *Vireo huttoni* Cassin

Rather common summer resident in the denser live oak brush and woods of southeastern Arizona, north sparingly to the Mazatzal Mountains and the Whiteriver-Fort Apache area, and west to the Santa Catalina and Pajaritos Mountains. Fairly common winter resident from Patagonia and the Santa Catalina Mountains west to the Baboquívari and even Quijotoa Mountains, north rarely to Salt River and perhaps the foot of the Natanes Plateau in Gila County (November 10 – ARP, wintering?); mostly in streamside trees and brush and in the lower oaks. Found uncommonly in early fall in the higher conifer forests adjacent to breeding areas. Rare west to the Bill Williams Delta, *November 24, 1953,* the Colorado River above Imperial Dam, *November 13, 1960,* and the Kofa Mountains, *September 25, 1956* and September 29, 1960 (all GM).

The olive-gray Hutton's Vireo is almost identical in coloration and pattern to the Ruby-crowned Kinglet, from which it can be told by its larger and non-black feet and bill. A prolonged mewing whinny, of typical vireo quality, immediately sets it off from the kinglet. The song is a monotonous repetition of a single phrase. This is a characteristic bird of the live-oaks, ranging from Vancouver Island to Guatemala; and the species would be a rewarding one for study. It is tame and easily observed.

Though it can be found all-year-round in the oaks of the Santa Catalina Mountains (Marshall), there is no question that this vireo is partly migratory in Arizona. You can see or hear it in your garden during migrations and the winter. For instance, in the Tucson Valley (Lower Sonoran Zone) it has been recorded from *August 7* (*1938* – ARP) and September 25 (1940 – Monson) to April 23 (1943 – A. J. Foerster) and April 26 (1942 – Jacot). Arizona specimens belong to the palest, grayest race of this widespread but easily overlooked little bird: *stephensi* Brewster.

312. BELL'S VIREO; ARIZONA VIREO; LEAST VIREO *Vireo bellii* Audubon

Common summer resident in dense low brush, especially mesquite associations along streams, up to the top of the Lower Sonoran Zone; absent from the bottom of the Grand Canyon, the Mexican border west of the Ajo region, and from the area of the confluence of the

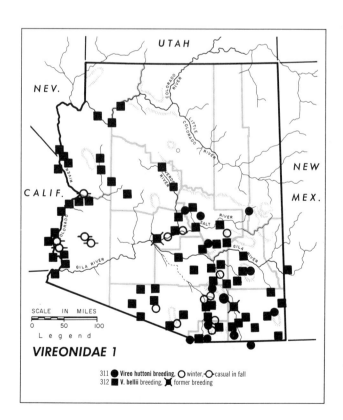

VIREONIDAE 1

311 ● *Vireo huttoni* breeding, ○ winter, ⊖ casual in fall
312 ■ *V. bellii* breeding, ✕ former breeding

Gila and Salt Rivers, but has been found on upper Colorado River (in recent years) to below Toroweap Valley, Grand Canyon National Monument (Dickerman, in 1953). Breeding range was formerly less discontinuous, as it bred at Phoenix in *1889* (V. Bailey). Scarce in recent years along Colorado River. Reduced or exterminated in some regions by cowbird parasitism. One winter specimen from Topock, February 7 to *March 7, 1951* (GM).

Bell's Vireo was named for John G. Bell, a well-known taxidermist who accompanied Audubon up the Missouri. This is our smallest vireo in Arizona, not much bigger than a kinglet, but with a relatively longer tail, which it waves about airily. Freshly molted birds have tinges of yellow and green on the flanks and back, respectively, but worn summer birds and juveniles are essentially white and gray, with wing-bars. A faint pale line runs from the bill backwards over the eye and down around half of the eye, leaving the lores dark. But the head shows no really strong contrasts. The song is most un-vireo-like and unmusical. The bird is constantly asking itself a question, then answering it: *Wee cha chu we chachui chee? Wee cha chu we chachui chew!*

Bell's Vireo is one of the principal victims of the increase of cowbirds (*Molothrus ater*) in recent years, being the only vireo that nests in the Lower Sonoran Zone, where cowbirds are commonest. While Lyndon L. Hargrave was living at Benson, his pair of vireos raised nothing but cowbirds for years, and finally the poor vireos disappeared. Two recent works on the breeding behavior and natural history, respectively, in the east are by Nolan (Condor 62, 1960: 225-244) and Barlow (Univ. Kansas Pub. Mus. Nat. Hist. 12, 1962: 241-296).

Like the Verdin and other birds which like mesquites, this vireo has spread out from the river bottoms as mesquites came in through abuse of the land; this has not extended its over-all distribution, however. It is authentically recorded in Arizona from March 8 (1910 along the Colorado River — Grinnell) and March 13 (1916 near Tucson — Howell) to *September 30 (1947* near Tucson — ARP), *October 9 (1935* at Madera Canyon — Gabrielson, US), and November 24 (1953 in Bill Williams Delta — Monson).

Arizona specimens belong to the race *V. b. arizonae* Ridgway, which is intermediate between yellow-green races to the east and a white-and-gray race to the west. Loss of the color tones by late spring (as in other vireos) is no doubt responsible for erroneous reports of *arizonae* in the Rio Grande Valley and of *pusillus* in the Colorado delta, Nevada, and southeastern California.

313. GRAY VIREO *Vireo vicinior* Coues

Fairly common summer resident in junipers of the Upper Sonoran Zone, in southeastern to northwestern Arizona, but absent from the Mexican border ranges west of the Chiricahua Mountains and from the area east of the Grand Canyon and north of the Little Colorado River and St. Johns. A rare migrant in Lower Sonoran Zone of southern and western Arizona during *September* and mid-*April* to early May (US, ARP, PH, MCZ, A. J. Foerster). Winters in small numbers in the mountains of Yuma County, and probably westernmost Pima County (not rare near Sonoyta in Sonora, *November 1955* — ARP), mostly south of the Gila River; one winter record from near Tucson, *December 31, 1949* (ARP). Erroneously reported from Hoover Dam area (specimens are *V. bellii*).

This is a very plain gray and white vireo with one wing-bar. The white eye-ring is complete but narrow. The tail, as in Bell's, is rather long and expressive. The song is like a Red-eyed or Solitary Vireo's but shortened. Though characteristic of the Upper Sonoran Zone, the exact habitat varies from place to place and includes mesquites along the northeast base of the Chiricahua Mountains, for instance. Gray Vireos usually prefer large junipers or chaparral with scattered trees. The summer range is relatively small, from southern California to west Texas; and the winter range appears to be still smaller, concentrating in central to northwestern Sonora. This vireo has a very strange call-note — a clear chatter.

The only arrival date available from the breeding grounds is April 1 (1885 — Scott) in the Santa Catalina Mountains. (The Gray Vireo must be very local there, for we have never encountered it — it would be looked for at the north base of these mountains.) It remains common in northwestern Arizona until about September 20; but the latest for the

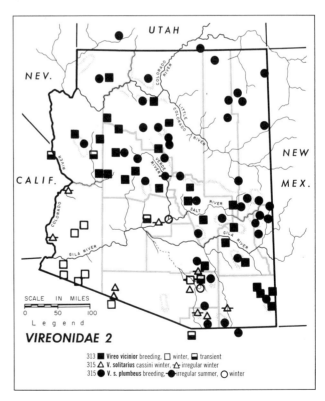

VIREONIDAE 2

313 ■ *Vireo vicinior* breeding, □ winter, ▬ transient
315 △ *V. solitarius cassini* winter, ⟁ irregular winter
315 ● *V. s. plumbeus* breeding, ⊝ irregular summer, ○ winter

north is September 16 (northeast of the San Francisco Peaks — Hargrave). It arrives on its wintering grounds as early as August 25 (1961, near Tule Well — Monson).

There is geographic variation in the Gray Vireo but it seems to be non-clinal, forming a mosaic, as in some of the other gray birds of the Great Basin and desert areas: the Plain Titmouse and the Bush-tit. Birds from the region of the type locality are pale, but darker ones occur both to the west (Hualapai Mountains and westward) and east (eastern Arizona).

314. YELLOW-THROATED VIREO *Vireo flavifrons* Vieillot

Three specimens: Chiricahua Mountains, *May 8, 1948;* near Mammoth, *August 17, 1948* (both ARP); and Bill Williams Delta, *October 10, 1953.* Also one record from Sierra Pinta, Yuma County, June 4, 1956 (last two by Monson).

315. SOLITARY VIREO; PLUMBEOUS VIREO; CASSIN'S VIREO *Vireo solitarius* (Wilson)

Common summer resident in Transition Zone throughout Arizona, and in the heavier vegetation of the Upper Sonoran Zone (locally and irregularly ranging down after breeding into cottonwood association of Lower Sonoran Zone). Common in migration statewide, principally in southwest in spring and mountains in fall. Winters in small numbers in willow-cottonwood association of Lower Sonoran Zone of central-southern and central Arizona, also in towns, dense mesquites, and casually in live oaks.

The two races of Solitary Vireo occurring in Arizona are so distinct that they can be recognized in the field. Indeed, it is not infrequent that one of each will be seen and heard singing in the same tree, during migration or winter. In this connection it must be remembered that vireos sing just about all the time, even on the nest. Our Arizona nesting birds are plain gray and white ("Plumbeous Vireo") but differ from the Gray Vireo in more extensive wing bars and eye marking. The latter is in the form of a broad question-mark starting at the bill, running back over the eye to encircle it, thus leaving only the lores dark. Most of our transients and winter visitants, however, show conspicuous yellowish on the sides and green on the back (formerly called "Cassin's Vireo"). The Solitary Vireo is not the only vireo that is solitary, since this is a characteristic of the family. Usually we see only one or two birds accompanying a flock of warblers or chickadees. The Solitary is the most frequently observed vireo in most of Arizona.

Vireo solitarius plumbeus Coues, our gray-and-white breeding race, is further distinguished in the hand by its large size, large bill, and broad white on the outer tail feather. It was formerly almost unknown away from its breeding range (one record, near Tucson, *May 3, 1884* — Stephens, AMNH); but starting in *1922,* with more careful search, it has been fairly common in both spring and fall at the mouth of the Verde River and at Tucson, where individuals winter rather regularly in the riparian cottonwoods and sometimes even in other shade trees. Southern Arizona spring migration is from early April to early May; extreme dates are April 9 (1943 in the Santa Rita Mountains — A. J. Foerster *et al.*) to *May 14* (*1937* near Redrock, New Mexico, the latest in lowlands — Paul A. Stewart, ARP). Fall migration in the south is chiefly during late September and October.

Northern Arizona occurrence is from the last of April to early October. Outside dates are April 30 (1934 at Flagstaff Hargrave) to *October 10* (*1948* in the Juniper Mountains — ARP). Not only do some birds move into the lowland cottonwoods right after breeding (for instance at Camp Verde and Montezuma Well, *July 19* and *August 8, 1916* — Taylor, US), but one had penetrated the Hudsonian Zone near the summit of the White Mountains, at about 10,800 feet altitude, by *August 13* (*1950* – ARP).

Vireo solitarius cassinii Xantus is the smaller yellow-and-green race, which is a common transient and uncommon winter resident in Arizona. It breeds along the Pacific Coast and to the north. Southern Arizona migrations are from late March through May and late August to October. (Extreme dates are March 24, 1940 in the Baboquívari Mountains — Monson, Phillips; *March 26, 1932* at Saguaro Lake on the Salt River — Hargrave, MNA; to *June 3, 1951* at Tucson — ARP; and again from *August 16, 1924* in the Chiricahua Mountains — Koelz, KANU; to *October 28, 1951* at Earp, California, opposite Parker — ARP). Northern Arizona migrations are in May and in August through September: *May 7* and *13* (*1936* at bottom of Grand Canyon — Grater, GCN; and near Flagstaff — Phillips, respectively); and again from *August 7* (*1934* in the Tsegi Canyons near Kayenta — Russell, UT) to *October 4* (*1934* at Eagar — Stevenson, ARP).

316. RED-EYED VIREO *Vireo olivaceus* (Linnaeus)

Rather rare *September* transient in deciduous vegetation in northern Arizona; casual in south, where there are specimens only from the Huachuca Mountains, supposedly on *May 20, 1895* (Lusk, *fide* Swarth), and from the Colorado River about 25 miles above Imperial Dam, *May 28, 1956* (GM). Accidental in and near Tucson, singing July 6, 1952, and May 17, 1958 (Phillips). Also taken in northern Sonora, *July, 1952* (ARP).

The Red-eyed Vireo is our largest vireo, but the size difference is not striking. It is very similar to both the Warbling and Philadelphia Vireos but its white superciliary is bordered above by a dusky margin, instead of meeting a uniformly colored crown. It also differs from the Philadelphia in having the yellow wash restricted to the sides and flanks, leaving the center of the breast white. The song (which has been heard in Arizona) is of the usual vireo type with single phrases, unlike the continuous warble of the Warbling Vireo. The coral red of the eye is not conspicuous and is present only in adults.

Vireo olivaceus olivaceus, which is the race occurring in the United States, is the northernmost and one of the dullest of the many forms. Others occur southward, chiefly in South America, whither all the races retire in winter.

317. PHILADELPHIA VIREO *Vireo philadelphicus* (Cassin)

Three fall specimens: near Tucson, *November 10, 1939,* and *October 7, 1953* (both ARP), and King Valley,

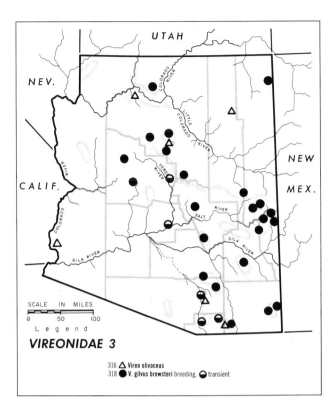

316 △ Vireo olivaceus
318 ● V. gilvus brewsteri breeding, ◐ transient

VIREONIDAE 3

south of Kofa Mountains, Yuma County, *October 27, 1954* (GM). Also a sight record in Tucson, September 28, 1942 (Alma J. Foerster). A fourth specimen lately taken south of Phoenix, *October 12, 1963* (Fr. Amadeo M. Rea, MVZ).

318. WARBLING VIREO — *Vireo gilvus* (Vieillot)

Rather common summer resident in the willows, maples, dense box-elders, and especially aspens from the Hudsonian Zone down, locally and irregularly, to the Upper Sonoran Zone; west to the Santa Catalina and Juniper Mountains. (Not in such adjacent Lower Sonoran places as St. Thomas, Nevada, and St. George, Utah, as stated by A.O.U. Check-list). Common in migration throughout the state. Observed February 23, 1945, Baboquívari Mountains (van Rossem) and February 17 near Phoenix (J. M. Simpson).

This is the plainest of our vireos: dull olive above and whitish below with a tinge of yellow on the flanks. There are no wing-bars or eye-ring and no pattern except a white superciliary. Warbling Vireos are common as migrants but nest only in the higher mountains. The migrations are very protracted; some birds are migrating through Arizona in every month from March to October. The breeding birds are partial to aspen groves, but they will accept other broad-leaved trees where sufficiently numerous within the coniferous forest. Indeed, Phillips found a nest in a beautiful green woods on Walnut Creek in the Juniper Mountains, essentially in the Upper Sonoran Zone — this being the only "eastern" looking woods he has seen in Arizona. Even on migration these vireos are not apt to be seen in conifers. Their song is a continuous warble, not broken into separate phrases, and it is often heard on spring migration.

Southern Arizona migrations are from *March 15 (1939* near Tucson, race *swainsoni* — ARP) to June 14 (1946, Topock — Monson), and again from early July (1957, lower Chiricahua Mountains — Lanyon), July 13 (1951, Bill Williams Delta — Monson); and *July 15 (1954,* Tucson — ARP) to *October 18 (1953* at Wikieup — ARP). Similarly extreme dates for northern Arizona migrations are from April 26 (1936, Betatakin Canyon — Wetherill) to May 23 (1937 at Keams Canyon — Monson); and from July 23 (1938 near Tuba City — Woodbury) to October 10 and 15 (1935, in Betatakin Canyon — Wetherill).

Vireo gilvus brewsteri (Ridgway) is our local breeding race of the mountains. This bird is larger than the other races, is slightly dark-crowned, and theoretically is grayish. It arrives very late in the spring, for the earliest that the species has been seen on the breeding ground in the high Chiricahua Mountains is on May 13 (1948 — Phillips *et al.*) — at a time, however, when there are still plenty of migrants, partly of this race, in lower, non-breeding areas to the southwest. For instance, Phillips and Marshall noted about 14 of the birds in the Sierra de los Ajos, Sonora, on *June 2, 1955* (WF, ARP). Even lowland migrants of *brewsteri* are rare before May 14, the earliest being *April 28 (1887* at Tucson — Brown, ARIZ). The migration route seems to be from the Sierra de los Ajos and Huachuca Mountains regions northeastward to Eagle Creek and east into New Mexico in spring. In fall, this race is virtually unknown as a migrant. In northern Arizona, all known records from the breeding area fall between *May 29 (1936,* at North Rim of the Grand Canyon — Grater, GCN) and *August 20 (1951,* near timberline on the San Francisco Peaks — Phillips, MNA).

Extreme rarity southeast of the Huachuca Mountains amounts almost to a profound gap in the breeding distribution of this race. Even in the Arizona border mountains there is some uncertainty as to its status. The implication that the species breeds "between 5000 and 7200 feet" in the Chiricahua Mountains (Tanner and Hardy, Am. Mus. Novit. no. 1866, 1958: 8) must be based on (1) migrants seen at low elevations and (2) failure to investigate the aspen groves! June specimens from the Santa Rita Mountains were thought not to be breeding, but again, there are some nice aspen groves high on the northeast slope of Mt. Wrightson, that hold the answer. Rancho la Arizona, in adjacent Sonora, is 'way too low for this species, whose supposed breeding there is unsupported by specimens in the Dickey Collection (LA — none of any race taken between *May 31* and *August 13, 1929).*

Vireo gilvus swainsoni Baird, the small Pacific coastal race, is the common migrant in Arizona. Theoretically it is also greener above and yellower on the flanks than is *brewsteri*. Migration periods are essentially those mentioned above in the account of the species as a whole, with allowances for a lesser span represented by actual specimens. (If any Arizona specimen approaches the eastern, light-crowned *gilvus*, it is likely to be that of Yaeger and Hargrave from Tempe, *May 5, 1940* — LLH #H296). An intermediate form, *leucopolius* Oberholser (small and gray) has so small a breeding area that its formal recognition hardly seems useful. Therefore we do not list our Arizona migrants of it.

WOOD WARBLERS
PARULIDAE

Wood warblers comprise a large and varied American family, of which may species are quite colorful; if not, then at least they usually show some yellow, red, or chestnut. Most are distinctly smaller than sparrows and move about actively in foliage, where with pointed slender bills they grasp insects. Of the Arizona kinds, most are transients or summer residents, wintering in Mexico. On the other hand, those of the eastern United States mostly winter in Middle and South America. Few species nest in the deserts, and those that do are confined to the rivers and washes. Most Wood Warblers have for call-notes a weak *sip* or metallic *chip*. Distinctive features of their songs, formed of high-pitched musical *chips* and *sweets,* had best be learned in the field. Unlike vireos they seldom sing until approaching the breeding grounds.

319. BLACK-AND-WHITE WARBLER *Mniotilta varia* (Linnaeus)

Rare winter resident near Tucson, at least since *1938* (ARP). Casual farther west and north; only three non-Tucson specimens, Huachuca Mountains, *May 27, 1936* (J. Stuart Rowley); Bill Williams Delta, *February 15, 1952* (GM), and Wickenburg, *November 22, 1959* (JSW); but two others from Boulder City, Nevada (Pulich, Phillips, ARP). There are sight records from the Salt River at Saguaro Lake (Hugh P. and Margaret Dearing) and from Parker (Monson). There are now five Tucson specimens, from five different years, plus sight records, which necessitate elevating this bird above the category of "casual."

Entirely streaked with the colors of its name, this warbler acts like a nuthatch by creeping around tree-trunks. In Arizona it is usually on willows and cottonwoods. It always arrives late at Tucson in the fall, inclusive dates being *October 20 (1954)* to *February 1 (1952* – both ARP). Extreme dates in the Bill Williams Delta are September 20 to *February 15* (both *1952* – GM). All southwestern birds are invariably immatures – in their first basic plumage.

320. PROTHONOTARY WARBLER *Protonotaria citrea* (Boddaert)

Two specimens: Tucson, *May 1, 1884* (Nelson, US) and Chiricahua Mountains, *September 8, 1924* (Kimball, MICH).

[WORM-EATING WARBLER *Helmitheros vermivorus* (Gmelin)]

[Hypothetical. One report only: Chiricahua Mountains, March 30, 1941 (Tom Kirksey).]

321. BLUE-WINGED WARBLER *Helminthophila pinus* (Linnaeus)

Casual: one record, Bill Williams Delta, *September 5, 1952* (GM).

322. TENNESSEE WARBLER *Helminthophila peregrina* (Wilson)

One specimen, Chiricahua Mountains, *April 7, 1925* (Kimball, MICH). A few recent sight records of migrants in the southeastern mountains. Also found in northern Sonora near Nogales, *October 14, 1954* (ARP).

Several of the sight records pertain to birds singing in spring. This bird certainly resembles the Warbling Vireo. When Marshall tried to collect one which was singing in deciduous trees near the summit of the Rincon Mountains (with a sling-shot), it cleverly traded places with a vireo at the last moment, and the Warbling Vireo specimen is all we have left to tell the tale.

323. ORANGE-CROWNED WARBLER; LUTESCENT WARBLER *Helminthophila celata* (Say)

Locally a summer resident in dense deciduous cover, chiefly willow thickets, in the Canadian Zone of eastern Arizona (White and Graham Mountains, and one locality in the Santa Catalina Mountains), possibly sparingly farther west (Mormon Lake). Common transient statewide. Winters uncommonly from Patagonia and Tucson west and northwest in dense brush; commonly along the Colorado River north to at least Davis Dam. Two midsummer records from southwestern Arizona: Kofa

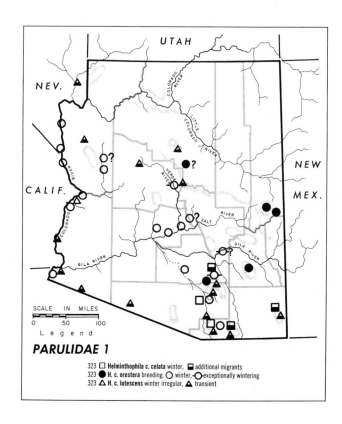

PARULIDAE 1

323 ☐ Helminthophila c. celata winter, ■ additional migrants
323 ● H. c. orestera breeding, ○ winter, -○- exceptionally wintering
323 △ H. c. lutescens winter irregular, ▲ transient

Mountains, June 25 and 26, 1956, and Colorado River 23 miles above Yuma, July 1, 1957 (both by Monson).

(The following was written by Allan Phillips in 1956, along with a Masked Bobwhite account, as samples for his projected book.)

DESCRIPTION: Length 4½ inches. A plain warbler without striking marks of any kind. A rather dark line through the eye, in some contrast to a paler superciliary above it, which is not white but of the same color as the throat. No other pattern on head, wings, or tail; breast often with a vaguely streaky appearance. The general color varies greatly with the sex, season, and subspecies; dull females of *H. c. celata*, in fall, are essentially pale gray below and olive-gray above, with rump slightly greener; while spring males of *H. c. lutescens* are green above, almost yellow-green on the rump, and nearly as bright yellow below as a Yellow Warbler. Most Arizona birds will be between these extremes, dull yellowish (never pure white) below and grayish green above. Worn breeding birds may show a tawny crown-patch.

FIELD MARKS: The slender pointed bill and plain head, with only a faint striping through the eye, identify the Orange-crowned Warbler. It has the usual *Helminthophila* habit of dipping the short, plain tail downward periodically, through a short arc.

VOICE: The usual call is a loud, sharp *chip* reminiscent of the Dark-eyed Junco. Its song, seldom heard away from the breeding grounds, is a quick tinkling series, the final half on a different pitch than the first and with a slightly harsher quality: *Weeee-ee-ee-ee-ee-chee-chee-chee-chee-chee-chee*. These loud, clear whistles are rather ringing without being of an explosive character, and carry fairly well. The song is sometimes preceded by a few slower notes.

A fairly common summer resident locally in dense willow thickets, and perhaps maple and even locally aspen groves of the high Transition and Canadian Zones. Common transient generally, abundant in southwestern Arizona. Winters fairly commonly, particularly along the low rivers of western and central Arizona, in streamside brush, thickets, and cottonwoods; locally in garden shrubbery of a dense type, such as privet hedges.

Southern Arizona migrations are from August 7 and 8 (*1951* in the Chiricahua Mountains — Marshall), *August 8* (*1940* near Patagonia — GM), and August 9 (*1953* near Parker — Monson), to at least October 22 (*1938*, Santa Catalina Mountains — Phillips); and at Overton, Nevada, to *October 29* (*1917* — Smith, AMNH). The spring movement is from February 18 (1955) and March 4–5 (1956 in the Sierra Pinta — Monson), to *May 26* (*1884* at Camp Verde — Mearns, US), and in the Sierra de los Ajos, northern Sonora, to *June 2* (*1955* — ARP). But aside from the southwest part of the state, the earliest spring migrants are recorded *March 15, 1939* (Tucson, ARP; Ajo Mountains, Huey, Trans. San Diego Soc. Nat. Hist. 9, 1942: 36).

Northern Arizona migrations are from *August 7, 1934*, on San Francisco Mountain (Stevenson, MVZ), August 12, 1936, in Betatakin Canyon, Navajo National Monument (Wetherill, *fide* Woodbury and Russell, Bds. Navajo Country), and August 16, 1931, at Flagstaff (Hargrave; banded), to October 12, 1938, at Keams Canyon (Monson, *fide idem.*), and from *April 25, 1948,* near Holbrook (ARP) to May 26, 1936, near Flagstaff (Phillips), if not later (see June 11 record above).

Subspecies: *H. c. orestera* (Oberholser) of the Rocky Mountains region is the common form in Arizona at most times and places, with status essentially that of the species. First arrivals of definite migrants in southern Arizona are *April 2* (*1939* in the Santa Rita Mountains — Jacot, PX) and *August 25* (*1924* in Chiricahuas — Kimball, MICH). Summer sight records for the North Rim of the Grand Canyon and Mormon Lake are merely presumed to pertain to *orestetra*; while the summer records for Yuma County cannot be assigned to any race. The latest northern Arizona date is *October 7* (*1956* at Grand Canyon Village — Geo. L. Shake and Hargrave, GCN). The latest southern Arizona migrant specimen is *October 20* (*1956* in the Chiricahua Mountains — E. Ordway, SWRS).

H. c. celata (Say) of the northeast is the scarcest of the three races in Arizona, but is apparently not rare as a transient from Tucson eastward. It also winters occasionally near Tucson (*December 29, 1946* — ARP) and Patagonia (*January 25, 1947* — ARP). Except for a returning migrant near Tucson, *February 24* (*1934* — Phillips, ARIZ), southern Arizona spring migration is from *April 14* (*1951* near Tucson — ARP) to *May 5* (*1956* in the Chiricahua Mountains — Phillips; ARP and SWRS), and (exceptionally?) to *May 26* (*1957* in the Chiricahua Mountains — ARP). Fall migration is from *September 12* (*1905* in the Huachuca Mountains — Marsden, MCZ) to *October 5* (*1884* near Oracle — Scott, AMNH). We could cite less typical specimens which would extend these periods.

H. c. lutescens (Ridgway) of the Pacific Coast is a fairly common transient, with migrations earlier than *orestera* except that the end of fall migration is equally late. In spring it is restricted to the southwestern part of the state, east to the Huachuca Mountains, Tucson, and Patagonia (Monson, GM), arriving earlier in the southwestern corner than elsewhere. In fall it covers much of Arizona, east to the Chiricahua (once, Kimball, MICH) and Huachuca Mountains, Camp Verde (Mearns), and Flagstaff (Phillips, MNA). Note, in all this, the extremely close parallel to the races of *Wilsonia pusilla*! Though wintering commonly in Sonora, only one typical *lutescens* has been taken in Arizona in winter: Parker, *December 11, 1946* (ARP). Southern Arizona migrations begin on the dates cited under the species (and *August 11, 1918*, in the Santa Rita Mountains — A. B. Howell, US). Fall migration extends to *October 21* (*1947* in the Ajo Mountains — ARP) and spring migration to *April 18* (*1885* at Fort Huachuca — Benson, US). A specimen from Tucson, *November 5, 1941* (GM), was probably a belated transient; while one from Tinajas Altas, southern Yuma County, on *February 16* (*1894* — Mearns and Holzner, US) was possibly an early spring transient.

In northern Arizona, the only specimens of *lutescens* are from Flagstaff, *August 23, 1947,* and *September 7, 1940* (Phillips, MNA).

[The remaining race, gray-streaked *H. c. sordida* Townsend of the channel islands, with a larger bill, was incorrectly reported in winter from Yuma on the basis of a soot-stained example of *orestera* — MICH.]

As a breeding bird in Arizona, the Orange-crowned Warbler is strangely restricted, being absent from the willow and gooseberry thickets of Flagstaff and the San Francisco Peaks, for example. Although specimens taken establish the fact that it does breed in Arizona, no nest has yet been found here. Its southern limit was in the willow bushes by a lovely, marshy meadow of iris and other flowers in the Santa Catalina Mountains. This delightful spot was the victim of land speculations, and at present is a house and parking lot in Summerhaven.

As a transient, the Orange-crowned Warbler is far less choosy. He will put in nearly anywhere, though he generally eschews conifers in favor of broad-leafed bushes and trees or even weeds. In late summer he is frequently out in a sunflower clump in some mountain park, or along a highway. Later you may often hear a loud noise in the cottonwoods of some river-bottom; look up, and you will see him in one of his most characteristic acts, searching the clumps of dry, crackling dead leaves for the insect larvae whose webbing holds them together and aloft. He is also given to searching the upper trunks and fluttering along the slanting limbs almost like a Bewick's Wren. Perhaps it is these unwarbler-like ways that help him survive winters much more rigorous than those to which most other warblers retire. Generally he minds his own business and keeps out of sight; but he is not shy, and will winter in the heart of Tucson almost every year if there are enough bushes and hedges for a modicum of privacy.

324. GRAY-HEADED WARBLER; RUFOUS-CAPPED WARBLER; NASHVILLE WARBLER; CALAVERAS WARBLER; VIRGINIA'S WARBLER *Helminthophila ruficapilla* (Wilson)

Common summer resident of low deciduous brush in the higher mountains throughout Arizona, chiefly in Transition and Canadian Zones; also breeds along low rivers north of Mogollon Plateau (Holbrook) in Upper Sonoran Zone. Descends locally (Santa Catalina Mountains) to foothills in July. Fairly common transient throughout Arizona, especially southwestward, in large brush and weeds and deciduous trees.

This is a slim warbler which has a leisurely up-and-down movement of the short tail. The head is gray with a white eye-ring and white or yellow throat strongly contrasting with the gray at the sub-malar line of demarcation. There is yellow both above and below the tail, and also more or less yellow on the chest, separated from the yellow crissum by a white belly. The northern races differ from *virginiae* in the extension of the chest mark over the breast and throat, and in a green wash over the back; whereas the back is plain gray in *virginiae,* our only nesting form. Since the green and gray forms can easily be distinguished in the field, we shall discuss their migrations separately.

We include here, as a single species, forms that have been accorded specific rank by other authors: the Nashville Warbler, Virginia's Warbler, and Colima Warbler. They all have the same song, call (*plisk*), tail-wagging, and habit of nesting on the ground, whereas *luciae* is a cranny nester (See Phillips, Anales Inst. Biol. Univ. Méx. 32, 1962: 364-365).

Geographic variation consists of a change from the green-and-yellow birds of the north, through the gray and less yellow race of the Rockies (and Arizona), to the large brownish-gray bird of the Chisos Mountains, Texas, and northeastern Mexico, which lacks yellow almost completely. Among the northern birds there is a cline, not very well marked, from east to west in increasing brightness, as in so many other species of wood warblers. Thus the Sierra Nevada breeders are the brightest, and Phillips thinks these contribute the bulk of our Arizona migrants.

This yellow-green race is called *Helminthopila ruficapilla ridgwayi* (van Rossem), known as the Nashville or Calaveras Warbler. It is a fairly common transient throughout Arizona. Here the spring and fall paths are complementary: in spring, mostly the southwestern and central lowlands; fall, mostly in the mountains of the east and northwest. Both migrations end abruptly and it is unusual to see the bird after the first few days of May and after the end of September. Spring migration in southern Arizona is recorded from March 27 (1943, California side of Imperial Dam; and 1955, Sierra Pinta, Yuma County — Monson) and March 27 and 31 (1948 and 1940, respectively, at Tucson — Phillips) to *May 12* (*1953* at Wickenburg — ARP) and May 15 (1910 at Pilot Knob, California — *fide* Grinnell). The only northern Arizona record is at Flagstaff, May 6, 1936 — Hargrave.

Fall migration records in southern Arizona are from *August 4* (*1918,* Santa Rita Mountains — Howell, US) to October 3 (*1953* at Tucson — ARP). Exceptionally late dates are October 9 (1953 near Prescott — Eleanor Pugh) and October 15–16 (1954, 27 miles south of Nogales, Sonora — Phillips). Northern Arizona records are from *August 7* (*1934,* San Francisco Mountains — Stevenson, ARP) to *September 27* (*1876* in the White Mountains — Aiken, AMNH).

Helminthopilla ruficapilla virginiae (Baird), known as Virginia's Warbler, is the gray breeding race of the Arizona mountains. It is also an uncommon transient, chiefly in spring, in the Lower Sonoran Zone west to the Pajaritos Mountains, Santa Cruz and Big Sandy Valleys. It is rarely seen before April 5 or after September 15. A remarkable concentration of 50 or more was seen in the mesquite thicket near Tucson about May 5, 1942 by E. C. Jacot and Guy Emerson! Extreme dates for southern Arizona migrations are from March 23 (in the lower Santa Catalina Mountains) to May 18 (1957 near Tucson – J. T. Bialac and Phillips); and again from July 18 (1938 near Tucson — Phillips) and *July 23* (*1946,* casual at Topock — GM) to *September 16* (*1905* in the Huachuca Mountains — Marsden, AMNH). Northern Arizona migrations have been recorded from April 5 and 16 (1953 near Flagstaff — E. Pugh) to May 10 (1929 at Phantom Ranch, bottom of Grand Canyon — Mrs. Pack, *fide* McKee) and *May 11* (*1933* near Bluff, Utah — *fide* Woodbury and Russell, UT), in spring. Fall dates are from July 29 (1939 near Winona, east of Flagstaff) to *September 25* (*1950,* 2 or 3 seen at Supai — Phillips, GCN).

325. LUCY'S WARBLER *Helminthophila luciae* (J. Cooper)

Abundant summer resident in dense mesquite and cottonwood-mesquite associations of the Lower Sonoran Zone, fairly common in ash-walnut-sycamore-live oak association of the Upper Sonoran Zone, in most of south-

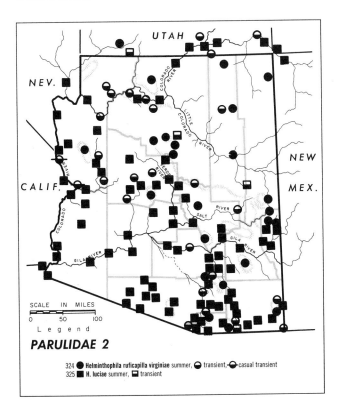

PARULIDAE 2

324 ● *Helminthophila ruficapilla virginiae* summer, ◐ transient, ◒ casual transient
325 ■ *H. luciae* summer, ◻ transient

ernmost Utah, April 28, (1931 — Greenhalgh, *fide* Woodbury, Condor 41, 1939: 161). Concerning Swarth's Miller Canyon birds, it is likely that the species now breeds there, owing to the increase of mesquite, for by 1947 it had become fairly common at and above Portal, a comparable habitat in the Chiricahua Mountains (Hargrave). In 1950, Phillips found about 15 on Date Creek on May 23, and a pair to each grove of big trees from Prescott to nearby Fort Whipple on May 29–30; this would be a brilliant increase since Coues' day, if only his confounded Fort Whipple reports, written apparently from very dim memory, could be trusted! Extreme desert localities are 5 miles southeast of Bouse (southeast of Parker), where Phillips noted a pair on May 7, 1952; Scott's Well and New Water Well (both north of the Kofa Mountains, May 26, 1955 and June 18, 1957, respectively — Monson).

326. PARULA WARBLER *Parula americana* (Linnaeus)

Rare fall and winter visitant across southern Arizona, with several records from the Tucson area (ARP) and casual ones west to the Colorado River (GM). Spring records are near Tucson, *March 26, 1938* (ARP); near Roosevelt Lake, May 30, 1952 (Dickerman); and Chiricahua Mountains, *May 25, 1957* (ARP). Though termed "casual" in A.O.U. Check-list, the period 1938–1953 produced three specimens, and two other individuals seen, near Tucson alone, and it bred in northern Sonora (*1952* — ARP and Marshall, WF).

The Parula is a small, intricately patterned warbler that is bluish-gray above, yellow below, and with the belly, wing-bars, eye- and tail-spots all white. The chest has a wash of cinnamon and, in the male of our northern form, a narrow black bar. The center of the back is yellowish-green.

The tropical forms, besides lacking the black on the chest, lack the white eye-spots and have a dusky border above the yellow throat. They live from central Sonora to South America. Both Paynter (Bull. M.C.Z. 116, 1957: 249-285) and Phillips (Anales del Inst. Biol. Univ. México 32, 1962: 365) consider all these Parulas to be conspecific. The fall and winter records extend from *September 16* (*1953* near Tucson ARP) to *January 5* (*1955* on the Colorado River 15 miles above Imperial Dam, in willow-arrowweed-camphor weed — GM).

All specimens and birds well seen in and about Arizona are of the typical northern race, *americana*. And except for the breeding pair in the Sierra de los Ajos, Sonora, all have been single birds. The lack of records from northern Arizona is not significant, for this region has been unworked in recent summers, and the species has been taken in western Colorado.

ern and central Arizona and along the entire Colorado River. Absent from deserts west of Growler Valley and from most of Phoenix area. In recent years has become scarce along lower Colorado River. Casual on and just north of Mogollon Plateau in migration.

Like the foregoing, Lucy's Warbler has a concealed rufous cap. It is a gray and white mite with a chestnut rump. There is no other pattern, and the gray upperparts shade into paler cheeks which do not therefore contrast with the eye-ring. The tail motion is more vigorous than in the preceding species. This is the only warbler which nests on the desert; the nest is behind slabs of bark, in sprouts on the trunk, or in cavities of trees. Juveniles have the rump buffy, not chestnut; and fall birds are warm buff beneath.

Lucy's and Yellow Warblers are by far the earliest non-wintering warblers to appear in spring, when by late March the territorial males are singing loudly from the tops of tall mesquites and cottonwoods. Lucy's song suggests the Yellow Warbler's, but is less emphatic and less organized.

Southern Arizona records are from March 10 (1910, three in the Chemehuevis Valley by the Colorado River — *fide* Grinnell) and March 11 (1939, several near Nogales — Dille) to *October 5* (at Valentine and at Tucson — ARP). More northern records are from April 16 (1937 in the far northwest at Pierce's Ferry — Hargrave) to August 10 (1938 at Lee's Ferry on the Colorado River — Woodbury).

This warbler is a very rare migrant away from its breeding grounds: Huachuca Mountains at mouth of Miller Canyon, April 8–12 (1902 — Swarth); near Flagstaff, *July 31* (*1934* — Jenks and Stevenson, ARP); and Kanab, south-

327. YELLOW WARBLER *Dendroica petechia* (Linnaeus)

Common summer resident in willows, cottonwoods, and sometimes sycamores, of Sonoran Zones almost throughout state, but peculiarly absent from some areas. Scarce to absent in recent years as a nesting bird along the lower Colorado River. Has bred once in willows of lower Canadian Zone on San Francisco Peaks (ARP–*morcomi*). Evidently locally reduced or exterminated by cowbird parasitism. Common migrant in all parts of Arizona except unbroken woodlands. Casual in winter along the Colorado River (Parker, two *morcomi*, *December 24, 1952* and *December 20, 1953* — GM).

Spring males are yellow below, on the sides of the head, and variably so on the top of the head. The breast and sides are streaked with chestnut. Spring females agree with males in having considerable yellow in the inner webs of the tail feathers; some resemble finely streaked males, but others vary greatly. Fall birds are never heavily streaked, and most of them are quite nondescript; even the yellow in the tail may be obsolete. The tertials are always pale-edged in fall and the eye-ring is pale, though seldom contrasted with its surroundings. These fall birds vary above from gray to greenish and from pale to dark; below from pale grayish to clear yellow. They are best recognized by their vigorous, sharp call, *chilp*. This warbler is another prime victim of the Brown-headed Cowbird in Arizona.

Spring migration of the Yellow Warbler is discussed by Phillips (Wilson Bull. 63, 1951: 130-131). It is under way in southeastern Arizona by mid-March, but begins nearly a month later in the lower Colorado Valley and in the late April in northern Arizona. It lasts through May, and in the northeast to June 10 (1933, common at Kayenta — Hargrave). The latest migrants in southern Arizona are June 9 (1956 in the Santa Rita Mountains — Bialac, Phillips, and Werner; see also *morcomi*). Wetmore (Condor 23, 1921: 63) noted birds south of Williams on July 8 and 14, 1918, which he suspected of initiating the fall migration. We, however, have no such early dates — the young are apparently driven from the breeding willow-cottonwood association in late June or July, when they become independent, but they do not seem to go far, at least at Tucson. The bulk fall migration begins at the start of August (earliest date July 29, 1934, on San Francisco Peaks — Stevenson). It continues (in the south and west) into early October. But the most interesting thing about the migration is the leap-frogging of northern populations past those of the south, for an understanding of which we must discuss the several geographic races. *Sonorana* is the first to enter and the first to quit Arizona, with no records after the first few days of September. May migrants are mostly *morcomi*. The early fall migration brings just about any race anywhere in the state in August; but the October migrants are mostly the dark northern birds of *rubiginosa* and *amnicola* persuasion, or are typical *morcomi*.

If *aestiva* and *petechia* are really conspecific, then they furnish an illuminating example of ecologic races, well distinguished, at least in males, by the chestnut head of the mangrove-inhabiting *petechia*. Good places to study contact between the two forms should be sought in Sonora, where the vast cottonwood forests of major rivers such as the Yaqui and Mayo may come close, at their mouths, to mangrove swamps.

But within the yellow-headed *aestiva* group, we find in general a cline in the western part of the continent from dark small birds in the north to large pale ones in the south. This cline is but weakly reflected in the east, where males are all heavily streaked on the breast and females are uniformly green and yellow. Females and immatures show great individual variation in the west. The darkest race of all is *rubiginosa*, of parts of Alaska; the next darkest, with crown and perhaps wing-bars lighter, is *amnicola* of the

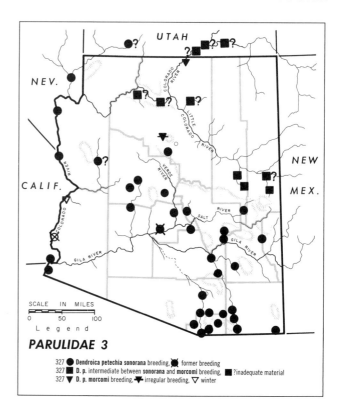

PARULIDAE 3

327 ● *Dendroica petechia sonorana* breeding, ◉ former breeding
327 ■ *D. p.* intermediate between *sonorana* and *morcomi* breeding, ■ ?inadequate material
327 ▼ *D. p. morcomi* breeding, ⤣ irregular breeding, ▽ winter

north; next is *morcomi*, breeding in north-central Arizona (of which *brewsteri* Grinnell is a synonym — all these are green-backed in the male.

Then there are two populations which are a paler, more yellowish-green above; one is the heavily-streaked small *aestiva* of the east; the other is the larger, finely-streaked breeding bird of northeastern Arizona, here referred arbitrarily to *morcomi*. Finally, the palest and yellowest race of all is *sonorana*, which breeds in the southwestern half of Arizona right across the state north to Oak Creek, Lake Mead, and possibly the Big Sandy.

[The following subspecies accounts are by Allan R. Phillips]

Dendroica petechia rubiginosa (Pallas) is apparently a statewide fall migrant except perhaps in the northeast corner. Most of the few specimens properly referable to this darkest race are from *September 24* (*1902*, Yuma — H. Brown, ARIZ; *1939*, Eagle Creek — LLH) to *October 7* (*1949*, Big Sandy River — ARP). The only spring specimen is a male from Yuma, *May 1916* (W. W. Brown, MCZ #325222), but there are others from nearby Bard, California, *May 12* and *29* (*1921* — Canfield, LA).

D. p. amnicola Batchelder is the widespread northern race to which Dr. Oberholser has referred about one third of the series of northern Arizona migrants (in MNA; see Hargrave, Condor 38, 1936: 121). These dark Yellow Warblers are widespread through northeastern Arizona in fall, and they occur south to Flagstaff (MNA). In spring however, they are rare, and recorded only in the southwest, from the lower Colorado Valley east to the Sonoyta Valley (Huey). *Amnicola* has been taken in northern Arizona from *August 2* (*1934*, Tsegi Canyons — Hargrave, MNA) to *September 12* (*1933*, grasslands east of San Francisco Peaks — Hargrave and Motz, MNA). There are specimens farther west from *August 14* (*1952*, Colorado River at southern tip of Nevada opposite Fort Mohave — ARP) to *October 9* (*1952*, Wikieup, Big Sandy River — ARP).

D. p. morcomi Coale breeds along the upper Colorado River as far south as willow groves extend — to Lee's Ferry (Dickerman, CU); once it bred casually in the high San Francisco Peaks (ARP). It is a very common transient generally, and the only two authentic winter records (above) pertain to it. Extreme dates for southern Arizona migrations are from *April 5* (*1915* at Laguna Salada, Baja California — Murphy, AMNH) and *April 17* (Salton River, Baja California — Mearns, US) to *May 29* and *31* (*1949* north of Wikieup and Juniper Mountains — ARP); and from *August 6* (*1940* at Tucson — ARP) to *October 7* (*1949* in Big Sandy Valley — ARP). When with misguided zeal Dr. Byron Cummings caused the Herbert Brown collection (ARIZ) to be relabeled, some Yellow Warblers got mixed, including one which would be the latest fall record for the state, *October 21* (*1888* at Tucson). If Phillips has correctly reassigned the label (away from an obvious spring bird) then this latest record is of *morcomi*. In any case, all the fall birds suspected of being mixed are good examples of this race. Northern Arizona migrants have been taken *May 20* and *27* at Flagstaff (Phillips, MNA) and from *August 3* at Tsegi Canyons (BNav) to *September 26* (*1950* at Supai — ARP).

The birds occupying most of Colorado and New Mexico are actually intermediate between *morcomi* and *sonorana*. Such birds extend into northeastern Arizona, occupying the north slope of the White Mountains region, probably as far as Tuba City (though we lack fresh-plumaged material from this area). Here they are recorded from *April 18* (*1952* at Shumway — Dickerman, ARP) to *August 29* (*1934* at St. Johns — Stevenson, ARP); still present in New Mexico south of Taos, on the Rio Grande, *September 3* (*1956* — ARP), and at Las Vegas, *September 4* (*1915* — Arthur Smith, MCZ). Such birds are very rare as migrants farther south, being recorded but once or twice: male taken near Tucson, *May 10, 1905*, and another "male," *May 14, 1905*, appears to be an unusually bright female of this type (both Kimball, MCZ).

Dendroica petechia sonorana Brewster breeds across the southern half of the state in four areas, shown on the map, whose discontinuities have probably been caused or heightened through parasitism by *Molothrus*. At Phoenix, where cowbirds are now so abundant, the early status of *sonorana* is unknown; but certainly there is no reason to suspect its occurrence since 1920. At Tempe Butte, not far away, a male was taken *May 29, 1933* (Yaeger, MNA), when the species was common; but none was seen later that year, and by 1940 it was definitely absent, being found no nearer than the Verde River, just north of Fort McDowell Indian Reservation (Yaeger and Hargrave). Thence it still ranges down toward Granite Reef Dam (Johnson, Simpson, and Werner), but is absent at the dam and below.

This race arrives on its southeastern Arizona breeding grounds long before the appearance of migrants of more northern races, and it leaves correspondingly early. Early dates are *March 9* (*1940* near Patagonia — GM); *March 10* (*1947* at San Bernardino Ranch — ARP) and several more there the next day (van Rossem, LA); *March 30* (*1952* below the Mazatzal Mountains — Dickerman, MIN); but none before April 2 (1886 — Mearns) at Camp Verde, April 1 (1899 — Herbert Brown) in the Colorado Valley, or before *April 14* (*1937* — Green, ARP) at Whiteriver. The latest fall records for the state are *September 2* (*1886*, Camp Verde — Mearns, AMNH) and *September 3* (*1902* at Yuma — Brown, ARIZ).

We cannot leave the subject of this bright and familiar bird without remarking that every village plaza in lowland Mexico has its one winter resident yellow warbler.

328. MAGNOLIA WARBLER — *Dendroica magnolia* (Wilson)

Casual transient, four records: Bill Williams Delta, October 5, 1949 (Monson); Topock, *November 11, 1951* (GM); Tucson, *November 6, 1955* (ARP); and Chiricahua Mountains, May 23, 1959 (George).

329. CAPE MAY WARBLER — *Dendroica tigrina* (Gmelin)

Accidental, one record: *1875* (J. A. Spring, Paris Museum), probably from Tucson. Also, one found at California end of Laguna Dam on Colorado River, *September 23, 1924* (Huey, SD).

John Arnold Spring was an accomplished Swiss teacher who taught in Tucson and made small collections in various fields of natural history in that immediate vicinity. The specimen is mounted and is in fall plumage.

330. BLACK-THROATED BLUE WARBLER — *Dendroica caerulescens* (Gmelin)

Casual transient: male found dead in Ajo Mountains, *April 30, 1955* (pickled specimen — Allan Thomas, ARIZ); one photographed in color in Chiricahua Mountains, May 5, 1955 (Mr. and Mrs. Harlan Eckler); one taken (other birds believed seen) in Chiricahua Mountains, *October 17, 1956* (Cazier, SWRS); and one seen east of Tucson, October 31, 1959 (Levy).

331. YELLOW-RUMPED WARBLER; MYRTLE WARBLER; AUDUBON'S WARBLER — *Dendroica coronata* (Linnaeus)

Common summer resident throughout boreal zones (less common in adjacent Transition Zone) except in the Hualapai Mountains and, apparently, central Arizona. Common migrant everywhere. Common winter resident of Lower Sonoran Zone and, locally, deciduous riparian Upper Sonoran Zone of southern, central, and western Arizona, except in driest desert portions; rare in winter in other parts of Upper Sonoran Zone in the south.

The Yellow-rumped Warbler is our most abundant and hardiest warbler, the only one wintering in any numbers, and the commonest one in summer in the high fir and spruce forests. Phillips has seen it flying over the highest ridges of the San Francisco Peaks. The pattern is rather intricate and variable according to sex, race and season. The dull basic plumage is the product of a complete pre-basic molt in young and old alike. In spring there is a prealternate molt which replaces almost all the feathers except the flight feathers and tail coverts; it turns the male, at least, into one of the handsomest and most contrastingly colored of warblers. Individuals in all plumages can be recognized by their yellow rump

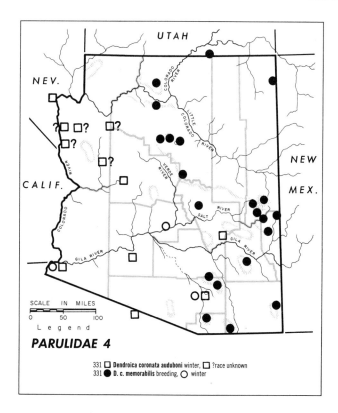

PARULIDAE 4

331 ☐ *Dendroica coronata auduboni* winter, ☐ ?race unknown
331 ● *D. c. memorabilis* breeding, ○ winter

patch, if not also by similar yellow on the crown and sides, pale throats, and white tail-patches that do not reach the tip of the tail.

Yellow-rumped Warblers are recorded in or near southern Arizona, at points away from the breeding grounds, from September to late May, but they are rare on the desert before late September. Extreme dates are August 23 (1950 in the Hualapai Mountains — Phillips) to May 31 (1949, Walnut Creek — Phillips) and to *May 29 (1953)* and June 3 (1955) low in the Sierra de los Ajos of northern Sonora (ARP — Marshall and Phillips). The latest date in the higher mountains is November 6 (1955, upper Bear Canyon, Santa Catalina Mountains — J. R. Stewart). Migrations always start earlier and last longer in the mountains than in the lowlands (except that a stray was seen at Parker, August 23, 1953 — Monson).

Northern Arizona migrations are from *April 12 (1937* at McNary — ARP) to May 25 (1937 at Hotevilla, Hopi Indian Reservation — Monson) and from August 26 (1937 in the Hopi Buttes — Monson) to *October 28 (1933,* Wupatki National Monument — Brewer, MNA), November 11 (1936 near Espero — Phillips), and November 12 (1935 at Springerville — Jacot).

Broad interbreeding, innumerable intermediate specimens, and similarity of song and habits indicate that the Myrtle and Audubon's groups of Yellow-rumped Warblers are conspecific (see Aldrich, *Birds of Wash. State.* 1953: 563; Phillips, Anales Inst. Biol. Univ. Méx. 32, 1962: 366). Slight differences in the timbre of the *chep* call (more of a *chup* in the Myrtle), number of rectrices with a white spot, and face pattern are hardly enough to cause us to fear some mysterious evolutionary force which is about to pry these birds apart! Geographic variation in the species as a whole consists of increase in amount of black (ignoring the black face patch of the male Myrtle in alternate plumage) and larger size to the south. The Myrtle Warbler group is northern, identified in winter plumage by a distinctive face pattern (absent in some females) of light superciliary and dark auriculars. There are breast streaks which are distinct against a whitish background; and two or three rectrices on each side have a white spot. The eastern race, *coronata,* is small and dark, with a brownish, almost ruddy suffusion over the back. In the west, *hooveri* McGregor is larger, lighter, and more grayish-brown. Intermediate birds from the Prairie Provinces are nearest to *coronata* but not typical (see Godfrey, Canadian Field Naturalist 65, 1951: 166-167).

In the contiguous but more southern Audubon's Warbler group, the winter basic plumage shows no cheek pattern, the unstreaked breast is clouded or washed with brown, and usually four or five outer rectrices on each side possess a distinct white spot. There is a cline from small birds with limited black (in alternate plumage) in the north and Pacific Coast areas — called *auduboni* (Townsend) — to the southern larger races with increased black, of which ours is *memorabilis* Oberholser, found from the Sierra Nevadas to the Rockies. The former is the prevailing winter bird in Arizona; the latter is the one which breeds here. The usual range of overlap in wing chord between the two races is 76.5 to 77 or 77.5 mm. in males, 71 to 72.5 mm. in females.

Dendroica coronata coronata is casual in Arizona, with one record from lower Aravaipa Canyon, 13 miles above its junction with the San Pedro River, on *April 14, 1940* (ARP). Although the color differences between *coronata* and *hooveri* are of no use at this season, the measurements (wing 71.5 mm., tail 56.9 mm.) are too small for the Alaskan race; they fit the Prairie Province *coronata* population. Sections of the testis are preserved on a slide, so that there is no question as to the sex.

Dendroica coronata hooveri is a rare winter visitant to Lower Sonoran Zone rivers and farms in western and southern Arizona, mostly to the west (Big Sandy and Colorado Rivers). It is a rare spring transient in eastern, central, and northern Arizona. The single fall record is from the Graham Mountains (*October 8, 1956* — Levy, US). No published record is verifiable, however, except in the A.O.U. Check-list. Sight records are generally erroneous, at least to the extent that they are based on a white throat, which shows up in many *auduboni* and intergrades. *Hooveri* has been collected in Arizona from *October* (above) to *April 14 (1952* in the Chiricahua Mountains — Marshall, WF) and *April 27 (1948,* a female so small as to approach *coronata,* from Shumway, north of the Mogollon Plateau — ARP).

For those timid souls who shrink from lumping Myrtle and Audubon's Warbler, we can present the occurrences of the intergrading specimens just as if they were a separate race. These extend from *November 26 (1954,* Ciudad Obregón, Sonora — ARP) and *January 7 (1940* on the Papago Indian Reservation — GM) to *May 2 (1937* at Tonto Natural Bridge — ARP).

Dendroica coronata auduboni is a common winter resident in Arizona. Fall arrival is somewhat later than *memorabilis,* the main migrations being from October to early November and April to May, in the south. Extreme dates there are from *September 27 (1928* on Walnut Creek, northwest of Prescott — Jacot, ARIZ) to May 25 (*1952* near Tucson — ARP) and later in the Sierra de los Ajos, Sonora, as noted above. Like so many other poorly reported ornithological data of the Lumholtz expedition to northern Sonora, the September 3 record is an absurd error. Northern Arizona specimen dates are from *September 24* to *October 5 (1934* and *1936,* respectively, both at Grand

Canyon Village — GCN), and *April 12* (*1937* at McNary—ARP).

Dendroica coronata memorabilis breeds in the highest mountains of Arizona, is an early fall transient, migrates in spring, and winters in parts of the state. No breeding birds have been found in the Hualapai, Juniper, Sierra Ancha, or Pinal Mountains. It has been taken away from the breeding grounds in southern Arizona from *September 4* in the mountains (Hualapai Mountains, *1951*–ARP) and *September 17*, in the lowlands (Wickenburg, *1948;* two — ARP) to *May 23* (*1950*, Date Creek — ARP) and later as noted above for the Sierra de los Ajos, Sonora. Specimens have been taken in northern Arizona between *April 12* (*1937* near McNary — ARP) and *October 28* (*1933*, Wupatki National Monument — Brewer, MNA). A male with enlarged testes taken on the San Francisco Peaks, *June 3, 1936* (Phillips, MNA), at the upper edge of the Canadian Zone, is to all appearances typical of *auduboni*, but it is probably an extreme variant of the breeding race there. Just below the Mogollon Rim, a female arrived north of Whiteriver on the exceptionally early date of *March 15* (*1937* — ARP).

332. BLACK-THROATED GRAY WARBLER — *Dendroica nigrescens* (Townsend)

Common summer resident in piñon-juniper woodland and other dense vegetation of high Upper Sonoran Zone, and apparently also Gambel oak thickets, of eastern Arizona, west to the Baboquívari and Bradshaw Mountains and the Grand Canyon region. Fairly common state-wide on migration. Winters uncommonly in cottonwood-willow and sycamore-mesquite associations in the Phoenix and Tucson areas and in the Baboquívari Mountains. One record from the lower Colorado Valley at Yuma, *February 18, 1940* (ARP), and also found in winter at Bard, California (Huey).

This handsome warbler has no bright colors, but doesn't need them. Two distinct song types are heard; one of them is wheezy and full of *z*'s, as is characteristic of the whole group of black-throated dendroicas, through the Hermit. It likes junipers and piñons, but on fall migration it ranges up nearly to timberline. Females have more or less white on the chin and upper throat. On the fall migration of this and other warblers see Phillips, Wilson Bull. 63, 1951: 134. In southern Arizona, near the southern end of its breeding range, this warbler has to be satisfied with oak encinal.

The Black-throated Gray Warbler was especially widespread as a winter resident in central and south-central Arizona in 1955–1956. You can usually find a single bird in a large mixed winter flock in the cottonwoods. Possibly this wintering is a phenomenon of recent years only; for instance, none was recorded in the Salt River Valley until after 1950 — in which case a *D. n. nigrescens* of *October 21, 1888* at Tucson (Brown, ARIZ) would be a very late migrant. Migration is mostly from mid-March to May and from late July to October in the south; from mid-April to May and mid-July to early October in the north.

Two races may be distinguished, according to Kenneth C. Parkes, on measurements and tail white. *Dendroica nigrescens nigrescens* is small and is the prevailing migrant and winter resident in Arizona. Specimen records from southern Arizona — it is rare in the north — extend from *August 29* (*1956*, Portal — George, SWRS) to *May 19* (*1940*, Tucson — Sutton, ARP).

Dendroica nigrescens halsei (Giraud), the breeding race, is characterized by longer wing and tail; it tends to have more white on the third rectrix. We have not had time to study collections other than ARP and those in Arizona; but this race appears to be very rare as a migrant after the beginning of August. Its only winter records are *February 6, 1938* at Tucson (ARP) and *February 23, 1957* in the Sierra Aconchi, Sonora (WGG). Migration periods are roughly outlined by southern Arizona specimens from *March 31* (*1931*, Huachucas — Jacot, ARIZ), *July 31* (*1939*, near Tucson — Lee W. Arnold, ARIZ), and *August 6* (*1936*, Whiteriver — Poor, ARP); it must surely stay later. In the north it has occurred from *May 2* (*1936*, south rim of Grand Canyon — Grater, GCN) to *September 3* (*1934*, Springerville — Stevenson, ARP). Two immature female migrants were taken on the upper San Francisco Peaks on *July 28, 1939* (Phillips and F. W. Loetscher, ARP), but no migrants have been taken away from Arizona breeding areas in spring.

333. BLACK-THROATED GREEN WARBLER — *Dendroica virens* (Gmelin)

Rare transient; though termed "accidental" in A.O.U. Check-list, there are already seven specimens, from late *March* to late *May,* and from late *October* to early *November;* localities represented are Huachuca Mountains (Lusk, *fide* Fisher), Tucson vicinity (ARP),

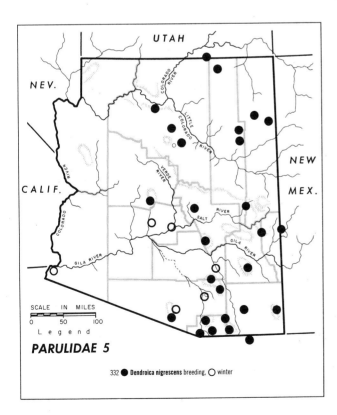

PARULIDAE 5

332 ● *Dendroica nigrescens* breeding, ○ winter

Grand Canyon at rim of Toroweap Valley (Seth B. Benson, MVZ), Ajo Mountains (ARP), Parker, and Bill Williams Delta (both GM). In addition, reliable sight-records come from Tucson, Parker, and Petrified Forest National Monument. Hybrids of *townsendi* X *occidentalis* (ARP) closely resemble this species.

This warbler has to be identified in the field by its low-pitched call and unstreaked back. As in the Black-throated Gray Warbler the call is a soft *tup*.

Fall records are notably late; an early fall sight record (L. L. Walsh, Auk 50, 1933: 124) is doubtless erroneous. Arizona specimens belong to the large-billed nominate race. The species seems most closely related to *nigrescens*. On their songs, see Stein (Living Bird 1, 1962: 61-71).

334. TOWNSEND'S WARBLER — *Dendroica townsendi* (Townsend)

Transient throughout Arizona, very common at higher altitudes, only fairly common in lowlands; scarce along eastern edge, and unrecorded in northeast, in spring. Two winter records: Picacho Reservoir (*March 1, 1959* —WGG) and Patagonia (*December 3, 1939* — ARP). Hybrids with *D. occidentalis* have been taken.

Townsend's Warbler is patterned exactly like the Black-throated Gray but the gray is replaced by green and much of the white by yellow. It differs from the other warblers of this group in its conspicuously darkened cheeks, separating two well-defined yellow lines above and below. Named for Townsend by Audubon, it came out in Townsend's account before Audubon published; so Townsend automatically received the priority — an accident, not an instance of self-glorification. (The same thing happened with the original description of Gambel's Quail, involving Nuttall and Gambel). The two hybrids with *occidentalis* are from Tucson, *September 16, 1953* and Las Playas, southern Yuma County, *April 28, 1956* (both ARP). Why do we not lump these two species if we so willingly did so for the Yellow-rumped Warblers? The differences here are that Townsend's and Hermit Warblers share a considerable breeding area in western Washington without much intermarriage, whereas Myrtle and Audubon's are geographically complementary and they intergrade extensively.

Southern Arizona migrations are from April 9 (Huachuca Mountains — Swarth) to *May 30* (*1918*, Wellton, Lower Gila Valley — Howell, US) and from August 3 (1948, Chiricahua Mountains — Arthur Aronoff) and *August 13* (*1937*, Salt River Valley — JSW; and 1936, Graham Mountains—Monson) to *November 12* (*1955*, Santa Rita Mountains — ARP). Northern Arizona records are *May 2* (*1937* at Mormon Lake — Phillips, MNA) and from *July 28* (*1939* on the San Francisco Peaks, several — ARP) to October 22 (1936 at Springerville — Phillips).

335. HERMIT WARBLER — *Dendroica occidentalis* (Townsend)

Abundant fall transient at high altitudes, scarcer at lower elevations, rare in Lower Sonoran Zone. Common spring migrant at all altitudes in southern Arizona west of the San Pedro River, north to Salome and the Mohave Mountains; rare farther east and north to Verde Valley, unrecorded then in the north. One winter record: Big Sandy Valley north of Cane Springs, *February 17, 1958* (Nancy Isham, Musgrove collection at Kingman).

The adult male Hermit Warbler with his triangular black cravat and all-yellow face is almost unmistakable. Females and young are identified by lack of streaking on the sides, and by the narrow, vague black line behind the eye. The back is rather gray. This species closely resembles Townsend's Warbler, and there are two well-defined hybrids from Arizona (ARP, cited under *townsendi*). One (immature male) resembles the Black-throated Green except for having a streaked back. The other is a brilliant spring male which appears like a Hermit, but has a broad yellow area below the black throat.

Like other warblers which breed in coniferous forest, such as the Yellow-rumped and Townsend's, the migration periods begin earlier, last longer, and consist of more individuals in the high mountains than in the lowlands. Authentic southern Arizona records are from April 1 near the summit of the Huachuca Mountains (1947 — Mr. and Mrs. H. L. Cogswell) to May 28 (also in the Huachucas — Swarth) and again from *July 18* (*1939* in the Santa Rita Mountains —ARP) to *October 14* (*1956*, several in Phoenix — Housholder, KANU), *October* 12 and *17* (*1952* in the Santa Catalina Mountains — Mary Jane Nichols and ARP), and October 25 (1958 at Martinez Lake above Yuma — Monson). The species usually disappears rather suddenly at the close of September.

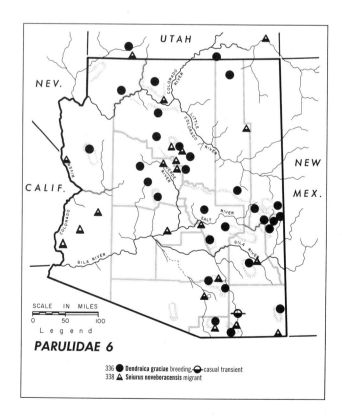

PARULIDAE 6

336 ● *Dendroica graciae* breeding, ○ casual transient
338 ▲ *Seiurus noveboracensis* migrant

Northern Arizona records are from *July 28* (*1939* on the San Francisco Peaks — ARP) to October 3 (1926 above Mormon Lake — Griscom) The most northern spring records are for the Mohave Mountains, *April 25* and *28, 1938* (Huey, SD).

[CERULEAN WARBLER   *Dendroica cerulea* (Wilson)]

[One was found dead at Boulder Beach on the Nevada side of Lake Mead, *June 6, 1954* (Grater, Lake Mead National Recreation Area Museum).]

336. GRACE'S WARBLER   *Dendroica graciae* (Baird)

Common summer resident of Transition Zone throughout the state. One record for Lower Sonoran Zone: St. David, April 19, 1939 (Monson).

The combination of yellow throat and chest, white belly, and black-streaked sides does not identify Grace's Warbler safely, since the female Townsend's may appear identical below. There has consequently been some confusion of these species. However, the cheeks, crown, and upper parts of Grace's are bluish-gray whereas Townsend's are greenish or blackish. Also the superciliary of Grace's is less pronounced, not extending so far back of the eye. Grace's Warbler is exceedingly partial to pines and does not commonly feed elsewhere. The song is a series of *chips* on one pitch followed without pause by a longer series at a higher pitch. Its plain quality suggests a Chipping Sparrow. Grace's Warbler was discovered by the young Dr. Coues at Prescott, named by him for his sister, and published by his mentor, Prof. Spencer Fullerton Baird. An unpublished study of its nesting was made by Lyndon L. Hargrave in the San Francisco Mountains. Like the several preceding *Dendroicae,* this has wing-bars and long white tail spots.

Extreme southern Arizona records are from *March 29* (*1917,* Chiricahua Mountains — Austin Paul Smith, AMNH) to *September 26* (*1948* in the Hualapai Mountains — ARP) and exceptionally to *October 29* (*1864* at Prescott — Coues, US). Northern Arizona dates are from *April 15* (*1937* at Sawmill — ARP) to September 27 (1926 at Mormon Lake — Griscom).

337. CHESTNUT-SIDED WARBLER   *Dendroica pensylvanica* (Linnaeus)

Casual in fall and winter. Two specimens: Wikieup, Big Sandy Valley, *October 4, 1952* (ARP); and Phoenix, one from November 11, 1954 to *January 9, 1955* (Margolin *et al.,* stolen by a cat). A gorgeous warbler seen October 25, 1957 at the home of William H. Woodin III, at the mouth of Sabino Creek, Tucson, has been identified as this species from its description in Marshall's field notes.

[PRAIRIE WARBLER   *Dendroica discolor* (Vieillot)]

[Hypothetical. One record near Tucson, December 7 and 8, 1952 (Tucson Bird Club; Phillips).]

[PALM WARBLER   *Dendroica palmarum* (Gmelin)]

[Hypothetical. One seen near Wikieup, Big Sandy Valley, January 29–30, 1951 (Phillips); two at Walnut Grove, Yavapai County, April 29, 1956 (Heidi McLernon); and one on California side of Colorado River about nine miles above Imperial Dam, September 22, 1942 (Monson).]

[OVENBIRD   *Seiurus aurocapillus* (Linnaeus)]

[Hypothetical. One seen at Walnut Grove, Yavapai County, May 4, 1955 (Heidi McLernon); and one at Boulder City, Nevada, June 16, 1954 (flew into a house — Mrs. A. West, *fide* Grater).]

338. NORTHERN WATER-THRUSH; GRINNELL'S WATER-THRUSH; ALASKA WATER-THRUSH   *Seiurus noveboracensis* (Gmelin)

Rather uncommon transient through whole eastern half of Arizona, and also west locally to Salome (WJS). Only two records farther west: Topock, Colorado River, September 21, 1951; and Colorado River about 30 miles above Imperial Dam, September 6, 1958. Also an old record for Nevada (presumably) part of Colorado River, *October* (Gambel), and reported in southwestern Utah (Zion National Park).

Hardly suggesting a warbler, this plain sooty brown bird with black-streaked whitish underparts walks along the shores of ponds and ditches. It teeters the rear of the body up and down like a Water Pipit or Spotted Sandpiper. Except for the streaks, its only pattern is a whitish or pale buffy superciliary stripe. Attention is most often attracted by its sharp call, though the bird itself is rather shy. Most of the migration dates in Arizona are compressed into remarkably short periods. An unusual concentration of about 12 birds was seen by E. C. Jacot and Guy Emerson on the Santa Cruz River south of Tucson about May 5, 1942.

Southern Arizona migrations are from *April 26* to *May 24* about Tucson (ARIZ, ARP) and from *August 27* (*1874* near Fort Crittenden — Henshaw, US; and *1884* near Tucson — Stephens, SD) to *September 16* (*1897* at Phoenix — Breninger) and exceptionally *October 31* (*1955* at Tucson — Anna E. Lawrence, ARP). Acceptable northern records are from *May 11* (*1933* near Bluff, southeastern Utah — Woodbury, UT) to May 22 (1937 at Keams Canyon — Monson) and from *August* 17 and *31* (*1934* at Mormon Lake — Hudspeth; Phillips, MNA) and *August 27* (*1947* at Flagstaff — Nora Kuhlman, MNA) to September 8 (1938 near Fort Wingate, northwestern New Mexico — Monson).

Geographic variation in color of fall plumage consists of a trend across the top of the continent from greenish-tinged back and yellowish underparts in New England (nominate race) to a more variable bird in Alaska, which is sooty-olive above and sometimes almost white beneath (*notabilis* Ridgway). Very dark-backed birds with medium yellow turn up both east and west on migration, and may be just variants. (This supposed dark race *limnaeus* Miller and McCabe from interior British Columbia is believed to consist partly of fresh, unfoxed *notabilis*.) Arizona birds, except perhaps the above *April* specimen, belong to *notabilis* as herein defined.

[LOUISIANA WATER-THRUSH *Seiurus motacilla* (Vieillot)]

[Not yet reported from within Arizona, though it probably passes over the southeastern edge in order to reach the Nogales area, where it winters regularly (four specimens, *mid-October* to *mid-December* — ARP; Marshall, WF) but uncommonly within 15 to 27 miles of the city, in Sonora.

Since our Northern Water-Thrushes practically lack yellow in spring, separation of the Louisiana bird would be difficult in the field. In the hand, *motacilla* would show a bigger bill, white superciliary, and white underparts contrasting with a buffy wash on the flanks. It also lacks spots in the center of the throat.]

339. KENTUCKY WARBLER *Oporornis formosus* (Wilson)

Accidental. One record, Ramsey Canyon, Huachuca Mountains, *May 23, 1959* (caught by a cat, ARIZ).

[CONNECTICUT WARBLER *Oporornis agilis* (Wilson)]

[Reported in A.O.U. Check-list as casual in Cochise County. Basis of this statement is unknown to us.]

340. MacGILLIVRAY'S WARBLER *Oporornis tolmiei* (Townsend)

Fairly common summer resident in *Ribes*-willow thickets of the Canadian Zone of the White and (very locally) San Francisco Mountains. Common migrant in brush throughout Arizona, except in unbroken forests.

J. K. Townsend discovered this bird and wished to dedicate it to his friend Mr. Tolmie, factor of the Hudson Bay Company. Audubon saw the same specimens and, disregarding Townsend's wishes, dedicated the species to his assistant in Scotland, William MacGillivray. This was his mentor, really, who took Audubon's crude writings, polished them, and added the technical details so as to make ornithological sense. Townsend's notice of the species, published in Philadelphia, appeared first, so that it bears the scientific name *tolmiei;* but this was either unknown to early ornithologists or they preferred to use Audubon's better-known work, which appeared almost simultaneously. As a result, Audubon's name *macgillivrayi* was current for many years and the common name, MacGillivray's Warbler, became firmly attached.

This is a plain green and bright yellow warbler with the head contrasting in some shade of gray or dull brownish. The throat in immatures is whitish, which may be tinged with pale buffy yellow but still contrasts with the bright yellow breast. Except for this line of the hood, the only pattern is a white arc on each eyelid, separated at front and rear. MacGillivray's Warbler prefers dense

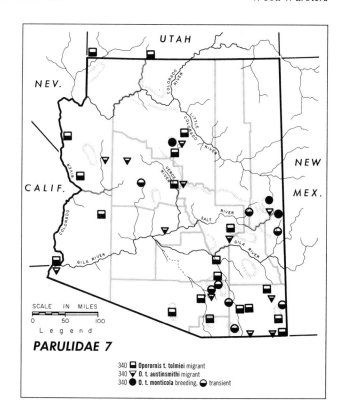

PARULIDAE 7

340 ▫ *Oporornis t. tolmiei* migrant
340 ▽ *O. t. austinsmithi* migrant
340 ● *O. t. monticola* breeding, ◐ transient

low brush; it descends to the ground, but only rarely does it rise into the lower limbs of trees. Only once, on the Big Sandy River, has Phillips heard it sing on migration. To one accustomed to the rarity of its congeners, the Connecticut and Mourning Warblers, in eastern North America, the number of MacGillivray's that may be seen in Arizona is a great surprise.

This is one of several warblers and other birds given as "summer residents" about Prescott by Coues on the basis, probably, of August migrants. Others include *Selasphorus rufus, Riparia riparia, Myadestes townsendi, Dendroica occidentalis, Wilsonia pusilla, Ammodramus sandwichensis, Spizella breweri,* etc. Surely there is no point in repeatedly citing such statements. Some birds are listed as "resident" on the basis of a single one taken, or (*Sturnella neglecta*) less than six seen.

MacGillivray's Warblers are very common transients in southwestern Arizona, and common in fall in brush from Lower Sonoran to Canadian Zone throughout the less densely wooded parts of the state, but apparently uncommon in spring in northeastern Arizona. Insufficient collecting, especially in the north, makes it difficult to tell the ranges and proportions of the races, but the greater abundance in southwestern Arizona in spring is probably due to the restriction of the main flight of typical *tolmiei* to that part of the state. Extreme dates for the migration in the south are from March 21 (1943 near Tucson — Jacot) to *June 6 (1943* at Benson — LLH) and *June 8 (1881* at Tucson —Stephens, MCZ); and again from *August 6 (1940* and 1949, Tucson — ARP, Phillips) to October 18–19 (1949 near Wikieup — Phillips), and near Tucson casually to November 6 (1950) and November 14 (1949 — Alma J. Foerster, Mary Jane Nichols).

Northern records are from April 25 (1948 near Holbrook — Phillips) to October 13 (1953 at Lake Mary —

Eleanor Pugh) and casually to *October 27* (*1931* near Bluff, Utah — BNav).

[The following subspecies accounts are by Allan R. Phillips.]

Geographic variation in *O. tolmiei* was described by Phillips, Auk 64, 1947: 296-300. There are two clines, one to duller back and rump eastward, the other to shorter tail northward. This last continues into the more northern and eastern Mourning Warbler, *O. philadelphia* (unknown in the western United States), with which *tolmiei* may be conspecific; in immature plumage the two are virtually identical except in the relative length of the tail; but adult *philadelphia* nearly always lacks white on the eyelids. Study of the races requires clean, unworn specimens from the various breeding grounds, cognizance of the long migration periods, and elimination of *Oporornis philadelphia* skins (apparently unheeded by Blake, Fieldiana Zool. 36, 1958: 559-560!)

The breeding race in Arizona is *O. t. monticola* Phillips, with the tail relatively longest, usually 56.5 mm. or more in males and not over 4.5 mm. shorter than the wing (chord). Back and rump are darker, grayer (less yellowish) green, and underparts average slightly paler (less chrome) yellow. Color of underparts is of little use in identification of races and tends to fade out in very old museum skins. This race apparently makes long flights between our mountains and Mexico, for it is rare in Arizona away from its breeding grounds. In fact, there are no such records at all in northern or western Arizona, and a mere handful elsewhere: *May 9* (*1925* near Paradise, Chiricahua Mountains —H. H. Kimball, MICH; color not quite typical) to *June 6* (as above); and from *August 29* (*1936* near Whiteriver, Navajo County — F. G. Watson, ARP) to *September 27* (*1947* near Tucson — ARP). The westernmost record is Prescott (*September 6, 1928* — Jacot, ARIZ).

O. t. austinsmithi Phillips differs only in its relatively shorter tail, rarely reaching 56.5 mm. in males and at least 4.5 mm. shorter than the wing. Breeding in the Rocky Mountains north of *monticola*, it migrates rather commonly west to central Arizona (Tucson, Phoenix, and Flagstaff regions), with single records west to the Big Sandy River (*May 29, 1949* — ARP; the singing bird) and Yuma (*October 3, 1902* — Herbert Brown, ARIZ). Southern Arizona dates are from *April 26* (*1922* at Tucson — H. H. Kimball, MICH) to *May 29* (as above) and *June 1* (*1949* in Juniper Mountains northwest of Prescott — ARP); and again from *September 2* (*1884* at Tucson — Herbert Brown, ARIZ) to early October (latest specimen *October 15, 1947*, at Tucson — ARP, was found on dissection to have suffered an injury to one wing). Northern Arizona dates (of specimens with defective tails, unfortunately) are from *August 23* (*1947* at Flagstaff — Phillips, MNA) to *September 23* (*1936* in the White Mountains — ARP) and *September 26* (*1933* east of the San Francisco Peaks — Hargrave, MNA; tail entirely missing).

O. t. intermedia Phillips is a somewhat variable race, birds breeding in California having longer tails (though seldom if ever as long as the longer specimens of *monticola*). Birds breeding in the interior of British Columbia are like *austinsmithi* (females may be indistinguishable), but males are brighter. *Intermedia* migrates throughout Arizona, in the last half of August and September and, in the south, from *April 3* (*1954* in Santa Rita Mountains —Keith S. Brown, Jr., ARP) to *May 31* (*1936* near Coolidge — A. H. Miller, MVZ). One taken *June 1, 1950* near Tucson (ARP) may be intermediate toward *tolmiei*. Extreme fall dates for the state are *August 6* (as above) to *October 7* (Tucson and Big Sandy River — ARP), with casuals at Caborca, Sonora, *October 29* and *November 1* (*1948* — Phillips and Amadon, ARP).

O. t. tolmiei is the brightest race, yellowest above, with a slight tendency (not useful in identification of specimens) to smaller size. Breeding in the Pacific northwest, it naturally does not occur in northeastern Arizona in spring, but (like most coast warblers) reaches northern Arizona in fall. Here it is recorded *August 5* (*1936* near the head of Beaver Creek Canyon, San Juan County, Utah — Russell, UT) and *September 8* to *September 26* (both *1933* on east side of San Francisco Peaks — Hargrave, MNA). In spring its eastern limits are the Chiricahua Mountains (H. H. Kimball, MICH), the Natanes Plateau north of Coolidge Dam (Goldman, US), and farther north the southwestern corner of Utah (Behle and Greenhalgh, UT).

Southern Arizona records are *March 24* (*1940* in the Baboquívari Mountains — ARP) to *May 15* (*1925* in the Chiricahua Mountains — H. H. Kimball, MICH), exceptionally *June 8* (as above); and *August 17* (*1948*, Feldman — GM) to *October 14* (*1925* at Bard — Huey, SD). A casual wanderer was collected by Kimball (MICH) near Paradise, Chiricahua Mountains, on *July 18, 1925!*

341. COMMON YELLOW-THROAT; YELLOW-THROAT *Geothlypis trichas* (Linnaeus)

Common summer resident at such reedy marshes as survive in the Sonoran zones of Arizona; and sometimes in dense, tall grass or weedy fields in the southeast. Probably breeds up to lower edge of Transition Zone in White Mountains region. Common migrant at weedy, brushy, and swampy places throughout the more open parts of the state from the Transition Zone (where recorded in fall only) down, even occurring in the most arid desert sections sometimes in spring. Males winter rather commonly (females very rarely) along the lower Colorado River, and locally east as far as Tucson and Safford.

The male with his black mask set off by white is unmistakable. Young males in fall have the black much obscured — virtually absent on the forehead. Females are very nondescript but have a yellow crissum which contrasts with the white belly and brownish or olive (to grayish in spring) flanks. Frequently the throat also is yellow, but it may be white. In any case there is distinct contrast at the submalar line but not elsewhere. This is a characteristic bird of marshes in North America, with a song *weechity weechity weechity,* a sharp but reedy call, and a stutter.

The spring migration, insofar as it can be distinguished from wintering, begins early in southern Arizona. At a pond near Tucson where Hargrave studied the birds in 1943, and where only males usually winter, he noted the arrival of a female on February 15. The species increased to about four on *February 22* and about 10 on March 2, with females at least as numerous as males. But almost all the females were gone by March 17. Spring migration has continued to June 7 (Tucson and Papago Indian Reservation — Phillips and Sutton). Fall migrants in the south are from *August 24* (*1951* at Boulder City, Nevada — Pulich, ARP) to November 11 (1938 near Tucson — Phillips). In the north, Common Yellow-throats are known to migrate at least between these dates: April 6 (1935 in the bottom of the Grand Canyon — Grater) to May 26 (1937 at Keams Canyon — Monson) and from *August 11* (*1947* near Flagstaff—Phillips, MNA) to September 30 (1938 at Keams Canyon — Monson).

The Common Yellow-throat is one of the least understood of United States birds — one which it would be well to study while the marshes last. For this serene environment is one of the inevitable victims of man's insensate desire to "improve" and "develop." The A.O.U. Committee on Bird Protection has learned of prairie marshes being drained for agricultural use, paid by one government agency, at the same time that the same individuals are being paid by another bureau to withdraw from cultivation their other lands! The only people who have been effective in conserving marshes are those whose economic interest gives their enlightened conservation instincts a solid footing: the duck-hunters and the gun and ammunition manufacturers.

There is geographic variation among yellow-throat populations in tone of general coloration, white on the forehead, and size. The forehead is white in the west, grayish white in the mid-west, and pale grayish in the east in males, but this also varies with season and wear. In the west, southern Arizona breeders are the brightest of all; they are yellowest on the back, least white below, and have a slightly larger bill. These are *chryseola* van Rossem, which breeds over most of Arizona on all the main river valleys except that of the Colorado. They are also transients in southern Arizona, but singing birds are breeders.

The next yellowest race is *occidentalis* Brewster (including *scirpicola* Grinnell). This arcs around Arizona to the north, breeding from southern California and the Colorado River up through Nevada, Utah, Colorado and northeastern Arizona, where it occupies the whole north slope of the Mogollon Plateau. Most of the Arizona transients and winter visitants are *occidentalis* or intermediates.

On the coast and farther north are birds which are dark and dull on the back and which usually have less yellow below and a smaller bill. Of these the best characterized race, but not occurring in Arizona, is the tiny, dark *sinuosa* Grinnell of central California, with its very small bill. This is the only one in which size differences are of any value. Two larger races of this dark group occur as transients in Arizona: *campicola* Behle and Aldrich (including tentatively *yukonicola* Godfrey and *alberticola* Oberholser) and *arizela* Oberholser. *Campicola* is the paler and grayer of the two, with grayer flanks, according to Behle and Aldrich. It is a very rare migrant through Arizona, with only nine specimens, some of which may be faded females of other races. *Arizela*, browner, darker, and a little browner on the flanks, nests on the Pacific Coast and is the commoner migrant of the two in Arizona. It possibly winters in western and southern Arizona.

You get an occasional variant of *chryseola* and *occidentalis* (and possibly also of other races) which is much yellower and includes yellow right through the forehead.

Geothlypis trichas chryseola, the summer resident over most of Arizona, is also a transient which has been noted off the wintering grounds from *March 25* (*1933*, Salt River

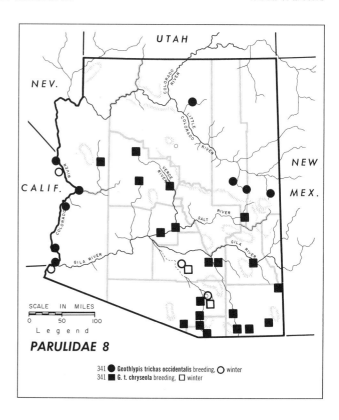

PARULIDAE 8

341 ● Geothlypis trichas occidentalis breeding, ○ winter
341 ■ G. t. chryseola breeding, □ winter

Valley — Yaeger, MNA) to *October 1* (*1956*, Chiricahua Mountains — Cazier and George, SWRS; this last is definitely a transient). In the Salt River Valley, young were still being fed in the nest on *August 8, 1958* (JSW). The only northwestern breeding colony is at Wikieup on the Big Sandy River; Behle (Condor 52, 1950: 193-219) was incorrect in supposing that the species breeds at Hackberry. This race winters — mostly males — as far north as Picacho Reservoir near Coolidge (*January 31, 1953* — ARP).

G. t. occidentalis breeds along the Colorado River and in northeastern Arizona; northern records are from *April 25* (*1948* at Holbrook — ARP) to *September 20* (*1936* in the White Mountains — Correia and Phillips, ARP). One breeding Springerville male is intermediate towards *chryseola* (Poor, ARP). This is our common transient in southern Arizona from *March 5* (*1938* at Tucson — ARP) to *May 17* (*1952* at Roosevelt — Mrs. A. Sanders and Dickerman, MIN) and again from *August 25* (*1893* in the Huachuca Mountains — Holzner, US) to *September 28* (in the Salt River Valley — JSW). It winters, mostly the males, north to Topock (ARP), including Tucson and Picacho Reservoir.

G. t. campicola is a rare migrant taken in southern Arizona from *March 20* (*1948*) to *June 3* (*1950*, both near Tucson — ARP) and again from *August 24* (*1951* at Boulder City, Nevada — Pulich, ARP) to *September 28* (*1884* at Tucson — Herbert Brown, ARIZ). In northern Arizona it has been taken only in September (*7, 1933*, grassland east of San Francisco Peaks — Hargrave, MNA; and *September 24, 1950* at Supai — Phillips, GCN). There is one winter specimen from Yuma (*December 28, 1902* — Herbert Brown, ARIZ).

G. t. arizela is chiefly a fall transient, with records in north limited to *September* (MNA) and those of the south falling between *August 28* (*1902* at Yuma — H. Brown, ARIZ) and *October 31* (*1948* at Caborca, Sonora — ARP). There are a few southern spring records from

April 21 (*1943* at Benson — LLH) to *May 17* (*1948* in northwestern Cochise County — ARP). The only winter record is at Picacho Reservoir, near Coolidge, on *January 31, 1953* (ARP).

342. RED-FACED WARBLER *Cardellina rubrifrons* (Giraud)

Common summer resident of Transition and especially Canadian Zones, west to the Santa Rita and Santa Catalina Mountains; less commonly north to the Mogollon Plateau and west, in recent years, to the Bradshaw Mountains and the mountains of the Flagstaff region. Migrant in adjacent Upper Sonoran Zone during cold springs; one lowland record on migration, near Tucson, *April 26, 1940* (ARP). Casual in foothills of Santa Catalina Mountains, February 22, 1956 (Housholder).

Imperfectly seen, this might look like a chickadee, with its gray upperparts and white underparts — until the red-marked head comes into view. No other Arizona bird has a head prominently patterned with red, black and white. An additional useful character, in flight, is the white rump. Actually the bird can feed just like a chickadee, out at the tips of fir and spruce branches. Unlike chickadees, however, it nests on the ground. It is found over a considerable part of Arizona, but in no other state except part of New Mexico.

The Red-faced Warbler was never found north of the Oak Creek area in the early days; Mearns took the eggs at Oak Creek in 1887. Above there, at Mormon Lake, the species began to be seen May 30, 1934 (Hudspeth) and June of 1942 and 1943 (H. and M. Dearing). In the San Francisco Mountains it was first seen on September 4, 1936 (on the top of Mount Elden —Phillips) and *July 17, 1937* (at Little Spring — Huey, SD). Mrs. Eleanor Pugh has found it fairly regularly there; all summer of 1952, and feeding young in 1953 at Mount Elden.

Southern Arizona records are from April 9 (1948 in the Santa Rita Mountains—A. O. Gross and James H. Veghte) to *September 21* (*1919* in the Chiricahua Mountains — Kimball, LA). Northern arrivals are *April 20* at Oak Creek (*1956* — Simpson and Werner, ARP), April 19, (1942 in the Magdalena Mountains, New Mexico — Monson), and April 29 (1953 on Mount Elden — Mrs. Pugh). The latest northern Arizona date is September 4 (above).

343. HOODED WARBLER *Wilsonia citrina* (Boddaert)

Accidental; one specimen, Patagonia, *July 13, 1947* (ARP); and singing male seen there, April 20, 1963 by Fern Tainter *et al.*

344. WILSON'S WARBLER; PILEOLATED WARBLER *Wilsonia pusilla* (Wilson)

Common to abundant transient throughout Arizona. There are three winter records: Tucson, December 3, 1944 to *January 27, 1945* (*pileolata* — Hargrave *et al.*, ARP); Colorado River about six miles above Imperial Dam, December 7, 1955 (Monson); and Phoenix, December 30, 1956 (Maricopa Audubon Society).

Wilson's Warbler is a small and very bright green and yellow bird. The yellow extends up onto the forehead and superciliary where it abruptly meets the cap, which is black in adult females and males, greenish in young females. The tail is frequently flipped above the level of the back. The call, *chep,* is decidedly harsher and throatier than the calls of most warblers. This is another warbler which is much more abundant in Arizona than in the eastern United States. And here it is more abundant in low trees, brush, weeds, and mesquites of open country than in the denser mountain forests.

Arizona migrations are principally from mid-April through May and again from mid-August through September. But numerous additional dates in the south extend these periods between extremes of March 3 (1943, Nogales — Dille) to June 6 (1960, Colorado River about 18 miles above Imperial Dam — Monson); and from July 31 (1938 — Phillips) to *November 6* (*1950* at Tucson — ARP). Migrations differ somewhat for the races, which we must now discuss.

In general, variation parallels that of the Orange-crowned Warbler, with dullest birds in the east as opposed to the brightest ones along the Pacific Coast. In both species there is a tendency for the largest specimens to be from the Rocky Mountains and for the smallest to be from the Pacific Coast.

Wilsonia pusilla pusilla is the dull eastern bird, a rare but probably regular May transient in southeastern Arizona and casual farther northwest. Typical specimens have been examined from near Tucson as follows: *May 11, 1917,* one female (C. I. Clay, Santa Barbara Museum #418 — burned up?); *May 16, 1886,* female (H. Brown, ARIZ); and *May*

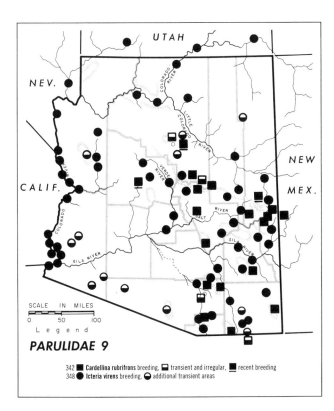

PARULIDAE 9
342 ■ *Cardellina rubrifrons* breeding, ◨ transient and irregular, ■ recent breeding
348 ● *Icteria virens* breeding, ◐ additional transient areas

19, 1940, female (ARP #616). A male from Prescott, *May 17, 1865* (Coues, US #40687) also appears typical, but is somewhat discolored, evidently by soot. Finally, a female was taken at Tonto National Monument, *May 30, 1962* (LLH).

The published records for Arizona are, except for Ridgway's reference (*Birds of North and Middle America*) to the above Prescott bird, based on intermediate specimens, which are quite common in May, and less so in August and September. But there is individual variation in breeding populations, and to record such variants as of other races will obscure the real migration routes. Most intermediates (dull *pileolata* in this case), have bright foreheads and lores which distinguish them from true *pusilla*.

W. p. pileolata (Pallas) is the very common and widespread race of the middle part of the continent. Brighter than *pusilla,* even dull variant females can be distinguished by their brighter yellow forehead and lores. In Arizona *pileolata* is a transient, usually abundant in the lowlands of southern and central Arizona, but apparently less numerous farther north. It migrates later than *chryseola* in spring, and arrives a little later in fall. Both seem to depart at about the same time, judging from the few October and November specimens available. Late May and June birds are all *pileolata,* as is the winter specimen cited above. Southern Arizona migrations are from *April 13* (*1885* at Fort Huachuca – Benson, US) to *June 3* (*1962* at Tonto National Monument – LLH); and again from *August 18* (*1892* at San Bernardino Ranch in extreme southeastern Arizona – Mearns, US) and *August 23* (*1953* at Gillespie Dam, Gila River – Dickerman, CU) to *November 6* (*1950* at Tucson – ARP).

Pileolata is the prevailing migrant in northern Arizona; it is abundant by *April 25* (*1948* at Holbrook–ARP) and continues through May, to return principally between August 24 and the end of September (with sight records for the species in October).

W. p. chryseola Ridgway breeds along the Pacific Coast of the United States. It is the brightest yellow of the races; also its back is more yellowish green than in the above forms. A rather common spring transient in southwestern Arizona, northeast to the Tucson Valley (and rarely to the mouth of the Verde River – LLH), it is more common and widespread in fall, when it has been recorded east to the San Francisco and Chiricahua Mountains. Migration begins earlier than in the other races, but continues equally late in fall in the south. All March, early April, and early August specimens are *chryseola*. The few northern Arizona records are *August 14* (not 31, *1889* on San Francisco Mountains – Merriam and Bailey, US; and the same in *1941,* south rim of Grand Canyon – J. R. Arnold, GCN) to *August 22* (*1932* and *1940* near Flagstaff – Yaeger, Hargrave and Phillips, MNA).

Southern Arizona migrations of *chryseola* are from *March 9* (*1948* at Canoa Ranch on the Nogales Highway –ARP) to *April 25* (*1939* at Bates Well, Organ Pipe Cactus National Monument – Huey, SD). Fall migration has been recorded from *August 1* (*1884* near Oracle – Scott, AMNH) to *November 5* (*1951* near Tucson – ARP).

345. AMERICAN REDSTART *Setophaga ruticilla* (Linnaeus)

Probably a rare summer resident on the Little Colorado River near Eagar, where a pair was discovered June 28 (H. H. Poor) and the breeding male taken *July 3, 1936* (Jenks, ARP), and a molting female was seen, August 12, 1950 (Phillips). In the extreme southeast and in the west it is regular, singly, in both spring and fall migrations, remaining into early June. Otherwise a casual transient, principally in the fall. There are recent winter records: one near Tucson from early November, 1955 to *January 25, 1956* (Lanyon *et al.,* ARP); and Colorado River about six miles above Imperial Dam, *December 20, 1958* (one of two taken), December 23, 1959, and November 4 and December 23, 1960 (GM; Bialac, Monson, *et al.*).

The red or yellow patches on the wing, sides, and basal parts of the outer tail feathers render the easy identification of this species at some distance. The adult males in addition have black head, sides, and upperparts. These birds dance actively about with their wings and tail spread, making themselves so conspicuous as to fool inexperienced observers into thinking they have seen a whole flock. The nest is in a fork or crotch of a small tree. This striking warbler was especially numerous and widespread in 1952–1953, when several were in the foothills above Tucson (Joseph Wood Krutch; one, ARP), one was taken on the Big Sandy River (ARP), at least two males in second pre-basic molt were found in northern Sonora in July (Marshall, WF), and another individual also wintered in northern Sonora (Phillips, CU). The peculiar migration routes in the southwestern United States are discussed by Pulich and Phillips (Condor 55, 1953: 99-100). Root's equating of Springerville with the well-studied San Francisco Bay region (Condor 64, 1962: 76) is a misleading way to compare a bird's status.

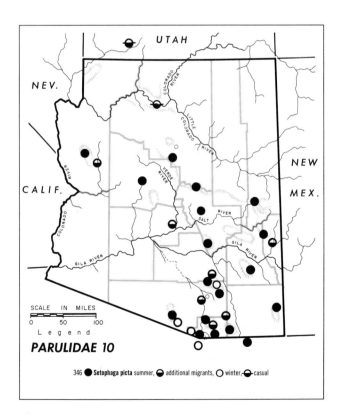

PARULIDAE 10

346 ● *Setophaga picta* summer, ◐ additional migrants, ○ winter, ◌ casual

A Redstart seen northwest of Tucson on April 7 and 11, 1953 (Whitney and Karen Eastman) is so early that it was possibly a wintering bird. Otherwise, dates of migrants extend from April 30 (*1905* near Yuma — Herbert Brown), May 4 (*1959*, Colorado River above Picacho, California — Monson) and May 20 to June 9 (*1956*, Santa Rita Mountains — Bialac, Phillips, and Werner); and again from *August 12* (*1884* near Oracle — Scott, AMNH) to *October 9* (*1955*, Colorado River about 15 miles above Imperial Dam — Monson). Additional records in adjacent Nevada include October 21 (Pulich and Phillips, *loc. cit.*)

346. PAINTED REDSTART — *Setophaga picta* (Swainson)

Common summer resident of the more heavily wooded parts of the Upper Sonoran Zone (higher live oaks, juniper, piñon, etc., usually mixed with pines) of all southern and central Arizona, west to the Baboquívari Mountains, and north to the Hualapai Mountains and the Mogollon Rim. Found post-breeding into the Transition and Canadian Zones of the same areas and rarely down to lower limit edge of live oak growth. Winters sparingly in the Pajaritos Mountains (and southeast foothills of Baboquívari Mountains, *December 18, 1960* — Mrs. Cristina Crossin, RSC), casually elsewhere in low canyons of the Upper Sonoran Zone of south-central Arizona. Rather rare spring migrant in southeast valleys, casually northwest to Salt and Big Sandy Rivers in fall. Accidental in Zion National Park, Utah (sight record only).

Like the American Redstart, the Painted Redstart prances with its wings and tail spread. The call suggests a Pine Siskin, and the song has a most pleasant lilting quality. The nest is placed on the ground, usually in a depression or cavity of a dirt bank. Though found in most of the mountains of Arizona, this gorgeous bird is hardly known elsewhere in the United States, save in adjacent southwestern New Mexico. The first warbler to appear in the higher oak regions, it starts to arrive in the middle of March and is common in the southern mountains in April.

Dr. Parkes (Wilson Bull. 73, 1961: 374-379) uses the generic name *Myioborus* for this redstart, meanwhile keeping the American Redstart in a monotypic genus *Setophaga*, allied with *Dendroica*. But within *Dendroica* there is much diversity, and perhaps some should be allowed in *Setophaga*, unless we are to "split" *Dendroica*, too. We feel that the point is not to separate our two redstarts but rather to include with them *Myioborus*, which is so obviously congeneric with *picta*. *Setophaga (Myioborus) miniata* is the Slate-throated Redstart, which has recently been taken as a casual in New Mexico. *Miniata* and *picta* are very similar in their song, eggs, habits, and ground nesting; it is *ruticilla*, with its *Dendroica*-like song, eggs, and tree nesting, which is the odd-ball. Our consulting oölogist, Richard S. Crossin, agrees with Parkes in deploring the lumping of *ruticilla* which we propose here.

Arizona migrations are from *March 14* (*1923* in the Huachuca Mountains — Jacot, MCZ) to *April 23* (*1925* in the lower Baboquívari Mountains — Bruner, ARIZ); and again from *July 10* (*1953* at Patagonia, a juvenile — Phillips and Dickerman, ARP) to *October 17* (*1883* near Oracle — Scott, AMNH). The earliest arrival in central Arizona is March 25 (*1935* — Monson) at Honeymoon Camp, Greenlee County. In late July these redstarts begin to ascend into the forests above their breeding grounds, but there is an early date of June 18 for Summerhaven, near the top of the Santa Catalina Mountains in 1947— (Nicholas M. Short). The nominate race, *picta*, occurs in Arizona. It shows more white in the flight feathers than do birds southeast of the Isthmus of Tehuantepec.

347. FAN-TAILED WARBLER — *Euthlypis lachrymosa* (Bonaparte)

One specimen, Guadalupe Mountains, Cochise County, *May 28, 1961* (Seymour Levy, US; see Auk 79, 1962: 119-120).

This is a strongly marked, redstart-like bird with a piercing *seep*. It is yellow and tawny below, and derives its name from white eye-markings. It stays on or near the ground.

348. YELLOW BREASTED CHAT; LONG-TAILED CHAT — *Icteria virens* (Linnaeus)

Very common summer resident of dense mesquite-willow-*Baccharis* and arrowweed associations along major rivers and at ponds in Lower Sonoran Zone; fairly common summer resident in deciduous brush along streams and in irrigated lands in Upper Sonoran Zone. Rare transient elsewhere, including the Transition Zone and the arid deserts of southwestern Arizona, and more often garden shrubbery in cities.

In size and shape this bird looks more like a tanager than a warbler. The yellow of the throat and chest ends abruptly against the white breast and brownish sides, as if cut off with a scythe. This is another bird with a white question-mark around the eye, as in the Solitary Vireo. Above it is plain olive. On their nesting grounds in the mesquites of Arizona, Chats are sometimes abundant, and they fill the air with a wide variety of calls, screeches, and stutters, both day and night. Occasionally one rises above the brush in an ecstatic love-song, clapping the wings loudly together above the back as it flies. There is another very strange display (Marshall) given by a male perched upright on a bare twig, in view of the female lurking overhead. With the coal-black mouth wide open, he leans slowly sideways past the horizontal to one side, then the other. Migrating Chats mostly keep out of sight and are silent, but one sang for four days, May 12 to 15, 1956, at Olive Road, Tucson — which again shows that migrants pause for several days (Phillips). Gilman (ms.) did not find the Chat breeding at Sacaton, so there may be a gap in the breeding range between the Colorado River and Tempe and the Santa Cruz River.

Southern Arizona records are from April 11 (1950, two at Topock — Monson) and *April 16* (*1938* near Tucson — ARP) and *April 15* (*1894* at Seven Wells, Baja California — Holzner, AMNH) to *October 12* (*1902* at Yuma — Brown, ARIZ) and *October 15* (*1892* on the San Pedro River at the Sonora border — Mearns and Holzner, US). The end of spring migration is indicated (in town where chats do not nest) by males found dead in Tucson, *May 18* (*1938* — H. Friedlander, ARP) and *May 21* (*1933* — Quaintance, ARIZ), and by one seen in Benson, May 26 (1943 by Hargrave).

Northern Arizona records are from May 8 to October 1 in the bottom of the Grand Canyon (1935 — Grater) and October 12 (1938, at Keams Canyon — Monson). Migrants were seen at Flagstaff as late as May 25 in the spring and as early as September 2 in fall (Mrs. Eleanor Pugh, in 1952).

The race of Yellow-breasted Chat occurring in Arizona is *auricollis* (Deppe). It is gray-backed, long-tailed, with more white on the malar area (less yellow) than the greener eastern race.

WEAVER FINCHES
PLOCEIDAE

349. ENGLISH SPARROW; HOUSE SPARROW *Passer domesticus* Linnaeus

Abundant resident of cities and towns, and rather common resident of large ranch headquarters, irrigated fields, etc., statewide. Became established in Arizona in the early 1900's, widespread by 1915.

This loud and messy bird shares with two other species, *Rattus rattus* and *Homo sapiens,* traits which make the three of them a blight upon the earth: omnivorous food habits, ability to colonize every corner of the world, and an inordinate capacity to procreate — in geometric progression. Since writing the above invective, Marshall concedes that the English Sparrow has estimable qualities, particularly of intellect. In a remarkably short time they associate an opening door with an impending volley from a sling-shot. Unlike other birds in a banding trap, they crouch still while the door is being opened, looking for egress past the trapper's hand. They quickly learn to reach into Bailey traps for seeds, without getting caught. There are also hidden capabilities of flight. Five or six times in the last 14 years at Tucson, Marshall has witnessed the chase of a Mourning Dove by an English Sparrow, during seasons when both are attracted to date palms for nesting. The dove is a very fast flier; nevertheless, the sparrow manages to "stay on his tail" (within about 8 inches) and will chase the surprised dove for a block or two! English Sparrows experiment with feeding innovations, including hawking after moths — without success to our knowledge. Of non-native birds we include this species and the Starling because they get into the countryside; whereas we omit the pigeon, *Columba livia,* from our list because it nests only in city buildings.

Everyone knows the male, but the female is another matter. She is a chunky sparrow with solid brown crown and a superciliary stripe which begins at the eye. English Sparrows, when not nesting in crannies of houses, build great globe-shaped nests of grasses and leaves in trees, with the entrance on one side. A native of Europe, this sparrow has been introduced almost throughout the world, and on purpose!

The English Sparrow arrived at Tucson in 1903–1904 (Howard, Condor 8, 1906: 67-68; Brown, Auk 28, 1911: 486-488); Tombstone, 1904 (Howard, *loc. cit.*); Flagstaff prior to 1907, and Williams in March 1907 (Wetmore); Benson prior to 1907 (Swarth); Casa Grande, 1907 (Florence M. and Vernon Bailey, FW files; Gilman, ms.); Sentinel, southwestern Maricopa County, 1907 (Dr. and Mrs. Bailey); Sacaton, late spring, 1908, and became established in the spring of 1910 (Gilman); Holbrook, Winslow, Fort Defiance, and Ganado prior to 1909, and Adamana and Navajo Springs by 1909 (Nelson, *fide* BNav); Oracle by winter of 1910–1911 (Harman, FW files; Brown, *loc. cit.*); Yuma about 1910 (Brown, *loc. cit.*); and common at Needles, California and one bird at Mellen (near Topock), Arizona, February, 1910 (Grinnell). It was doubtless established in all cities and towns by 1910, though ornithologists were lacking to report the birds.

In 1913 the English Sparrow was found at Parker and Wickenburg (common — Goldman), and at Roosevelt and ranches on Pinto Creek near there (Goldman and Willett, FW files); in 1914 it was abundant at Safford and common at a ranch at Bonita, foot of the Pinaleno Mountains (Holt, FW files); it was common in 1915 at Springerville (Goldman) and generally abundant throughout Arizona by 1916–1918. In the Salt River Valley it was common before Vic H. Housholder's arrival in 1919. It was seen around the stables and hotel at Grand Canyon Village as early as 1933 (Mr. and Mrs. Crockett), but in the absence of farming activities it seems never to have become common there. The spread of the English Sparrow was thus evidently along the railroads from the east, for the birds built their nests in box-cars. They were not introduced directly into Arizona.

In Arizona, the English Sparrow preserves its northern hemisphere reproductive cycle, nesting in spring and not in later summer and fall. Young at Olive Road, Tucson, were able to fly by March 9, 1947 (Alma J. Foerster). Around the first of *June, 1963,* Marshall collected young at Tucson (KANU) that were already well into the first prebasic molt, with patches of black appearing on the chest of males!

MEADOWLARKS, BLACKBIRDS, AND ORIOLES
ICTERIDAE

The icterids are a diverse family including terrestrial, arboreal, marsh inhabiting, parasitic, territorial, and colonial species. Some are highly gregarious; and members of this family form mixed flocks numbering millions. It has been calculated that the total population of birds in the United States and Canada is six billion, but it would not be surprising to learn that there are practically that many blackbirds alone.

Meadowlarks differ from other icterids in their cryptic pattern of streaked brown upperparts; all others of the family have black in the male. In the orioles this is relieved with yellow, orange, or even scarlet. The male Bobolink, on the other hand, is the only passerine with entirely black underparts and extensive light markings dorsally — a reversal of the usual rule of counter-shading. Icterids are another Pan-American family, but few species range northward to Canada or even the northern tier of states. Some of the tropical caciques and oropendolas are of extremely large size, for passerines.

350. BOBOLINK *Dolichonyx oryzivorus* (Linnaeus)

A small breeding colony near Show Low in 1937 is the only one recorded in the state (Poor, ARP). Otherwise, a rare fall transient in western Arizona; though called "casual" by A.O.U. Check-list, five specimens were taken near Wikieup, *October 2 to 10, 1952* (ARP); another was heard on the Big Sandy River in late September, 1949 (Phillips); and one was seen in the Bill Williams Delta, September 14, 1954 (Monson).

The male is described above, in the alternate plumage only. The female and winter male plumages are streaked, and these birds look like overgrown sparrows with spiky tail-feathers. This species is popularly known in the southeastern United States as the Rice-bird. It has long been famous for its peculiar fall migration. Most of the birds congregate then along the Atlantic Coast marshes, though the breeding range extends across to the west side of the Rocky Mountains and into the Great Basin. From Florida, most of them seem to fly non-stop to South America. Small numbers, to be sure, have been found in Louisiana in fall. But the species remains unknown anywhere in Mexico outside of the Yucatan Peninsula. Where the few go that migrate through western Arizona is a complete mystery. The Bobolink is an open country and farmland bird which nests on the ground.

MEADOWLARKS

Genus *STURNELLA*

The coloration of meadowlarks on the back is a wonderfully intricate pattern of buffs, browns, and black pencillings which conceals them perfectly against dead grass. The head is broadly striped with buff and dusky. The yellow of the underparts is varied by a black V across the chest. Rising or settling in the grass, meadowlarks display prominently the white sides of the short tail. Their flight is a series of rapid wing-beats followed by a short glide; the wings appear to be set far back because of the rather long neck, long tertials, and short tail. They inhabit grasslands, grassy parks in the pines, and fields; in winter they may even be found in rather unpromising desert: They build domed nests under tussocks of grass.

There are two species of meadowlarks whose breeding distributions overlap and are rather intricately entwined in Arizona. In some places they adjoin, in others they change hands from one year to another. In wet years one or the other species will breed in fresh grass of areas usually uninhabitable; and conversely, they may absent themselves from traditional breeding areas during drought. These two species are the Eastern Meadowlark (*Sturnella magna*) and the Western Meadowlark (*Sturnella neglecta*). Unlike the situation revealed by Dwain Warner's studies and those of Wesley E. Lanyon in the east, there is no consistent ecologic distinction between the two species here in Arizona; for instance, *magna* occupies the high prairies of the White Mountains and *neglecta* those of the San Francisco Mountains. On the plains out north of Williams, in 1907, Alexander Wetmore (then working for the railroad) found *magna* in summer; when he went back in 1918 only *neglecta* was there. Then Wetherill and Hargrave found nothing but *magna* in 1938; and Wetherill was very surprised, on returning to Williams in 1949, to note only *neglecta!*

But the most interesting thing about the meadowlarks is that in Arizona, where they overlap so broadly, they are much more similar in general coloration, pallor, and size than anywhere else in their respective ranges. This is a brilliant example of convergent evolution, by which the cryptic color pattern must correspond with the color of our local grass and soil in both birds. In other words, of all the many varied races of *magna*, that which occurs in Arizona (*lilianae*) most closely duplicates *neglecta*, whose distribution it here overlaps. This should confound the adherents of the somewhat shaky doctrine of "character displacement," for in no respect are there greater differences in other characters – in voice, color, or habits – which this concept demands should be exaggerated in the area of overlap, so as to facilitate species recognition. Their voices are just as diagnostic, to man and bird, where they occur alone, as here in Arizona. By now you will correctly have guessed that the best means – and the only one in the field – by which the two species can be identified is the call and song. Finer details of color pattern in *S. m. lilianae*, by which Arizona specimens can be identified, were worked out in 1951 by Phillips; they are fairly unique among the various races of that species, so that differentiating traits recommended in other areas do not apply here. The main point is that the third from the outside rectrix in *lilianae* is essentially white, whereas it is mostly dark in *neglecta*, being white only along the shaft on the inner web.

351. EASTERN MEADOWLARK; ARIZONA MEADOWLARK; TEXAS MEADOWLARK *Sturnella magna* (Linnaeus)

Fairly common summer resident of grassy plains and fields in southeastern, central-southern, and northern Arizona. Its centers of abundance are the Altar, Sonoita, San Rafael, and Sulphur Springs valleys, the high prairies of the White Mountains, the grasslands just north of the Mogollon Plateau forests, and the valleys near (especially north of) the Juniper Mountains. It does not extend north of the Little Colorado Valley and the Grand Canyon nor west beyond Coconino County as far as now known. During winter, it is found in varying abundance in southern Arizona west to at least Arlington in the Phoenix region and to the Organ Pipe Cactus National Monument, but is rare in the north, except perhaps at Springerville.

The Eastern Meadowlark is distinguished by its call and song, all of which are high-pitched. The call is a harsh *dzzht* and the song is made up of five or six clear plaintive whistles. In the hand in Arizona, the third rectrix from the outside is almost entirely white. Minor characters are a white malar area, and upper tail coverts tending to be solidly dark along the shafts, or to have dark streaks along with the bars; and deeper buff flanks and crissum.

In northern Arizona, in the San Francisco Mountains-Flagstaff region, the Eastern Meadowlark occurs principally from March to October, with a winter (?) specimen for *February 14 (1934* — Hargrave, MNA). Lower down, at Springerville, both species occurred up to *January 8 (1936* —Jacot, CU and ms.), after which only *neglecta* was seen. The latest date in the high White Mountains prairies is *October 12 (1936* — Correia, ARP), where collecting activities were carried on into November.

In southern Arizona, at Tucson, which is a *bona fide* non-breeding area, the earliest record is of a single bird which brazenly uttered a sharp *dzzht* at our passing car on *September 30 (1956* — Wm. G. George, ARP). The latest spring date there is *March 5 (1939* — ARP).

This species normally occupies its center of abundance to the exclusion of *S. neglecta,* and sometimes occurs in small numbers elsewhere in fields adjacent to those occupied by *neglecta.* The two do not normally associate, nor do they obey any regular pattern of distribution. The most that can be said is that our local race, *Sturnella magna lilianae* Oberholser (poorly differentiated from *hoopesi* Stone of southern Texas), breeds on the elevated grasslands of southeastern Arizona, as on the Sonoita Plains; whereas *neglecta* is the bird of the alfalfa fields below, as around Phoenix.

352. WESTERN MEADOWLARK *Sturnell neglecta* Audubon

Common summer resident in the grassy parts of northern and central Arizona (except most places where *magna* occurs), and locally in irrigated valleys almost throughout Arizona except the southeast, beyond Tucson. Nested at least once in Sulphur Springs Valley in an exceptionally wet year (1941). In wet years it nests also in well-vegetated desert areas of southwestern Arizona. In the northwest it apparently breeds in eastern Mohave County, but not anywhere farther west in the Kingman region; we cannot believe it a "common permanent resident in the . . . desert areas" of adjacent Nevada. Common in grassy and semi-grassy areas in migration. In winter common in grassy parts of the Sonoran zones and farmlands of southern and western Arizona, rather uncommon in grassy or cultivated ponderosa pine openings and in northeastern Arizona farmlands and grasslands.

The voice of the Western Meadowlark is at a lower pitch than that of the Eastern. The call is a more liquid *djert* and the song, after a few opening whistles, bursts into a rich liquid bubbling. The third rectrix from the outside, as seen in the hand, is dark along the edges, with a median white stripe occupying about half its width. Males and many females show an extension of the yellow throat onto the malar stripe or even above. But in winter specimens, it is necessary to ruffle the feathers to see this since only the bases are yellow in some individuals. The upper tail coverts are barred in a ladder-like fashion with black, which does not join broadly along the shaft. The head stripes are less black than in the Eastern Meadowlark. Phillips worked out these differences of tail pattern applying only in the west, for eastern races of *magna* have a tail like *neglecta* (but are dark-bodied). In general color and pallor the two species are identical in Arizona.

In occasional years, heavy winter or spring rains on the desert will bring up a stand of grass which will become inhabited with Western Meadowlarks "from out of nowhere." Such was the case on the desert between Needles and Vidal Junction, California in 1952 and in the Growler Valley and Las Playas areas of southern Yuma County in 1958 (all Monson). In the Sulphur Springs Valley in *1941,* two sets of eggs and parents were collected on *April 13* (WJS). This latter is the only breeding record for southeastern Arizona, except for a pair or two which will breed some years in the farms beside Binghampton Pond, Tucson (Jack R. Hensley, Univ. of Arizona, term paper for ornithology class).

The Western Meadowlark is recorded away from its breeding grounds in southern Arizona, usually in large flocks,

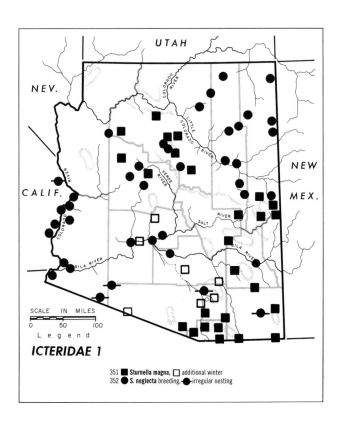

from *September 27* (*1885* east of Baboquívari Peak — H. Brown, AMNH) to at least April 15 (1939 near Tucson — Phillips); singles and unidentified meadowlarks remain a little later. First migrants appear in the Colorado Valley by mid-September (Monson). In northern Arizona it occurs all year at Flagstaff, but increases in late February or March, indicating a migration at that time.

Most specimens of Western Meadowlarks in Arizona collections belong to the interior nominate race, *neglecta.* A northwestern race has been described as *confluenta* Rathbun, which is more heavily marked with black. Since the black pattern of meadowlarks is confined to the interior parts of the feathers, the wearing off of the lighter edges will make a bird appear darker as the seasons progress past the assumption of basic plumage in fall. Therefore this race is not safely identifiable after December or January, and several worn specimens, previously ascribed to *confluenta,* have to be ruled out. The only Arizona specimens that seem properly referable here are from Springerville, *November 10, February 23,* and *March 21* (*1935* — Jacot, ARP); near Tucson on *April 9, 1943* (C. T. Vorhies and Kittie F. Parker, ARIZ) and *April 5, 1957* (Jack R. Hensley, ARIZ); and Potholes, California on *October 12, 1902* (Herbert Brown, ARIZ).

353. YELLOW-HEADED BLACKBIRD
Xanthocephalus xanthocephalus
(Bonaparte)

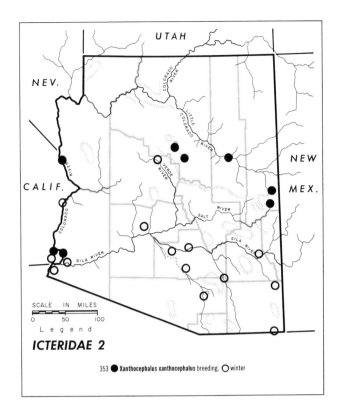

ICTERIDAE 2

353 ● *Xanthocephalus xanthocephalus* breeding, ○ winter

Breeds in small colonies at reedy lakes on and north of the Mogollon Plateau, and along the lower Colorado River. Common in migration at marshes and cattle pens, and in smaller numbers at lakes, fields, and stock tanks throughout the state. Winters abundantly, nearly all males, in marshes and adjacent farmlands across southern Arizona.

The Yellow-headed Blackbird is one of our least mistakable birds. In the male, of Robin size, the body is almost all black and the head and chest are bright yellow. Females and young have the yellow limited to the face and throat, the latter being almost white in some young females. As with other blackbirds, the male is considerably bigger than the female. A peculiar feature about the plumage of these birds is the yellow anal circlet of bright feathers which contrasts with the black underparts. When feeding in cow pens, they hold the stern way up while prying at and shifting cow manure with the bill and feet; hundreds of the winking anal spots thus revealed make a rather peculiar spectacle — there should be some display function for this characteristic.

Adult males are the first to appear in southern Arizona, in July; yet they make up the bulk of the wintering population also. For the females, though arriving later, have farther to go; they comprise about 80% of the wintering swarms near the southern limit of the range, in Jalisco and Nayarit. This species and Cassin's Sparrows are the most conspicuous of the early July migrants into the southeastern Arizona and southwestern New Mexico lowlands. Yellow-headed Blackbirds nest in colonies in cattails or tules, where the males try hard to sing, with discouraging results. They also gather into enormous throngs to feed in stockyards. There is a roost estimated at 100,000 west of Peoria. On February 22, 1947, Wm. X. Foerster, Pulich, and Phillips tried to count the roost at Picacho Reservoir and wound up with a rough estimate of 253,000 blackbirds — mostly Yellowheads.

Southern Arizona records are from June 26 at Sacaton (Gilman) and *July 2* (*1918* at Bates Well south of Ajo— A. B. Howell, US) to June 1 (1952 at Imuris, Sonora — Phillips and Yaeger) and June 7 (1939 near Tucson — Phillips). At San Bernardino Ranch, in the extreme southeast, E. R. Tinkham and Phillips noted three birds on July 7 (1947) and about 25 the following day — obviously new arrivals. Northern Arizona occurrence is from April 3 (1953 at Marshall Lake near Flagstaff, two seen — Mrs. Eleanor Pugh) to October 18 (1934 at Grand Canyon National Park — McKee).

354. REDWINGED BLACKBIRD
Agelaius phoeniceus
(Linnaeus)

Common to abundant resident in marshes and irrigated farmlands of Sonoran zones throughout the state, feeding in farmlands and roosting in marshes during winter, when numbers augmented by birds from north. Occasionally found in migration at stock tanks and wells on the desert. Common summer resident at marshy lakes of the Transition Zone and higher in northern Arizona; rare and irregular in winter in Transition Zone.

The name describes the male, except that his red is confined to the shoulder (lesser wing coverts). The female is dusky, streaked with brown above and white below. She has a prominent pale superciliary. Redwings gather into large flocks but are often in little groups also.

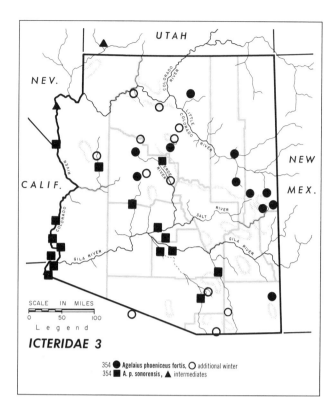

ICTERIDAE 3

354 ● Agelaius phoeniceus fortis, ○ additional winter
354 ■ A. p. sonorensis, ▲ intermediates

In winter the sexes flock separately, and even in May, flocks of young males may be found in some areas of Sonora where the species is not known to nest. Phillips has taken wandering females in Arizona in late May and late July, so that odd birds may turn up almost anywhere at any time. Flocks have even appeared in mid-winter at Grand Canyon Village, a most unsuitable-looking environment (Harold C. Bryant). The Redwing is adaptable, and where there are no cattails it will nest in willows, mesquite, or elderberry trees. It may be that practically all of them wait until just before the summer rainy season to nest; certainly the nesting seems late in Arizona as compared with other areas, but there is much variation. The earliest eggs were found April 28 (at Sacaton — Gilman), and our earliest young out of the nest are *May 30* (*1952* at Imuris, Sonora — ARP) and June 3 (1933 at Binghampton Pond, Tucson — A. H. Anderson).

The following account of subspecies by Allan Phillips was edited by Robert W. Dickerman:

In Arizona two types of Redwings may occur: one with the female so inconspicuously streaked as to appear black, and the male's epaulet entirely red; the other with the breast of the female streaked with white, and the male's red epaulet bordered with buff. Otherwise, males of all races are alike except for bill-shape, only partly reflected in their respective females. As van Rossem correctly pointed out (Condor 28, 1926: 215-230), the bill of the first-year male is shorter and thicker than that of the adult male; and all wing-length differences in Arizona are matters of age only. First-year males also show less buff on the middle secondary coverts.

A. *Black female type*

Agelaius phoeniceus californicus of the Great Valley of California represents this group. It has been found close to Arizona at Calipatria, California (*October 21* and *November 8, 1921* — Kalmbach, US), and should be looked for along the Colorado River.

B. *Striped female type*

This group varies in the darkness and richness of the female's coloration. Southern nesting populations are generally the palest.

1. *Agelaius phoeniceus nevadensis* Grinnell, with dark females and uncertain validity, nests in the northwestern Great Basin. We cannot identify the males. But females, determined with reference to standards from the interior of southern British Columbia, do winter in Arizona. Although not common, they are widespread, even to Yuma (*December 29, 1902* — Clay Hunter, ARIZ 659). Southern Arizona occurrence is from *October 7* (*1952* at Wikieup — ARP) to *April 14* (*1946* at Alamo — van Rossem, LA) and *April 16* (*1886* at Camp Verde — Mearns, US). The only northern Arizona female is from Grand Canyon Village sewer basin, *November 9, 1956* — Hargrave and Dilley, GCN).

2. *Agelaius phoeniceus neutralis* Ridgway, breeding in southwestern California, has the female as dark as *nevadensis* but with a heavier bill. A female specimen is from Supai, *November 20, 1912* (Nelson, US 239868).

3. *Agelaius phoeniceus fortis* Ridgway (*utahensis* a synonym) is the commonest race in Arizona. The female is pale and the bill is short and thick. *Fortis* breeds over the northeast half of the state and winters commonly south and west (we have no record of pure *fortis* west of Wikieup and Sonoyta, Sonora). Migrants and strays have been taken at non-breeding sites in summer and in the Transition Zone in winter (GCN, MNA). A *fortis* was taken at Flagstaff on *December 13, 1931* (Hargrave, MNA), and a female banded from the same flock was caught alive at Cottonwood, on the Verde River, December 8, 1934, by C. Parrish. Migrants to southern Arizona have been found from *October 3* to *May 11* (*1952–1953* at Wikieup — ARP). Status at Camp Verde has changed as in the Flicker (Mearns, US). In the 1880's there were no Redwings in summer; by *1916* there was a breeding population of mixed *fortis-sonorensis* — presumably derived from Flagstaff and Peck's Lake, respectively.

4. *Agelaius phoeniceus sonorensis* Ridgway (corrected spelling from *sonoriensis*) has a long, thin bill and the female is pale dorsally, especially in her reddish-browns. This is the common breeding Redwing at Tucson, Phoenix, and the lower Colorado River. There is a breeding colony on the Big Sandy River from which Phillips has with difficulty obtained a specimen in fresh fall plumage, owing to the hordes of migrating *fortis* there. Although *sonorensis* is mostly resident, it spreads south in winter sporadically to Mazatlán, Sinaloa! Specimens from areas where it did not breed are from *October 23* (*1889* at Calabasas — V. Bailey, US) to *February 25* (*1885* at Camp Verde — Mearns, AMNH). The specimen formerly considered to be the type of *sonorensis* is very pale above and does belong to the southwestern Arizona race. It was taken at the old location of Fort Grant, on the lower San Pedro River. But the correct type locality is Mazatlán (see Deignan, U. S. Nat. Mus. Bull. 221, 1961: 572).

Sonorensis included the birds breeding along the Arizona-Sonora boundary, according to van Rossem. Phillips finds them extremely variable, giving the impression of *fortis* in the east and of *nayaritensis* Dickey and van Rossem farther west (series from Imuris, Sonora — ARP). Fresh fall specimens taken before the arrival of migrants are necessary for working out their affinities.

5. *Agelaius phoeniceus caurinus* Ridgway from coastal Washington and British Columbia is the same as *sonorensis* ventrally and in bill shape, but is much darker on the top surface. There are two female stragglers: Sonoyta, Sonora (*January 12, 1884* — Mearns and Holzner, US 132620) and San Pedro River, "Texas" = east of the Santa Catalina Mountains (*January 28, 1886* — Scott, AMNH). These differ from northwest coast specimens only in their less buffy head and neck, due of course to fading by our famous Arizona sunshine.

355. ORCHARD ORIOLE *Icterus spurius* (Linnaeus)

Two specimens, Chiricahua Mountains, *September 2 and 8, 1956* (Wm. G. George, SWRS and ARP). Also a few recent sight records, mostly of fall migrants, from the upper San Pedro Valley east.

Except for its throaty call and louder song (heard once at Hereford — Hargrave and Phillips), this species is a small edition of the Hooded Oriole in all but the adult male plumage.

356. HOODED ORIOLE *Icterus cucullatus* Swainson

Common summer resident of large mesquite, palm, walnut, and sycamore associations of the Sonoran zones across southern Arizona, north in recent years to below Davis Dam, Kingman, and the south base of the Mogollon Plateau. In migration occurs on deserts of extreme southwestern Arizona. Four winter sight records: Tucson, December 23, 1951, January 26, 1952 (both Phillips), and December 1959 (Anders H. Anderson); and Sierra Pinta, December 27, 1957 (Monson). Statement that race *nelsoni* winters at Tucson (A.O.U. Check-list) is not based on any specimen known to us (mentioned because it is often the eastern race of a species which gets to Arizona as a casual winter visitor).

Male orioles are our only medium-sized birds with conspicuous yellow and black plumage, except for the Western Tanager. Orioles differ from tanagers in their long hanging nest. The females of our species are much duller than the males, the female Hooded being olive-green above (cheeks paler) and dull yellow below. All Arizona orioles have white wing-bars or wing patches. The orange head of the male Hooded Oriole separates his black bib from the black back and wings. The rump and underparts are orange-yellow, and the tail is entirely black. First year males resemble the female, but have a black throat; they nest in this plumage. The call is a weak *eek,* and the song is rather thin and high for an oriole. The Hooded Oriole is more slender than our other orioles, and its bill is slender and very slightly

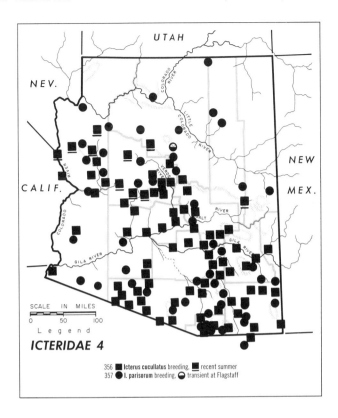

ICTERIDAE 4

356 ■ Icterus cucullatus breeding, ■ recent summer
357 ● I. parisorum breeding, ○ transient at Flagstaff

curved down. This species nests by preference in sycamores, mesquites, hackberries, or palms; occasionally even under the eaves of houses. But it is uncommon or rare in cottonwoods, where largely replaced by the Northern or "Bullock's" Oriole. Hoodeds nest two or three times a year with very indifferent success, as they are the principal target of the Bronzed Cowbird; neither are they immune to parasitism by the formidable Brown-headed Cowbird. The nest, shallow for an oriole, is of grasses and fibers sewn to the underside of a palm leaf; it is of a pale straw color.

Here are some miscellaneous notes from Phillips' diary of July 22, 1947, at 113 Olive Road, Tucson: "Swarm of insects appears during shower at 14:45 hours; as shower ends, many English Sparrows, 3 Hooded Orioles, 1 male Cardinal, a few House Finches, and 1 Arizona Crested-Flycatcher feed on them in the yard. The Hooded Orioles are a male, female, and full-grown juvenile. The female sings repeatedly with her bill full of flies before flying over to the palm at the adjacent south end of 115 Olive Road!"

The Hooded Oriole has extended its range northward spectacularly during the present century. Phillips noted one at Fort Whipple on May 29, and a pair every hundred yards along Date Creek, May 23, 1950; it had not been found in either place by Coues. A nest was found at Union Pass, Ute Mountains, and a pair seen at Hell Canyon, north of Prescott on May 28 of the same year. The species was seen at Hackberry, east of Kingman, at the end of May, 1949, and at Walnut Creek, Juniper Mountains, just after that. A straggler to Gambel oaks of the Transition Zone, Hualapai Mountains, was noted on May 27, 1950; and a male adult was at Fort Apache on April 17, 1952 (all Phillips). None had been found earlier at these places, and in particular, none at Fort Apache, the most intensively studied of these localities, through at least 1937. The steady pace of northward progress of the species parallels that in California (reaching West Berkeley by 1946 — Marshall) and has an important bearing on its racial taxonomy — see below. Concerning the march on Prescott, H. H. Keays did not know the bird there in the 1880's. He came back in the 1920's and collected one for a California museum because he was so surprised to see this species in his boyhood haunts.

Aside from the wintering records cited in the first paragraph and for a late, unmolted (sick?) adult male from Tonto National Monument on *October 30* (*1960* — Norman Messinger, SWAC) the principal period of occurrence in Arizona is from the last third of March to the beginning of October. Tucson records extend from March 17 (1916 — A. B. Howell) to October 16 (1955 — Anders H. and Anne Anderson). A male was seen in the Kofa Mountains, March 15 (1955 — Monson). Earliest spring arrival at Yuma is March 21 (1905 — Nelson and Goldman, US); and to the north: March 24 (1933, common — Yaeger) at Granite Reef Dam near Mesa; April 1 (1956 — Dale S. and Jeanne King) at Globe; and *April 12* (*1952* — Dickerman, CU) below the Mazatzal Mountains. Last seen at Tempe October 11 (1952 — Yaeger).

Of the races of *Icterus cucullatus,* we can first dispose of the more eastern nominate *cucullatus,* not yet taken in Arizona. It is brighter than any western race, and in the female, the flanks are contrastingly gray. There is a winter female from Granados, Sonora (ARP).

Then in the Cape District of Baja California is the long-billed form *trochiloides* Grinnell, though its earliest name is probably *californicus* Lesson, by analogy with other French names of the period (for instance the types of the Gilded Flicker and Costa's Hummingbird which were undoubtedly from the Cape). This is discussed by Phillips (Anales Inst. Biol. Univ. Méx. 33, 1963: 353-355). The type of *californicus* was in Paris but is no longer extant. It cannot be identified from the original description: "bec . . . fortement recourbé, très-acéré" because no bill measurement is given. Judging from its rate of northward extension, however, the Hooded Oriole would not have been in upper California at that time; therefore *trochiloides* and *californicus* probably apply to the same population.

California must have been colonized from northern Baja California or the southeastern deserts, however, and not from the Cape. For the California and Arizona birds are identical racially. (If there is a California race, it should migrate through southwestern Arizona, whence we have picked up no off-color birds.) Although currently thought of as two races on the basis of van Rossem's splitting in 1945, the California bird is merely the sooted town version of the Arizona race, *nelsoni* Ridgway. Nor is the Sonora race any good. Black not reaching to the eye — van Rossem's main point — is apparently due entirely to the make of the skin. In conclusion, there is a western race, *nelsoni* Ridgway, exclusive of the Cape, which is characterized by paleness, short bill, and by not being contrastingly gray-flanked in the female; and all Arizona specimens to date belong here.

357. SCOTT'S ORIOLE *Icterus parisorum* Bonaparte

Common summer resident of the live oak and yucca associations in the mountains of southeastern and central Arizona, and in Joshua trees from Wickenburg region northwest. Smaller numbers breed in the piñons of most of northern Arizona, and in beargrass (*Nolina*) and *Yucca* in southwestern Arizona, almost to the Colorado River. Occurs as an uncommon migrant in the valleys of southeastern Arizona, and casually at Flagstaff (juveniles, *August 1* and *5, 1947* — Phillips, MNA). May winter irregularly from Baboquívari Mountains west to Organ Pipe Cactus National Monument, but no specimens.

The adult male Scott's Oriole is a clear yellow and black oriole without any orange tints. He has entirely black head and upperparts, with yellow sides of the tail basally (tip all black). Females and immatures are dusky olive-green above and they lack any gray or brown tones; rather, they are washed with green. The best field mark is the dusky cheek. One-year-old males and old females have a black throat, and sometimes their entire head is black. The call is a nasal *aan* but the song is of rare beauty — a clear piping suggestive of the Western Meadowlark. Scott's Orioles favor areas with yuccas for nesting and these are particularly common in the live-oak belt of the foothills and higher grasslands. Lt. Couch dedicated this species to Gen. Winfield Scott, as *Icterus scottii,* but it was later found that Bonaparte had previously named it in honor of the Paris brothers of France — owners of an early ornithological collection.

Southeastern Arizona migrations are from *March* 9 and *10* (*1940* near Patagonia — Monson and Phillips, GM) to May 1 (*1907* at Benson — A. P. Smith, AMNH); and again from July 19 (at Sacaton — Gilman, ms.) and *September 4* (*1938* near Tucson — ARP) to October 13 (1944 in the mountains northwest of Nogales — Loye Miller and van Rossem). By mid-August the birds become very hard to find and remain so for a month. A vagrant was at Bill Williams Delta, August 3, 1954 (Monson). (It should be noted that both the dates October 26, 1909, for this species and October 8, 1885, for *I. cucullatus* cited by Oberholser, Bird-Lore 25, 1923: 389 and 244, are erroneous.)

Scott's Oriole shows a partiality, in summer, for areas with tall, tree-like yuccas (over a meter tall) in some quantity, when it nests at all in the Lower Sonoran Zone. Thus it is normally absent from the Tucson Valley save on migration, nesting only in the surrounding mountains and yucca-grasslands. Nonetheless it has probably nested irregularly at Indian Dam, in cottonwood-mesquite-willow association; here a male was seen May 25, 1947 (Erle D. and Virginia M. Morton), and two pairs (of which a male sang once) on April 15, 1956 (Tucson Bird Club); unfortunately the spot was not revisited later in either summer. On the other hand, a male flying over Binghampton Pond, on the other side of town, May 18, 1953 (Dickerman), was with little doubt a belated transient.

In some years, at least, Scott's Orioles seem to arrive earlier in southwestern than southeastern Arizona. Thus a male was singing 21 kilometers west of Gila Bend, March 6,

1947 (van Rossem); and a male seen later the same month at headquarters of Organ Pipe Cactus National Monument (Phillips, van Rossem) was said by Supt. Supernaugh to have been present for weeks, possibly since January. On the east slope of the Baboquívari Mountains a male was singing February 23, 1945 (Loye Miller, van Rossem), but none was noted there before late March by either Bruner or Phillips in earlier years.

Northern Arizona records are from May 17 (1929 in piñons of Slide Canyon, Kaibab Forest — McKee) to September 1 (1914, Apache County — Ligon).

358. NORTHERN ORIOLE; BALTIMORE ORIOLE; BULLOCK'S ORIOLE   *Icterus galbula* (Linnaeus)

Breeds commonly in cottonwood-willow association of the Sonoran Zones except along parts of the Mexican boundary. On migration prefers the same broad-leaved trees in open country, and ranges throughout the state, especially in Lower Sonoran Zone. There are a very few winter records, including specimens from Parker, *February 1, 1947* (*parvus*, GM) and near Phoenix, *November 21, 1957* (*bullockii*, R. Roy Johnson, JSW) and *November 25, 1955* (*bullockii*, Simpson and Werner, PX).

This is our only oriole with orange in the tail (adult males). The female has gray flanks and whitish belly, and is quite grayish above, unlike any other Arizona oriole except the very rare Scarlet-headed. She is also distinguished by her orange or yellow cheeks, in our races. Some adult females, probably the old ones, become quite bright beneath, with a narrow black throat stripe almost like the male's. The male is wholly black above except for the orange rump and a large white patch in the wing, which flashes conspicuously in flight. The nest is a long hanging bag which appears blackish due to the use of considerable amounts of horse-hair. It is commonly hung from cottonwoods and mesquites. The call is a harsh scolding and the song is a clear, brief piping.

This oriole rarely hosts *Tangavius,* the Bronzed Cowbird. But a female was feeding figs to two juvenile *Tangavius* at Benson on August 5, 1943; and a young *Tangavius* fledged there (with three young orioles), July 10, 1949 (both Hargrave).

Breeding distribution along the Arizona-Sonora border is a spotty affair, for the birds absent themselves as nesters from suitable habitat at Guadalupe Canyon and the Chiricahua Mountains, yet they do breed at the San Bernardino Ranch and also south of Naco and around Nogales. At other parts of the extreme south they apparently breed irregularly. Thus at Patagonia they were seen on June 19 and July 13, 1947, but none was found in June of the following year (Phillips), although they were near Nogales at that time.

On migration there is a flight of the race *parvus* in March and early April through the southwest, eastward to the San Pedro Valley. These birds are bound for the Pacific Coast.

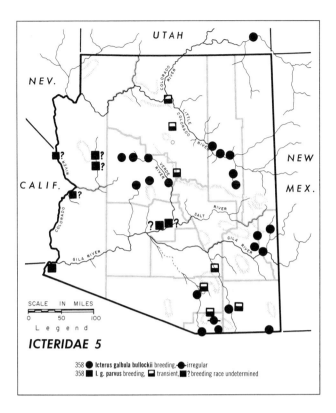

ICTERIDAE 5

358 ● *Icterus galbula bullockii* breeding, ◆ irregular
358 ■ *I. g. parvus* breeding, ◻ transient, ■? breeding race undetermined

None of the breeding population of *bullockii* seems to reach Arizona until about mid-April. The birds remain widespread in southern Arizona until late September, after which there are stragglers in October and even until *November 4 (1953* at Tucson, and later in northern Sonora — ARP). Over-all extreme dates for the migration of the species in the south are from March 3 (1946 near Tucson — Vorhies and Tucson Bird Club) through May and to June 10 (1940 at Menager's Dam, Papago Indian Reservation — Sutton and Phillips); and again from June 26 (1938 on the Saguaro National Monument — Phillips) through September and to the *November* date cited above. Strays have even been seen at Phillips' home in Tucson, June 14, 1955, and June 8 and 20, 1950. Fall migrants appear there regularly beginning sometime between June 27 and July 7.

Northern arrivals are usually in early May (at Snowflake — Albert J. and Louise Levine), but some earlier migrants have been noted from April 19 (at Flagstaff — Hargrave). Spring migration there continues to *May 26 (1933* — Hargrave, MNA; a male with enlarged testes measuring 9 x 6 mm., collected on the desert east of the San Francisco Peaks, of the race *parvus!*) and about June 8 (1953 at Point of Pines, Ft. Apache Indian Reservation — Lorin Haury). Fall records are from July 25 (1939 near Flagstaff — Phillips) to September 10 (1914 in Apache County — Ligon, FW files).

Aside from the spectacular difference in the head and wing coloration of the adult male, geographic variation in this species consists chiefly of rather tenuous size differences; adult males show these more than do the females and immatures. This is another case, as in the Ruby-crowned Kinglet, Orange-crowned Warbler, and Wilson's Warbler, of the largest birds being from the intermountain states, and the smallest being from the Pacific Coast.

The eastern United States race, *galbula,* has the entire head and chest black in the adult male, and is somewhat smaller than *bullockii,* with which it intergrades in Oklahoma (Sutton, Auk 55, 1938: 1-6, and color plate; also

Sibley and Short, Condor 66, 1964: 130-150). *I. g. galbula* also differs in the wing and tail pattern, and females are usually deeper orangish than *bullockii*. It is only a hypothetical member of the Arizona avifauna. One was seen on the Colorado River about 18 miles above Imperial Dam on September 22, 1956 (Monson), and a specimen was taken in Sonora, 26 miles south of Nogales on *October 12, 1954* (ARP).

In the west, all birds north of Durango look alike; the males are characterized by their completely orange underparts, except for a narrow black stripe down the midline of the throat. Two western races have been made out, differing only in size. The smaller is *Icterus galbula parvus* van Rossem, which breeds in coastal California and east apparently to the lower Colorado Valley. Not enough specimens of the breeding populations of the Big Sandy Valley and the Phoenix area are on hand to settle whether or not they, too, belong to the small race or are intermediate. As noted above, there is a small flight of this race through southern Arizona in March and early April. And, true to form, van Rossem picked a member of this flight for his type of *parvus*: *April 1* in eastern San Diego County. He has probably exaggerated its distinctness, for there is much overlap between these and *bullockii*. If *parvus* be recognized, it must include the *May 26* male from desert east of the San Francisco Peaks (mentioned above) and an immature female from the southeast boundary of Grand Canyon National Park, Upper Sonoran Zone grasslands, on *August 18, 1935* (McKee and J. R. Arnold, GCN). This gives a rather odd migration pattern. Other eastern *parvus* (AMNH) are from Camp Verde, *April 29, 1886, May 1, 1888,* and *May 22, 1884* (Mearns) and from Benson, *April 1, 1907* (Austin Paul Smith). Also there are *March* birds from Patagonia and Tucson (ARP) and one in *April* at Mammoth (ARIZ).

The larger race, *bullockii* (Swainson) is the one which arrives in south-central Arizona in mid-April, and to it all the preceding accounts apply, unless specified otherwise.

359. SCARLET-HEADED ORIOLE — *Icterus pustulatus* (Wagler)

Rare to casual winter visitor (and transient) to the Tucson region, though hardly "accidental," as it has been found *December 19, 1886* (H. Brown, ARIZ), October 30 to *December 26, 1948* and *March 19* and *July 31, 1952* (all ARP). Also one in Sonora south of Nogales, *December 19, 1956* (Dickerman and Phillips, ARP).

Superficially, the male of the Scarlet-headed Oriole resembles the Hooded Oriole, though it is brighter about the head and has a narrower black throat patch. It is also a heavier bird, with a thicker, straight bill. The back is streaked black and orange, not solid black as in the Hooded. Females and immatures of our northern race (*microstictus* Griscom) are much like the female Northern Oriole and indeed have been misidentified as such in collections. In the hand, the tail is longer and the mandible is black with an abruptly contrasted pale bluish base, whereas the Northern's bill is slender and with the mandible dull bluish gray to the tip. In the field, Phillips is impressed with the contrast of the bright orange malar area with the duller chest and sides, but he would not recommend field identification of these females in Arizona without a shotgun. The calls, especially the long chatter, are rather similar to those of the Northern Oriole, but there is also a weaker, thinner call somewhat suggestive of the Hooded. As we go south, females become brighter and brighter until finally, at the Isthmus of Tehuantepec, they are male-plumaged.

360. RUSTY BLACKBIRD — *Euphagus carolinus* (Müller)

Rare fall and winter visitant to at least western, central, and southern Arizona; records include eight specimens (Camp Verde, Picacho Reservoir, Tucson, Globe, Benson, Bill Williams Delta, Topock, Tule Well in southern Yuma County — AMNH, JSW, ARP, LLH, US, GM). Their dates range from *October 24* (*1887,* Camp Verde — Mearns) to *March 7* (*1962,* Benson — Seymour Levy).

This blackbird is like a small Brewer's; but the female in autumn has a pale superciliary contrasting with dark, rusty-overlaid upperparts, and a whitish eye. Both sexes would appear slightly smaller than Brewer's Blackbirds, should the two species be seen together. In Arizona, however, the Rusty Blackbird occurs singly, along the edges of rivers or reservoirs. Such lone blackbirds should be scrutinized with care, though the males can hardly be identified in the field after the rusty edges have worn off their upper-parts in early winter.

361. BREWER'S BLACKBIRD — *Euphagus cyanocephalus* (Wagler)

Common summer resident in the vicinity of willows and in well-watered farmlands on and just below the Mogollon Plateau and in the Springerville region; scarcer in Chuska Mountains region, on the South Rim of the Grand Canyon (irregularly?) and on the Kaibab Plateau. Unknown farther west. Transient throughout state, generally uncommon except in farmlands. Abundant winter resident, principally in and near farmlands, south and west of the Mogollon Plateau; also winters sparingly at lakes and farms of Mogollon Plateau and northward. Two summer records in Lower Sonoran Zone: California side of Havasu Lake, *June 12, 1947* (GM), and Phoenix, *July 26, 1953* (Dickerman, ARP). Also three birds in low Upper Sonoran Zone near Tuba City, *July 6, 1936* (Phillips, MNA).

This is an all black, or in the female all gray, bird of Robin size found in open fields, and meadows in the pine forest. It is abundant in stockyards, but does not favor cattails and tules except for roosting at night. Close-up the male is seen to be handsomely glossed with purple and green, with a striking white eye. Females almost invariably have brown eyes, and are completely unstreaked and without any pattern except a vague superciliary in fresh fall plumage. In winter Brewer's Blackbirds need farmlands or marshy lands. They come into town, possibly just to get gravel, only at Parker, Needles and Benson. They do not feed on lawns here as they do on the Pacific Coast.

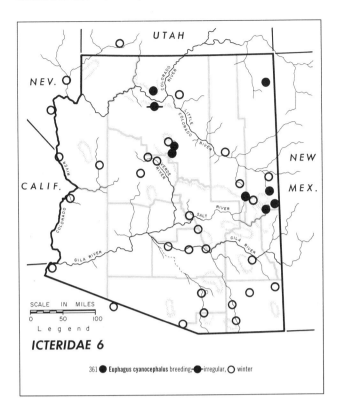

ICTERIDAE 6

361 ● *Euphagus cyanocephalus* breeding, ◐ irregular, ○ winter

Southern Arizona records are from August 22 (1887 at Camp Verde — Mearns), "August" (in the Boulder Dam region — Grater), and *September 4* (*1941* at Salome — WJS) to May 13 (1905, when still common about Yuma — — Herbert Brown), *May 13* (*1927* near Patagonia — *fide* Swarth), May 18 (1950 at Topock — Monson), and May 23 (1931 near Tucson — Vorhies).

Northern Arizona migrations are from March 3 (1936) and March 18 (1939 at Flagstaff — Phillips, Hargrave; Kassel) to May 15 (1935 at Kayenta — Wetherill); and again from *September 9* (*1934* at Babbitt Tank, west of Winslow — Hargrave, Russell, MNA) to November 14 (1938 at Flagstaff — Kassel).

A peculiar generalized blackbird found on the San Pedro River nine miles south of Mammoth, *March 14–15, 1928* (Law and Brooks, US) is thought by Phillips to be a hybrid between this species (or *Quiscalus*?) with *Xanthocephalus*. Robert W. Dickerman, on the other hand, thinks it is *Cassidix X Agelaius*. At any any, Munro (Rep. B. C. Prov. Mus. Nat. Hist. and Anthr. for 1954, 1955: B84) reports a female Yellow-head displaying to males of Brewer's Blackbird. Brooks' statements (Auk 54, 1937: 160-162) that it had "organs much enlarged ... very near breeding condition" are directly contradicted farther on by gonads "of the same size as in the two Red-wings," much smaller birds that "were not breeding at that time!" No blackbirds breed in March in Arizona, to our knowledge.

We have not seen material to judge the validity of the coastal race *minusculus* Grinnell, but it ought to be good by analogy to other species whose populations differ in migratory habits. Coastal Brewer's Blackbirds are essentially sedentary, whereas those of the interior are highly migratory. Should *minusculus* be valid, on the basis of its smaller size, then several wintering birds from southwestern Arizona could be assigned to it: Yuma, *November 7, 1905* (part albino, H. Brown, ARIZ); Sonoyta, Sonora, *January 20, 1894* (Mearns and Holzner, US #132625); and Sacaton, *February 24, 1914* (E. R. Adams, US #241154). These three males measure in wing chord 126.8-127.5 mm. and in tail, 95.2-96 mm. The male from Phoenix, *July 26*, could also be called *minusculus*. On the other hand, equally small males (ARP) are from St. Johns, *November 17*, and Shumway, *January 14*, in the northeast, where migrants from the coast are most improbable.

362. BOAT-TAILED GRACKLE; GREAT-TAILED GRACKLE *Cassidix mexicanus* (Gmelin)

Locally resident in recent years in southeastern Arizona as far west as Phoenix; other breeding stations include Safford, San Bernardino Ranch east of Douglas, upper San Pedro Valley, and whole Santa Cruz Valley. There are records from Alamo Canyon, Ajo Mountains, *May 14, 1939* (Huey, SD — *nelsoni*), and Apache Lake, *May 10, 1952* (R. W. Dickerman, CU — *monsoni*).

This is the largest of our blackbirds, with corresponding greatest difference in size between the sexes. The long tail is somewhat folded or keel-shaped and seems to sag out behind in flight. It appears that the bird has to beat its wings briskly in order to keep from being dragged to the ground! Both sexes have white eyes (in the adults) and resemble the Rusty and Brewer's Blackbirds in coloration. There is a variety of calls like clanging gates, squeaks, squeals and other unearthly outpourings. Their loudness is apt to attract attention. The grackle is of particular interest in Arizona on account of the nearly simultaneous invasion of the state

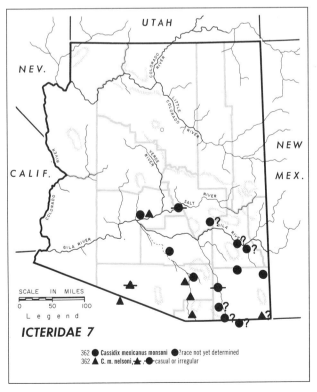

ICTERIDAE 7

362 ● *Cassidix mexicanus monsoni* ●? race not yet determined
362 ▲ *C. m. nelsoni*, ▲-, ●- casual or irregular

by two quite different races in the late 1930's. They are now well established at Tucson at dairies and at Randolph Park, where there is also a large roost.

About 1950 a breeding colony settled at Kinsley's (Arivaca Junction), right on the main highway to Nogales, despite the human crowds that flock to those resort ponds. It was summer resident (mid-April to mid-September) through the 1950's there, but has recently begun to winter as well. Another favorite roost now is a cattail-choked pond by the State Agricultural Inspection Station at San Simon (5 specimens of *monsoni, 1956* — ARIZ, ARP). The bulky nests of grackles are hard to find, hidden as they are in the densest tamarisks, cottonwoods, and exotic conifers. For a map and comments on the Arizona invasions of this species, see Phillips, Condor 52, 1950: 78-81.

Somewhat of a furor invests the taxonomy of *Cassidix* since Selander and Giller (Condor 63, 1961: 29-86) claim to discern two species overlapping in eastern Texas. Although their two kinds of males have different call-notes, they will mate with anything. The authors suppose that the females will accept only the proper male, but made no actual observations on females! If these grackles are so selective, then this is a matter for some surprise in view of the unconventional *amours* in this group of blackbirds. It would perhaps be helpful to have data on individual specimens to supplement Selander and Giller's various amalgamated statistics.

At Tucson, the situation is no less complicated (on a racial level). For while it is perfectly clear that a large race entered Arizona from the east, and that the small one came north from Sonora, they have now met at Tucson to produce a population which seems to exceed in variability the combined norms for both (Selander, Marshall, Johnston, *et al.*, University of Texas Collection; not seen by us).

Cassidix mexicanus monsoni (Phillips) is the large eastern race (minimum tail-length of first-year female 130 millimeters), in which the adult female is dark. This was probably the first grackle to reach Arizona — at Safford in May 1935, and May and June of 1936, breeding in the latter year (Monson). As *nelsoni* had not then arrived at Tucson, farther southwest, it seems likely that these birds had spread west from the Rio Grande and Lordsburg colonies, which were already established in New Mexico. This probability is enhanced by the taking of a male *monsoni* at Benson on *June 6, 1943* (LLH), and of other typical birds near Randolph on *February 4, 1947* (ARP) and on the Río Bavispe in northeastern Sonora, at Colonia Morelos, on *June 21, 1953* (ARP). At Tucson the first *monsoni* was taken (female, with a male *nelsoni* and 6 other males) *January 31, 1945* (ARP); nesting began in 1957 or before. The above straggler from Benson remains our only specimen from the San Pedro Valley. Here we have occasionally seen birds from May (at Fry) on, in the upper part of the valley. At Palominas, near the Sonora border, Hargrave and Phillips found two nests (one with young, the other apparently with eggs) high in cottonwoods at a pond, July 26, 1953. The birds had left before the next visit, in September. Joseph A. Wilcox, a nearby rancher, had seen the last birds at his open fields and bare "lake" on August 28 — a surprisingly early departure. We have always received the impression, in the field, that the San Pedro Valley grackles are *monsoni*.

C. m. nelsoni (Ridgway) is a smaller bird, with pale-colored female plumage, nesting in Sonora. This appeared at Tucson as a summer visitant from April to August of 1937 and 1938. It reappeared in 1940, bred again, and established itself. For a time it was mostly a summer resident, but was also a permanent resident in small numbers thereafter; it was recorded *March 17, 1941* and January 26, 1942 (Vorhies, ARIZ). Even now it is more common from April to September than during the winter. The mixed populations now extend north to Phoenix (JSW). East of the Santa Cruz Valley, the only record is a female (worn; measurements seem to be *nelsoni,* but color is not typical) from San Bernardino Ranch, in extreme southeastern Arizona (Hargrave, Brandt Collection, Univ. Cincinnati). Unfortunately, no unworn material from this ranch is available.

363. BROWN-HEADED COWBIRD; DWARF COWBIRD; NEVADA COWBIRD *Molothrus ater* (Boddaert)

Common summer resident of the less densely wooded parts of the Sonoran zones, especially in southern and central Arizona; scarce in piñon-juniper areas and straggler (breeding?) in adjacent ponderosa pine. Common transient south and west of the Mogollon Plateau. Winters commonly in southern Arizona lowlands where there are livestock concentrations, notably at Parker, Yuma, Willcox, and in the Phoenix and Tucson areas.

"Cowbird" is a contraction of the name "cow-pen bird," bestowed upon this species by the early colonists because of its habit of swarming around the cattle. This species is the commonest Arizona cowbird, and until the present century it was the only one. Our local breeding race is the smallest Arizona blackbird, hardly bigger than a sparrow. The male has a deep coffee-brown head and black body, whereas the female is uniform grayish-brown. Sometimes she is vaguely streaked on the breast with darker, and the throat may be whitish, particularly in the northeastern race. The bill is short, deep, straight-edged, and sparrow-like. The calls are musical squeals.

It is important to know the colors of the juvenile, so as to distinguish it from its step-brothers and foster parents. It is grayish brown, with buff edges to all the feathers, and this buff predominates on the underparts, which are thus streaked and speckled with dark. The pale edgings give a characteristic scaled appearance to the wing. Juvenal males in August and September pass directly into a plumage like their fathers'! They are still distinguishable, however, by some light juvenal feathers retained in the under wing covers. This cowbird is reputed to have but one molt a year, the prebasic, which simply means that nobody has ever looked at spring specimens! James T. Bialac and William G. George collected a large series near Tucson, in *March, 1958* (ARIZ, ARP) in order to prove that several races roosted together; it was at once evident (Phillips) that there is an extensive prealternate molt of males and females alike. This involves the head, neck, down feathers in the adjacent apteria, the legs, and rarely the rump and crissum.

This bird is famous for its parasitic habits, known since the days of Alexander Wilson; it is particularly detrimental to, and has affected the distribution of,

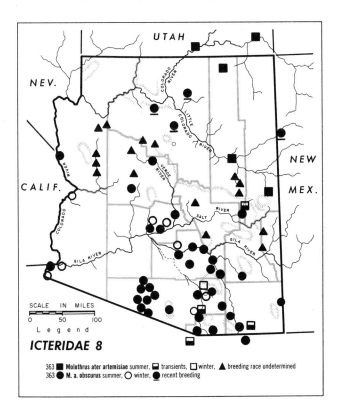

ICTERIDAE 8

363 ■ Molothrus ater artemisiae summer, ⊡ transients, ☐ winter, ▲ breeding race undetermined
363 ● M. a. obscurus summer, ○ winter, ● recent breeding

Bell's Vireo and the Yellow Warbler. It has become much commoner in recent years in Arizona, and also has spread into the northeast section and Grand Canyon Village. Apparently it was rare in winter at Tucson even as late as 1927 (Swenk and Swenk).

Dr. Dwain Warner has induced juvenal females to incubate by shooting them full of progesterone. Adults stoutly refuse.

The fall migration begins early, in late July. This has led to some confusion regarding the breeding range, particularly in the southeastern United States. The Brown-headed Cowbird is recorded away from its winter range in southern Arizona from March 18 (1947) until October 5 (1949; both in Bill Williams Delta — Monson) and October 16 (1951 at Havasu Landing, California — Monson). Northern Arizona summering is mostly from late April (at Kayenta — Wetherill and Hargrave; also Holbrook — ARP) to "September" (in the Grand Canyon Region — Grater). But generally, cowbirds are gone from northern Arizona by mid-August. An exceptional record is of a flock of six to eight birds of both sexes in the cow-sheds at Red Lake, north of Williams, on December 1, 1956 (Hargrave)!

Present evidence indicates that the proportions of the different races in winter varies from year to year and from place to place. But this should be checked by systematic collecting at the various roosts. Favorite roosting places are in cattails, tall oleander hedges, and date palms.

Molothrus ater obscurus (Gmelin) is the smallest race of the species and has a thin bill. It breeds over Arizona generally except in parts of the northeast, and it is the predominant race in winter. Of 60 birds taken from the March, 1958 roost at Tucson, 73% were good *obscurus*, a few seemed intermediate, and the rest were large. *Obscurus* is also the race of a flock (five specimens, April to May, 1957 — Hargrave GCN) which appeared unexpectedly at the mule corrals at Grand Canyon Village about April 10, 1957, grew to sixty (nearly all males), and remained at least into June (Louis Schellbach — specimens measured for us through kindness of Merrill D. Beal, Chief Park Naturalist).

It is likely that the present overlap of *obscurus* and *artemisiae* in the northeastern part of the state is a result of a northward range extension by *obscurus* in recent years. But there is too little prior exploration for documenting this. As yet, there are no intermediate specimens at hand, and the first *obscurus* north of the Mogollon Plateau were taken in the early *1930's* (MNA).

North of us, and in parts of northeastern Arizona, is the biggest race, *artemisiae* Grinnell, of the same coloration as the above. It seems to interdigitate with *obscurus* in an odd checkerboard of breeding occurrences, without both having been found in the same place as yet (unless this is implied in the obscure statements of Johnson and Richardson, Condor 54, 1952: 359 — do both breed in southern Nevada?). *Artemisiae* also occurs as a transient and winter visitant in southern Arizona, some birds arriving in August (and possibly late July, for this is the race of which Phillips took an example at Tepic, Nayarit, on *July 30!*).

Molothrus ater ater occurs irregularly in winter west to Tucson. It differs from *artemisiae* in its thicker bill, which is rather arched at the base. The female has a whiter throat, contrasting more with the chest, than in the two western races. In this race of the eastern United States it is a temptation to say that the female is also darker, but soot is the inevitable adornment of eastern birds. *Ater* is slightly smaller than *artemisiae* in wing and tail lengths. All three races were found in the March, 1958, roost at Tucson.

364. BRONZED COWBIRD; RED-EYED COWBIRD *Tangavius aeneus* (Wagler)

Common summer resident (starting early in 20th century) of irrigated areas and sycamore canyons in southern and central Arizona, west to Organ Pipe Cactus National Monument and Wickenburg, north to the Phoenix and Globe regions, and east to the San Pedro Valley and extreme southwestern New Mexico; rarely farther northeast and west to the Colorado River. Winters, usually in small numbers, at Tucson.

This is a Redwing-sized all-black bird whose heavy, moderately long bill is slightly curved along its upper profile. Seen close, the red eye of the male is conspicuous. Females are slightly smaller; in our race they are plain gray, lacking the brownish tinge of the female Brown-headed Cowbird. The eye of the female is reddish or orange except in very young birds. The Red-eyed Cowbird is parasitic, like the Brown-headed, but is far more choosy, depositing a high percentage of its favors among the Hooded Orioles, with lesser amounts to the Summer Tanager and Northern Oriole. There is an instance of an egg taken with a set of three of the Yellow-breasted Chat, near Tucson on *June 7, 1959* (Patrick J. Gould, ARIZ). The egg is quite different from the Brown-headed's speckled white egg, being plain very pale blue. The nestling is almost solidly slate gray with

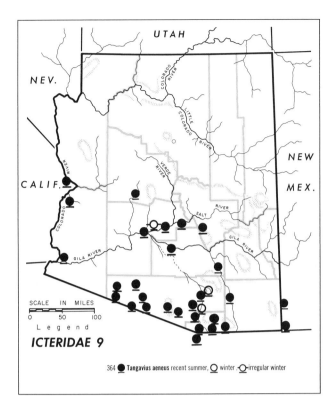

364 ● *Tangavius aeneus* recent summer, ○ winter, ⊙ irregular winter

ICTERIDAE 9

only narrow pale edgings. In the hand, the outer primaries of adults show a peculiar hook on the inner web. The male has a neck ruff which is raised during an elaborate display, accompanied by quivering wings and peculiar whistles like a creaky cart. The female also has a ruff, and in southern Texas and southward she is colored entirely black, like the male. This has led authors of field guides, who are unfamiliar with subspecies, to describe the Arizona bird incorrectly as having similar sexes.

The Red-eyed Cowbird appeared suddenly in Arizona in *1909*, or at least was discovered simultaneously that year at Tucson (Visher) and Sacaton (Gilman). It was found regularly thereafter and probably began to winter within a few years, since one was taken at Tucson as early as March 6 (*1916* by Thomas K. Marshall — Ivan Peters collection, Tucson). The first mid-winter record was in January 1928 (Ellen C. Rogers, report to FW files); however, there are still hardly any winter records away from Tucson (seen once near Phoenix February 23 — R. Roy Johnson). Even in northern Sonora, very few winter.

The Red-eyed Cowbird turned up in the Patagonia Mountains in *1910* (Herbert Brown, ARIZ). It was traced west to the Altar Valley in 1917 (W. L. Dawson) and to Sells on the Papago Indian Reservation in *1918* (A. B. Howell, US); north to Wickenburg in *1927* (US), and to Saguaro Lake in 1935 (Hargrave); and east to the San Pedro Valley in 1929 (Gorsuch). It reached Guadalupe Canyon, New Mexico by *1947* (ARP).

Aside from the wintering birds (which were particularly numerous at Tucson in early 1958 and late 1963) and for an early arrival by a male in eastern Tucson on March 10 and 11 (1953 — Bruce Cole, Mrs. Erle D.

Morton), adults are generally found in Arizona only from late April through July, rarely into early August. But the young are being fed to the end of August by the unfortunate Hooded Orioles, and remain to mid-September. Our sexually dimorphic race is *milleri* van Rossem. Extreme dates in western Arizona, where it is still rare, are April 14 (1952) to July 30 (1950 — both near Parker, Monson).

TANAGERS *THRAUPIDAE*

This is a large tropical American family of which three species regularly reach Arizona, but only one goes on into Utah and beyond. Ours are sedate medium-sized birds larger than sparrows; their bills are rather heavy, with curved edges. Adult males are red-bodied or, in the Western Tanager, red-headed with yellow body and black wings and tail. Females are generally olive above and yellowish below. They differ from female orioles in their heavier, less tapered bills, and most species lack wing-bars. The nests are saucer-shaped, placed high among the branches, and not suspended.

365. WESTERN TANAGER; LOUISIANA TANAGER *Piranga ludoviciana* (Wilson)

Common summer resident of the Transition and boreal forests throughout the state except from Prescott region northwest, where lacking. Common, sometimes abundant, transient throughout Arizona, some fall birds arriving in low desert areas early in July, soon after the last

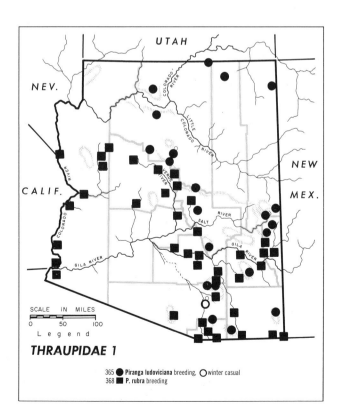

365 ● *Piranga ludoviciana* breeding, ○ winter casual
368 ■ *P. rubra* breeding

THRAUPIDAE 1

spring migrants depart. One winter record, Tucson, *February 11, 1945* (ARP).

This is our only tanager with white wing-bars. The call is a soft *prit-it* and the song is rather wheezy but musical and Robin-like. This is a close relative of the Scarlet Tanager, and elsewhere hybrids have been taken between these two. It is a very abundant species in May flights of some years, even across the deserts. The protracted migrations, suspended only in the last ten days of June and the first ten of July, have led to the erroneous impression of nesting in places where this does not really occur, for instance at Cajón Bonito Creek, in the northeast corner of Sonora, and Rancho la Arizona (van Rossem). Even on June 25 (1951), two males were seen together in Arizona cypresses of Bear Canyon, Santa Catalina Mountains, far below the nesting area at the summit (Phillips). Actually the species breeds nowhere in Sonora, and the fir forests at the top of the Huachuca Mountains mark its southernmost outpost. Nor does it breed in the Chisos Mountains of Texas (in that state it nests only in the Guadalupe Mountains of the New Mexican border), nor as low as 6600 feet in the Chiricahua Mountains (as implied by Tanner and Hardy, Am. Mus. Novit. 1866, 1958: 9).

We have defined above the short midsummer period of no migration, which seems slightly longer in northern Arizona (June 11 to July 16). It remains to furnish early spring and latest fall dates. These are, for southern Arizona, March 12 (1955) and April 13 (1928 at Tucson – Mrs. Helen Cosper, Ellen C. Rogers) to October 18 (1956 in the Kofa Mountains – Monson) and *October 20 (1954* near Tucson – ARP). The species is hardly ever seen now before the last days of April, however (the next earliest recent date is April 17, 1957, at Gadsden, below Yuma – Monson) or after October 5th, despite several mid-April reports prior to 1920. Northern Arizona records extend between April 26 (1936 in Betatakin Canyon, Navajo National Monument – Wetherill) and *October 9 (1884* on the Fort Apache Indian Reservation – Mearns, AMNH), and casually to *October 23 (1963* in the bottom of the Grand Canyon – Norman Messinger, GCN).

366. SCARLET TANAGER *Piranga olivacea* (Gmelin)

Casual. Three records: Tucson, *May 18, 1884* (H. Brown, ARIZ); Otero Canyon, Baboquívari Mountains, March 2 to 22, 1932 (Phillips); and Wikieup, Big Sandy Valley, *October 19, 1949* (ARP).

This species is known in Arizona only from females and immatures. They closely resemble other plain-winged tanagers but are definitely green above, and the males show black in the wings.

367. HEPATIC TANAGER *Piranga flava* (Vieillot)

Common summer resident in dense oaks, fairly common in pines and large piñons, throughout Arizona, except

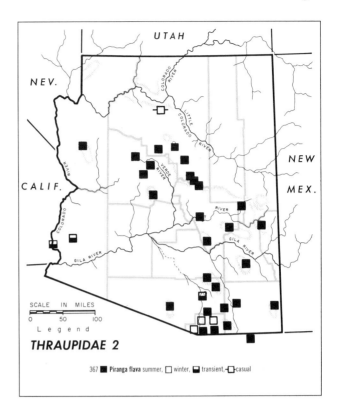

THRAUPIDAE 2

367 ■ *Piranga flava* summer, □ winter, ■ transient, ⊡ casual

in north and northwest, where local. Breeding range includes Baboquívari and Hualapai Mountains. Winters rather regularly (but rarely) in Santa Cruz County, from the Sonoita Valley west to the Pajaritos Mountains (ARP and Marshall, WF). Has been recorded on migration in lowlands only six times: near Tucson, *September 11, 1938* (small flock, ARP), October 2, 1940 (Monson), and *October 24, 1958* (WGG); in Tucson, May 14, 1953 (Phillips); in the Castle Dome Mountains of Yuma County, *June 6, 1959* (GM); and Colorado River about 18 miles above Imperial Dam, *November 18, 1960* (GM). A report from the extreme northwest (Virgin River) is unsubstantiated.

In this mountain species the liver-red color of the male is varied only by a gray cast to the back and cheeks. Both sexes have a definitely black maxilla and a bluish-gray mandible. The call is a *tchup* suggestive of the Hermit Thrush. Lowland reports are mostly erroneous.

Southern Arizona migrations, outside of the extreme south-central wintering area, are from *April 3 (1932* in the Huachuca Mountains – Jacot, ARIZ) to May 14 (1953) and again from *September 11 (1938*, both Tucson – Phillips and ARP) to October 25 (1907 – Swarth, in the Huachuca Mountains). Northern Arizona records are from early May at Flagstaff (Hargrave) to *October 11* or later (*1851* in the San Francisco Mountains – Woodhouse). Though regular near Flagstaff prior to the Second World War, it was practically limited thereafter to the foot of Mt. Elden, where found regularly by Mrs. E. Pugh. Arizona birds belong to the northwestern race *hepatica* Swainson, characterized by large size and dull colors.

368. SUMMER TANAGER; COOPER'S TANAGER *Piranga rubra* (Linnaeus)

Common summer resident in most of the willow-cottonwood association of the Lower Sonoran Zone, rather uncommon in other broad-leafed cover (sycamore, walnut) of the Sonoran zones, south and west of the Mogollon Plateau. Rare transient, April-May and end of July to September, in the Sonoran zones away from its breeding grounds. In winter has been found in the Phoenix and Tucson areas; also sight records for Patagonia and for near Needles, California.

In the Summer Tanager, the bright red of the male extends uniformly over the head and back. Females and young vary among shades of brown, gray, yellow, and buff; but again, the cheeks do not contrast with the rest of the head. The field guides tell us that the bill is yellowish brown, but this is true only during the breeding season. The bill becomes darker in fall and is quite dusky in young birds. The call is a sharp *kit-it-up,* more staccato than in the Western Tanager. This is our only nesting tanager in the lowlands. According to Mearns (ms.), the Summer Tanager was very rare in the Camp Verde region from 1884 to 1888, but it was common by 1916 (FW files).

Piranga rubra rubra is the small, dark, small-billed race of the eastern United States. It is the only form known to winter in Arizona, where it also occurs in migration, but is very rare. Specimens are from the Santa Rita Mountains, *April 18, 1889* (MICH); Phoenix, *November 16, 1957* (R. Roy Johnson, JSW); south of Phoenix, *November 22, 1963* (Fr. Amadeo M. Rea); below the Mazatzal Mountains, *April 20, 1952* (Dickerman, CU); Tucson, *December 17, 1938* (William Homer Brown and Phillips, ARP); and Rancho la Arizona, Sonora, *May 23, 1929* (J. T. Wright, LA). There are intermediates from Tucson (*April 22–* MVZ); Bard, California (*September 23 –* SD); and Reserve, New Mexico (*August 25 –* MICH).

Piranga rubra cooperi Ridgway, of the west generally, has the reverse characters: it is large, pale, and with a long heavy bill. Breeding in Arizona, this race has arrived as early as *April 13* (*1915* at Bard, California – Kimball, AMNH; another specimen catalogued as taken *April 9* cannot be found) and *April 14* (*1940* at Tucson – GM). Immatures leave in the fall ahead of the adults and they are rare after mid-September. The latest date for an immature is *September 24* (*1925* at Bard, California – May Canfield, SD). Adults remain commonly into early October, when the males are either much commoner than the females, or are more easily seen. The latest verifiable *cooperi* is *October 26* (*1958* near Wickenburg – JSW). This behavior, of adults remaining later than juveniles, is similar to that of the Gray Vireo, Arizona Crested and Olivaceous Flycatchers, and Northern Oriole.

Van Rossem probably had faded summer birds in worn plumage when he gave his extravagant description of "*hueyi*" from the Colorado Valley. For despite their apparent isolation, unworn birds from this population are indistinguishable from those at Tucson – the beautiful fall series of Laurence Huey from Bard, California (SD) has been consulted. More to the point is a separation of the Big Sandy Valley population, whose juveniles are rich ochraceous below, dark brownish above, with very little yellow (ARP); and perhaps that of Imuris, Sonora (65 kilometers south of Nogales, Arizona), wherein the females are very orange, especially on the crissum and tail (ARP). In working with these specimens it is important not to compare immature males and females together, for the immature male is brighter – paler and yellower – than the females, which are darker and browner.

GROSBEAKS, FINCHES, SPARROWS AND BUNTINGS
FRINGILLIDAE

This is the largest family in Arizona and one of the most diverse. Most species have heavy conical bills, but the Crossbill has pointed curved mandibles which cross at the tip. Fringillids include such familiar birds as sparrows, buntings, canaries and various finches except the weaver finches. There are three convenient groupings – subfamilies – in Arizona. The first is *Cardinalinae,* a New World group which can scarcely be distinguished structurally from the tanagers, being brightly colored, and with a very deep bill which in most has a downward curved culmen (upper profile of the maxilla). These are mostly arboreal or brush-inhabiting birds, included on our list from the Cardinal through the Dickcissel. The second subfamily, the *Carduelinae,* including those from the Evening Grosbeak through the goldfinches, is mostly Old World, though wide-spread, consisting of short-legged, short-necked, small-eyed birds that go about in large flocks harvesting buds and seeds (often in treetops). They are fond of salt. Many species show red on the head or more or less yellow elsewhere. There is current dissension as to whether the *Carduelinae* really belong in this family. The third subfamily, the *Fringillinae* or *Emberizinae,* from the towhees to longspurs, consists mostly of plain, streaked brown sparrows, including the European buntings. Traces of chestnut or rufous are usually the only bright colors visible. They feed mostly on the ground; and many of them scratch with both feet at the same time.

AMERICAN GROSBEAKS AND BUNTINGS
CARDINALINAE Subfamily

369. CARDINAL *Cardinalis cardinalis* (Linnaeus)

Common resident (reports of altitudinal movements unconfirmed) of the taller and denser Lower Sonoran brush of southern and central Arizona, west to Bates Well on the Organ Pipe Cactus National Monument, the Gila Valley above Gila Bend, and Wickenburg; and north (mainly along rivers) to the foot of the Mogollon Plateau in several places. In recent years, also a small

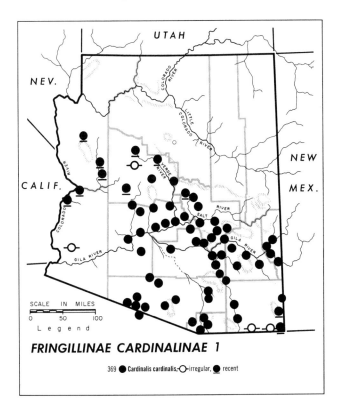

FRINGILLINAE CARDINALINAE 1

369 ● Cardinalis cardinalis,-○-irregular, ● recent

early explorers (Heermann, Kennerly, *et al.*). Stephens and Henshaw found it in few places. In the 1880's it was already present at the mouth of San Carlos River, Globe, Payson, and the Agua Fria River (Mearns), and south to the Santa Cruz and San Pedro Valleys. Mearns (ms.) first saw the Cardinal at Camp Verde on November 7, 1885, and found none in summer until *1887,* but it was rather common there by 1916 (FW files). By 1954 it had extended up the Verde to the mouth of Sycamore Creek (M. A. Wetherill) by *1946* to Parker (GM; van Rossem, LA); by 1950 to Date Creek (Phillips), where not found by Coues; and by 1947 it had become established at Guadalupe Canyon (van Rossem and Phillips). The Cardinal was seen near Bisbee in 1953 (Dr. Guy G. Gilman) and was reported as visiting San Bernardino Ranch in 1947 (van Rossem and Phillips), but was still not established at the latter place in 1948 (Phillips, *et al.*). Arizona and northern Sonora specimens belong to the race *superbus* Ridgway; they are pale, large, very bright red, and with a whitish chin in most females.

370. PYRRHULOXIA *Cardinalis sinuatus* (Bonaparte)

Rather common summer resident in dense Lower Sonoran brush of southeastern Arizona, from the San Bernardino, middle San Pedro, and Santa Cruz valleys west over most of the Papago Indian Reservation to the Ajo region; in 1958 and 1959 was found westward to the Mohawk and Castle Dome Mountains of Yuma County (GM); still present in the Growler Valley, western Pima County, in 1962 (Monson). Winters over same range (except in parts of the San Pedro Valley—Hargrave) and also in the Patagonia area and other

population along the Bill Williams and Big Sandy Rivers that extends to the Colorado River in the Parker area (van Rossem, Condor 48, 1946: 247-248 and Monson, Condor 51, 1949: 265). Inhabits Upper Sonoran Zone at Tonto Natural Bridge (1955 — Phillips, Mary Jane Nichols) and Oracle (Scott, Sowls; nest and eggs in oak — Gould, ARIZ). Still very local in the southeast. Occasional at Prescott in winter; also a specimen from the Castle Dome Mountains of Yuma County, *February 24, 1956* (GM). Has considerably extended its range in past 75 years, both in Arizona and elsewhere.

The all-red male with his black bib is unmistakable. Cardinals resemble Pyrrhuloxias in having a prominent crest and long tail; both of these, plus the underside of the wings, are red. The female Cardinal differs from the Pyrrhuloxia in its straight-edged reddish bill. (Young of both species have blackish bills.) Cardinals are sedentary and live even in cities. The female sings antiphonally with the male. An old female, as in tanagers, may rarely become quite red. Territories and ecological overlap of Cardinals and Pyrrhuloxias at Tucson have been studied by P. J. Gould (Condor 63, 1961: 246-256).

The alleged Colorado River record from the 19th Century is an error. Bischoff's itinerary, supposed to be lost, was published long ago (Wheeler, *Prelim. Rep. Expl. & Surv. principally in Nev. & Ariz. in 1871,* 1872: 15-16, 62-70, 73) and shows that he was near the Pinal Mountains and traveling toward Tucson on the date the bird in question was collected.

This striking species, a favored cage-bird among the Mexican native population, was not reported by any of the

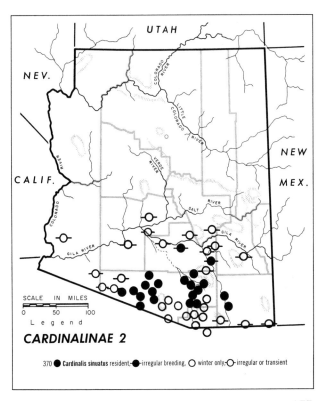

CARDINALINAE 2

370 ● Cardinalis sinuatus resident,-●-irregular breeding, ○ winter only,-○-irregular or transient

areas higher than the breeding range (*i.e.*, Baboquívari Mountains). Has wandered north to Gila Bend (Howell, FW files), south of Phoenix at Komatke (*1964* — Fr. Amadeo M. Rea), Sacaton (Gilman, ms.), near Superior (regular? — Bischoff), the lower San Pedro Valley (Monson), and San Carlos (Jacot, Phillips). Coues' report from the Yuma district is to be disregarded.

Closely related and very similar to the Cardinal, the Pyrrhuloxia nevertheless shows some interesting points of difference. It gathers in winter into groups of a dozen or so; and it appears regularly at that season in places near the oak zone, where it does not breed, such as Patagonia. The bill is brownish-yellow. It is somewhat snubby, shorter than a Cardinal's and down-curved along the culmen, like a parrot's. The Pyrrhuloxia is a trim, slender, handsome bird with relatively long, thin crest and tail. Altogether it is more aristocratic-looking than the Cardinal. The lovely deep red of the male's breast is handsomely set off by his gray sides. These two species occur together in the same habitat in much of south-central Arizona, where they have been studied by Gould (Condor 63, 1961: 246-256).

Contrasted with the steady expansion in breeding range of the more sedentary Cardinal, extensions, retractions, wanderings, and migrations are shown in the more mobile Pyrrhuloxia. Banding in the mesquite bosque on the San Xavier Reservation just south of Tucson (Gould, *op. cit.*) shows that the same individuals are present there both summer and winter, though in winter they form flocks which move past territorial boundaries. The migration of birds (from what source?) into the upper drainage of the Santa Cruz River and west to the Baboquívari Mountains is regular and occurs every winter. There are over a dozen records, mostly from late November to early March, in this region, indicated by the open circles on the accompanying map. In September, migrants appear in sections of Tucson where none nests (Phillips). Then there are temporary extensions of breeding area; for instance, north to nine miles east of Casa Grande, where a pair nested only in one year, about *1924* (eggs collected by Dwight D. Stone). In this same period Pyrrhuloxias were seen by Mr. and Mrs. Crockett at Mammoth (June 15, 1924) and the north side of Phoenix (one male, early 1925), their only records north of Tucson. Finally, there are the wanderers, or pioneers, whose efforts have so far led to no increase in the nesting area, as shown on the map around to the north and east of the breeding range.

Arizona specimens belong to the race *fulvescens* (van Rossem); they are large, with brownish-gray back (not slaty).

371. COMMON GROSBEAK; BLACK-HEADED GROSBEAK; ROSE-BREASTED GROSBEAK; ROCKY MOUNTAIN GROSBEAK *Pheucticus ludovicianus* (Linnaeus)

Common summer resident of Transition Zone and, in southern Arizona, of moist and high Upper Sonoran woodland, especially about creeks and deciduous

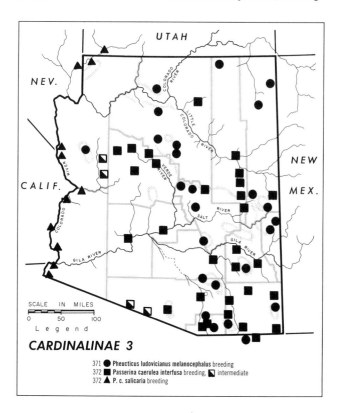

CARDINALINAE 3

371 ● *Pheucticus ludovicianus melanocephalus* breeding
372 ■ *Passerina caerulea interfusa* breeding, ◨ intermediate
372 ▲ *P. c. salicaria* breeding

thickets. Common transient in most parts of state below Canadian Zone, rarely higher, with records practically spanning the summer. "Common" in an orchard at upper edge of Lower Sonoran Zone near Kingman, June 24–28, 1902 (Stephens)! The above applies to the western form. The rose-breasted race of the eastern United States is a casual transient or summer visitant to Arizona, where all records are for the southeastern and extreme western parts; also an accidental winter visitant on the Salt River near Mesa, *November 28, 1958* (Werner, JSW).

This is a chunky, stubby-billed bird varied with black, white, and various shades of brown according to the sex. Males are chiefly black on the head and upperparts. In the eastern race, rare in Arizona, the underparts and rump are white and there is a rose-red inverted triangle under the black throat. Our usual western form is tawny on the rump and breast, with pale yellow belly.

The wholly black and white wings and tail are the hallmark of males that are more than one year old. Females are sparrow-like in color but have dark cheeks in prominent contrast to the pale superciliary. The crown is also highly contrasted. The tawny-orange of the male is reduced in the female and absent in some young birds. The yellow under wing coverts are replaced by red in males of the eastern race. The call is *pik* and the loud song is sometimes given in flight — also freely on migration. This grosbeak reappears regularly in the lowlands on or soon after the 4th of July, thus being one of the forerunners of the fall migration. Spring migration lasts well into June.

Spring migrants of the Common Grosbeak arrive in southern Arizona in late March (29 in Growler Mountains—Monson) and early April (7–9 — Bruner, Phillips, Foerster *et al.*) and they have reached Grand Canyon Village in the north by April 27 (McKee). Fall birds have been seen as late as September 25 (at Supai — Phillips) and exceptionally October 27 (1950 at Grand Canyon Village — Ruby Chase) in the north; in southern Arizona the last date is October 21 (1931 in the Baboquívari Mountains — Phillips).

Profound geographic differences are shown in the Common Grosbeak — between the eastern rose-breasted form and the western black-headed, brown and yellow bird. But they interbreed extensively in South Dakota and Nebraska, according to Swenk (Nebraska Bird Review 4, 1936: 27-40) and West (Auk 79, 1962: 399-424). Their voices and biology seem identical.

Pheucticus ludovicianus ludovicianus, the eastern rose-breasted race, is found mainly in May and June, with one at Kanab, Utah on April 26, 1935 (C. Greenhalgh). There are but four specimens, aside from the winter record cited above: Huachuca Mountains, supposedly on *June 29, 1894* (Lusk, CAS) and about *September* or *October 1929* (W. W. Brown, CLM); Parker, *June 27, 1953* (GM); and Grand Canyon Village, *September 22, 1963* (Messinger, GCN).

Western birds, formerly known as the Black-headed Grosbeak, are thought to comprise two races on the basis of bill size. The difference is scarcely enough to separate populations, let alone to maintain that there is an altitudinal difference! Actually the small-billed lowland birds are just migrants on the way to the Pacific Coast, where they breed. If recognizable, they may be called *Ph. l. maculatus* (Audubon). Small-billed birds are uncommon transients in western Arizona.

Pheucticus l. melanocephalus (Swainson), with larger bill, breeds in the mountains of Arizona, and to it belongs the overall species account above. The bill must require at least six months to reach full size. Yet an adult female taken at Big Saddle Camp, Kaibab Plateau, *June 28, 1925* (Hall, US), has a very small bill, but is doubtless a variant of the breeding population.

372. BLUE GROSBEAK *Passerina caerulea* (Linnaeus)

Fairly common summer resident of willow, cottonwood, and moist mesquite-farmland associations of the major valleys of the Sonoran zones. Occasional elsewhere, even on the Mogollon Plateau, including one specimen, White Mountains, *September 12, 1936* (ARP). One valid winter record: Parker, *February 18, 1951* (GM).

This is an overgrown bunting, *Passerina,* with which its song, calls, and tail-flick all agree. Its bill is greatly swollen, particularly at the base. The adult male is deep blue, appearing black at a distance, with a cinnamon wing patch. Females are nondescript solid brown. Young males in their first year are variously patched with blue, usually on the head. In eastern Arizona this is one of the last summer residents to put in an appearance — around *May 18–20* at Tucson for the first males (extreme arrival is May 3) — and they are still nesting in September. Numbers and distribution in Arizona seem reasonably constant.

Racial difference among the birds nesting in Arizona and California is confined to the size of the bill. *Passerina caerulea salicaria* (Grinnell) is the small-billed race which breeds from California east at least to the Colorado River Valley, as pointed out by van Rossem (*Birds of Sonora,* 1945) and as exemplified by the series from the Yuma area (ARIZ, LA). Birds of the Big Sandy Valley and Organ Pipe Cactus National Monument region are variable. *Salicaria* arrives earlier than the eastern *interfusa:* April 21 (1948 — Pulich), and regularly by the last days of April. It remains until the end of September (and in 1946 to October 13 — Monson). The winter bird from Parker belongs in this race.

P. c. interfusa (Dwight and Griscom), with the large bill, has arrived in south-central Arizona as early as *April 29* (*1889* at Phoenix — V. Bailey, US). Migration in central Arizona precedes that at Tucson and farther east, but is later than that of the race *salicaria;* for Mearns took specimens as early as *May 6* (*1887* at Camp Verde — AMNH). This grosbeak has remained as late as October 31 in fall (of 1939 at Hereford — Monson).

Northern Arizona records are from June 10 (1915) to *September 19* (*1934*), both at Springerville — Goldman, FW files; Stevenson, ARP.

373. COMMON BUNTING; LAZULI BUNTING; INDIGO BUNTING *Passerina cyanea* (Linnaeus)

Rather uncommon summer resident in willow associations of the Sonoran zones of central and northeastern Arizona, south and west to the bottom of the Grand Canyon, Prescott, Camp Verde, and possibly the Mazatzal Mountains and Sierra Ancha. Common transient in brush and tall herbaceous vegetation throughout the less densely wooded parts of the state, records nearly spanning the summer. Winters rarely in southern Arizona from Hereford west and northwest to the Nogales and Phoenix regions (but not west of Santa Cruz Valley); in recent years only, and apparently increasing.

The above applies to the western form, known heretofore as the Lazuli Bunting. The eastern race, formerly known as the Indigo Bunting, which has crossed with the Lazuli at least at Flagstaff, nests in central Arizona, and is a rare migrant and accidental winter visitant.

The Common Bunting is the one we most usually see in Arizona. The male is all blue or, in our more numerous race, he has a salmon-buff chest band and white belly. The female is solid brown above and on the head, with pale underparts overlaid by brown in fresh fall plumage. Until 1917 only this western race was known in Arizona, but more recently there have been several records of the eastern race, principally at seasons when it would be suspected of nesting. As yet, only one definitely intermediate adult male specimen has been taken in Arizona (Flagstaff, *June 20, 1957* — ARP). A protracted attempt by R. Roy Johnson to study

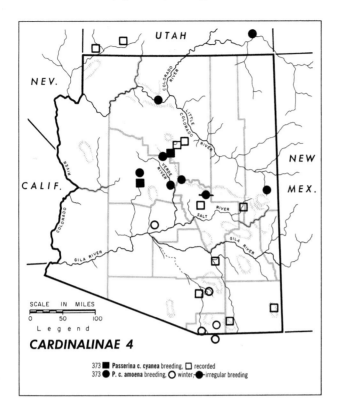

CARDINALINAE 4

373 ■ Passerina c. cyanea breeding, ☐ recorded
373 ● P. c. amoena breeding, ○ winter, ⬤ irregular breeding

from *May 10* to *30,* and again from *July 13* to *September 15* (Monson; ARP; Harter, SD; Phillips, MNA, respectively). Though the *May 30* specimen was already in breeding condition at Tucson, it is unlikely that this race nests south of the Salt River.

P. c. amoena (Say), in which the male is sky-blue, salmon, and white as described above, is identified in the female by the whitish tips of the secondary coverts and by the unstreaked pale breast covered by a tawny wash. Southern Arizona migrants are from March 14 (1933 at Diversion Dam on the Salt River — T. S. Roberts) to June 2 (1939 at Tucson — Monson and Phillips), and again from June 16 (1927 in the Santa Rita Mountains — *fide* Swarth) to October 10 and 31 (1939 at Hereford — Monson). Northern Arizona migrations are from April 24 (1935 in the bottom of the Grand Canyon — *fide* McKee) to May 31 (1936 near Flagstaff — Phillips), and again from June 30 (1937 at Keams Canyon — Monson) or early July to September 24 (1926 at Flagstaff — Griscom). Supposed breeding records south of the Gila Valley are probably erroneous, and its "breeding" in the Boulder Dam region, Nevada (Grater; Linsdale, Pac. Coast Avif. 23, 1936: 117-118) requires confirmation.

The first record that probably represents wintering of *amoena* was March 4, 1939, near Tucson (Allan Brooks), unless a male taken at Tempe, *November 10, 1932* (Yaeger, MNA) was wintering there. The first mid-winter records are *December 3, 1939* near Patagonia (ARP); January 1, 1940 at Nogales (Dille); and *February 22, 1943* near Tucson (LLH). There were unusual numbers wintering in 1955-1956; and in the winter of 1957-1958, there were flocks up to 50 around the edges of the mesquite bosque south of Tucson (Marshall).

possible crossing in the Sierra Ancha was frustrated for lack of birds. In the Flagstaff, Oak Creek, and Fort Apache areas, such studies might be more successful.

The explosion of the eastern Indigo race into Arizona was part of a general expansion in the past thirty years, as it has also appeared in Tarrant County, Texas (Pulich) and there is suggestive evidence of breeding at certain points in Mexico in the mid-1950's (Dickerman, Phillips). The sudden appearance of one race (Indigo Bunting) inside the range of another (Lazuli Bunting) is of course a most unusual occurrence and would probably have gone unnoticed had there been a large population of the latter to absorb the newcomers. In central Arizona this was not the case, as the Lazuli is sparse and local there. R. Roy Johnson wrote in the summer of 1962 that "there is still one male Indigo and no Lazulis in the Sierra Ancha," where he had spent several summers as a fire lookout.

Passerina cyanea cyanea is all deep blue in the male; females are brown, with tips of the secondary coverts only slightly paler, and they are vaguely streaked on the breast. A rare migrant in recent years, this race is now nesting at least in the Oak Creek and Prescott areas and possibly on the Sierra Ancha and along White River. Lone males have been seen in southeastern and central Arizona in summer, and one was seen near Tucson, December 6, 1957 (Marshall). Of non-breeding females there are only two records: Flagstaff, September 15, 1940 (Phillips); and Chiricahua Mountains, *May 26, 1957* (ARP). The preponderance of adult males recorded from Arizona would suggest that the dull females have been generally overlooked. Migration is

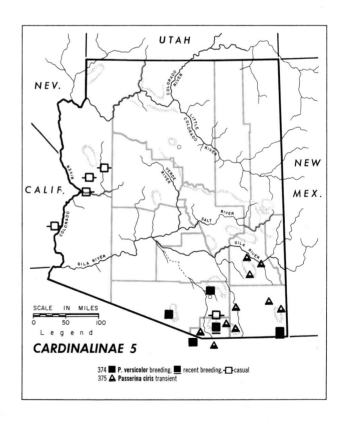

CARDINALINAE 5

374 ■ P. versicolor breeding, ■ recent breeding, ☐ casual
375 ▲ Passerina ciris transient

374. VARIED BUNTING; BEAUTIFUL BUNTING
Passerina versicolor (Bonaparte)

Locally an uncommon summer resident (breeding in July?) in low thorny thickets of the higher Lower Sonoran Zone in foothill canyons of southern Arizona (west perhaps to the Ajo area), north to the Santa Catalina Mountains, with the Baboquívari Mountains the center of abundance. There are casual records for near Patagonia, July 14, 1884 (Stephens, SD) — where it has bred since 1962; the east side of the Mohave Mountains near the Colorado River, *October 27, 1949* (ARP — our only specimen of *pulchra*); the Bill Williams Delta, September 20, 1952 (Monson); and Wikieup, October 1 to 6, 1952 (Phillips). Also, 15 or more seen at Blythe, California side of Colorado River, in February, *1914*, including two specimens, *February 8* and *9* (Hornung, LA).

Despite the beautiful colors in the picture books, the male merely appears all black. (With luck you may see a flash of red or of blue in the crown). The female is solidly brown and nondescript. She is a trifle stubbier-billed than our other buntings, with a more regularly curved culmen, but this is visible only in the hand. This bunting has a sharp call like MacGillivray's Warbler, rather louder than the notes of its allies, and we have never heard it utter the lisping buzz of the Common Bunting and Blue Grosbeak.

There is a strange scattering of records in fall and winter well to the northwest and west, whence the only two females available belong to different races, judging from the fact that one is much paler and grayer than the other. (The pale bird from the Mohave Mountains resembles the Baja California race, *pulchra* Ridgway, whereas the Hornung specimen from Blythe is very brown, like the Sonora form *dickeyae* van Rossem). The species is extremely local in Arizona, one of the most local of all Arizona birds; for instance, in the Santa Catalina Mountains area it is known only from the lower part of Espereso Canyon. Although the species arrives early in May in northern Sonora, our first record for Arizona is June (Sutton, Phillips and Hargrave, Auk 58, 1941: 265-266). It remains well into September, at least in Sonora. The above October records are considered exceptional.

Racial variation is evident only in females and young, as noted above. *P. v. dickeyae* is our breeding race.

375. PAINTED BUNTING
Passerina ciris (Linnaeus)

Formerly a fairly common fall transient (not "casual," as stated by A.O.U. Check-list) in extreme southeastern Arizona, north to the Gila River and west to the Graham Mountains and the Nogales area. Not found in any numbers since 1884 (H. C. Benson, MVZ), and seldom recorded since 1914 (Holt, US): one in Pajaritos Mountains, July 27, 1933 (Campbell, MICH), one in Chiricahua Mountains, August 11, 1956 (Ordway, SWRS), and one 10 miles south of St. David, August 17, 1959 (Levy, US). Still common in Sonora south of Nogales, *October 1954* (ARP), but no ornithologist worked in southern Arizona that season.

The Painted Bunting is an abundant bird in northern Sonora at times and probably is in southern Arizona too, if anyone would go out and look. The male is an unbelievable hodge-podge of purplish blue and yellowish green, varied with bright red underparts and rump. The female is green, yellower below; but many immatures are gray and buff, with scarcely a hint of green. The Painted Bunting remains in northern Sonora near Nogales until mid-October, when the prebasic molt is completed and the birds are quite different from the ragged, grayish urchins that arrived in August from the east. This is an interesting additional example of a "molt migration," supposed to happen only in ducks, but known also in the Western Kingbird, which molts in Sonora during August.

This bunting is recorded in Arizona from *June 18* (at the Graham Mountains — the Ernest G. Holt specimen cited above) and *July 12* (*1902* in the Huachuca Mountains — Swarth, CAS) to *September*.

The apparent decrease of the Painted Bunting in Arizona can have no connection with the prohibition of keeping it caged, as the largest numbers were found by Henshaw in country remote from the few settlements of the early 1870's. The only fully adult male taken in Arizona was the Graham Mountains bird, which shows no evidence of having been in captivity.

376. DICKCISSEL
Spiza americana (Gmelin)

Uncommon fall transient (not "casual," as implied by A.O.U. Check-list) in Lower Sonoran valleys and farms in the southeast; casual elsewhere, with no records for northeastern corner as yet. All authentic records are between *August 8* (Mearns, AMNH) and October 16 (Mr. and Mrs. A. H. Anderson).

The Dickcissel is a strange bird of uncertain relationships. As we see it in Arizona in fall it resembles nothing so much as a female English Sparrow, tinged with yellow. The prominent superciliary and streaked back distinguish it from the female Blue Grosbeak, found in the same habitat. It feeds mostly in fields, returning at any alarm to the adjacent mesquite, along with hordes of lesser sparrows.

Dickcissels were fairly regular in Arizona up to *1892* (Mearns), and have been since 1947 (Wilmot Road Cienaga, Tucson — Anders H. and Anne Anderson). Between those times they were encountered only near Patagonia on *September 24, 1927* (Mailliard, CAS) and near Springerville on *September 20, 1937* by Ed McClintock (two taken at two places — ARP).

FINCHES AND OLD-WORLD GROSBEAKS *CARDUELINAE* Subfamily

377. EVENING GROSBEAK *Coccothraustes vespertinus* (W. Cooper)

Rather uncommon and erratic summer resident about deciduous vegetation in the Transition Zone of eastern and northern Arizona, west to the Grand Canyon (both rims), the Bradshaw Mountains, the Sierra Ancha, and the Santa Catalina and Santa Rita Mountains; possibly also the Hualapai Indian Reservation (immature, October 8, 1948 — ARP). Rare and irregular transient and winter visitant in the wooded parts of the Sonoran zones throughout the state, in flight years reaching even Yuma at least once (May 6, 1902 — Herbert Brown). Also generally rare in winter in and near its breeding territory, but common locally in flight years.

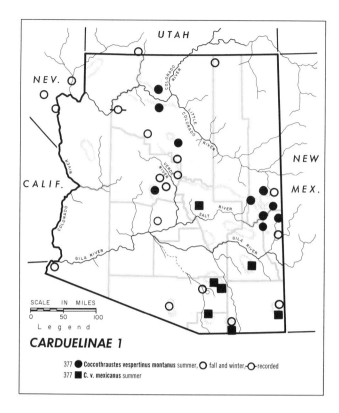

CARDUELINAE 1

377 ● *Coccothraustes vespertinus montanus* summer, ○ fall and winter, ⊖ recorded
377 ■ *C. v. mexicanus* summer

Although this species is to be expected in firs and spruces, in Arizona we have no knowledge of its occurrence above the upper limit of ponderosa pine forests, within which it prefers to feed on buds and seeds in Gambel's oak, maple, and New Mexican locust. Its loud *peer* can be heard afar, as is characteristic of flocking birds which feed in tree-tops and fly for long distances. This is the only chunky, olive-yellow bird with black-and-white wings in Arizona. The enormous bill turns apple-green in the spring.

The Evening Grosbeak's erratic occurrence, long flights, and occasional spectacular irruptions constitute an excellent example of behavior typical of its subfamily, Carduelinae, and is of course connected with the vagaries of food supplies. Independence of insect food permits some cardueline to nest at odd times, but this does not appear to be true of the Evening Grosbeak. The few nesting records for Arizona are in early summer, and no juveniles have been seen in winter. Our earliest date for a nearly full-grown juvenile is June 2 (*1940* on the North Rim of Grand Canyon — Behle, GCN).

Evening Grosbeaks are rare in the Transition Zone between November and April, except for occasional large numbers (January 30 to April 27, 1943, and particularly March 4 to April 18, 1940) in the mixed Transition and Upper Sonoran Zones of the South Rim of Grand Canyon.

Every few years, if not annually, some Evening Grosbeaks appear below the pine belt, with a concentration of records from late April to early June and from mid-October to mid-December. Principal flights (largest numbers or farthest from forest) have been in 1885–1886 (Mearns), 1931, 1935–1936 and 1938–1939 (northern Arizona only), 1950–1951, and 1955–1956. Only on this last invasion were the birds present all winter in the lowlands. Flocks of six appeared near Prescott on October 13 (Gallizioli) and at Sedona October 15 (Douglas Rigby). By *December* the birds were common near Phoenix and Tucson (ARP), and they remained into May near Prescott (Gallizioli).

Although present as supposed winter visitants or transients (which we do not doubt) in the Charleston Mountains of Nevada by October 7 (1931 — van Rossem, Pac. Coast Avif. 24, 1936: 51), dates in Arizona at points remote from breeding grounds are only from October 12 (1931, Baboquívari Mountains — Phillips) to June 7 (1884, Upper Sonoran woodlands of Copper Canyon, near Camp Verde — Mearns) and "several times" in June (1944, Pepper Sauce Canyon near Oracle — Jacot). Except for the 1955–1956 flight, the only records at such points between December 13 and April 24 are March 15, 1934, fourteen at Kiet Siel Pueblo, northwest of Kayenta, northern Navajo County (Wetherill); March 31, 1962, flock at feeding station in northwest Tucson (Marshall); January 11, 1948, two eating mistletoe on Hassayampa River below Wickenburg (Housholder); and December 1950 to January 23, *1951*, at a convent garden in downtown Tucson (Marshall *et at.*, WF).

There are three races of the Evening Grosbeak and all have been taken in Arizona. Our commonest is *montanus* (Ridgway), of which *californicus*, *warreni*, and probably *brooksi* (all Grinnell) are considered synonyms (Phillips, Anales del Inst. Biol. 32, 1962: 369-370). This is the largest-billed race and the female is heavily washed with brown. Except for the Santa Catalina Mountains and south, its status is essentially that of the species in Arizona, as given above. The question of the validity of *brooksi* does not appear to be fully settled. Some competent taxonomists maintain *brooksi* as a darker race of the northern Rocky Mountains, with the female less washed with yellowish (*fide* Kenneth C. Parkes). On this view, some winter birds in Arizona would be *brooksi*. Dark coloration seems to us, however, to be a matter of fresh plumage and occasional soot- or grease-discoloration.

C. v. mexicanus Chapman is the race breeding from the Santa Catalina and Chiricahua Mountains (and apparently the Sierra Ancha — JSW) southward into Mexico. It differs from *montana* only in the more slender, though equally long, bill. This makes for difficulty in distinguishing individual specimens because of overlap from the White Mountains southward. Even at Flagstaff thin- and thick-billed birds may be found mated (Phillips, MNA). We still lack midwinter specimens except from the Chiricahua Mountains (Mark Raue, SWRS).

The most interesting race is nominate *vespertinus*. Long known as a bird of western Canada, it suddenly appeared in the northeastern United States in the winters of 1886–1887 and especially 1889–1890. During the 20th Century, such invasions became increasingly frequent and widespread, involving ever more birds and territory. They began to nest in mid-century east to Quebec! This great expansion was, of course, far more easily detected in the eastern third of the continent, where no Evening Grosbeaks had occurred previously, than in the west. Nonetheless it appears to be affecting Arizona as well. Two females from the 1955–1956 flight have been verified as *vespertinus* by Kenneth C. Parkes. These are from the Southwestern Research Station at Portal, Chiricahua Mountains, *March 19* (Ellen Ordway and Jim Hand, ARP) and Indian Gardens on Oak Creek, *April 20* (ARP). *Montanus* was taken the same year at these and other points. *Vespertinus* differs from *montanus* in its shorter, stubbier bill. Color differences are slight, the principal one being the grayer body of the female.

378. RED CROSSBILL; *Loxia curvirostra*
MEXICAN CROSSBILL   Linnaeus

Irregularly common resident of the more extensive Transition and Boreal Zone forests. May occur in piñons in winter. Rare and irregular in the lowlands (see the racial accounts below).

The Red Crossbill epitomizes cardueline traits, for it breeds in winter, with snow draped around the nest — when and where abundant cone crops permit. Just to confuse ornithologists and oölogists, an occasional odd pair may nest in late spring or late summer, so that some breeding activity may be noted at almost any time. Nevertheless, in western North America at least, the principal breeding season is in early winter. The lovely songs of male crossbills enliven the frozen forests from November on. Though only one actual nest has been found in Arizona (Santa Catalina Mountains — RSC), there is a fine account of winter colonies in Colorado by Bailey, Niedrach, and Baily (Denver Mus. Pictorial no. 9, 1953: 1-62).

The name crossbill is derived from the manner in which the tips of the mandibles overlap each other, thus permitting the bird to open pine scales with the full force of the powerful muscles which close the jaw. In the tree-tops where flocks of crossbills spend much of their time when not coming to the ground for water or salt, they show no color or pattern for identification, but they can easily be identified by the loud *kip-kip-kip* uttered just before and during their undulating flight.

The fledgling crossbill looks exactly like a female Cassin's Purple-Finch — streaked brown and white. Young females soon molt into a dull olive garb, which color they retain throughout life; there is a touch of yellow on the rump. Occasional females, possibly old birds, acquire a dull orange tinge. Young males, on the other hand, often molt directly from the streaked juvenal to a red first basic plumage, distinguishable from fully adult males only by the olive edgings of the flight feathers. (Of course, most of these birds show a few streaked juvenal feathers in the middle of the belly for some months.) Like other red birds, as we have mentioned of L. L. Short's work on the Flicker, red and yellow pigments are easily interchangeable. Though many ornithologists consider yellow or "xanthochroic" males to be immatures, one often finds red males molting into a yellower plumage. This is presumably dependent upon the diet at that particular time of year. Yellow crossbills are commoner in some regions than others (particularly in Alaska and Nayarit), probably due to dietary differences.

The normal male Crossbill's red is a lovely pale rosy hue (described by Griscom as a distinct race, *benti*) when freshly molted. It becomes increasingly deep scarlet with wear and tear, and more brick-red with dirt. Griscom, who describes the eastern race as brick-red, modestly shrinks from referring to his own publication of its custom of dancing "up and down with quivering wings in evident enjoyment" of the smoke from a brick chimney (Bull. M. C. Z. 66, 1938: 531)! Actually the small eastern race, unknown from Arizona, is paler than *bendirei* when clean. Conversely the larger northeastern race, *pusilla*, is darker than our *benti*. Both eastern races have heavier, straighter bills than do our western forms. In the west, crossbills show marked geographic variation in size. This reaches a maximum in central and northern Mexico and southern Arizona; that race, *stricklandi* Ridgway, is not only long-winged, but also has an enormous head and bill, some individuals weighing nearly twice as much as the tiny northwest coast *sitkensis* Grinnell; there are also measurable differences in most of the long bones (Hargrave). The next largest race, *benti* Griscom (of which *grinnelli* Griscom is a synonym, and *bendirei* Ridgway the former and perhaps correct name) is somewhat paler than our other races. The name *bendirei* is now applied to the small race formerly called *minor* in the west. This confusion is due to the equivocal measurements of the lectotype of *bendirei*, which Phillips has examined. Measurements of our four races are given by Bailey, Niedrach, and Baily (*op. cit.*). Also adding to the confusion is the crossbill's propensity to wander — in whole tribes — to territories occupied by other races, there to nest and to constitute the only example in birds of more than one race breeding in the same area. This is a distinction shared only with Man himself and the Roof-Rat (excluding dubious cases in the Brown-headed Cowbird and English Sparrow and more respectable "open ring" forms).

In spite of all this confusion and irregularity, much is now known of the status and movements of Crossbills in Arizona, especially at Flagstaff, thanks to the efforts of Hargrave, Phillips, and Wetherill over many years. These observations show that Crossbills are not "absent from any one locality for years at a stretch" in Arizona (Griscom, Proc. Boston Soc. Nat. Hist. 41, 1937: 133); rather collectors are absent! The birds occur every year, or nearly every year, though as stated above quite irregularly, both as to the number and the race present.

Benti is the race normally found at Flagstaff and in northern Arizona generally; probably it occurs there annually. It visits southern Arizona nearly as often as *stricklandi*, though larger numbers of the latter have been taken. The only desert record of *benti* is at Willow Beach, below Hoover Dam in the northwest, *November 14, 1938* (Grater; lost, not seen by us).

Stricklandi occurs only irregularly north of the border ranges. In flight years, however, it has spread as far north as Oregon (*1920* – US); there are records for all of Arizona except the west and the extreme northeast, where very little collecting has been done. It has been taken in central or northern Arizona every few years, despite the scarcity of collectors there, and has occurred in numbers northwest to Williams (Wetmore, personal collection) and the Hualapai Indian Reservation (Pine Springs – Mearns). There is not a single desert record.

Bendirei (of Griscom and the A.O.U.) is not a common bird in Arizona, yet has been taken at many scattered points throughout the state, except the central and southwestern deserts. In northern Arizona it seems to occur somewhat more often than *stricklandi,* but in smaller numbers. Nor is there any positive proof of breeding of typical *bendirei* nearer than southern Colorado (Monument, El Paso County – Breninger, US). In the Tucson valley, however, it is a different story; here *bendirei* is the dominant race among the few birds that appear, as there are specimens taken *January 25, 1956* (ARP) and *December 30, 1960* (Streets and Gould, ARIZ); only *bendirei* and *sitkensis* are known to have occurred here. Since 1960 hardly a winter goes by without crossbills flying over Tucson (Marshall) or visiting bird-baths there (H. K. Gloyd).

Sitkensis is the most erratic and aberrant race that reaches Arizona, where it doubtless does not breed, and where it has never been seen feeding in native pines. It seems unusually tame for a crossbill. These tiny birds (for crossbills) take one look at the large cones of our ponderosa pines and decide they had better feed on sunflower and other weed seeds or upon the small cones of Aleppo pines planted at Tucson. Nevertheless, one collected from a small flock in piñons near Springerville, *September 9, 1908* (C. Birdseye, US), had evidently descended from higher country, for its stomach was "full mainly of spruce seeds." One was taken from a flock of five at Parker on *August 23, 1953* (GM) – a far cry from the normal home in the cold, dripping firs and hemlocks of coastal Washington, British Columbia and Alaska!

The main flight to Arizona was in 1950–1951. Two were taken from a flock of four in weeds beneath cork-bark firs and small aspens on the San Francisco Peaks, *August 30, 1950* (ARP). Soon after, on September 18, a flock of twelve appeared at Flagstaff (Wetherill), from which five taken *September 21* (ARP) proved to be *sitkensis*. Crossbills reached the Tucson valley by November – an unprecedented occurrence; two secured from a small flock, *November 6* (ARP) proved to be *sitkensis* also. They were seen in the valley until January 19, 1951 (Wm. X. and Alma J. Foerster, J. A. Munro).

In summary, *sitkensis* has been taken in Arizona in three years which correspond to major flights elsewhere. Probably it reaches us in every major irruption; lack of collectors in Arizona is the likely explanation of the fact that we have no records for 1878–1879, 1898–1899, 1913–1914, 1920, 1922–1923, and 1941–1942. In Arizona, it occurs in any Life Zone from Lower Sonoran to Hudsonian. It arrives in late summer, and is evidently gone by mid-winter. It is usually in flocks of 4 to 12 birds, and apparently does not associate with our usual races. The two specimens from the Great Basin region thus far mentioned in print were both called *"bendirei."* These are the above Springerville bird and one from Utah in *June* (!), *1950* (Selander, Condor 55, 1953: 158-160). Selander's excellent series, and other Utah birds (UT), were kindly lent by Dr. Behle.

379. COMMON PURPLE-FINCH; "PURPLE FINCH"; CALIFORNIA PURPLE-FINCH *Carpodacus purpureus* (Gmelin)

Irregular fall and winter visitant, usually in small numbers, in central and southern Arizona east to Camp Verde, Oracle, and the Huachuca Mountains (and exceptionally New Mexico). Two taken on South Rim of Grand Canyon, *December 22, 1934* (Grater, GCN, MNA), and one seen at Supai, September 23, 1950 (Phillips) – the only northern Arizona records.

The Common Purple-Finch is a nondescript streaky brown bird without prominent coloration in wing or tail. Adult males, more than one year old, are a lovely rosy color, which gives rise to the misnomer "Purple-Finch." This color is deeper and mixed with brown above; below it covers the breast, but gives way to unstreaked white on the lower belly and under tail coverts. (Both Cassin's Purple-Finch and the House Finch have brown streaks on the belly or under the tail.) The female (or young) can hardly be identified by sight alone. The heavy head and relatively large bill, with the short notched tail, would identify her as a *Carpodacus,* were it not for the fact that young crossbills share this appearance. All these birds except the House Finch likewise share a pattern of a broad dark stripe behind the eye, set off by pale streaks above and below. The Common Purple-Finch must be recognized by its call, a light *tic*. It is the species of this stripe-headed group most likely to be found in the Lower Sonoran deserts, but habitat is not an infallible clue.

Invasions of Common Purple-Finches in Arizona occur in the same winters as unusual numbers of Lawrence's Goldfinches and others, as in 1950–1951. That winter the Tucson Valley received visits of Evening Grosbeaks, both Common and Cassin's Purple-Finches, Red Crossbills, three species of Goldfinches plus the Pine Siskin, various jays and nuthatches. In addition other northern birds came into the nearby mountains; and there was a flight of Acorn Woodpeckers to the west base of the Baboquívari Mountains (Marshall).

The usual winter occurrence in Arizona is from October through March, with the latest record being April 23 (1955, Madera Canyon, Santa Rita Mountains – Marshall). The largest flights were apparently those of 1885–1886, 1934–1935 and 1947–1948.

The taxonomy of the Common Purple-Finch is at present somewhat uncertain. Almost all Arizona and southwestern New Mexico birds are currently referred to the race *californicus* Baird, a relatively small plain race with an olive wash in the females and young. This race is said to breed all along the Pacific Coast of the United States. The birds of coastal Washington and southwestern British Columbia however, have been separated as a somewhat darker race, *rubidus* Duvall, which may be recognizable. If so, many of our birds will become *rubidus*. There is still a question to what extent post-mortem changes ("foxing") and soot

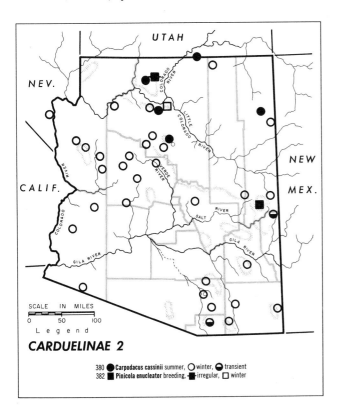

CARDUELINAE 2

380 ● Carpodacus cassinii summer, ○ winter, ◐ transient
382 ■ Pinicola enucleator breeding, ■- irregular, □ winter

discoloration are involved. The more contrastingly patterned eastern race, *purpureus,* is of casual occurrence: one immature female near Tucson, *January 6, 1956* (ARP; specimens of *californicus* were taken in Arizona the same winter).

380. CASSIN'S PURPLE-FINCH; CASSIN'S FINCH *Carpodacus cassinii* Baird

Common summer resident in boreal forest openings of the Kaibab Plateau; decidedly uncommon and apparently irregular in Transition and Canadian Zones elsewhere in northern Arizona, and no definite nesting record. Irregularly abundant winter visitant in high Upper Sonoran and, in southern Arizona, open Transition Zone generally, and very irregularly into Lower Sonoran Zone as at Tucson, Camp Verde, Wickenburg, Salome, Kingman, the Big Sandy Valley, the Kofa Mountains, and even near Tule Well (Yuma County, *November 8, 1960* – GM). On the Mogollon Plateau it is chiefly a transient, though sometimes wintering.

Except for its call, a three-syllable *chidilip,* Cassin's Purple-Finch is virtually a carbon copy of the Common Purple-Finch. In the hand, however, it is seen to have thin dark streaks on the under tail coverts, and it is somewhat larger. While it is usually a mountain bird, identification on that basis is unsatisfactory. All these *Carpodacus* and the very similar juvenal Crossbill have been found on the south rim of the Grand Canyon, while there have been flights of *cassinii* into the lowlands in 1885–1886, 1950–1951, and 1960–1961.

Authentic southern Arizona records are from *October 4 (1951,* in the Hualapai Mountains – ARP) and *October 18 (1958,* summit of Santa Catalina Mountains – George, ARIZ) to May 17 (1948 in the Chiricahua Mountains – Brandt, Hargrave and Phillips) and *May 20 (1927* near Patagonia – Swarth, Proc. Calif. Acad. Sci. 18, 1929: 323). Northern Arizona migrations are principally in late October to early November, and again in March to April. Some Arizona specimens would be *C. c. vinifer* Duvall if geographic races can be recognized; they are slightly darker than our usual bird.

381. HOUSE FINCH; *Carpodacus mexicanus* LINNET (Müller)

Abundant summer resident in the less dense vegetation of the Sonoran Zones, especially about towns and ranches; in recent years has spread to towns and ranches in the Transition Zone. Irregularly reaches even Canadian Zone (Kaibab Plateau). In winter it withdraws into lower valleys, and spreads out commonly to the deserts of western and southwestern Arizona.

The House Finch has various chirping calls none of which resembles those of the previous two species. The song is a clear warble which would be as beautiful as those of the Purple-Finches and Red Crossbill, did it not end in a harsh buzzy note, in Arizona. Mexican birds, which show various local song dialects — some of them most polished — sing quite differently. As its name implies, the House Finch is a familiar and tame bird, roosting in great numbers in vines and trees in town, where some also nest. There are few places in the desert where House Finches do not occur; nests may be found in chollas far from any farms or human habitations. Unlike Purple-Finches, most young males gain their red plumage at the first prebasic molt. House Finches differ slightly in shape from their relatives, having a stubbier, more curved bill, more slender body, and less notched tail. The female has hardly a suggestion of the head pattern so prominent in other species, and the male shows broad brown streaks on the flanks as well as the under tail coverts. Like other red birds, occasional yellow variants occur; males turn yellow in captivity (by the molt); and the red of wild males turns brighter, less rosy, with wear.

There are reports of this species in the Transition Zone for every month except January; but it is rarely seen there except in summer (May to early October). Even then it is limited to certain towns, ranches and lakes, and apparently is not of annual occurrence. The records suggest that it is less rare there since 1938. In Upper Sonoran Zone in the north (Wupatki National Monument), it is common from late March to early October only (J. W. Brewer, Jr., David and Corky Jones). A peculiar event was their return in numbers to the lower Santa Catalina Mountains near Oracle "late in February" (Scott, Auk 4, 1887: 197).

Although there are striking geographic differences in the red coloration, these show up far to the south of us, in Mexico. All Arizona birds are referred to *C. m. frontalis* (Say), a race abundantly supplied with synonyms. This northernmost race of the species has diffuse red combined with relatively large size.

382. PINE GROSBEAK
Pinicola enucleator (Linnaeus)

Fairly common and doubtless resident in the boreal forests of the White Mountains. Was present on the Kaibab Plateau June-July, 1929 (Rasmussen, Vorhies, Mrs. Bailey) but not seen there since. There are also two winter records for the South Rim of the Grand Canyon, *December 15, 1950* (Phillips, GCN) and *January 6, 1957* (Hargrave, GCN).

Though the adult male, whose rosy head and chest contrast with a gray belly, is a handsome bird, most Pine Grosbeaks are rather nondescript clear gray birds like a shortened, chunky Gray Jay with a touch of olive-yellow or dull reddish about the head. The call bears some resemblance to that of a Cassin's Purple-Finch, but is more musical. Though descending to feed in lush meadows at times, Pine Grosbeaks are usually seen perched on the topmost spire of a tall spruce.

The population of the White Mountains was separated as *P. e. jacoti* Jenks. Study of a large series (ARP) by Allen J. Duvall indicates that this population, despite its isolation, is only about 40% separable, and all Arizona birds should be called *montana* Ridgway.

383. ROSY FINCH
Leucosticte tephrocotis (Swainson)

A few sight records, and eleven specimens taken at two localities *November 27* and *December 26, 1956* (Hargrave, GCN); all from the South Rim of the Grand Canyon.

Rosy Finches are dark, unstreaked brown birds that feed on open ground. Unless flying, they do not look "rosy." Hargrave found them feeding in a compact flock of forty to fifty birds, by the roadside where snow did not cover all the weed seeds. Rosy Finches nest in rocky parts of the Arctic-Alpine Zone, reaching their southern limit in northern New Mexico in summer. Because of their ground-feeding habits, the northern races must of necessity migrate in fall, but the southern ones usually move only a short distance southward and downslope. Rosy Finches are recorded in Arizona from November 8 or 9 (1919 — Taylor, FW files) to early January (N. Dodge) and March 23 (1930 — McKee, exceptional?).

All Hargrave's specimens are referable to the Great Basin race *atratus* Ridgway, known by its sooty (less rusty brown) body, especially dark in the male, though hardly black. (The common name of this race, sometimes regarded as a species, is "Black Rosy Finch.") Taylor's, Dodge's, and McKee's birds were first believed to be a paler race, *tephrocotis* (which has been reported taken not far north in Zion National Park, Utah); but after examining Hargrave's specimens, McKee is not certain that his bird was different. Thus, the recognition of any race besides *atratus* in Arizona must await the capture and careful comparison of specimens.

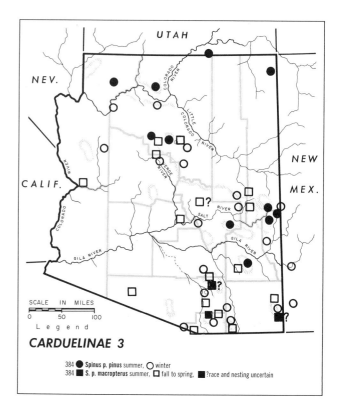

CARDUELINAE 3

384 ● *Spinus p. pinus* summer, ○ winter
384 ■ *S. p. macropterus* summer, □ fall to spring, ■? race and nesting uncertain

384. PINE SISKIN; PINE FINCH
Spinus pinus (Wilson)

Common summer resident (though hardly any nests recorded) from the Transition Zone to timberline on the Kaibab and Mogollon Plateaux, including the San Francisco and White Mountains; rather uncommon to rare as a presumably breeding bird south to the Chiricahua and Santa Rita Mountains, and probably to Mt. Trumbull in the northwest. Winters more or less commonly in weedy fields and river valleys almost throughout the state, but rather sparingly in the southwest and along the Colorado River, sometimes appearing in flocks in the Sonoran Zones by *August 10* (ARP) and remaining to June 7 (1917 near Tucson — Dawson).

The Pine Siskin is about as nondescript as a female Purple-Finch or House Finch, but is distinguished by its small size, slender form, sharply pointed bill, and the yellow in the spread wings and tail. Both the amount and depth of color of the sharp streakings vary greatly, and so does the amount of yellow at the bases of the flight feathers. The characteristic call is a shrill, rising whine, *shreeeee,* and there are other weak, high calls.

The difficulty involved in determining the breeding range stems from the prolonged and erratic movements, as well as the absence of actual nest records except those of Sutton in the Santa Rita Mountains (Auk 60, 1943: 349). Thus there are numerous records of flocks far from any suitable coniferous forest throughout May, and single birds have lingered at Sunnyside, at the lower edge of the pines in the Huachuca Mountains, to June 22 (1951 — Marshall) and *June 25 (1937* — Jacot, ARP). This leaves scarcely six weeks during which we may suspect (but cannot prove) that all are on the breeding grounds. It is unusual, however, to see Pine Siskins in the desert valleys before mid-October, the above August records being the only one earlier than September 29 (1961, Salt River Canyon — Marshall).

Geographic variation in Arizona birds is slight in comparison to the great changes shown over the entire range of the species, leading to an unstreaked green and yellow bird with a black cap in the male in Chiapas and adjacent Guatemala. Most Arizona birds, particularly in summer, are of relatively small size and heavy streaking (*pinus*). Those which apparently breed in the border ranges, and which flood northward later, are slightly larger and sometimes show a reduction in the streaking of the underparts together with an increase in the amount of yellow in the flight feathers. These are *S. p. macropterus* (Du Bus). The two races occur during winter over much the same territory, sometimes in the same flock. The larger flocks, however, in certain parts of Arizona (Whiteriver and Flagstaff regions, where considerable winter collecting has been done) are predominantly *macropterus*. This race occurs there from late November to mid-April at least. Southern Arizona records well away from the forests are *August 10* (as above) and *October 24* (*1947*, Ajo Mountains — Tinkham, ARP) to *May 21* (*1917* in the lower Chiricahua Mountains — Austin Paul Smith, AMNH).

The small northern race, *pinus*, has been taken away from the forests in southern Arizona from *October 28* (*1884* near Oracle — Scott, AMNH) to *May 21* (*1935*, Eagle Creek near Morenci — Jacot, ARP), and *June 25* as above.

385. AMERICAN GOLDFINCH; PALE GOLDFINCH   *Spinus tristis* (Linnaeus)

Irregularly a rather common winter resident in deciduous trees and weedy fields of the Sonoran Zones, sometimes remaining to early June. Transient on the Mogollon Plateau; seen recently in summer there (Phillips) and in Hualapai Mountains (Musgrove), but no summer specimens. Older summer records inside Grand Canyon and on Navajo Mountain (Utah) require substantiation.

The most common of the clear, high-pitched calls of the American Goldfinch, is a *per-chic-o-ree* given in undulating flight. They lack the buzz of the Pine Siskin and the plaintiveness of the Lesser Goldfinch. The brilliant alternate plumage of the male is seldom seen in Arizona; instead, our birds are usually in the dull basic plumage, which is similar to the much commoner Lesser Goldfinch but without any green and with a pronounced cinnamon wash on the flanks. Yellow is limited in this plumage to the head, the underparts being mostly whitish.

Although considered "evidently a rare species in Arizona" by Swarth and others, the American Goldfinch is a regular winter visitor and is seen frequently from about mid-October through April in the lower valleys. Winter records are few on and near the Mogollon Plateau and along the lower Colorado River. Extreme dates for northern Arizona are from October 8 (1941, Lava Canyon, Grand Canyon — McKee) to June 4 (1937, Keams Canyon — Monson) and *June 7* (*1915*, Springerville — Goldman, US). Southern Arizona records are from September 11 (1938 near Tucson

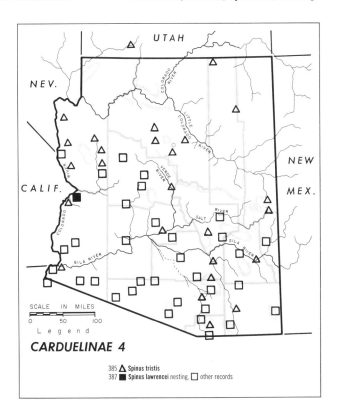

CARDUELINAE 4

385 △ Spinus tristis
387 ■ Spinus lawrencei nesting, □ other records

— Phillips) and *September 30* (*1952*, Wikieup — ARP) to June 9 (1959 near Globe — Harold A. Marsh).

The above account refers to the usual race in Arizona, *pallidus* Mearns. This is the largest and palest of the species. The small, dark coast race *salicamans* Grinnell is represented by a single female from Parker, *May 17, 1948* (GM) as kindly identified for us by Dr. Alden H. Miller. The medium-sized dark eastern race, *tristis*, with reduced white portions of the wing and tail, occurs rarely in eastern Arizona; an occasional bird from Tucson and Tempe strongly approaches *tristis*, but the only perfectly typical specimen is from the upper Black River area, White Mountains, *November 17, 1936* (ARP).

386. LESSER GOLDFINCH; GREEN-BACKED GOLDFINCH; ARKANSAS GOLDFINCH   *Spinus psaltria* (Say)

Fairly common summer resident in deciduous trees and brush (especially willows and cottonwoods) in the Sonoran and (locally) Transition Zones throughout Arizona; also found locally in live oaks. Irregularly common post-breeding visitor and winter resident in the Sonoran Zones, less common in extreme northeast; in late summer visits Canadian Zone (Kaibab Plateau), and may winter irregularly in Transition Zone. Breeding season remarkably prolonged or irregular.

The whistled calls of this smallest goldfinch are extremely plaintive and easily imitated to decoy the bird passing high overhead. The song is Canary-like, as in other members of the genus. During its undulating flight, white areas are seen on both the tail and wing of the Lesser Goldfinch. The most interesting thing about the bird is the odd span of nesting times, at least in southern Arizona.

A juvenile from Patagonia, *February 18 (1952* — J. A. Munro, Royal Ontario Museum of Zoology) was judged to have hatched in the first half of January. A nest was being built on March 17 near Tucson (Howell, Condor 18, 1916: 212). Nests with eggs in the first half of April were found near Tucson (1953 — Eliot Porter) and near the mouth of Aravaipa Creek (1962 — RSC). A pair began a nest at Phantom Ranch, bottom of the Grand Canyon, on February 13 (1935), commenced incubating on April 12, and fledged the young on May 10; a new nest was started on May 20 (Mrs. Clara Shields, *fide* McKee). A fully-grown young was taken in the Ajo Mountains, Organ Pipe Cactus National Monument, *May 21 (1947* — ARP). Then besides normal summer nestings, there are several in the fall. A nest was being built near Patagonia on September 14 (1947 — Mr. and Mrs. Wm. X. Foerster, Phillips); a young bird about two weeks out of the nest was seen near Camp Verde on October 19 (1884 — Mearns); and young being fed by an adult were taken at Benson on *November 5 (1945* — LLH).

Aside from sporadic late *November* to January records, Lesser Goldfinches have been found in and near the pines of northern Arizona from April 4 (1931 — Hargrave) to October 23 (1952 — Mrs. Eleanor Pugh), both at Flagstaff. Southern Arizona migrations are not clear because of the widespread breeding and wintering.

Except for the Springerville region, all Arizona birds are currently referred to *S. p. hesperophilus* Oberholser, green above, and therefore formerly known as the Green-backed Goldfinch. Mexican males, on the other hand, are black-backed. Unfortunately the type of *psaltria* was collected from an intermediate population in Colorado, where both back colors occur in males, with no proof that these are age differences, as has been assumed. Considering that black-backed males occasionally crop up at Springerville and Patagonia (ARP), it seems more likely that this is simple polymorphism, with two color phases. Two races, but not more than two, can certainly be distinguished (only in males) from northern Mexico and the United States on the basis of the striking difference in back color, and these should take the two oldest names: *psaltria* (the type of which is green) for the green-backed birds of California, Arizona, Sonora, and the partly-green populations of Colorado and New Mexico, and *mexicanus* Swainson for the wholly black-backed populations farther south. Incidentally, the black-backed male from Patagonia was shot at a nest under construction and fell into it. The female arrived with a bill-load of material, threw out the male, and continued to build.

387. LAWRENCE'S GOLDFINCH *Spinus lawrencei* (Cassin)

Irregularly common transient and winter visitant in weedy areas and at watering places of the Sonoran zones across southern Arizona, north to the Prescott region. In 1952, at least, nested in western Arizona, near Parker (nest taken — G. W. Bradt, ARIZ). Specimens in breeding condition taken at Wickenburg, *May 12 (1953* — ARP), and in Castle Dome Mountains, *May 6, (1955* — GM). A flock of five seen along Colorado River 18 miles above Imperial Dam, June 2 (1958 — Monson).

The call of this handsome finch is of two tinkling notes in succession, the second a minor third below the first. The yellow in the wing is on the outside and much easier to see than in the Pine Siskin, from which it further differs in being without streaks. The body is gray, and the male has a yellow chest patch and black all around the bill. Our few summer records suggest that this little nomad may be extending its breeding range eastward from California. Except for these, the localities on the map represent its wintering flights to Arizona, which are not regular, but occur on the average of one year in two.

The normal wintering span of the flocks here is from October to April, with extreme dates in late *September (23, 1961,* at Tucson — RSC) and May 20 (1925, in the Baboquívari Mountains — Bruner). Like our other goldfinches, these birds feed in flocks on sunflowers, thistles, and weeds in open fields. They used to roost in privets by Phillips' window in Tucson until the leaves would fall, in January.

SPARROWS AND OLD-WORLD BUNTINGS
FRINGILLINAE Subfamily

388. GREEN-TAILED TOWHEE *Pipilo chlorurus* (Audubon)

Rather common summer resident in low deciduous brush of the Transition and boreal zones of the White and San Francisco Mountains and the Kaibab Plateau.

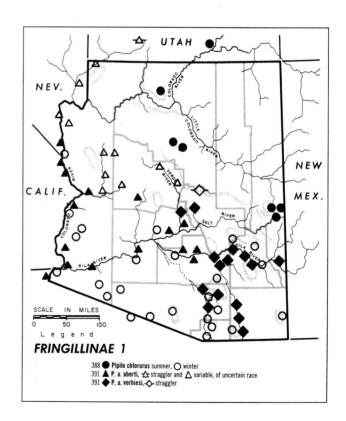

FRINGILLINAE 1

388 ● *Pipilo chlorurus* summer, ○ winter
391 ▲ *P. a. aberti*, △ straggler and △ variable, of uncertain race
391 ◆ *P. a. vorhiesi*, ◇ straggler

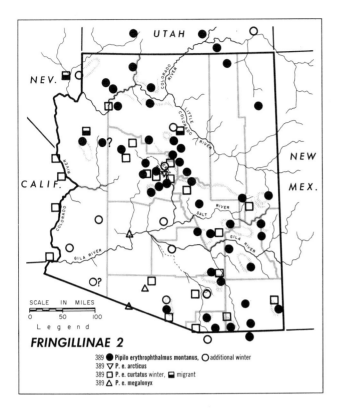

FRINGILLINAE 2
389 ● *Pipilo erythrophthalmus montanus*, ○ additional winter
389 ▽ *P. e. arcticus*
389 □ *P. e. curtatus* winter, ▨ migrant
389 △ *P. e. megalonyx*

Common transient in dense brush throughout Arizona; and winters fairly commonly in low, weedy brush of the Lower Sonoran Zone and adjacent areas of the Upper Sonoran Zone of central and central-southern Arizona; in most winters scarce in extreme west, southwest, and southeast.

This small green and gray towhee has a double mewing call, a lively song, and seeks its food on the ground by scratching backwards with both feet simultaneously. This peculiar manner of the bird's scratching itself down through layers of dead leaves and mulch is here encountered for the first time, as it is a unique trait in the subfamily Fringillinae. The principal means of livelihood for members of the genus *Pipilo* and the Fox Sparrow, it can also be seen occasionally in the Vesper, White-crowned, Lincoln's and Song Sparrows as well as in the Mexican Junco. Leaves are kicked out to the rear, and the sound of steady scratching is the best clue to the presence of such full-time scratchers as the Green-tailed Towhee, Rufous-sided Towhee, and Fox Sparrow. Also encountered here for the first time is the yellow underside of the wrist, which crops up in fringillines such as the Grasshopper, Botteri's, Cassin's and Sage Sparrows; in none of these is it as brilliant and extensive as in the Green-tailed Towhee, however.

The migration period in northern Arizona is from April 12 (1932 at Flagstaff, banded — Hargrave) to May 28 (1946, Wupatki National Monument — Davey and Corky Jones, *et al.*); and again from August 22 (1936, Betatakin Canyon, Navajo National Monument — Wetherill) to October 22 (1935, Flagstaff — V. R. Kiessling, also a trapped and banded bird). Just below the Mogollon Rim, in Oak Creek Canyon, one was seen at Indian Gardens on April 7 (1956 — Bialac).

Extreme records in southern Arizona are from August 10 (1874, Fort Bowie — Rutter, *fide* Henshaw) and *August 25 (1936,* Tucson — Jacot, ARP) to June 3 (1956, J. W. Hardy in the Chiricahua Mountains) and *June 6 (1893* near Tucson—Price and Wilbur, Stanford University Coll.). The conspicuous influx of migrants is in April, with the earliest record in suburban Tucson on March 19 (1954— Bruce Cole). Much more pronounced than in fall, this spring migration brings Green-tailed Towhees into city gardens, where an individual will sing, rest and feed for several days to a week, showing that migration is a leisurely affair. The last of two birds in Phillips' yard at Olive Road, Tucson, stayed a week and then left into the teeth of a storm (April 27, 1953), without proper respect for the principles of migration stated in texts. In mesquites outside of Tucson, the birds are extremely abundant at this time and the males sing, fight, and chase each other incessantly in anticipation of staking out territories in the north.

389. RUFOUS-SIDED TOWHEE; SPOTTED TOWHEE; SPURRED TOWHEE; MOUNTAIN TOWHEE; NEVADA TOWHEE
Pipilo erythrophthalmus (Linnaeus)

Common summer resident in dense broad-leafed brush of the Upper Sonoran Zone, ranging locally into the Transition Zone. Winters commonly in the Upper Sonoran foothills of northwestern, central, and southern Arizona, fairly commonly in brushy canyons and river valleys of Lower Sonoran Zone of southeastern Arizona; less commonly westward to the Colorado River, where it is a rather rare transient, and even scarcer in winter in most years. Apparently rare to uncommon in winter in the northeast.

In the United States and Canada male Rufous-sided Towhees are black birds with small white spots on the back, wings, and corners of the tail; they have chestnut flanks, and a white belly. Females, according to the race, resemble the males or are variously paler, more gray or brown. Brown females are almost unknown in the west, however. All ours have red eyes, brighter in adults. Spending its life scratching leaves under dense brush, this towhee only occasionally ascends into view in order to mew or give its breezy song, which suggests the name, Towhee. In Arizona we have a resident population in the mountains and wintering birds from the north.

Extreme wintering records for non-breeding southern localities are from *September 22 (1930* at St. Thomas, Nevada— van Rossem, LA), and September 28 (1955 at Molino Basin in the lower Santa Catalina Mountains — Lanyon) to May 5 (1957 in Tucson — Ivan Peters), May 6 (1961) and May 11 (1955, both south of Tucson — Marshall). But it is rare in the lowlands after the first week of April.

In the north, records well away from breeding areas are from *September 23* (*1933* at Wupatki National Monument — Hargrave, MNA) to April 25 (1948 near Holbrook — Phillips). Wanderers sometimes appear not far from the brushy breeding localities in the Flagstaff-Mormon Lake region in late summer, the earliest being *July 25* (*1939* — ARP).

Arizona birds are chiefly of the race *montanus* Swarth. Females of this form are nearly as black as males, except on the rump. Contrary to Swarth's opinion, this race is not entirely sedentary in the mountains, but occurs regularly in the valleys and lower mountains in winter, west even to the Castle Dome Mountains of Yuma County (GM). In summer there are numerous pairs of Rufous-sided Towhees living in the net-leaf oak chaparral on the south face of the summit of the Santa Catalina Mountains and also in willows along the creeks. But in midwinter few are found there. Whether these birds simply move down-slope, or contribute to the distant records, cannot be told without extensive banding and much more luck than usual. For the percentage of recoveries of small passerine birds away from the point of banding is infinitesimal.

Because the identification of male *montanus* is risky, we have no certain fall arrival date of this race in the Arizona lowlands, from which we lack adequate specimens of females. Female *montanus* has been taken in the valleys from *October 13* (*1954* south of Nogales, Sonora — ARP) to *April 9* (*1939* near Tucson — ARP). In northern Arizona there are no satisfactory data.

The north-central race *arcticus* (Swainson) is characterized by a female in which the head and back is pale brown. It is a rare or casual winter visitor to Arizona, recorded only from the female taken at Camp Verde on *January 7, 1888* (Mearns, US #235589). It has also been taken at Provo, Utah, *November 30, 1872* (Henshaw, US #63510), so it should be looked for carefully in the eastern half of Arizona in winter.

The Great Basin race, *curtatus* Grinnell, is equally common as a winter resident in Arizona with our own *montanus*, but with the female grayish instead of blackish. It was described from measurements of breeding males of Nevada, which is unfortunate in view of the principal geographic variation being confined to coloration of females. Our birds have accordingly been identified in comparison with females from southern interior British Columbia. Extreme dates of typical females in southern Arizona (besides the Nevada September 22 bird cited above) are from *September 26* (*1949* below the Hualapai Mountains—ARP) to *March 24* (*1917* in the Chiricahua Mountains — Austin Paul Smith, AMNH). Northern Arizona records, pertaining to migrants only, are from *September 23* (*1933*) to *April 5* (*1934* on the northeast slope of the San Francisco Mountains region — Hargrave, MNA).

California races differ from *montanus* in being still darker, particularly on the flanks, and with reduced white spotting. Several southern Arizona specimens are more or less intermediate, but reasonably typical specimens of the large-footed *megalonyx* Baird are from Arlington, *November 22, 1918* (Austin Paul Smith, AMNH #368229) and Sonoyta, Sonora, *January 16, 1894* (Mearns and Holzner, US #132611). Both are immature males. The Sonora bird is altogether too dark-sided and heavy-footed for *curtatus*, as recorded by van Rossem (*Birds of Sonora*, 1945). An adult male from near Ventana Ranch, Papago Indian Reservation, *March 21, 1947* (ARP) likewise has dark flanks and large feet and appears referable to *megalonyx* (though van Rossem believed it to be *falcinellus* Swarth).

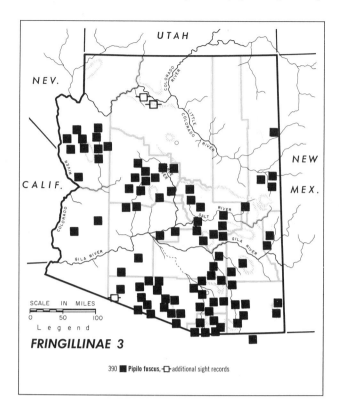

FRINGILLINAE 3

390 ■ Pipilo fuscus, ▫ additional sight records

390. BROWN TOWHEE; *Pipilo fuscus* CANYON TOWHEE Swainson

Rather common resident in scattered low but dense brush in high Lower Sonoran and low Upper Sonoran zones of most of southern, central, and northwestern Arizona. Restricted to rocky hills and desert mining camps westwardly. Found west to the Black and Mohave Mountains of western Mohave County, the Kofa Mountains, and the Ajo region west sparingly to Papago Well. Island populations may be found on the north side of the White Mountains region from Springerville to Lupton. Three sight records along rims of Grand Canyon, from Village westward (McKee, H. C. Bryant, and Grater).

This towhee appears all brown as seen against the ground, which it matches in color, and over which it runs and hops between places where it scratches for food. At close range a black dot is seen on the chest. (In the hand, the back of fall birds is actually gray.) Visitors from California are surprised to notice its entirely different voice from the California bird — a hoarse *hic-cup* call note and a pleasant chipping song. Not only a permanent resident wherever found, the pairs moreover are remarkably sedentary, remaining mated for life upon circumscribed territories (Marshall, Condor 62, 1960: 49-64). There is a long nesting season from March to September. A nest with eggs was taken in February (RSC). Unlike the birds in California, it is not found in city gardens; indeed, Tucson is the only large city it even approaches in its Arizona distribution. It is entirely absent (except in the mountains) from the Phoenix region, where instead, the Abert's Towhee is a garden bird.

It is remarkable that so sedentary a bird should not show more spectacular geographic variation. The race *mesoleucus* Baird covers all Arizona and most of New Mexico. This is the palest, most white-bellied, most brightly rufous-capped race of the species. Interestingly, some of the specimens from the Harquahala Mountains are darker than normal, but not in a sufficient proportion to merit a racial name. Thus we do not recognize "*relictus* van Rossem" at the dire risk of reducing taxonomy "to its narrowest and most sterile application."

391. ABERT'S TOWHEE *Pipilo aberti* Baird

Common resident in dense undergrowth of the willow-cottonwood and large mesquite associations of the main rivers of the Lower Sonoran Zone in southern and western Arizona. The Upper Sonoran Zone report ("Fort Whipple") is an error.

Abert's Towhee is a more cinnamon brown than the Brown Towhee; it lacks the chest spot, but shows a light-colored bill framed in black feathers of the front part of the face. In the hand, fall adult males are seen to be suffused ventrally with a beautiful pinkish tone. In many details of voice (pattern of calls and songs but not timbre), egg color, and familiarity as a door-yard bird in Phoenix, this species is more an Arizona counterpart of the California Brown Towhee than is *mesoleucus*. A riparian bird, the Abert's Towhee approximates or overlaps the distribution of the Brown Towhee only on parts of the upper San Pedro River (Gould), Aravaipa Creek (Phillips) and on the lower Santa Cruz River and its tributaries in the immediate vicinity of Tucson, where over-lapping populations were studied by Marshall (Condor 62, 1960: 49-64). Like the Brown Towhee, in this species the pairs remain mated for life on definite territories, except for an occasional divorce, as studied in banded birds (Marshall, *op. cit.*); and the nesting season is as long or longer. At Sacaton, Gilman (ms.) found eggs from February 28 to September 4. Peaks of nesting activity in spring and again in fall follow rains at those times (Marshall, Proc. XIII International Orn. Congress, *in press*).

For a recent discussion of geographic variation and the snarled nomenclature of the Abert's Towhee see Phillips (Anales del Inst. Biología Univ. de México 33, 1963: 357-367). In brief, the pale cinnamon race, *aberti*, ranges from the Colorado Valley and Desert east to at least the Phoenix area, with the exception of the Big Sandy Valley, where the population is variable. The somewhat darker and more grayish-brown eastern race, *vorhiesi* Phillips, ranges in the Tucson region and probably along the upper Gila and San Pedro Rivers down to their junction.

392. LARK BUNTING *Calamospiza melanocorys* Stejneger

Common to abundant winter resident in brushless, weedy, or barren-looking parts of the Lower Sonoran Zone of southeastern Arizona; scarcer and irregular westward but common some years, even to southern Nevada, usually reaching the Colorado River only in fall. Apparently a fairly common transient in eastern Arizona in open Upper Sonoran Zone from Holbrook to Fort Apache but rather rare farther west or north. A *May 6* record for extreme southwestern Utah (Behle, Condor 44, 1942: 231) is probably exceptional.

In late summer compact flocks of stocky, fluffy, heavy-billed finches arrive upon the most barren grounds of southern Arizona. They are soon in their basic, female-type plumage, heavily streaked with blackish both above and below, following their annual prebasic molt. They feed on the ground, where they bounce along by hopping with their short feet. In April there is a complete prealternate molt in both sexes; the females retain their previous pattern but the males become wholly black except for a prominent white patch on the folded wing. (Other plumages have dull buffy wing-patches). In May the males are beginning to sing a song which in concert sounds just like a singing flock of Brewer's Sparrows — with much trilling, tinkling and buzzing — but is much louder.

The usual occurrence in southern Arizona is from August to May, with extreme dates near or south of Tucson from *July 9 (1940 – ARP)* and July 25 (1933 – Vorhies *et al.*) to May 22 (1947 – Alma J. Foerster, Junea W. Kelly, *et al.*) and May 24 (1951 – Wm. X. Foerster). East of the Chiricahua Mountains one was seen July 23 (1957–Levy). By July 31 to August 1, 1951, perhaps 200 were near Douglas (R. B. Streets). In northern Arizona the species is a migrant only, principally in September and April;

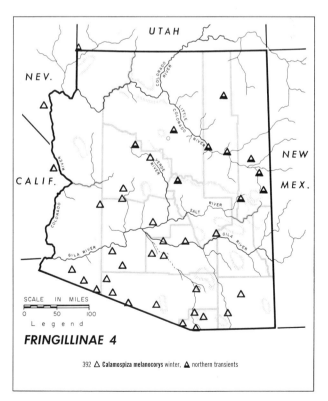

FRINGILLINAE 4

392 △ *Calamospiza melanocorys* winter, △ northern transients

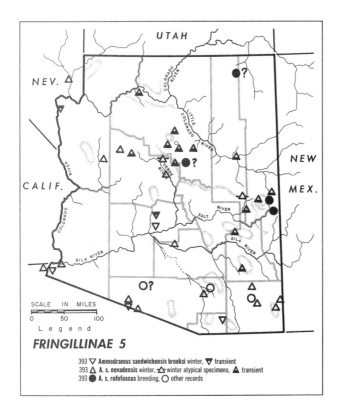

FRINGILLINAE 5

393 ▽ *Ammodramus sandwichensis brooksi* winter, ▼ transient
393 △ *A. s. nevadensis* winter, ⟁ winter atypical specimens, ▲ transient
393 ● *A. s. rufofuscus* breeding, ○ other records

extreme dates are August 25 (1934 southeast of the Petrified Forest—Stevenson) to September 29 (1931 at Wupatki National Monument) and again from *March 4* (*1934* in the same region — Hargrave, MNA) to May 13 (1937 near Keams Canyon — Monson). Apparently the abundant southern Arizona winter residents do not cross northern Arizona, but arrive from eastward. There are records just below the Mogollon Rim at Whiteriver from *August 8* to *23* (*1936* — Watson and Poor, ARP), however.

393. SAVANNAH SPARROW; LARGE-BILLED SPARROW *Ammodramus sandwichensis* (Gmelin)

Fairly common summer resident locally at lakes and moist fields on and just north of the Mogollon Plateau in the White Mountains (including Springerville area), also possibly Mormon Lake (and formerly Kayenta?). Common transient at lakes, ponds, marshes, and in fields and level grassy spots throughout the state. Common to abundant in winter in irrigated fields of the lower Colorado Valley and elsewhere in irrigated valleys, grassy swales and plains, and along bodies of water in Lower Sonoran Zone throughout southern and western Arizona. Also found locally and rarely in winter in weedy fields and grassy edges of lakes and ponds in northern Arizona, casually into Transition Zone (near Lakeside, *December 18, 1936* — ARP). Late summer straggler to lakes near Yuma (large-billed form *rostratus*): one authentic record, *August 15, 1902* (Brown, ARIZ).

The Savannah Sparrow is a small sparrow, streaked above and below, with a large wing and powerful flight, for it inhabits windy prairies and open damp fields. Often there is yellow visible over the eye, especially in summer. The tertials are long, to protect the wing from abrasion against the grass; the tail is slender, short, and somewhat notched because the rectrices are pointed in this genus. There is usually, except in the large-billed Mexican races, an extensive prealternate molt in April or May of the body and (in some central Mexican birds, at least) the tertials and central rectrices. The call is a shrill *sip*. All these traits distinguish the bird from other common sparrows which are streaked on the chest, such as the Song and Lincoln's Sparrows.

Migrant and wintering Savannah Sparrows occur in southern Arizona from *August* (*nevadensis;* also seen twice on August 24 in Bill Williams delta — Monson) to *May 8* (*anthinus*), May 11 (1950) and May 22 (1952; both in Topock-Needles area—Monson). Northern Arizona migrations are from August 20 (1936 near Flagstaff — Phillips) to *November 1* (*anthinus*), and again from February 10 (1938 at Ganado, Apache County — Monson) and *February 21* (*nevadensis*) to May 10 (1937 near Keams Canyon — Monson).

Geographic variation in this complex species can be presented synoptically as follows, for the races occurring in Arizona:

A. First we may divide out the relatively thick-billed, plain-backed, salt-marsh races of the Gulf of California, of which only one has wandered into Arizona; it is *rostratus* (Cassin), referred to at the end of the opening paragraph.

B. The small-billed, streaked-backed races can be divided variously:

1. With heavy black chest streaks and dark coloration above; only two races in Arizona:
   a. The breeding bird, *rufofuscus* (Camras), which differs from *anthinus* in blacker head with less buffy cast, black streaks underneath (and possibly slightly darker, with heavier bill). It may be a synonym of *brunnescens* Butler of Mexico. It breeds in the White Mountains district. (Birds formerly bred at Kayenta, but the race is not known; one from Mormon Lake, *August 5, 1933* — Phillips, MNA, is too worn for racial identification.)
   b. The San Francisco Bay race, *alaudinus* (Bonaparte), an accidental, with one record (Tucson, *January 1, 1938* — ARP). This is smaller than the above and its black streaks are a little heavier. Identification of the specimen is confirmed by Dr. Alden H. Miller.

2. Intermediate; streaks often black and heavy, but otherwise the bird is pale and small: *brooksi* (Bishop) of the humid northwest coast. Rare in Arizona.

3. With narrower, often less blackish chest streaks and usually paler coloration above and on head:
   a. *Nevadensis* (Grinnell) is like *brooksi* but larger, seldom having heavy black streaks below, and is like *brooksi* in its thin bill, whose depth is less than one-half the length. (Lack of a yellow superciliary is not a good racial character, because it varies.) Breeding in the Great Basin and Great Plains, it is by far the commonest transient and

a common winter resident in Arizona. Most of the old reports of *"alaudinus"* belong here.

b. *Anthinus* (Bonaparte) is like *nevadensis* but is darker and browner above, less grayish. It is similarly large and thin-billed. In fall it is quite buffy on the head. It is also a very common winter resident in Arizona, breeding in the interior of Alaska and northwestern Canada. One specimen approaching the larger *P. s. sandwichensis* is reported from Arlington, Gila River (*fide* Peters and Griscom).

c. All the rest of the races have thicker bills and none occurs in Arizona, although *crassus* (Peters and Griscom), which breeds in southern coastal Alaska, is to be looked for in southwestern Arizona in winter.

Ammodramus sandwichensis rufofuscus has been taken only twice off the breeding grounds in Arizona: at Tucson, *February 6, 1887* (Brown, ARIZ) and Sulphur Spring, Cochise County, *March 16, 1895* (Osgood, AMNH). There are no satisfactory migration dates on the breeding grounds, since the latest specimen available is *September 9* (*1934* near Springerville — Stevenson, ARP).

Although *brooksi* is restricted in winter to "western California to central Baja California" by the A.O.U. Checklist, there are the following records in Arizona: Willow Beach below Hoover Dam, *October 3, 1938* (Grater, Lake Mead National Rec. Area collection); Yuma, *November 17, 1902* (Clay Hunter, ARIZ); Quitovaquita, *November 29, 1939* (Huey, SD; Trans. San Diego Soc. Nat. Hist. 9, 1942: 373); northwest and west of Phoenix, *October 17, 1953* (ARP), *February 22, 1957* (Housholder, ARP), and *September 27, 1958* (Simpson, JSW); and Sonoita, *December 9, 1935* (Dille, ARP).

Southern Arizona migrations of *nevadensis* are from *August "14"* (*1884* at Camp Verde — Mearns, US and ms.; actually the 13th) and *August 26* (*1873*) to at least *October 26* (*1874*, both at Fort Apache — Henshaw, US); and again from *March 9* (*1937* near Fort Apache — ARP) to *May 2* (*1886* at Tucson — H. Brown, ARIZ). Northern migrations are *August 26* (*1934* west of Winslow — Russell and Hargrave, MNA) to *September 21* (*1934*, common at Springerville — Stevenson, ARP) for the fall; spring records are between *February 21* (*1937* west of Winslow — Hargrave, MNA) and *April 26* (*1948* near Holbrook — ARP).

Anthinus occurs as a transient and winter resident in southern Arizona from *September 24* and *May 8* (H. Brown, ARIZ, AMNH). In northern Arizona it does not winter and is found principally from late *September* through *October* to *November 1* (Hargrave, MNA), and again from the end of *February* through *March* (specimens in ARIZ, ARP, MNA). Farther west near Prescott there is a specimen taken *May 7* (Jacot, ARIZ). Since *anthinus* breeds farther north, it arrives in Arizona a month later than *nevadensis* in fall. Yet it is also a laggard in spring. The winter range parallels that of *nevadensis* (see map), extending north to San Carlos and Wikieup (ARP), lower Oak Creek (Jacot, ARIZ), and formerly (locally) Cottonwood Landing under the present Lake Mohave in Nevada (*fide* Linsdale).

394. GRASSHOPPER SPARROW *Ammodramus savannarum* (Gmelin)

Fairly common resident in dense grassland and alfalfa fields of San Rafael and Babocomari valleys, the upper end of the Sonoita Valley, and possibly the Altar Valley, Coyote and Quinlan Mountains region, in southeastern Arizona. Has bred (possibly irregularly) at Fort Grant, and perhaps once (1916 — H. T. Jackson and Taylor, FW files) at Camp Verde. Fairly common winter resident in dense grass (usually mixed with low brush) of southeastern Arizona, west to the Papago Indian Reservation; uncommon and irregular farther west, even to the Colorado River where it is mainly a very rare migrant; but one winter record years ago at mouth of Bill Williams River (Kennerly and Möllhausen, US — not on Big Sandy River, as stated in A.O.U. Check-list).

An exceedingly handsome bird in the hand, the Grasshopper Sparrow as seen when flushed from its grassy haunts is an obscure grayish-brown streaky-backed sparrow with a short, thin tail. If it happens to alight in a bush, the face and chest show light ochraceous, and a buff crown streak can be seen. The chest may be finely streaked with rufous, but never with black. The name is derived from its insect-like buzzing song.

There are three brownish races in the United States (excluding the blackish birds of the Florida Everglades). In the east, *pratensis* (Vieillot) has a heavy bill, dark color, and is rather small. It is mentioned here because a single specimen from Gardner Wash, east base of the Santa Rita Mountains (*September 28, 1958* — Bialac, ARIZ) shows these characters. *Pratensis* would not be expected in Arizona because *perpallidus* (Coues) breeds east all the way to southern Illinois and adjacent Indiana (US).

A. s. perpallidus, with opposite characters, is the common Arizona race, except locally in the extreme south.

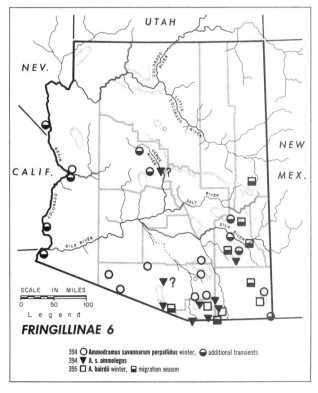

FRINGILLINAE 6

394 ○ *Ammodramus savannarum perpallidus* winter, ● additional transients
394 ▼ *A. s. ammolegus*
395 □ *A. bairdii* winter, ■ migration season

Nesting well to the north, northeast, and locally west of Arizona, it is here found as a transient and winter resident; and to it applies the entire opening account except that pertaining to breeding areas. Southern Arizona records are from *August 7* (*1940* near Elgin — GM) and August 15 (*1876* on Ash Flat, Graham County; abundant, this race? — Aiken) to at least *March 26* (*1885*, Ash Creek south of Prescott), *April 17* (*1885* near Gage, southwestern New Mexico — both Mearns, AMNH), and *May 7* (*1934* opposite Fort Mohave in Nevada—Linsdale and Compton, MVZ).

Ammodramus savannarum ammolegus Oberholser is the breeding race of southern Arizona whence several specimens (ARP) indicate winter residency there too, although most move south into Sonora for the winter. It is paler still and has more rufous than the foregoing, with this color extending all the way to the upper tail coverts and usually cropping out as streaks on the breast. It occurs regularly only within a restricted area of grassland in south-central Arizona; beyond this there is only the juvenile from Fort Grant (*July 28, 1874* — Henshaw, US), and possibly the sight record for Camp Verde (above), where there are no specimens and where the species has not been seen since!

395. BAIRD'S SPARROW *Ammodramus bairdii* (Audubon)

Until about 1878 an abundant transient and doubtless winter resident in the grasslands of southeastern Arizona, north to northern Graham County (Aiken); until about 1920 decidedly uncommon but still a winter resident about the bases of the Chiricahua and Huachuca Mountains. Now apparently much rarer, recent records coming only from beneath these and the Santa Rita Mountains, south of Bowie (*April* — WJS), and the Sonoita Plains. One record from near Eagar, southern Apache County, *October 14, 1934* (Stevenson, ARP). Ranges west along Mexican border to Altar Valley (Sasabe area — Stephens and ARP); but unrecorded as far northwest as Tucson.

Now very rare in Arizona and confined to the densest stands of grass, Baird's Sparrow must be carefully distinguished in the field from the common Savannah Sparrow. Instead of getting up when flushed and flying away high like the Savannah Sparrow, it flutters along in a zig-zag (like some Grasshopper Sparrows) low over the grass and dives in farther on. Its sharp and narrow black streaks, forming a necklace on the chest, and extensive ochre (not yellowish) on the pale side of the head and coloring the median crown stripe are distinctive, if only a good look at it could be had. Its rarity now, in contrast to its former abundance in Arizona, can only be attributed to the destruction of grasslands in its northern breeding range. Grasslands and riparian woods have always been neglected by the conservation movement, which concentrates on preserving mountain forests. Yet many of our most unique and highly specialized North American animals and birds are obligatory members of this dynamic and instructive grassland community. Portions of it should be preserved for study, let alone for its beauty and serene aspect.

Baird's Sparrow decreased sharply between 1876 and 1880, in all probability, as the many excellent ornithologists who

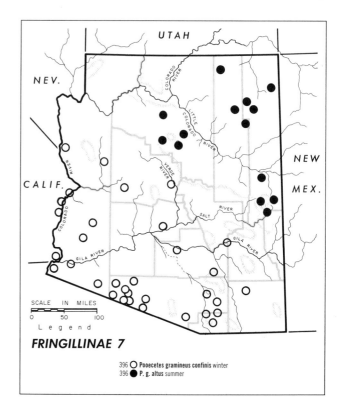

FRINGILLINAE 7

396 ○ *Pooecetes gramineus confinis* winter
396 ● *P. g. altus* summer

collected in Arizona in the years between 1876 and 1902 reported exactly three birds (Altar Valley, *August 29* and *30, 1884* — Stephens; Fort Huachuca, *February 24, 1887* — Cahoon, MCZ). Arizona records were from *August 15* (*1876*, Ash Flat — Aiken, Colo. College) to at least May 3 (base of Huachuca Mountains — Swarth; sight record?). An interesting study of Baird's Sparrow was made by Cartwright, Shortt, and Harris (Trans. Royal Canad. Inst. 21, 1937: 153-197).

396. VESPER SPARROW *Pooecetes gramineus* (Gmelin)

Fairly common summer resident in dry grasslands from high Upper Sonoran Zone to low Canadian Zone along and north from the Mogollon Plateau. Common migrant in open country generally. Winters commonly in weedy fields and grassy areas in the Lower Sonoran Zone of southern and (locally) central Arizona, northwest to Congress, Salome, and the Verde and Big Sandy Valleys.

In migration and winter the Vesper Sparrow usually feeds in small, loose flocks. It is a streaked bird above and below, and has a white outer tail feather, whitish eye-ring, and pale sides of the neck (setting off the darker cheeks). Occasionally the rufous of its lesser wing coverts is revealed beneath the scapulars, which gave it its old name of Bay-winged Bunting.

Southern Arizona records are from August 24 (1885 at Camp Verde — Mearns; 1932 near Picacho, Pinal County — L. L. Walsh) to May 2 (1948, Tucson — Phillips *et al.*); and May 20 (1946 on California side of Parker Dam — Monson). Northern Arizona migrations are from February 28 (1932, banded at Flagstaff—Hargrave) to May 6 (1934), and again from August 28 (1933; both at Mormon Lake —

Phillips) to mid-October, exceptionally to November 6 (1936 — Phillips) and November 12 (1935 — Jacot; both near Springerville).

As in the Grasshopper Sparrow there are four races in the United States, and again the eastern bird is dark and heavy-billed. It is *gramineus*, which does not reach Arizona. The next darkest race, likewise unknown here, is the rather rare northwest coast *affinis* G. S. Miller. It is small, with a buffy wash below. The pale, large bird of the Great Plains and Great Basin is *confinis* Baird, our common transient and winter visitant. Its northern Arizona migrations are from *August 29* and *September 11* to *October 3*; and in spring from *March 9* to *April 13* (all specimens in MNA and ARP). It is present in southern Arizona from *September 10* and *13* to *April 7* (specimens in ARP, MIN, and ARIZ, respectively).

Pooecetes gramineus altus Phillips (in press) is the same large size as *confinis*, of the same darker brown back as intergrades between *affinis* and *confinis* in central Washington, but without the buff below of *affinis*. This race breeds in northern Arizona (and adjacent states), where it is recorded from *April 13* to *September 17* (MNA, ARP). Specimens from southern Arizona are few, the majority apparently passing over without stopping. Dates are *September 1* to *October 1*, twice in *February*, and *March 27* (ARP, ARIZ, LLH).

397. LARK SPARROW *Chondestes grammacus* (Say)

Locally a fairly common summer resident in brushy grasslands from high Lower Sonoran Zone to Transition Zone in eastern Arizona, west to Nogales, Prescott, and perhaps Grand Canyon regions (also possibly in Mt. Trumbull area), but distribution between these points not continuous. Also nest found in 1939 at Quitovaquito, western Pima County (Huey). Common migrant in eastern Arizona, uncommon west of Baboquívari and Aquarius Mountains and the Phoenix region. Winters commonly in weedy farmlands and fairly commonly in grass-brush associations of the Lower Sonoran Zone of central-southern Arizona and the Lower Colorado Valley, north and east to the Santa Cruz valley and the Phoenix area, sparingly to the San Pedro and (casually?) Verde Valley. Has wintered irregularly in grassland at Fort Huachuca (common in late *February, 1887* — Cahoon, MCZ).

The Lark Sparrow is one of the most distinctive and easily recognized sparrows, not only because of its handsome, patterned plumage, but also owing to its clear metallic *chip*, given as it flies overhead. The tail corners are broadly white as it alights. Wintering flocks, like Brewer's Sparrows, often express their well-being in late morning concerts; but the Lark Sparrow's melodious notes are interspersed with throaty gurgles. Because they sing away from the breeding grounds, and migrate both late and early, their breeding range is not easily determined; but in southeastern Arizona it is certainly spotty.

Southern Arizona migrations are from the last half of March through May, and mid-July through September; but some individuals linger into June (9 and 16 near Tucson — Marshall, Phillips) and return by early July (sometime between 5th and 7th at Globe — Swarth; 9th at Tucson — Marshall).

Satisfactory northern Arizona records extend from April 15 (above Fort Apache at Chimo Spring — Dickerman and Phillips) and *April 16* (Flagstaff — Phillips, MNA) to October 3 (near Grand Canyon Village — Grater). Breeding here is more general, so that the extent of the migrations is difficult to determine. Birds seen near Tuba City, east of the Grand Canyon, July 4 and 6 (Phillips), may have been early fall transients. All Arizona specimens belong to the pale western race *strigatus* Swainson.

398. TREE SPARROW *Spizella arborea* (Wilson)

Uncommon winter resident in brushy and weedy parts of the Transition Zone, and of Upper Sonoran rivers and farmlands, on and northeast of the Mogollon Plateau. Ranges south and west to Moenave, east of the Grand Canyon (Monson); Flagstaff; and the upper Black River in the White Mountains (ARP). Casual to the San Carlos area, Gila Valley, *January 11* and *22, 1937* (ARP) and to southern Nevada (Pulich, ARP); once seen inside Grand Canyon (McKee). An old report near Tucson is unlikely and unsubstantiated.

This is the only small native sparrow to be seen in most parts of northern Arizona in winter, where it associates with flocks of juncos. It is unstreaked whitish below, with a single dusky spot in the center of the chest. The solid rufous crown, rufous around the shoulders and rufous (not dusky) line back from the eye easily distinguish it from other streaked-backed sparrows at this season, as does also the yellow base of the mandible. With us, it appears partial to small willows and the adjacent weeds.

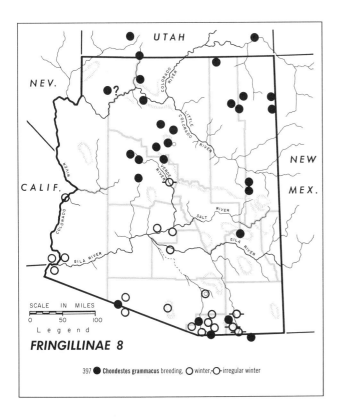

FRINGILLINAE 8

397 ● *Chondestes grammacus* breeding, ○ winter, -◐- irregular winter

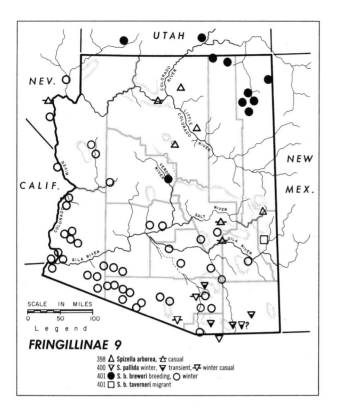

FRINGILLINAE 9
398 △ Spizella arborea, ⩟ casual
400 ▽ S. pallida winter, ▼ transient, ⩡ winter casual
401 ● S. b. breweri breeding, ○ winter
401 □ S. b. taverneri migrant

Extreme dates at Flagstaff are October 30 to March 16 (Kassel, Plateau 13, 1941: 66-67). Arizona birds are of the pale western race *ochracea* Brewster.

399. CHIPPING SPARROW — *Spizella passerina* (Bechstein)

Abundant summer resident in open parts of the Transition and boreal zones, and rather common in open wooded parts of Upper Sonoran Zone, of northern Arizona, west and south to the Hualapai Mountains, the Prescott region, Payson, and Whiteriver. Also breeds locally in Upper Sonoran woods of the Huachuca and Chiricahua Mountains; supposed June or breeding records farther west not substantiated. Migrates throughout the state. Winters abundantly in oak-grasslands (Upper Sonoran Zone) of central-southern Arizona, and commonly in farmlands and the moister parts (grassy or wooded) of Lower Sonoran Zone throughout southern Arizona, north less commonly to Davis Dam (once to Lake Mead, Nevada — Mrs. Nora Poyser), the Big Sandy Valley, and the lower Salt and Gila River valleys. May winter casually in Transition Zone of White Mountains (specimens, late *November 1936* — ARP); but a lone specimen from Bridgeport, Verde Valley, *February 24, 1931* (Jacot, ARIZ), does not justify the statement (A.O.U. Check-list) that it winters at Camp Verde.

This is one of the tamest, most familiar, and also one of the commonest sparrows at many times and places, particularly in association with Yellow-rumped Warblers and Bluebirds in parks in the pines in late summer. It is also one of the earlier sparrows to start moving south in the fall, before molting (this applies both to juveniles and adults). The gray rump and narrow black line through the eye are good identifying marks at any time. Other more publicized traits vary with the season (such as the rufous crown and black bill — showing in spring and summer only).

General occurrence in southern Arizona is from August to late May. Extreme dates away from breeding areas are July 19 (1946, Bill Williams Delta — Monson) and July 22 (1958 at Martinez Lake above Yuma — Monson) to May 23 (1943 near Tucson — Hargrave *et al.*), May 24 (1960 in Kofa Mountains — Monson), and June 7 (1957, Guadalupe Canyon in the extreme southeast — Levy). According to Coues it lingered at Prescott in 1864 "until latter part of November"; while arrival near Whiteriver, below the Mogollon Rim, was on *March 8* and *27* (*1937* — ARP).

Northern Arizona migrations are from March 1 (1933 — Barbara H. McKee) and March 11 (1937 — Natt N. Dodge; both at Grand Canyon Village), and March 3 (1953 near Flagstaff — Eleanor Pugh) to May 24 (1933, banded at Wupatki National Monument — Hargrave); and again from August 3 (above Lee's Ferry in Utah) to November 6 (1938 at Flagstaff — Kassel). An exceptional flight was noted on the upper Black River, White Mountains, *November 23, 1936* (ARP). Usual occurrence is from late March to late October.

The common Arizona race is *arizonae* Coues; to it all the above applies. Under this name we include all the pale populations of western North America; the proposed races *stridula* Grinnell and *boreophila* Oberholser do not seem usefully separable, a conclusion reached with the assistance of Kenneth C. Parkes.

The darker eastern race, *passerina*, is probably only casual in Arizona. One of the three specimens from the Black River, *November 23, 1936* is of this race (ARP). Other specimens from the White Mountains regions are from Basin Lake, *September 1, 1937* (A. W. Sanborn, ARP; juvenile) and *September 13, 1934* at Springerville (Stevenson, ARP). Farther west there are accidental specimens from Navajo Mountain, Utah, *August 16, 1935* (Russell, GCN) and Yuma, *November 1, 1902* (Herbert Brown, ARIZ).

400. CLAY-COLORED SPARROW — *Spizella pallida* (Swainson)

Rare transient, irregularly (or formerly?) more common, from Sonoita Valley east. Occurs sporadically west to Tucson, and once to the Altar Valley (January 29, 1956 — Phillips and M. J. Nichols). Specimens have been taken in the upper San Pedro Valley (ARP), at Fort Crittenden (Henshaw, US), Tucson (ARIZ, ARP), Elfrida (in Sulphur Springs Valley — GM), and probably near Bisbee (Robinette, formerly AMNH; not seen by us). Found wintering in some numbers just inside Sonora in the San Rafael Valley (ARP). Some published records are erroneous.

The erroneous records are partly Brewer's Sparrows and partly juvenile Chipping Sparrows, which implies that this little bird is hard to distinguish. Many collectors have had the experience of shooting suspected Clay-coloreds and finding them to be *breweri*. The head of *pallida*, however, has a definite pattern of contrasts, with a brown cheek-patch (bordered above and below by dark but not black lines) in contrast to the gray neck, which separates the brown and buff tones on the head from those on the back and on the sides of the chest. There is also more contrast around the crown, resulting from the wider, heavier black streaks on the sides of the crown, which cause both the narrow central pale streak and the pale superciliary to stand out in greater contrast than in Brewer's Sparrow (which often has definite suggestions of such a pattern, but not the full contrast of much black in the crown, gray nape, and brownish, black-streaked back). The Clay-colored is distinctly the handsomer of the two. When in doubt, your bird is a Brewer's!

The few Tucson records are for various scattered dates from *September 10* (WGG) to March 2 (Phillips and Mrs. H. B. Donnan) in four different years but chiefly 1955–1956 and the fall of 1959. Farther southeast in Arizona eight specimens are concentrated in the period *September 1* (Henshaw, US) to *September 10* (ARP), with one record on *March 29* (GM).

401. BREWER'S SPARROW — *Spizella breweri* Cassin

Common summer resident in sage and other tall, dense brush of Upper Sonoran Zone in the northern part of the Navajo Indian Reservation. A colony also in Lower Sonoran Zone, Camp Verde to Fossil Creek, in 1880's (Mearns). There are single summer records for Flagstaff, June 14, 1936 (Hargrave) and Vail (near Tucson), *July 2, 1940* (GM), but none for "Fort Whipple" (Prescott), where erroneously stated to breed in A.O.U. Check-list. Common migrant in open parts of the Transition and Sonoran Zones statewide. Abundant winter resident in Lower Sonoran Zone (except in drier open areas, with purer creosote, of the extreme southwest; and except in uncultivated portions of the Colorado Valley); rarer in northern edge of Lower Sonoran Zone but winters to near Kingman (ARP) and to southern Nevada.

Brewer's Sparrow lacks the contrasting gray rump patch of the Chipping Sparrow, and it usually lacks the definite facial pattern and buff tones of the Clay-colored. In winter in southern Arizona it prefers the open desert, leaving the brushy or grassy foothills to its close relative the Chipping Sparrow. Its call is the same high *chip* which gives the Chipping Sparrow its name. Its delightful bubbling, trilling, and wheezing songs are given in concerts all winter and spring. It is the only common bird in creosote deserts about Tucson.

Its usual occurrence in southern Arizona is from August to mid-May. Extreme dates, excepting those mentioned above, are *July 31* (*1939*, north of Douglas — GM; and 1949 at Sonoita — Phillips) to May 28 (1950 at Fort Mohave — Phillips). It is recorded in non-wintering localities beginning *March 27* (*1937* at Fort Apache — ARP), and lasting until November 10 (1951 at Cazador Spring, south base of Natanes Plateau — Phillips).

Northern Arizona migrations are from March 20 (1936 — J. W. Brewer, Jr.) and March 31 (1942 — Corky R. Jones) to *May 24* (*1933* — Hargrave, MNA; all at Wupatki National Monument); and again from *August 1* (not July 29, as stated by Jenks and Stevenson, Condor 39, 1937: 41) to at least October 12 (Woodhouse, Monson).

The supposed breeding of this bird in the Huachuca Mountains (Willard, Condor 10, 1908: 206) is an error, pertaining actually to the Chipping Sparrow (*fide* Jacot, who visited the same spot).

All Arizona specimens save one belong to the nominate race (including that published as *taverneri* by Stevenson, Condor 40, 1938: 86-87). *Spizella breweri taverneri* Swarth and Brooks has broader black streaks above and a darker gray breast. (Streaks on the chest, supposed to be a racial trait, are merely remnants of the juvenal plumage; and differences in the bill may be seasonal). Our single *taverneri* is from Honeymoon on Eagle Creek, near Morenci, *May 23, 1935* (Jacot, ARP). Intensive efforts farther west have failed to secure a single specimen of this race, which breeds above timberline in the mountains of western Canada.

402. BLACK-CHINNED SPARROW — *Spizella atrogularis* (Cabanis)

Fairly common summer resident in Upper Sonoran chaparral across Arizona from east to northwest below the rim of the Mogollon Plateau, west to Hualapai Mountains; south very locally to the Chiricahua and Mule Mountains (GM), and possibly the south end of the Huachuca Mountains (but not in Sonora, as stated in A.O.U. Check-list). Once seen at Flagstaff, August 5, 1947 (Phillips). Fairly common winter resident locally in scattered brush of high Lower Sonoran and low Upper Sonoran hillsides of central-southern Arizona, from the foot of the Natanes Plateau and the Santa Catalina Mountains west to the Ajo Mountains. Also winters in most years to the higher mountains of Yuma County,

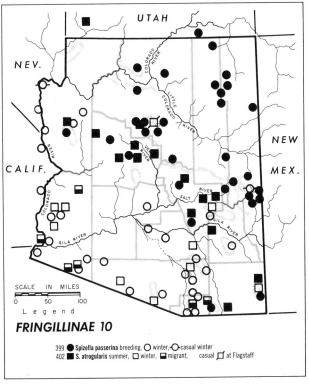

FRINGILLINAE 10

399 ● *Spizella passerina* breeding, ○ winter, -○- casual winter
402 ■ *S. atrogularis* summer, □ winter, ▨ migrant, casual ⌑ at Flagstaff

where seldom common, however. East of the San Pedro Valley known as a wintering bird only in the Chiricahua Mountains, where locally resident. Casual transient to Colorado River and lower mountains and valleys of Yuma County; unrecorded in open valleys of the southeast, and in the northeast.

The Black-chinned Sparrow combines the smooth gray body and head color of our local juncos with the trim form and streaked back of other *Spizellae*. It associates with these in winter, when it does not show a black chin. This publicized field mark is present only in males, following the pre-alternate molt in late winter. The Black-chinned Sparrow is quite a local bird, and the observer can live in Arizona for many years without seeing it unless he goes to the right places in the chaparral. The center of its abundance is the extensive chaparral of the Tonto Basin and Prescott regions. When wintering birds are once located, as in Molino Basin of the lower Santa Catalina Mountains, they can be seen time after time in the same bushes on subsequent visits. Although both the breeding and winter ranges occupy extensive parts of Arizona, only in the Chiricahua and possibly the Pinal Mountains do they approach each other closely.

Black-chinned Sparrows are found on the breeding grounds from early April to October 10, and on the wintering areas from early September to mid-April, normally. In 1955–1956 an extraordinary flight developed in the low mountains of extreme southwestern Arizona, where the bird has been otherwise unknown. This was a time of unusually beneficial rain. A specimen was taken there as late as *April 28 (1956 — GM)*, a date equalling the latest elsewhere (April 28, 1942, at Sahuaro Lake, Salt River — Hugh and Margaret Dearing).

The normal Arizona race is *evura* Coues, a large, pale form. Darker birds occur both in Mexico and on the Pacific Coast, the latter being smaller as well. Specimens taken in the 1955–1956 flight, mentioned above, are not typical of either *evura* or the coast race, *cana* Coues (including *caurina* Miller); but most of the *cana* are May birds, and the paler appearance of southwestern Arizona skins may be due to fuller feather-edgings, as well as greater fading in the desert sun. The tail measurements are far below those of *evura*: adult male, 68.8 millimeters; first year birds showing some wear, 63.6 (male) and 64.4 (female). The question is made still more difficult because the only fall specimen (the adult) is a partial albino! It seems that color is less reliable, under these circumstances, than size, and these three birds from the Kofa Mountains and Tule Well are tentatively referred to *cana*.

Genus *AIMOPHILA, SENSU LATU*
by Allan Phillips

403. RUFOUS-WINGED SPARROW *Aimophila carpalis* (Coues)

Common resident locally in mixed bunch-grasses and thornbrush of the Lower Sonoran Zone in central-southern Arizona, from near Oracle and the Tucson region west across the Papago Indian Reservation as far as Ventana Ranch and Menager's Dam. Formerly more common and less local, but presently again extending its range, at least to southeastward, where it appeared

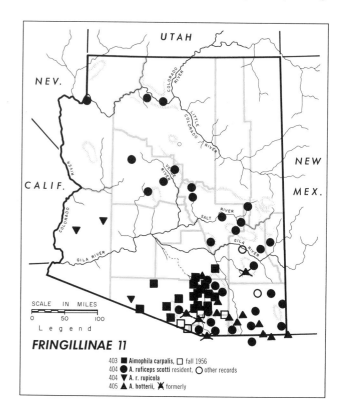

FRINGILLINAE 11
403 ■ Aimophila carpalis, □ fall 1956
404 ● A. ruficeps scotti resident, ○ other records
404 ▼ A. r. rupicola
405 ▲ A. botterii, ✗ formerly

in 1956–1957 in unexpectedly high places.

This is a trim sparrow somewhat like a summer-plumaged Chipping Sparrow at first sight. The rufous on the wing is not conspicuous in life, because this patch is hidden under the scapulars. But the bird is easily recognized by its head pattern of two thin black "whiskers" on each side, a rufous stripe through the eye, and a crown that is rufous at all seasons. Its most distinctive and startling feature is the high-pitched and penetrating *seep*. Songs vary in form, but always end in an accelerating series of metallic *chip*s. Habitat features common to most places where this sparrow occurs are flat terrain, some desert hackberry bushes in which to build the nest, some chollas, some grasses, and some bare ground. Despite its reputation as a will-o'-the-wisp, it is a tame bird, easy to study, and its nests are easily found. Birds at all seasons, and nests, are easily visible in the edges of bushes, not hidden away in dense grass as are other *Aimophilae* and their nests. The bird is noted for its late summer rainy season nesting and for its peculiar molts (Phillips, Wilson Bull. 63, 1951: 323-326).

Bendire discovered the Rufous-winged Sparrow near old Fort Lowell, Tucson, in "the early part of June," 1872, when several nests were found in small mesquites and the bird was considered common; thus Coues launched the new species from Bendire's specimens and letters. Bendire later (Orn. and Ool. 7, 1882: 121-122) enlarged upon this first notice, stating that he had discovered the species on June 10, had found eggs from "about June 14" to September 1872, and that the birds were then "very common on the ridges bordering Rillito Creek" in a mixture of mesquite, thornbush and tall bunch-grass. Meanwhile Henshaw (*Rept. Geog. Surv. W. 100th Merid.* 5, 1875: 291-293) also found it "in abundance" at Fort Lowell in early September, 1874, but "only among the mesquite thickets." Twenty specimens were taken, *September 9 to 12*. He gave an account of its habits and quoted from Bendire. Its abundance at Fort Lowell at this time may also be judged by a letter from Henshaw to Merriam, October 25, 1874, stating that the species was "exceedingly difficult to capture" (Nelson, Auk 49, 1932: 406).

Brewster (Bull. Nuttall Orn. Club 7, 1882: 195) reported that Stephens found it, from *mid-April to June, 1881,* "sparingly about Tucson and Camp Lowell. It inhabited the mesquite thickets, keeping closely hidden in the bunches of 'sacaton' grass, from which, when flushed, it flew into the branches above." A nest and three eggs were taken May 25. Stephens (Auk 2, 1885: 228) later traced the species west, in northern Sonora, to the Altar Valley, August 12, 1884. Scott (Auk 4, 1887: 203), found it common at times on the east side of the Santa Catalina Mountains at 3000 to 4500 feet altitude, but less common in summer; he found no nests. The latest specimen of the era of discovery was apparently the male taken by Herbert Brown at Tucson, *February 7, 1886* (ARIZ).

For almost thirty years, no reliable record was made in Arizona until Goldman (FW files) saw several on the slopes of the Coyote Mountains in early September, 1915. Soon A. B. Howell (Condor 18, 1916: 212-213) called attention to its absence about Fort Lowell, where in fact it was apparently wiped out by 1884. Dawson (Jour. Mus. Comp. Ool. 2, 1921: 33; ms.) recorded the Rufous-winged Sparrow from the desert west of Sells, *June 15, 1917,* where C. I. Clay took a pair (Santa Barbara Museum, examined by Phillips). Luther Little took a lone male at Sells, *July 10, 1918* (US). Fourteen years were to pass between Little's record and the next. Swarth (Proc. Calif. Acad. Sci. 18, 1929: 328) considered the species extinct in Arizona due to overgrazing. He also regarded it as characteristic of his "Eastern Plains Area" east of the Santa Rita Mountains, where it had not occurred.

The 1931 A.O.U. Check-list mentioned the species as occurring in Arizona "formerly." Van Rossem (Trans. San Diego Soc. Nat. Hist. 6, 1931: 299) questioned Swarth's theory of overgrazing, stating that this sparrow was very adaptable and favored especially the cholla cactus and mesquite associations. He did not question its extinction in Arizona. Moore (Proc. Biol. Soc. Wash. 45, 1932: 233) then recorded four specimens taken at Fresnal, west base of the Baboquívari Mountains, in *June, 1932.* The following winter, as Moore informed Phillips (*in litt.*), he revisited Fresnal and secured six additional specimens (all that were seen). Van Rossem (Trans. San Diego Soc. Nat. Hist. 8, 1936: 144) averred that "this species is a typical *Spizella* in almost every respect and why it has been considered an '*Aimophila*' is incomprehensible"; it "occupies, normally, very much the same ecologic niche as *Amphispiza bilineata*." He attempted to correct "an erroneous assumption of rarity," despite his failure to secure a single specimen in more than two months of intensive collecting. Moore (Condor 48, 1946: 117) then presented "factual data" on the same Fresnal birds, deemed in retrospect to have been "actually *abundant.*" Meanwhile Sutton and Phillips (Condor 44, 1942: 64) traced the species to its western limits, on the Papago Indian Reservation.

Mr. Edouard C. Jacot rediscovered the species at Tucson, *April 27, 1936,* the first record there in over fifty years! On October 8, 1938, Phillips found a singing bird a few miles away, and on *October 9,* he saw at least six; two adults taken were still in the midst of the annual molt, and a young bird retained a number of the feathers of the juvenile plumage. A pair was again seen there February 25, 1939. On *February* 19 and *25, 1939,* two or three were seen and one taken at another locality, half way between those originally found by Jacot and by Phillips. Later many others were found, here and elsewhere (see map).

Professor J. J. Thornber, of the University of Arizona Department of Botany, informed Phillips that those areas near Tucson where this species survived have never been subjected to heavy grazing. In 1901 they were meadows of *Bouteloua rothrocki*. Limited grazing sufficed to bring in *Aristida* to replace the gramas, but there was always a grassy cover. This is proof that overgrazing caused the decrease of this species; here in isolated spots that escaped the general abuse of the early 1890's and early 1900's were colonies of Rufous-winged Sparrows, while in the rest of the valley (whence came the early reports of abundance) the species was absent. Given the bird's requirement of grass with scattered bushes, such disappearance was inevitable.

If this species required simply mesquite or cholla cactus and was very adaptable, as van Rossem thought, we should find it greatly increased in Arizona with the spread of those two plants into the grasslands under the influence of overgrazing. Instead we find just the opposite. The Rufous-winged Sparrow obviously requires a heavier grass cover than *Aimophila bilineata*. The latter species is still, and always was, widely distributed in the deserts of southern Arizona.

The great irruption of the fall of 1956 brought first records to such points, at altitudes above the previous Tucson Valley range, as Canoa Ranch south of Continental (Phillips, Bialac), Gardner Wash north of Sonoita (Marshall), south of Arivaca (Levy), and a stray high in Madera Canyon, Santa Rita Mountains (James M. Gates; color photograph examined)! Likewise one appeared then in Phillips' yard; and another on the saguaro desert east of Tucson on February 17, 1958 (Wm. X. Foerster). At present they seem to be established around Continental and Arivaca Junction (Marshall).

404. RUFOUS-CROWNED SPARROW; SCOTT'S SPARROW
Aimophila ruficeps (Cassin)

Common resident of open, grassy and rocky Upper Sonoran Zone hillsides of southern Arizona, north sparingly to parts of the Mogollon Rim region (summer only?) and west to the Kofa (GM) and Ajo Mountains. It is also found sparingly along most of the Grand Canyon, where its range and status are poorly known.

The Rufous-crowned Sparrow has much the same head-pattern as the preceding species, but the back is not black-streaked. Instead, the streaks are rufous-brown bordered with gray, unusual in sparrows. The eye-ring is conspicuously white.

The Rufous-crowned Sparrow, like the following two species, nests, feeds, and spends most of its time on the ground, rarely alighting higher than a low bush even when curious or alarmed for the safety of its brood. A favorite vantage point for singing is a big rock, from which it scuttles down into the grass. Its characteristic call is *deer-deer,* and its song a series of *chip*s which changes pitch two or three times in the course of the outburst. Only by watching the birds flush from the edge of the Catalina Highway at dawn will one get an idea of the great numbers of Rufous-crowned Sparrows.

There is still no good evidence of migration in this sparrow, other than short "weather movements" downward in snowy weather and fluctuations at its lower limit (the mouth of Sabino Canyon, near Tucson) from year to year. Usually we do not find them there, but they were present regularly from the mid-1940's to about 1953. There are, to be sure, few October-to-March records north of the Gila Valley, but one was found dead inside the Grand Canyon, February 28 (*1939* — McKee), and others were banded there in January 1937 and March 1939. One was seen in Sulphur Springs Valley south (?) of Willcox, April 18, *1895* (Osgood, Condor 5, 1903: 149), and there have been occasional unsubstantiated reports at Tucson in winter or migration. But the only actual Lower Sonoran Zone specimens are from the Gila River (if accurately labeled) long ago: near Geronimo, below Safford, *August 17, 1876* (Aiken, Colorado College Pubs., Studies Ser. 23, 1937: 29; probably not "common"); and at or near same place, *September 11–12, 1873* (Henshaw).

As in the Rufous-winged Sparrow, we leave the discussion of the strong influence of rainfall upon the breeding season to the continuation of Bent's *Life Histories.*

Most Arizona birds are of the pale, light-crowned, large northern race *scotti* (Sennett). Those of the southwestern mountains (see map) are slightly darker above and are currently known as *rupicola* van Rossem, though they may represent merely a northwestern extension of the range of *simulans* van Rossem, with which they have never been compared.

405. BOTTERI'S SPARROW
Aimophila botterii (Sclater)

Rather uncommon summer resident (no nest yet found within state) from near the southeastern corner of Arizona west to near Sonoita, usually in giant sacaton grass; also found in Oracle region in *1940*. Formerly much more common, especially before 1895, when it ranged west to the Altar Valley and north to Fort Grant.

Unless a Botteri's Sparrow is kind enough to sing, it can hardly be identified in the field. In fact, it can hardly be seen! Only the territorial males get up high enough in a bush to be detected, and when they stop singing the species simply disappears, so that its winter range is almost wholly a matter of conjecture! This is true of no other Arizona sparrow, and very few of our other birds (Elf Owl, Flammulated Screech-Owl, and Allen's Hummingbird) are as poorly known in winter.

But if you are lucky enough to find a Botteri's Sparrow, you won't see much. It is as "plain as a mud fence," with no conspicuous pattern anywhere: plain whitish below, dull gray-faced, and streaky above (black, rusty brown, and grayish). The crown is mostly blackish (heavily streaked) with a pale median line hardly noticeable except on the forehead, as so often occurs in sparrows. Flushed close under foot, it seems an ordinary sparrow with a rather long, dark tail. The song resembles those of the Rufous-winged and Black-chinned Sparrows in ending in an accelerating series of *chip*s, but opens with a few soft *clip*s and thin, Horned Lark-like *chleep*s as we have heard it in Arizona. (In Mexico, at least, the opening is more variable.) In any case, it is not what you would call music.

The clouded history of Botteri's Sparrow in Arizona is outlined by Monson (Auk 64, 1947: 139-140). Briefly, between *1893* (Mearns, US) and *1932* (A. Walker, CLM) there was but a single record, of two birds in *1903* (Breninger, F). This was due to failure to search the sacaton at the right time, since Monson found them so widespread later. Extreme Arizona dates are from *May 17* (*1888* in the Huachuca Mountains — Price, MCZ) and *May 23* (*1940* near Elgin, Santa Cruz County — GM) to *October 7* (*1893* in the Sulphur Springs Valley at the Sonora border — Mearns, US; not October 17, as once given by van Rossem, *Birds of Sonora,* 1945: 277).

It is likely that, prior to the general overgrazing of the 1880's and 1890's, this sparrow was less restricted ecologically and more widespread in the lush grasslands. Otherwise Henshaw, Nelson, and Stephens would not have found it in so many places.

The pale, reddish Arizona race is *A. b. arizonae* (Ridgway), as shown by the recent revision of Webster (Condor 61, 1959: 136-146).

406. CASSIN'S SPARROW
Aimophila cassinii (Woodhouse)

Common post-breeding summer visitant, and locally and probably irregularly a fairly common winter resident, of the more extensive tall grass areas of the Lower Sonoran Zone in southeastern Arizona. Ranges west to the Coyote and Baboquívari Mountains (exceptionally to west of Growler Mountains in *fall of 1959,* and Monument 180 on Mexican border, *August 25, 1961* —GM), and north to the Gila River; also found at Camp Verde, *July 21, 1916* (Taylor, US). Has not been found nesting in Arizona or adjacent areas, for it is absent from early May through June.

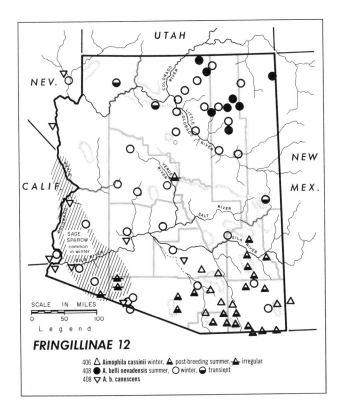

FRINGILLINAE 12
406 △ Aimophila cassinii winter, ▲ post-breeding summer, ▲ irregular
408 ● A. belli nevadensis summer, ○ winter, ◐ transient
408 ▽ A. b. canescens

In habits and appearance, Cassin's Sparrow is exactly like Botteri's; but its lovely song of four clear, sweet whistles (the second prolonged and trilled or quavering) instantly identifies it at a distance. The astonishing thing about this song is that it is given so fervently, often on the wing, by males in full breeding condition all through July and August in Arizona, where the species never nests! The most any female has ever been known to do here is to construct part of a nest! For an account of this extraordinary situation, see Phillips, Auk 61, 1944: 409-412. No other bird here (or elsewhere, to our knowledge) shows this behavior away from its breeding grounds. None shows as clearly that breeding seasons and areas are not to be determined by the condition of the male alone, and that a bird can leave its breeding range with testes still at their maximum development!

The contrast between the conspicuousness of the species when singing and its secretiveness at other times renders uncertain the timing of its migrations. Winter birds must surely be overlooked in many cases before their behavior starts to change. Migration evidently continues into September or October and starts again in March or April. The only real certainty, however, is that authentic Arizona specimens are all from *June 29* near Elgin, Santa Cruz County, and *July 6* in the Sulphur Springs Valley near Lowell to *May 6* near Tucson (all *1939-1940* — GM).

407. FIVE-STRIPED SPARROW *Aimophila quinquestriata* (Sclater and Salvin)

Known only from a specimen taken at west base of Santa Rita Mountains, *June 18, 1957* (Binford, MICH; see Auk 75, 1958: 103).

The Five-striped Sparrow is a bird of brushy, rocky semi-desert slopes (high Lower Sonoran Zone, chiefly) of parts of western Mexico, where its seasonal status remains poorly known. It breeds regularly north in Sonora to the Imuris region, about 70 kilometers south of Nogales (ARP). It is like a darkened Black-throated or Sage Sparrow, with a slaty color spreading over the sides and flanks. The 5 white stripes are the superciliaries, malar stripes, and a more-or-less distinct median stripe from the chin back onto the throat. The dark cheeks and throat give a strong resemblance to the Black-throated Sparrow, as does the reddish tinge on the back. Beneath the slaty color, however, may be seen (in the hand) the pattern of the Sage Sparrow. Its habits and song resemble those of the Black-throated Sparrow, but the juvenal plumage is unstreaked (Phillips, Anales Inst. Biol. Univ. México 32, 1962: 389-390).

While we would expect this sparrow to turn up somewhere in the desert hills from the Baboquívari Mountains west to the Ajo Mountains, its discovery at the mouth of Madera Canyon, which we had visited almost annually in early May, was a great surprise. It can hardly be more than accidental there. The specimen belongs to the well-marked pale northern race *septentrionalis* van Rossem.

408. SAGE SPARROW; *Aimophila belli* BELL'S SPARROW (Cassin)

Common to abundant summer resident of open sagebrush on the Navajo Indian Reservation north of the Rio Puerco Valley, west to Echo Cliffs and the Hopi Buttes (and possibly in sage areas farther west?). Summer reports elsewhere probably due to confusion with juvenal *bilineata*. Winters commonly on open ground with sparse brush in southwestern and western Arizona, and in Upper Sonoran grasslands of northern Arizona; scarcer farther east and south. Also occurs in winter in dense salt-bush stands of Colorado Valley. Casual fall transient on the high prairies of the White Mountains (October 10, 1936 — Phillips).

After the last few species, the Sage Sparrow is a relief. He runs around waving his long black tail in the air above his back, so anyone can identify him who can tell a sparrow's beak from a wren's. Even better, he does this on open ground with good visibility. You can actually *see* his pale gray tones; the white lore-spot, eye-ring, and underparts; and the dusky spot in the center of the chest. In most of Arizona he is only seen in winter, and favors us with nothing more musical than a tinkly junco-like call.

Southern Arizona records are from September 15 (1946 near Parker — Monson), and September 16 (near Phoenix — Simpson, Werner) to March 26 (1955, west of Sierra Pinta, Yuma County — Monson). The Sage Sparrow is recorded away from breeding areas in northern Arizona from September 22 (1926, west of Grand Canyon village at the entrance to Cataract Canyon — Griscom) and September 25 (1922 on the northeast side of the San Francisco Peaks — Swarth) to *March 3* (*1934*, same place — Hargrave, MNA; still common) and April 9 (1937 at Kayenta — Monson).

The commoner Arizona race, to which all the above applies, is *A. b. nevadensis* (Ridgway) of the Great Basin region. This is a large, pale race with fine streaks on the back and thin black moustache. A similar but smaller race, *canescens* Grinnell (usually equally pale, though the exceptional Picacho Reservoir specimen, ARP, is darker) breeds in south-central California. It ranges into western Arizona in winter, where it seems nearly as common as *nevadensis*. Since, of course, males are decidedly larger than females, determination of the race usually depends on careful determination of the specimen's sex. Satisfactory records of *canescens* cover only the period *October 19* to *March 2*, whereas *nevadensis* has been taken from *October 1* (*1949* on Big Sandy River — ARP) to at least *March 16* (*1886* near Camp Verde — Mearns, U. S.; apparently listed as March 26 in Mearns' catalogue, *fide* L. L. Short).

409. BLACK-THROATED SPARROW; DESERT SPARROW *Aimophila bilineata* (Cassin)

Common summer resident of scattered low brush or cactus in arid Sonoran zones throughout the state, scarcer west of the Gila Bend and Organ Pipe Cactus National Monument areas in the south. Winters rather commonly in scattered Lower Sonoran thorn-brush north to the Gila and lower Salt River valleys (casually north of Prescott, December 13, 1960 — Phillips), thence to the Hoover Dam region, but sparsely west of the Castle Dome Mountains, seldom reaching the extreme lower Colorado River, and absent from brush-less grassland in the southeast. Rare transient (and wanderer?) on the Mogollon Plateau.

As its former name told us, the Black-throated Sparrow is the "Desert Sparrow" *par excellence*. Given a few chollas and a small wash with other brush, he is at home in the barren creosote mesas around Tucson, which he enlivens in spring and summer with his musical tinkling song. Adults are unmistakable, with a handsome, full black throat (rounded behind) and dark head relieved by white malar and superciliary stripes and a small white crescent on the lower eyelid. But young birds, even into September or October, lack the black throat, being white below streaked with dusky. We suspect that they are often misidentified as Sage Sparrows due to their black tails with just a little white on the outer feather. The tail, however, is not raised above the level of the back.

Most northern Arizona records are from the last of *March* to mid-August, rarely later; but a specimen was taken on the northeast slope of the San Francisco Mountains region, *September 8* (*1933* — Hargrave, MNA). We cannot but feel that the several reports later in autumn should have specimen support. Just west of northern Arizona, it was seen at Supai on September 23 (1950 — Phillips). We can find no specimen of Sage Sparrow from Flagstaff on July 28, 1889, as reported by Merriam, and suspect that this report was based on a juvenal-plumaged *bilineata* already migrating south.

Southern Arizona migrations are from *March 2* (*1888*) and *March 18* (*1885* at Camp Verde — Mearns, AMNH) to "about the end of April" (1902 or 1903), and from late July (1902) to *September 30* (*1913*, all in the Huachuca Mountains — Swarth; Howell, LA). On the Santa Rita Experimental Range several appeared and were singing by July 23 (1955 — A. H. and Anne Anderson)! The last seen at Camp Verde were on September 27 (1885 — Mearns).

A. H. Miller (Proc. XII Intern. Orn. Congress 2, 1960: 521) claims exaggerated geographic variations in the breeding season of this sparrow. On the peninsula of Baja California, even in the years of extreme rainfall, there is no real evidence that it nests "from October to June" (see van Rossem, Trans. San Diego Soc. Nat. Hist. 10, 1945: 237-244; Banks, Condor 65, 1963: 309-310). Instead it varies among spring, summer, or fall according to the year or locality. In Arizona the breeding season is by no means so curtailed as stated by Miller. Our earliest records are of a nest being built in the Sierra Pinta, Yuma County, on March 29 (1957 — Monson) and a family of at least two full-grown young east-southeast of Essex, California, May 13 (1952 — Monson, Phillips). Nesting does not end in June, despite the heat, but continues well into the summer. Young in juvenal plumage are often seen through September in northwestern Arizona, and were still present farther south in the Ajo Mountains October 25 (1947 — Phillips). A nest with eggs was found on the Kofa Game Range, Yuma County, on August 26 (1955 — Monson). Near Tucson, presumably in 1872, Bendire found "quite a large lot" of nests and took three eggs on August 25, fresh eggs on September 3, and a nest (doubtless with eggs) on September 14, and found it "still laying" eggs on September 13, if not later (*fide* Coues, Amer. Naturalist 7, 1873: 323-324). We do not yet know how many broods are raised by a single pair in a season, however.

Arizona birds are all referred to *A. b. deserticola* (Ridgway). They are large (long-tailed), with a moderate white spot on the outer tail-feather. Most of them are quite reddish-backed; but grayer birds occur locally (southern Papago Indian Reservation), and seem indistinguishable from *opuntia* (Burleigh and Lowery) of Texas and nearby re-

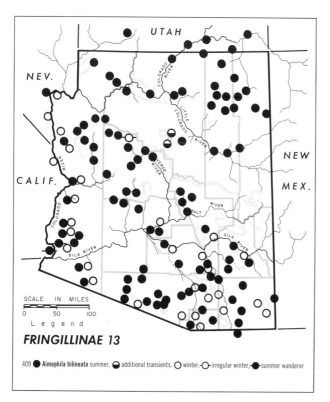

gions. It seems to us that in this species, geographic color-variations are too haphazard to be usefully named except in their broadest outlines, and that the number of races could profitably be reduced to about five.

Genus *JUNCO*
by Allan Phillips

410. BROWN-EYED JUNCO; DARK-EYED JUNCO. Includes SLATE-COLORED JUNCO; CASSIAR JUNCO; WHITE-WINGED JUNCO; OREGON JUNCO; SHUFELDT'S JUNCO; THURBER'S JUNCO; SIERRA JUNCO; PINK-SIDED JUNCO; GRAY-HEADED JUNCO; and RED-BACKED JUNCO (part)   *Junco hyemalis* (Linnaeus)

Very common summer resident in Boreal Zone forests of the Mogollon and Kaibab Plateaux. Fairly common summer resident in the adjoining Transition Zone, and on the Coconino Plateau and in the northeast. Reported sparingly south in former years to Coronado Mountain north of Clifton (Goldman, FW files), the Natanes Plateau, the Sierra Ancha, and the Mazatzal and Bradshaw Mountains; but no nest yet found in any of these ranges. Abundant transient throughout, scarcer in Lower Sonoran Zone. Abundant winter resident of open forests and woods of the Upper Sonoran Zone and above; rather common in the moister, more brushy areas of the Lower Sonoran Zone; also common in wet years in southwestern desert mountain ranges.

Brown-eyed Juncos are rather small ground-feeding sparrows which have no streaks except in juvenal plumage. There is usually a gray or black "hood" over the head and breast, marked off from the pure white belly. The lateral rectrices of the otherwise dusky tail are conspicuously white in flight. The song of most races is a monotonous trill, though of pleasant musical quality unlike the dry buzz of the Chipping Sparrow. But the breeding form of Arizona has more varied songs, approaching in that respect the songs of Yellow-eyed Juncos to the south. The call is a light snapping sound, *tic*, which may be run into a series. Territorial in summer, the species forms flocks in winter. These are a joy to behold in Arizona, as they generally include representatives of populations from all over the north — easily distinguished through binoculars, or with the unaided eye in those birds which are enticed to the picnic table, by their prevailing colors. As the flock drifts by you will see birds with gorgeous pink or buff sides, others with blue-gray hoods, some with black hoods, some slaty; and if you are very fortunate you may see a pale gray bird with a white bar on the wing coverts.

Like the Flicker, the various races of Brown-eyed Juncos intergrade with each other and accordingly constitute one species. Also like the Flicker, the races differ so much in color that ornithologists have been (and many still are) loath to unite them, despite the patent biological evidence, including interbreeding on a large scale wherever and whenever possible. In the early and mid-1930's Hargrave and Phillips made a banding study of juncos at the Museum of Northern Arizona, Flagstaff. Not only did they note influx and departure of races, but they succeeded in capturing and banding seven or eight out of the nine races known to occur in Arizona — all but *thurberi* and possibly *simillimus* — plus some intermediates.

On the first map we show the distribution of our breeding race, *dorsalis*, together with that of the geographically complementary Arizona (Yellow-eyed) Junco, a distinct species which differs in song, behavior, and color of soft parts. Alden H. Miller (Univ. Calif. Publ. Zool. 44, 1941: 210) stood on Aztec Peak of the Sierra Ancha, a breeding haunt of the Brown-eyed Junco, and could see Pinal Peak, the home of breeding Yellow-eyed Juncos to the south. "This 37-mile break in distribution is one of great significance; it divides the genus into two groups, geographically complementary, between which there is at present no interbreeding." Aztec Peak is of further interest in that Dark-eyed Juncos have not consistently occupied it in summer, possibly in connection with increasing aridity in the southwest. *Dorsalis* perhaps nested there continuously up into the

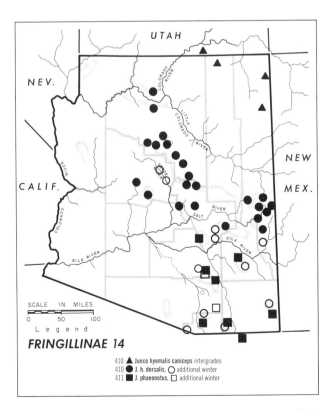

FRINGILLINAE 14

410 ▲ *Junco hyemalis caniceps* intergrades
410 ● *J. h. dorsalis*, ○ additional winter
411 ■ *J. phaeonotus*, □ additional winter

1930's for *June* specimens were taken in *1917* and *1931* (Swarth and Miller, MVZ). Later, including the decade of the 1950's, the bird was absent at the several visits of Simpson and Werner, and during R. Roy Johnson's studies on the ornithology and botany of the range (Master's Thesis, Univ. Arizona Library, 1960). Finally, on July 19, 1962, Johnson found an adult feeding young in the exact spot where Miller had seen the birds in 1931, and where Johnson had always looked for them. He writes: "It snowed three inches up there May 28 and is the lushest that I've seen in the years that I've been going up there to look for birds."

The second map shows the distribution of those wintering races (*hyemalis, simillimus, thurberi,* and *aikeni*) which are local or uncommon in Arizona. The remaining races are found throughout most or all of the state as transients and winter residents, and do not require a map.

The United States races are distinguished in the following synopsis, beyond which we give their Arizona status and dates of occurrence:

Junco hyemalis hyemalis (Linnaeus). "Slate-colored Junco." With dark hood, concave posterior border of hood (where it meets the white belly); a uniform bird in which the hood, back and sides are all the same dusky gray (brown-washed in females). Breeds across the northern part of the continent, wintering chiefly in central and eastern parts, where it is the only common junco except for the larger *carolinensis* in the southern Appalachians.

J. h. henshawi Phillips (for this and *simillimus* see Anales Inst. Biol. Univ. Méx., 32, 1962: 372-377). "Cassiar Junco." Black hood in male, with concave border; the back, often brown-washed, is contrasted with head; sides paler gray than head, in female washed with buff but not solidly so. Breeds in interior northern and central British Columbia and southern Yukon.

J. h. simillimus Phillips. "Oregon Junco." Black hood in male with straight (or slightly convex) border; back reddish-brown, sides orange-buff. Breeds on the northwest coast of Washington and Oregon. (The term "Oregon" or "Pink-sided Junco" is often applied to the several races from this one to *mearnsi*, collectively.)

J. h. oregonus (Townsend). "Oregon Junco." Like *simillimus* but darker, with a very black hood in the male and a richer, redder brown back. It breeds in coastal British Columbia and has been incorrectly credited to Arizona. Two intermediates toward *oregonus* have been taken, in the Canelo Hills, Santa Cruz County, *December 2, 1935* (Dille, ARP) and White Mountains, *September 29, 1936* (Correia, ARP).

J. h. thurberi Anthony. "Thurber's" or "Sierra Junco." Hood with straight margin, black in the male; back tan or cinnamon, sides pale orange-buff, contrasted with the head. Breeds in the Sierra Nevada of California. One Yuma male, wing 74.5 mm. (Herbert Brown, ARIZ), has a somewhat smaller size and richer coloration that make it closer to the sedentary race *pinosus* Loomis of the central coast of California.

J. h. shufeldti Coale. "Shufeldt's Junco" (part); "Montana Junco." Same as *simillimus* but slightly duller, grayer-brown back, slightly paler head and rump, and longer wing and tail. An interior race, breeding in southern British Columbia and the western part of the northern Rocky Mountains of the United States. (The slaty, not black, headed *montanus* Ridgway is included here, following A. H. Miller's conclusions.)

This and all the foregoing have sexual dimorphism, with the female paler and browner than the male. (Thus females of dark or gray races will resemble males of lighter or browner races.)

J. h. mearnsi Ridgway. "Pink-sided Junco." This and all the following have the sexes alike. *Mearnsi* is somewhat larger than the foregoing. Hood bluish-gray (in field), back grayish-brown, sides broadly pinkish-cinnamon (much brighter than back). Breeds east of range of *shufeldti* in northern Rocky Mountains of United States and in Cypress Hills, Saskatchewan. (*Annectens* Baird is not considered to be a stable population, following A. H. Miller; but it too is a handsome bird — a *mearnsi* with the reddish back of *caniceps*. It breeds in northern Utah and winters in Arizona.)

J. h. aikeni Ridgway. "White-winged Junco." Large size. Hood, back and sides all the same shade of pale gray. White wing-bars normally present (these may occur, exceptionally, in other races). Breeding in the Black Hills of Wyoming and South Dakota, it is a sort of dull relative of *mearnsi*.

J. h. caniceps (Woodhouse). "Gray-headed Junco." The weakly set-off hood is pale gray and slightly concave; back chestnut-rufous, sides pale gray. Breeds in central Rockies of Colorado and Utah.

J. h. mutabilis van Rossem. "Nevada Junco." Like *caniceps* but variably intergrading toward *thurberi*. Resident in mountains of southern Nevada and southeastern California. Not yet found off its breeding range.

J. h. dorsalis Henry. "Red-backed Junco." The only race with bill large and black (bluish below). Otherwise like *caniceps* but throat whiter. Breeds in northern Arizona

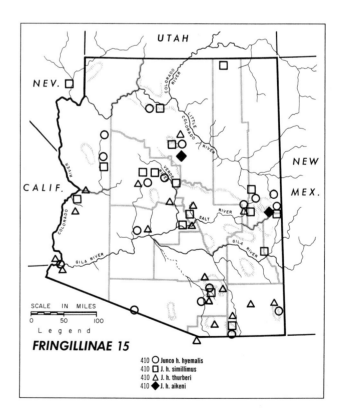

FRINGILLINAE 15

410 ○ Junco h. hyemalis
410 □ J. h. simillimus
410 △ J. h. thurberi
410 ◆ J. h. aikeni

and most of New Mexico; migration very reduced, compared to *caniceps* and northern races.

All the above forms except *dorsalis* have a pink bill. All are of the usual small size except *mearnsi*, which is somewhat larger, and *aikeni* and *dorsalis* which are large and fill the hand while being banded. All the northern and coastal forms down through *shufeldti* have a hood and dark wings usually without a white bar. They are winter visitors to Arizona, and their females are paler and browner than the males. There are slight differences in measurements and in the amount of white on the tail. The deep reddish tones in the back of such races as *shufeldti* become bleached and grayer with wear, a change probably accentuated in those wintering in sunny Arizona.

The following synopsis gives the status and dates of occurrence of the various races in Arizona:

Junco hyemalis hyemalis. Very uncommon but regular winter resident in eastern Arizona, rarer to the west, but ranging all the way to Yuma (H. Brown, ARIZ); see map. Usually single birds with flocks of other races. Recorded in and near northern Arizona from *November* 1 and 2 (*1956* at Grand Canyon — Willard E. Dilley and Hargrave, GCN) to *April 8* (*1937* north of Whiteriver — ARP). Extreme southern Arizona records, *October 7* (*1952* at Wikieup — ARP) and April 16 to *May 1* (*1938* near Tucson — ARP), are probably exceptional by 3 or 4 weeks. Also one banded by Hargrave, who had already had extensive experience with the many races of Junco, at Flagstaff, April 15, 1938.

J. h. henshawi. Fairly common winter resident (never in large numbers) in northeastern and central Arizona, less common farther south and west but reaching Yuma (three records: *1932, 1940* and *1955* — last two GM). Outnumbers *hyemalis* 4:1 or 5:1 at Flagstaff and in the White Mountains region, and males outnumber females by a like proportion, except when the males are leaving in February and March; but of course females are much less distinctive, so that many are surely overlooked in the swarms of *shufeldti*. The length of stay of each race in Arizona corresponds to the distance of its breeding grounds; thus we expect them to arrive in fall in the order *caniceps, mearnsi, thurberi, simillimus, shufeldti, henshawi,* and *hyemalis,* and to reverse this order in their departure. Each race is naturally most conspicuous just before the arrival of the next, except for *shufeldti,* whose great numbers tend to conceal all other races from late October to late March. Records of *henshawi* in and near northern Arizona are from October 8 (1936, banded at Flagstaff — Hargrave) to April 15 (1952 at Chimo Spring above Fort Apache — Dickerman and Phillips). In southern Arizona, where it hardly outnumbers *hyemalis,* it is recorded from *October 27* (*1942* at Imperial Dam, California — GM; peculiar pale headed female) to *March 31* (*1937* at Whiteriver — ARP).

J. h. simillimus. Status not clear because indistinguishable afield; but apparently fairly common in southern and western Arizona. Northern Arizona specimens, mostly atypical, range from *September 22* (*1934* on the South Rim of Grand Canyon — Grater, GCN) to *April 8* (*1937* below the Mogollon Rim north of Whiteriver — ARP). Southern Arizona records are from *September 8* (*1924* on Mingus Mountain near Prescott — Jacot, ARP) and *October 1* (*1952* at Wikieup — ARP) to *April 11* (*1888* at Camp Verde — Mearns, AMNH) and *May 2* (*1929* at Tucson — Hine, F).

J. h. thurberi. Apparently rather common in southwestern Arizona; uncommon to fairly common east to Prescott, the Salt River Valley, and even the Chiricahua Mountains (Austin Paul Smith, AMNH). Rare in the Fort Apache area (ARP), and probably merely casual at Flagstaff (*December 3, 1931* — Hargrave, MNA). Though nearly all Arizona specimens are from *October 29* to *April 28,* these dates are probably not representative of its stay in southwestern Arizona, where little systematic collecting has been done. Late records are *May 12* in northern Sonora (according to A. H. Miller, Pac. Coast Avif. 33, 1957: 384) and *May 18* in the Huachuca Mountains (Hine, F). The fact that all the last three dates cited were in *1929* suggests irregularity in spring departure, however. The next latest date is *April 27* (*1956* at Papago Well in southwesternmost Pima County — ARP); here again it is noteworthy that a belated female, probably of this race, was seen in the Kofa Mountains, Yuma County, on May 16 of the very same year (Monson). Excluding these two years, we have no record later than *April 19* (*1884,* near Prescott — Mearns, US).

J. h. shufeldti. This race is the abundant winter bird of Arizona, especially in the northern woods. It is perhaps predominant in the sporadic appearances of juncos in numbers in the valleys farther south. Thus at least 100 (including *caniceps*) were seen at Binghampton Pond, Tucson on December 26, 1956 (M. J. Nichols); and "thousands" were in the Verde Valley, supposedly driven down by the storm of March 2, 1886 (Mearns). They evidently lingered on in Beaver Creek near Camp Verde, where he found them "still very numerous" April 2 and saw "a number" April 26, 1886. It was again "quite common" in the valley October 17, 1887, and juncos thought to be of a *shufeldti*-like form were "very abundant" in the Beaver Creek locality four days earlier! These October and April flocks must have included a considerable proportion of *mearnsi* and probably other races. *Shufeldti* has been taken in and near northern Arizona from *September 22* (*1936* in the White Mountains — ARP) to *April 8* (*1937* near Whiteriver — ARP); and similar birds have been seen at Flagstaff from September 16 (1939, banded — Wetherill) to April 22 (1936 — Phillips). Again, as in *thurberi,* we have unusually late records in a single year (1935 in this case); these are *April 28* near Bluff, southeastern Utah (UT; referred to *simillimus* in BNav), and May 8 (in the Kayenta region — Wetherill). Southern Arizona records of *shufeldti* are from *October 1* (*1952* at Wikieup — ARP) to *April 8,* as above.

J. h. mearnsi. Common in east, especially the White Mountains, but much scarcer westward, and apparently rather irregular west of the Prescott and Ajo Mountains regions. Recorded west to southeastern corners of Nevada (Kaolin — Austin Paul Smith, AMNH) and California. Recorded in northern Arizona from September 23 (1936 near Flagstaff — Hargrave) to May 8 (1937, Keams Canyon — Monson); in southern Arizona from *October 10* (*1904* in the Huachuca Mountains — Kimball, AMNH) to May 11 (1953 high in the southern part of the Aquarius Mountains, southeastern Mohave County — Phillips).

J. h. aikeni. At least seven found at Flagstaff and in the White Mountains, *November 21, 1936,* to *February 1937* (ARP, MNA, LLH; see Hargrave, Phillips, and Jenks, Condor 39, 1937: 258-259). No valid record since that remarkable flight.

J. h. caniceps. The juncos breeding in the northeast, north of the Little Colorado and Zuni Rivers, are nearest this race but not typical (see BNav); those of the Grand Canyon rims are variably intermediate. Otherwise, *caniceps* is a very common member of the winter flocks of juncos in eastern and central Arizona, but rare west of Prescott and the Ajo Mountains regions. Recorded westward, in fall only, to Hualapai Mountains (Phillips), Topock (GM), Bill Williams Delta (Monson), and at Bard, southeasternmost California (Huey, Condor 28, 1926: 44). In wet years, not uncommon in winter in Kofa Mountains, as in 1955–56 (Monson). Recorded off its breeding grounds in northern Arizona from September 17 (Hargrave) to May 14 (Phillips; both birds banded at Flagstaff in 1936). Southern Arizona records are from September 27 (1952, only record for Hualapai Mountains — Phillips; 1953, top of Santa Catalina Mountains — Marshall) to May 24 (1957 in Santa Rita Mountains — Levy) and exceptionally to *June 5* (*1892* at Fort Huachuca — Loring, US) and *June 6* (*1957* in the southeastern corner of Arizona in Guadalupe Canyon — Levy, US).

J. h. dorsalis. Our breeding birds are of this race, except in the northeast as mentioned above. Typical *dorsalis* is a rare migrant south of the valleys just below its breeding range. Of seven supposed migrants from Arizona and western New Mexico listed by Miller (Univ. Calif. Publ. Zool. 44, 1941: 210), only one is presently confirmable (Huachuca Mountains — Kimball, MCZ), though one (Chiricahua Mountains) we have not seen. On the other hand, there are valid published records not listed by Miller from the Chiricahua Mountains, *March 26, 1881* (Stephens, MCZ; see Brewster, Bull. Nuttall Orn. Club 7, 1882: 195) and Peppersauce Canyon near Oracle, *April 7, 1885* (Scott, AMNH; see Scott and J. A. Allen, Auk 4, 1887: 201-202). Three specimens were taken in one winter (*October 20, 1956*, to *January 5, 1957*) in the vicinity of Cave Creek alone, in the Chiricahua Mountains (Ordway, George, Cazier; SWRS). Though a nearly typical *dorsalis* was taken at Camp Verde, *October 8, 1885* (Mearns, AMNH), records farther from the breeding range are all from *October 20*, as above, to April 27 (1952 in the Santa Rita Mountains — Dorothy E. Snyder *et al.*). Proper search would probably yield records each winter.

This junco seems a fine example of an altitudinal migrant, at first glance. During severe winters, at least, it occurs commonly in the Verde and upper Gila Valleys (San Carlos area), just below the Mogollon and Natanes Plateaux. It is notable, however, that these valleys lie to the south of the breeding range. Young juncos in late summer may assemble a very short distance below the main breeding grounds but still in the pines (north slope of San Francisco Peaks — Hargrave); but there are no records of *dorsalis* for the adjacent, lower Wupatki National Monument; nor did Jacot see any in the winter of 1935–1936 about Springerville, Shumway, or farther north on the north slope of the White Mountains region. (Since he knows juncos well, and no other dark-billed, pale-throated junco occurs there, its detection would have been easy). Thus the available evidence indicates that these birds move downward in a southerly direction only. The southwesternmost record is in the Pajaritos Mountains west of Nogales (ARP).

There are notable concentrations in the Transition Zone around the San Francisco Peaks from late July to early October, consisting mostly of young birds that have descended from higher forests where the species nests more abundantly. Again a migration through the pines becomes obvious from February 25 (1937, Walnut Canyon National Monument — Wetherill) to April 23 (1936, Flagstaff — Phillips); snowstorms in late March produce numerous new, unbanded birds in the banding traps.

411. YELLOW-EYED JUNCO; ARIZONA JUNCO; MEXICAN JUNCO *Junco phaeonotus* Wagler

Abundant resident in Transition and boreal zones of southern Arizona, north and west to the Santa Rita, Santa Catalina, Pinal and Graham Mountains. Also winters in adjacent Upper Sonoran Zone, but rarely reaching its lower edge. One migrant station outside of breeding range: Whetstone Mountains, *September 26* to October 5, *1907* (Austin Paul Smith; AMNH, examined).

In plumage this junco is an exact replica of *Junco hyemalis dorsalis* of the brown-eyed species, with its blue-gray head and body, and bright reddish-brown back. Here the resemblance ends, for its beady eye is a rich yellow, which also tinges the feet and bill except for the black top of the bill. Its varied song is usually in three parts of contrasting pitch and rhythm. Each male has several song patterns in his repertoire, but the observer soon learns to recognize them all as belonging to this species by their sweet, thin quality suggestive of a wood warbler. The gait is a peculiar shuffle, between a hop and a walk. The birds do not join their dark-eyed cousins, but stay to themselves, in groups not exceeding family size. All these traits, as well as the total lack of interbreeding and complete geographic separation in the breeding season, proclaim this as a species distinct from *Junco hyemalis*, the Brown-eyed Junco. They are common to all the races of *Junco phaeonotus*, which ranges as a year-round resident from southern Arizona to Guatemala.

Yellow-eyed Juncos descend below the forest at times, especially in severe weather, and have been noted there from *September 18* (*1952* in Molino Canyon, lower Santa Catalina Mountains — ARP) to May 15 (1957 at the Southwest Research Station, Chiricahua Mountains — Cazier). The Arizona race is *palliatus* Ridgway, the palest and northernmost form of the species.

413. WHITE-CROWNED SPARROW; GAMBEL'S SPARROW *Zonotrichia leucophrys* (Forster)

Very common transient in all brushy places. Abundant winter resident in weeds about tall brush, principally in farmlands and large washes, throughout the Sonoran Zones, though less numerous north of the Mogollon Plateau and east of Tuba City. Leaves Mogollon Plateau and rims of Grand Canyon by early December. Casual (not known to breed) near timberline in the White Mountains, *July 11, 1936* (ARP); has been seen casually in summer in Lower Sonoran Zone, where it lingers to *early June* (F, Stanford, LLH).

These abundant flocking birds arrive in Arizona in fall, just after a complete molt into fresh, basic plumage of two kinds. Birds hatched the summer of the same year are in their first basic plumage, with crown stripes brown and buffy. Those older than one year are in the definitive basic plumage, wherein the crown is decorated with alternating stripes of black and white. In spring, birds at your feeding station can be seen to molt, emphasizing the head pattern in old birds, and completely transforming the immatures into a dress of adult aspect, their first alternate plumage. All winter the birds forage in flocks and spend considerable time in community singing.

Two well-marked races are easily distinguished afield in Arizona; both of them breed far to the north. The first is *Zonotrichia leucophrys leucophrys* (Forster) in which the light superciliary goes forward with no darkening past the eye to the bill, which is small and yellowish. This gives the bird a wide-eyed, placid appearance.

We follow Todd (Auk 70, 1953: 370-372 and plate 10) in recognizing that the type of *leucophrys* is a white-lored bird, albeit from a Hudson Bay population which is polymorphic in this regard, but in which the preponderance of specimens have white lores. This is our abundant winter resident — probably the most abundant bird in the irrigated valleys of Arizona, from the end of September to mid-April. After this it rapidly diminishes as the other migrant form increases. Extreme dates in the south are *September 19* (*1949*) to *May 17* (*1940*, both at Tucson — ARP). Northern Arizona migrations are principally from late September to mid-November, and again in late April. Extreme dates there are *September 14* (*1889*, Grand Canyon — US; *1953*, Flagstaff — Wetherill, MNA) to *May 10* (*1929*, bottom of Grand Canyon — V. Bailey, US).

Zonotrichia leucophrys oriantha Oberholser is our second race, principally a migrant, whose large pinkish-brown bill and darkened lores (black in adults) give it an imperious mien. Look for this bird in the last half of September and in May, when few *leucophrys* are about. In occasional winters, such as *1952–1953* (ARP), 1960–1961 and 1961–*1962* (netted by Marshall; RSC) *oriantha* remains in numbers through the winter at Tucson; ordinarily only one or two birds are seen in the whole valley at that season.

Southern Arizona migrations are from *September 9* (*1928* at Wagner, south of Prescott — Jacot, ARIZ; *1947* at Wickenburg — ARP) to October 20 (1949 at Wikieup–Phillips); and October 30 (1952, Parker Dam — Monson); and again from the end of February to *June 6* (*1893* at Tucson — Price and Wilbur, Stanford Univ.). There are a few sight records (Tucson, San Pedro River, Tempe) in late June and July or early August, but the subspecies is unknown in all these cases. The July bird from the summit of the White Mountains is *oriantha;* it is an adult male which was singing at a little spring, where it was collected by Hustace H. Poor (ARP). We have not found this species there again in summer.

Nigrilora Todd, the eastern race of the black-lored type, is darker than *oriantha*, but cannot be distinguished from it in the field. Four specimens from Arizona are considered by Kenneth C. Parkes to be this race, in the sense that the races are currently divided, on the basis of color of the lores. They are from Nogales (Dille #978, ARP), Mesa (LLH #H396), Eagar (Stevenson #1063, ARP) and Topock (GM #208).

412. WHITE-THROATED SPARROW
Zonotrichia albicollis (Gmelin)

Rare winter resident in recent years in Lower Sonoran brush in valleys and foothill canyons (casually in Upper Sonoran Zone) across southern Arizona, except the drier deserts, west and north to the Colorado River (GM, and north to Nevada — Gullion) and the mouth of the Verde River (ARP).

The White-throated Sparrow is considerably more rusty above than the much more abundant White-crown, with which it is often associated. Both species have white throats, but that of the White-throated is sharply set off by dark gray edges all around. The yellow spot on the lores is also helpful.

Since its discovery in Arizona in *1939* (GM), there have been records at increasingly frequent intervals, reaching a peak in the winters of *1952–1953* and 1953–1954 when several were found each winter, especially around Tucson. We have no records between 1956 and 1961. In March 1962 one was at a feeding station on Marshall's Towhee study area in the mesquite woods south of Tucson for more than a week, along with a Golden-crowned Sparrow, a Harris' Sparrow, and both races of White-crowns! The numerous specimens and sight records for Arizona fall between *October 29* (*1949*, mouth of Verde River–ARP) and April 21 (1955, Tempe — Hugh Hanson) except for an extraordinary occurrence near Portal, Chiricahua Mountains, in early June (Mary Belle Keefer).

414. GOLDEN-CROWNED SPARROW
Zonotrichia atricapilla (Gmelin)

Very rare winter visitant in recent years in southern Arizona; specimens from Topock (GM), the Ajo Mountains (ARP), near Tucson (ARP; Marshall, ARIZ), and an older one from Potholes, California (*April 18, 1916* — Huey, LA); most recently from the lower Santa Catalina Mountains (RSC). Sight records from Tucson (Phillips; Marshall) and the Colorado River six miles above Imperial Dam (twice — Monson). Two records on migration in northern Arizona: Springerville, April 25, 1953 (Dickerman and Phillips), and near Anita, south of the Grand Canyon, *October 8, 1956* (LLH).

This species, in the first basic plumage which we usually see in Arizona, is like a dull young White-crown, but has a dark bill and the appearance on the breast of being very dirty. The golden crown stripe is inconspicuous.

The species was first taken within Arizona in *1947* (Ajo Mountains, *October 24* — ARP). Southern Arizona records extend from this date, through the winter, and to May 5 (1959, Tucson — Marshall; adult in prealternate molt).

415. HARRIS' SPARROW
Zonotrichia querula (Nuttall)

Four records, south to Gila Valley: Sacaton, *March 16, 1913* (Gilman); Moenave, Navajo Indian Reservation, February 19, 1937 (Monson); Tempe (banded by Mrs. Birchett) November 22 to December 23, 1947; and

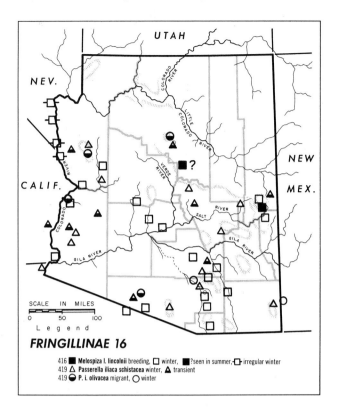

San Carlos, *November 9, 1951* (ARP). Also one at Tucson March 11-18, 1962 (Marshall).

The buff on the light parts of the head of the immature (which is the only plumage we normally see) is almost as rich as that of the Baird's Sparrow. Like the Golden-crowned Sparrow, it is slightly larger than our other *Zonotrichiae*. There is a congregation of dusky streaks forming a cravat on the breast.

416. LINCOLN'S SPARROW — *Melospiza lincolnii* (Audubon)

Summer resident, perhaps not uncommon, in willows of higher elevations of the White Mountains; possibly also at Mormon Lake. Common migrant in dense low cover bordered by grass or weeds, probably throughout state. Winters rather commonly in dense brush, reeds, and farm hedgerows of the main Lower Sonoran Zone valleys of southern and western Arizona, east and north to near Patagonia, the lower San Pedro Valley, Phoenix, the Davis Dam area, and casually to Boulder City, Nevada (but not, as stated by A.O.U. Check-list, at Flagstaff nor on San Francisco Mountain).

The song of Lincoln's Sparrow is nothing less than brilliant, as heard in summer in the willow thickets of the White Mountains meadows. The common call is thin and buzzy, like a bunting's. Another is a sharp *tep* exactly like the Fox Sparrow's. There is a buffy wash across the chest with its narrow black streaks.

Presumably a genus should possess distinctive traits of morphology and behavior. Such can be said of the grass-dwelling *Ammodramus*, with its long tertials and slender, pointed rectrices and of *Calcarius* with its long hind claw. But if you had an albino of any species from *Aimophila* through *Passerella* (on our list) you would be hard-put even to identify it to genus. For these genera seem to lack structural characters, being based on color and relative lengths of primaries. They differ structurally no more than do the various races of Song and Fox Sparrows among themselves. The White-crowned Sparrow has hybridized with the Song Sparrow, and the White-throated Sparrow several times with the Dark-eyed Junco (Dickerman, Auk 78, 1961: 627-632). Rather than follow Linsdale's, and later Miller's very commendable recommendation of sinking *Melospiza* in *Passerella*, why not simply take the oldest name for the entire group, which is *Emberiza* or *Junco*? Doubtless most ornithologists will find this rather gamey fare, and they cling to the hope that anatomical and "ethological" revelations will eventually resolve the execrable state of fringilline genera!

In March, migrating Lincoln's Sparrows seem to be behind every thick bush; were they in flocks like White-crowned Sparrows, we would call them abundant. Southern Arizona migrations are principally in late September to October and March to April. Outside dates are from *August 24* (*1956*, Chiricahua Mountains — George, SWRS) and *September 4* (*1938* near Tucson — ARP) to *November 9* (*1885*, Camp Verde — Mearns, US); and again from *February 7* (*1957* — ARP) and February 9 (1953 — Dickerman and Phillips; both in the city of Tucson where none winters) to May 27 (1934 at Tucson — Phillips).

Migrations in northern Arizona are from August 26 (1947 and 1954 at Flagstaff — Phillips) to November 15 (1947 at Holbrook — Phillips); and again from February 19 (1937 at Moenave, east of the Grand Canyon—Monson) to May 22 (1936 near Flagstaff — Phillips). From the above it will be obvious that Miller and McCabe (Condor 37, 1935: 145) and the A.O.U. Check-list exaggerated the "limits of the winter period . . . arbitrarily set at November 1 and March 15"!

The above applies to the nominate race *lincolnii*, of which *alticola* Miller and McCabe is considered a synonym by Wetmore (Birds of Southern Veracruz, Proc. U. S. Nat. Mus. 93, 1943: 339) and Phillips (Jour. Ariz. Acad. Sci. 1, 1959: 28). (Several earnest ornithologists have foundered on McCabe's color descriptions or have left museum work in despair.)

On the Pacific northwest coast is a small race, *gracilis* (Kittlitz), characterized by a pronounced buff overcast on the head and nape. Otherwise as Wetmore said (*loc. cit.*), "Color differences are not discernible except to note that there is considerable individual variation." Here again we suspect more rapid fading in Arizona than in the coastal fog belt, and only three specimens are small enough to qualify as *gracilis*: March 7 and *April 7, 1910* (Stephens and Dixon, MVZ, *fide* Jenks) from the Colorado River at the foot of the Needles and below Cibola, respectively; and a very small male taken east of Flagstaff on *February 28, 1937* (Wetherill, MNA).

417. SWAMP SPARROW — *Melospiza georgiana* (Latham)

Rare and irregular winter visitant in southeastern Arizona west to Tucson, casually west to the Colorado River in southern Arizona (Bill Williams Delta, *November 28, 1952* and *December 23, 1953* — GM). One record from northern Arizona: Tuba City, December 19, 1936 (Monson). The statement that it occurs only "casually" in Arizona (A.O.U. Check-list) is an underestimation. From Tucson alone there are four specimens (LA, LLH, ARP, WF); and twice it has been thought that four individuals were simultaneously present at Sabino Pond and Rillito Creek (R. B. Streets; A. B. Howell). There are additional sight records near and south of Tucson, and additional specimens from Hereford (San Pedro River; two — ARP) and the New Mexico part of San Simon Cienega (Marshall, ARP).

This is like a dark Song Sparrow without streaks below. It is found in reeds and cattails. In Arizona we have not seen the rusty-capped alternate plumage, our birds being dark-crowned. A westward surge began in 1948 (Fleetwood, Auk 68, 1951: 112-113) and several birds were seen in Arizona, Sonora, and adjacent New Mexico during 1951-1953, the same winters when White-throated Sparrows appeared. Since then, few have been noted. Seasonal occurrence has been from November 1 at Kinsley's (Arivaca Junction, 1952 — Phillips) to *February 14* (*1943*, Tucson — LLH).

Geographic variation, as currently recognized, consists of a cline of darkening color from northwest to southeast. All Arizona specimens are the pale northwestern *ericrypta* Oberholser except one, which is the dark *georgiana*: Bill Williams Delta, the *November* bird cited above (GM).

418. SONG SPARROW *Melospiza melodia* (Wilson)

Locally common resident in reed-sedge-brush types along major permanent rivers in southern and western Arizona, east to Safford and to San Bernardino Ranch in extreme southeast; also summer resident along the upper Colorado River tributaries (Grand Canyon region) and on permanent brush-lined streams of Upper Sonoran and Transition Zones on and adjacent to the Mogollon Plateau (chiefly in the White Mountains region). Rather common winter resident locally at reedy ponds, brushy streams, and farmlands with brushy, weedy edges in the Sonoran Zone valleys and even in low Transition Zone where it does not breed, and probably higher. Very rare transient elsewhere. Greatly reduced as a resident bird in the southeast in the present century because of habitat restriction.

This chunky denizen of extensive brush and cattails beside permanent running water has the liberal streaking of the throat and breast concentrated into a central chest spot. It and the Common Screech-Owl are the most geographically variable of all birds of North America. But in the Song Sparrow, the profound variations are confined to the resident populations of the Pacific Coast and the Mexican Plateau; inland and eastern United States birds are more migratory and show a lesser degree of racial development. Nevertheless, in Arizona the field observer can easily make out three types, comprising groups of races, and he would do well to record them as such, because of the interesting migration data so obtained. First there is the desert type (*saltonis*, *fallax*), very pale, with the tail and streaks both above and below of reddish-brown. The mountain Song Sparrow is the second type (*montana*, *fisherella*, *juddi*) with narrow streaks of black and a gray-brown tail. The third, so rare in Arizona that specimen verification would be desirable, is of northwestern birds (*merrilli*, *rufina*), very deep rufous or sooty above, and heavily marked below with broad chestnut streaks. (Heavily black-streaked California birds should be looked for, since one was taken in Sonora — "*mailliardi*," ARP).

Arizona is an outpost in the wide summer distribution of the Song Sparrow, for there are no breeding populations east of us, at this latitude, except in extreme northern New Mexico. There is an enormous gap in its breeding range, from the southeastern United States through New Mexico and down the eastern half of Mexico. If the Song Sparrow can nest along the desert rivers of Arizona, Sonora, and torrid Baja California, why should it eschew the Rio Grande, the lower Mississippi, the Sabinas?

Breeding populations of south-central Arizona Song Sparrows have led a precarious existence, depending as they do on water. Mankind still multiplies there, along the Santa Cruz River drainage, heedless of his destruction of water resources and the grasslands which could replenish the rapidly dwindling fossil supply. Song Sparrows (of the race *fallax*) were the first victims of the fate which is in store for man himself. In the 1880's they were abundant, outnumbering *montana* in winter. They bred along the Santa Cruz to at least *1895* (Herbert Brown, ARIZ), but apparently not long thereafter. The last of the Sonoita Creek population was seen by Phillips in 1947 — there is still running water, but cattle have eaten and trampled out all the herbaceous vegetation of this so-called "bird sanctuary." The only surviving colonies in natural habitat south of the Gila River are at Feldman, Mammoth, and San Bernardino Ranch. The birds also thrive at man-made Picacho Reservoir. These last, and the population along irrigation canals at Phoenix, may have contributed to a recent come-back of the bird at Tucson; Marshall noted a singing male in an irrigation overflow on the San Xavier Indian Reservation in the summer of 1961, and Richard S. Crossin discovered nesting (RSC) both in that and the following summer beside a reeking slough near Cortaro, just northwest of Tucson.

Southern Arizona migrants of northern populations of the Song Sparrows appear from *September 18* (*1938* — ARP) to *April 19* (*1886*, Brown — ARIZ) and May 2 (1943 — J. R. Simon *et al.*; all near Tucson). Fall migration was "at its height" near Camp Verde on October 13 (1887 — Mearns) and the earliest spring records are February 16 and 20 (1932 in the Baboquívari Mountains — Phillips).

Northern Arizona fall migration is from late August (1889 at San Francisco Mountain — Merriam) to at least October 12 (1936 at Oak Springs, Defiance Plateau — Monson). Spring migrants are recorded from *March 6* (*1937* near McNary — ARP) to April 25 (1948 near Holbrook — Phillips).

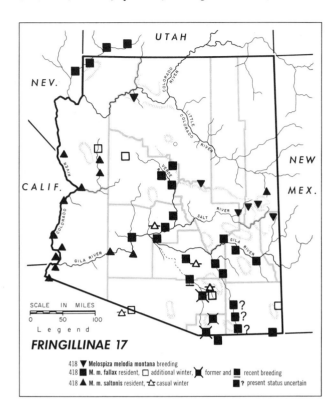

FRINGILLINAE 17

418 ▼ Melospiza melodia montana breeding
418 ■ M. m. fallax resident, □ additional winter, ◪ former and ■ recent breeding
418 ▲ M. m. saltonis resident, △ casual winter ■? present status uncertain

Melospiza melodia montana Henshaw is the breeding race of northern Arizona (west apparently to Supai, west of the Grand Canyon). It is also by far the most common transient and winter resident statewide; the above accounts of migration apply to *montana*. In the hand it is seen to be pale though sharply black-streaked, thin-billed, and with the upper tail coverts grayish-brown streaked with black. The head pattern is in gray and black; it is the least rufescent of our usual Arizona races.

M. m. fallax (Baird) is of the pale "desert type" and is very rufous, especially upon the crown and unstreaked upper tail coverts. As noted above, its streaks are normally all reddish-brown. The light areas of the side of the head are clear gray. It nests diagonally across the state from northwest to southeast. Though primarily a resident of Lower Sonoran Zone streams, it ascends Oak Creek to the level of the lower pines (ARP), and probably is the race for which there are sight records up the West Fork into pine and fir country. *Fallax* is commonly found in winter throughout its breeding range; but additionally some migrate, so that it is a rare winter resident in southern Arizona non-breeding localities, east to the New Mexico border (San Simon Cienega – ARP). The melancholy history of Tucson area birds, recited above, applies to this race.

M. m. saltonis Grinnell is a weakly characterized desert race differing from *fallax* in being slightly paler and smaller. The crown is not as deep chestnut as is that of *fallax*, and the light areas of the side of the head are white. It is the palest race of the entire species, in keeping with trends seen in other variable species along the lower Colorado Valley. In some years it is a rare winter resident east as far as Tucson. But there was a big flight to the Rillito River there in mid-winter of 1915–1916 (*December-January* – A. B. Howell and Huey, LA). A recent record is that of Lyndon L. Hargrave, who took one at the old Midvale Pond, Tucson, on *February 6, 1943*. Not only is this bird a very good example of the lower Colorado River race, but it could not be a variant of the local bird, which was extirpated at that time! Such records extend from late *October* (in adjacent Sonora – ARP) to late *February* (ARP, LLH).

M. m. juddi Bishop of the northern Great Plains is similar to *montana* in its sharp black streaking, but has a swollen bill. The light areas of the side of the head are clear buff. It is represented in Arizona by two typical specimens taken in winter; near Tucson by Howell and Huey (LA) and Ute Mountains, Mohave County by Jon Coppa (ARIZ).

M. m. fisherella Oberholser of the Great Basin is darker, more heavily streaked, and with a larger bill than *montana*. The several Arizona specimens, more or less typical, have been taken from *November* (LLH, JSW) to *March 30* (*1956*, Colorado River about 25 miles above Imperial Dam –GM #437).

M. m. merrilli Brewster is the reddish-brown race of Idaho, which is streaked above and below with deep reddish brown. It is considerably darker than *fallax*. The ventral streaks are blurry and they cover more of the underparts than do those of the above races. The three Arizona specimens are from Quitovaquita, Organ Pipe Cactus National Monument, *November 28, 1939* (Huey, SD); Wikieup, *October 1, 1952* (ARP); and Mammoth, *November 24, 1960* (RSC). Its main winter range lies farther west and northwest. The report at Fort Apache is a misidentification.

M. m. rufina (Bonaparte) is a large and very duskybrown bird whose blurry but dense ventral streaks are on a grayish background (instead of white below, as in all the above races). It has a long slender bill and breeds in coastal British Columbia and adjacent Alaska, where it is mostly resident. Wanderers had previously been found south to Washington, and it was a surprise to see one at a lake near Continental, *January 10, 1963* (Marshall, sling-shot, ornithology class, and faithful swimming dog "Donkey," ARIZ).

419. FOX SPARROW   *Passerella iliaca* (Merrem)

Local winter resident, usually in very small numbers except in the Hualapai Mountains (and in some years, the Kofa and adjacent mountains), in dense Sonoran Zone thickets and chaparral in western and southern Arizona, south barely into Sonora. Rare transient along the lower Colorado River and other principal watercourses and canyons, and in the San Francisco and White Mountains regions. Accidental in western Arizona as late as *May 15* (*1956*, Kofa Mountains – GM) and as early as August 29 (1938 below Hoover Dam – Grater).

The Fox Sparrow is the largest of our streaked and spotted fringillines, almost the size of a Rufous-sided Towhee. In coloration it is much like a Hermit Thrush. Unlike the other sparrows which have streaks on the chest, it lacks a superciliary stripe, its crown and face being plain. The base of the bill is yellow. Beneath the light malar area is a chain of dark spots, becoming larger and denser posteriorly until those on each side join below, covering the breast and gathering into a dark central blob. All these spots are in the shape of inverted V's. The Fox Sparrow scratches in dead leaves under dense bushes, working both feet simultaneously and scratching a long time in a single spot. The usual wintering period in Arizona is from late September to March.

Fox Sparrows nest clear across the northern part of the continent and extend southward in the west to the Rockies of Colorado and Utah and to the high mountains of southern California. In this vast area they show pronounced geographic variations, rivaling those in the Song Sparrow. Their study has constantly been thrown off the track by failure of investigators to take into account seasonal wear, which causes the back to become grayer after the fresh fall coat begins to wear away. But more profound is postmortem foxing which over the years reddens museum specimens. Thus some races have been described from recent gray breeding birds as compared with old, red winter specimens. There are three groups of races; each group can be identified in the field:

A. Eastern birds, tail and spots reddish-brown, tail short.

1. *Passerella iliaca zaboria* Oberholser breeds in northern and eastern Alaska. Even the back is conspicuously streaked with reddish-brown, and the overall prominence of foxy reddish tones in the plumage gives the name Fox Sparrow to the species. It is represented in and near Arizona by six specimens, including the *May* Kofa Mountains bird cited above. The others are from near Tucson, doubtless in *fall of 1875* (Spring, Paris Museum); near Fort Huachuca, *November 1, 1889* (Stejneger, US); one at Sabino Ranch, Tucson from December 26, 1949 (Richard H. Reed) to *January 10, 1950* (ARP); Topock, *October 10, 1952* (GM); and Santa Cruz, Sonora, *December 22, 1954* (ARP). (The typical race, *iliaca*, is still more reddish; it has not been taken as far west even as Minnesota —*fide* Dickerman.)

2. *P. i. altivagans* Riley breeds in interior British Columbia. It is practically unstreaked on the back and is not as red as the above. Six reasonably typical Arizona specimens are from the Huachuca Mountains, allegedly in *April, 1917* (W. W. Brown — whose data on labels are the terror of all right-minded systematists — AMNH); Wikieup, *October 19, 1949* (pale and reddish, toward *zaboria* — ARP); Hualapai Mountains, *October 3, 1948* (ARP); Topock, *October 17, 1953* (GM); south of Tucson, one present at feeding station from March 7 until *March 19, 1960* when it ran into a towhee net and died of shock (Marshall, ARIZ); and Mammoth, *November 12, 1961* (RSC).

B. Southwestern group; rather pale uniform gray upperparts which contrast with dull reddish-brown wings and tail; long-tailed (tail usually longer than wing), spots dull. These birds predominate in Arizona.

1. *P. i. schistacea* Baird, breeding in the Rocky Mountains and Great Basin, is the only race regularly occurring in Arizona, to which the opening account applies (except for the Sonora and *May* Kofa Mountains birds). It is common in the Hualapai Mountains from early October on. *Canescens* Swarth is a synonym, as shown by an excellent August series from the White Mountains, California (LA); so is *swarthi* Behle and Selander, judging from its position right in the middle of the range of *schistacea* and from examination of a few specimens kindly loaned by Dr. Behle.

2. *P. i. olivacea* Aldrich, from the mountains of Washington, is a weakly differentiated race, slightly darker on the posterior upper parts than *shistacea*. If it be recognized, then we would refer here a scattering of Arizona specimens (GM, ARP), shown on the map, of dates ranging from *October 3* to *December 31* in the south and west, and *October 1* on the San Francisco Peaks (all ARP).

3. California races, generously over-split, darker and with large swollen bill. Two specimens from southwestern Arizona were kindly identified by Alden H. Miller as *P. i. megarhynchus* Baird: Ajo Mountains, *October 24, 1947* (ARP) and Sierra Pinta, Yuma County, *December 12, 1955* (GM).

C. Dark coastal races with uniform dull brown backs. *P. i. townsendi* (Audubon) from Sitka area, taken once in Chiricahua Mountains, *November 28, 1914* (van Rossem, J. E. Law collection, *fide* Swarth, U.C.P.Z. 21, 1920: 146-147, 199). This was quite a strain on Swarth, who was loath to admit that coastal and Californian birds of any species should migrate into Arizona. The only place where groups B and C might intergrade would be in passes through the Cascade Mountains in southern British Columbia and Washington. Such an intermediate, presumably *fuliginosa* Ridgway x *olivacea*, is a specimen taken at the Southwest Research Station, Chiricahua Mountains, *September 29, 1956* (Cazier and Ordway, ARP).

420. McCOWN'S LONGSPUR — *Calcarius mccownii* (Lawrence)

Rare and irregular (formerly abundant) winter resident in grassy plains and valleys of eastern Arizona; recorded west formerly to the Altar Valley, the Gila Valley south of Phoenix, the Agua Fria River east of Prescott, and the northeast slope of the San Francisco Mountains. Since 1922 recorded only from the Sonoita-Elgin-San Rafael Valley area and near Bowie, in Santa Cruz and Cochise Counties.

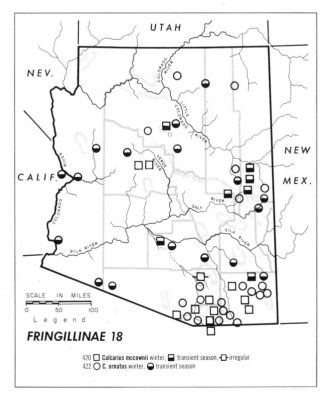

FRINGILLINAE 18

420 ☐ *Calcarius mccownii* winter, ■ transient season, -☐- irregular
422 ○ *C. ornatus* winter, ● transient season

Longspurs in dull winter plumage show no distinctive patterns or field marks except in flight. The spread tail of the McCown's shows a broad black **T**, framed laterally in white. The high-pitched call is a single note.

McCown's Longspur was probably equally common as the Chestnut-collared in Arizona in early times and was more widespread, for Mearns took 40 specimens in the central part of the state from *1885–1888* (US, AMNH, ARP). It still "fairly swarmed" at Sulphur Springs in 1895 (to April 6 – Osgood), but decreased rapidly thereafter. The last record for many years was at Cochise, Sulphur Springs Valley on *January 20, 1902* (three specimens – Robinette, formerly US). In *1922* Swarth took one and saw "a few others" in the San Francisco Mountains. For Monson's rediscovery of the bird at Elgin, in the fall of 1939, see Condor 44, 1942: 225. Seymour Levy saw three near Bowie late in 1961, and collected one from a flock of several hundred on a dry lake bed there on *March 7, 1962* (US – the Chestnut-collareds were out in the grass). Seasonal records in southern Arizona extend from early October to early April.

421. LAPLAND LONGSPUR — *Calcarius lapponicus* (Linnaeus)

Apparently very rare, only two acceptable specimens: Meteor Crater, Coconino County, *November 15, 1947* (ARP); and Colorado River about seven miles above Imperial Dam, *November 18, 1955* (GM). Also two seen at latter point, November 11, 1956 (Monson); and has been seen in winter in western part of Navajo Indian Reservation, where it may be regular (Monson). Specimens found in other places (Phoenix, Petrified Forest) were probably brought there by automobiles coming from undetermined regions, perhaps out-of-state, and the latter record should be deleted from A.O.U. Check-list.

Longspurs in winter are gregarious birds of open grasslands, where they can be identified by their distinctive calls and tail-patterns. Like the Horned Lark and Pipit they have a long hind claw, the "larkspur," and long tertials to protect the flight feathers from wear against the grass.

The Lapland, McCown's, and Smith's now seem extremely rare in Arizona. But who gets out in northeastern Arizona in the dead of winter to look for them? Rather than discuss them further we recommend that the observer arm himself with a good field guide, a shotgun (a 12 gauge double-barrel, complete with .410 and .22 auxiliaries), and a deft pair of ears, and that he go out and seek the birds in order to add to our understanding of their status and numbers in the state.

Arizona specimens of the Lapland Longspur belong to the pale western race *alascensis* Ridgway.

422. CHESTNUT-COLLARED LONGSPUR — *Calcarius ornatus* (Townsend)

Abundant winter resident in the grasslands of southeastern Arizona, and fairly common migrant and winter resident in grasslands farther north, even up to the Canadian Zone (snow-cover permitting). In fall migration occurs in small numbers (singly or in flocks of five or less) more or less regularly, usually on open land near water, nearly statewide west to the Colorado River.

At present the only common and regularly winter visiting longspur in the state, the Chestnut-collared can be identified by the extensive white on each side of the tail, black-and-white spread wing, and by its two-syllable call, *chútup*.

Southern Arizona records are from *September 24 (1874*– Rothrock, US, *fide* Henshaw) near Mount Graham and September 25 (1952 on the California side of Havasu Lake – Monson) to May 3 (1902 below the Huachuca Mountains – Swarth). Northern Arizona records extend from September 18 (1922 on the northeast slope of the San Francisco Mountains – Swarth) to *March 14 (1888* near St. Johns—Swinburne, location of specimen unknown). This species was still on the snowy high Canadian-Transition Zone prairies of the White Mountains on Phillips' and Correia's last visit, *November 18 (1936* – ARP).

423. SMITH'S LONGSPUR — *Calcarius pictus* (Swainson)

One specimen: White Mountains, *April 24, 1953* (ARP). No winter record, but possibly not as scarce in Arizona and northern Sonora as this single specimen would imply.

Smith's Longspur has a thin multiple call-note, uttered three or more times in succession. It is the only longspur completely pale buffy underneath in the alternate plumage. The Arizona specimen is in worn basic plumage, however.

INDEX

Abbreviations, museums viii
Abundance categories x
Accipiter 19-20
Accipitridae 19
Accipitriuae 19
Actitis 34
Aechmophorus 6
Aegolius 54-55
Agelaius 165-167, 171
Aimophila 198-203
Aix 14
Ajaia 8
Alaudidae 93
Alcedinidae 67
Alectoris 30
Amazilia 65-66
Ammodramus 156, 192-194, 208
Amphispiza (see *Aimophila*) 201-203
Anas 12-14
Anatidae 8
Anatinae 12
ANHINGA, *Anhinga, Anhingidae* 4
ANI, Groove-billed 46
Anser 10-11
Anseriformes 8
Anserinae 9
Anthus 137-138
Aphelocoma 104-105, 106
Apodidae 58
Aquila 24
Archilochus 62-63
Ardea 5-6
Ardeidae 5
Arenaria 33
Asio 54
Asyndesmus 71-72
Atthis (see *Selasphorus heloisa*) 64
Auriparus 113
AVOCET, American 36-37
Aythya 15-16

BALDPATE 13
Bartramia 34
BECARD, Rose-throated .... x, 76
       Xantus' 76
BITTERN 7
    American .... xvii, 7
    Least 7
BLACKBIRD 162
    Brewer's 170-171
    Red-winged *165-167*, 171
    Rusty 170-171
    Yellow-headed *165*, 171
BLUEBIRD, Azure 131-132
    Chestnut-backed 132
    Eastern 131-132
    Mexican 84, *132-133*, 196
    Western .... xiii, *132*, 196
BOBOLINK 88, 162, *163*
BOBWHITE .... xiii, 28-29
Bombycilla 138-139
Bombycillidae 138
BOOBY, Blue-footed, Brewster's, Brown 4
BRANT 10
Branta 9-10
Breeding range 139-140
Breeding seasons , ... 3, 40-41, 43-44, 93, 98
       139-140, 141, 183, 188, 198, 200, 201, 202
Bubo 51, 53-54
Bucephala 16

BUFFLEHEAD 16
BUNTING 176, 179
    Bay-winged 194
    Beautiful 181
    Common *179-180*, 181
    Indigo 179-180
    Lark 191-192
    Lazuli 179-180
    Painted 181
    Varied 181
BUSH-TIT .... xiii, *111-113*, 133, 144
BUTCHER-BIRD 140
Buteo 21-23
Buteogallus 24
Buteoninae 21
Butorides (see *Ardea virescens*) 5
BUTTERBALL 16

Calamospiza 191-192
Calcarius 208, 211-212
Calidris 35
Callipepla 29
Calothorax 62
Calypte (see *Archilochus*) 62-63
Camptostoma 92-93
Campylorhynchus 119
Cannibalism 20
CANVASBACK *15*, 16
Capella 33
Caprimulgidae 55
Caprimulgus 55-56
CARACARA, *Caracara* 25-26
Cardellina 159
CARDINAL *176-177*, 178
Cardinalinae 176
Cardinalis 176-178
Carduelinae 176, 182, 183
Carpodacus 184-185
Casmerodius (see *Ardea alba*) 6
Cassidix .... xi, *121-122*, 124
Catastrophes 2
CATBIRD .... xi, *121-122*, 124
Cathartes 18
Cathartidae 18
Catharus 127-131
Catherpes 120
Catoptrophorus 34
Centrocercus 28
Centurus 69-70
Certhia 115-116
Certhiidae 115
Chaetura 59
Changes (*see* Historic changes)
Charadriidae 32
Charadriiformes 32
Charadrius 32-33
CHAT, Yellow-breasted *161-162*, 173
Chelidonias 40
Chen (see *Anser*) 11
CHICKADEE 102, 159
    Black-capped 109
    Mexican 109
    Mountain 109-110
Chlidonias (see *Chelidonias*)
Chloroceryle 67
Chondestes 195
Chordeiles 56-58
CHUKAR 30
Ciconiidae 7
Ciconiiformes 5

Cinclidae	116
Cinclus	116
Circinae	25
Circus	25
Cissilopha	105-106
Cistothorus	119-120
Clangula	16
Coccothraustes	182-183
Coccyzus	45
Colaptes	68-69
Colinus	28-29
Collocalia	58
Colorado River Valley	xiii, xv
Columba	40-41
Columbidae	40
Columbigallina (see *Columbina*)	43
Columbina	43
Commonness	x
CONDOR, California	19
Conservation	xiii-xvii, 28-29, 194
Contopus	89-91
COOT, American	17, 24, 30-31, *32*
Coragyps	18-19
CORMORANT, Double-crested, Farallon, Mexican, Neotropical, Olivaceous	4
Corvidae	102
Corvus	106-107
Cotingidae	76
COTTONTOP	29
Coturnicops	31
COWBIRD, Bronzed	x, 167, 169, *173-174*
Brown-headed	143, 150, 167, *172-173*, 183
Red-eyed	172-173
CRANE	5, 30
Sandhill	30
CRESTED-FLYCATCHER, Arizona	xv, *80-81*, 82, 176
Great	80
Crocethia	36
CROSSBILL, Red	xi, 176, *183-184*
Crotophaga	46
CROW	102, 107
Common	xi, *107*
CUCKOO	45, 46
Yellow-billed	*45*, 57, 91
Cuculidae	45
CURLEW, Hudsonian	34
Long-billed	*33-34*, 36
Cyanocitta	*103-104*, 106
Cyclarhis	142
Cygnus	9
Cynanthus	65, 66
Cypseloides	58
Cyrtonyx	29
DABCHICK	2
Dates	x, 17
Dendragapus	27, 28
Dendrocopos	74-75
Dendrocygna	11-12
Dendroica	149-155, 156, 161
Dichromanassa	6
DICKCISSEL	176, *181*
Diet (effect of)	68, 183
DIPPER	116
Dolichonyx	163
DOVE	40
Common Ground	43
Inca	43-44
Mourning	20, *42-43*
Rock	40
White-winged	*41-42*

DOWITCHER, Long-billed	36
Short-billed	35
Dryocopus	69
DUCK (see also Baldpate, Gadwall, Golden-eye, Mallard, Merganser, Pintail, Scaup, Teal, etc.)	181
Black	12-13
Black-bellied Tree	x, 11-12
Florida	13
Fulvous Tree	11, *12*
Gray	13
Mexican	12
Mottled	13
New Mexican	12
Ring-necked	15
Ruddy	*17*, 32
Wood	14
DUCKS, Diving	15
Dumetella	121-122
DUNLIN	35
EAGLE	21
Bald	24-25
Golden	24
Ecology (see Habitats and Life Zones)	
EGRET, American	6
Brewster's	6
Common	5, 6
Great	6
Reddish	5, 6
Snowy	5, 6
Egretta (see *Ardea thula*)	6
Elainiiae	19
Emberiza	208
Empidonax	17, 81, 82, 85-89, 91,92
Eremophila	93-95
Ereunetes	33, 35, 36
Erolia	35
Erosion	xiii-xvii
Eudocimus	8
Eugenes	65
Euphagus	170-171
Eupoda	33
Euthlypis	161
Evolution	xi, 13, 47, 68-69, 163
Falco	26-27
FALCON	19, 25
Aplomado	xi, xiii, 26-27
Peregrine	*26*, 44
Prairie	26
Falconidae	25
FINCH (see also Goldfinch and Purple-Finch)	176
"Black Rosy"	186
Cassin's	185
House	27, 141, 184, *185*
Pine	186-187
Rosy	186
Fires	xiii
FLICKER	xi, xv, 27, *68-69*, 75, 166, 168, 203
Florida (see *Ardea caerulea*)	6
FLYCATCHER	77, 102, 127
Acadian	89
Arizona Crested	xv, *80-81*, 82, 176
Ash-throated	81-82
Beardless	92-93
Buff-breasted	xiii, 85, *86*, 90,131
Coues'	*90*, 91
Crested	80
Dusky	85, 87
Gray	85, *86-87*, 142
Hammond's	85, 87

Least	87-88
Nutting's	82
Olivaceous	82-83, 176
Olive-sided	90, 91
Pale-throated	82
Scissor-tailed	80
Silky	138
Sulphur-bellied	80
Traill's	xv, 85, 89
Vermilion	77, 91-92, 93
Western	xi, 85, 86, 88-89
Wied's Crested	80-81
Willow	89
Wright's	87
Yellow-bellied	89
Forest, boreal	ix
mesquite	xvi
Fort Huachuca	xiii
Fregata	4
Fregatidae	4
FRIGATE-BIRD	4
Fringillidae	176
Fringillinae	176, 188
Frost flight	135
Fulica	32
GADWALL	13
Gallinago (see *Capella*)	33
Gallinula	32
GALLINULE	30
Black	32
Common	32
Florida	32
Purple	31-32
Gavia	1
Gaviidae	1
GEESE	8, 9
Gelochelidon	39
Geococcyx	45-46
Geographic variation	x-xi, 183
Geothlypis	157-159
Glaucidium	51-52
GNATCATCHER	133
Black-tailed	124, *134*
Blue-gray	133-134
Plumbeous	134
Western	133-134
GOATSUCKERS	55
GODWIT, Marbled	36
GOLDENEYE, American	16
Barrow's	16
Common	16
GOLDFINCH, American	187
Arkansas	187-188
Green-backed	187-188
Lawrence's	184, *188*
Lesser	187-188
Pale	187
GOOSE, Blue	11
Cackling	10
Canada	*9-10*, 11, 12
Hutchins'	10
Ross'	11
Snow	11
Tule	11
White-fronted	9, *10-11*
GOSHAWK, American	xi, 19-20
Mexican	23
GRACKLE, Boat-tailed	x, 171-172
Great-tailed	171-172
Grasslands	xiii
GREBE	1, 9
Eared, Least, Mexican, Pied-billed, Swan, Western Horned	1-2
GROSBEAK	176
Black-headed	178-179
Blue	*179*, 181
Common	xi, 178-179
Evening	*182-183*, 184
Pine	186
Rocky Mountain	178-179
Rose-breasted	178-179
GROUSE, Blue	27-28
Dusky	27-28
Sage	28
Gruidae	5, 30
Grus	30
Guiraca (see *Passerina caerulea*)	179
GULL	32, 37, 127
Bonaparte's	38-39
California	38
Franklin's	38
Glaucous-winged	38
Heermann's	39
Herring	38
Laughing	38
Ring-billed	38
Sabine's	39
Thayer's	38
Western	38
Gymnogyps	19
Gymnorhinus	107-108
Habitats	ix-xi
Haliaeetus	24-25
HARRIER	25
HAWK	19, 127
American Rough-legged	23
Black	22, 24
Broad-winged	22
Cooper's	20, 26
Duck	26, 27
Ferruginous	23
Fish	25
Gray	23
Harlan's (*see* Red-tailed H.)	21
Harris'	23-24
Marsh	25, 27
Mexican Black	22, 24
Pigeon	27
Red-bellied	21
Red-shouldered	21
Red-tailed	18, *21*, 23
Rough-legged	23
Sharp-shinned	20
Sparrow	27, 84, 133
Swainson's	22, 23
White-tailed	22-23
Zone-tailed	22, 24
Helminthophila	146-149
Helmitheros	146
HERON	5
Black-crowned Night	6-7
Common	5
Great Blue	5, 6, 7
Great White	5
Green	5-6
Little Blue	6
Louisiana	6
Snowy	6
Tricolored	6
Yellow-crowned Night	7

Hesperiphona (see *Coccothraustes*)	182-183
Hesperocichla	127
Hibernation	56
HIGH-HOLE	68-69
Himantopus	37
Hirundinidae	95
Hirundo	97-100
Historic changes	xiii-xvii
HONKER	9-10, 10
Huachuca Mountains	xiii
HUMMINGBIRD	59-62, 127
Allen's	60, 63, 64, 200
Anna's	x, 60, 61, 62, 63
Black-chinned	61, 62, 65
Blue-throated	60, 61, 65
Broad-billed	60, 61, 65, 66
Broad-tailed	61, 63, 65
Bumblebee	64
Calliope	60, 61, 64-65
Costa's	60, 61, 62-63, 168
Heloise's	60, 64
Lucifer	60, 61, 62
Morcom's	64
Rivoli's	60, 61, 65, 66
Ruby-throated	61, 63
Rufous	60, 63-64, 156
Salvin's	65-66
Violet-crowned	x, 61, 65-66
Western species, key to	60-61
White-eared	60, 61, 66
Hybrids	63, 65, 66, 72, 154, 171, 175, 208
Hydranassa (see *Ardea tricolor*)	6
Hydroprogne	39-40
Hylocharis	66
Hylocichla (see *Catharus*)	
Hypothetical species	x
IBIS	5, 8
Glossy, Scarlet, White, White-faced, White-faced Glossy	8
Wood	7-8
Icteria	161-162
Icteridae	162
Icterus	167-170
Intelligence	102
Introductions	29, 30, 162
Invasions (*see* Migration)	
Iridoprocne (see *Tachycineta bicolor*)	96
Irruptions (*see* Migration)	
Italics	x
Ixoreus	127
JACKSNIPE	33
JAEGER, Long-tailed, Parasitic, Pomarine	37
JAY	71, 90, 102
Arizona	104, 105
Black and Blue	105-106
Blue	103
California	104
Canada	103
Florida	104
Gray	103
Long-crested	103-104
Mexican	104, 105
Piñon	103, 107-108
Rocky Mountain	103
San Blas	105-106
Scrub	x, xiii, 103, 104-105
Steller's	103-104, 105
Texas	104
Woodhouse's	104-105

Junco	203-206, 208
JUNCO, Arizona	189, 203, 206
Brown-eyed	x, xi, 93, 147, 203-206
Cassiar	203-205
Dark-eyed	x, xi, 93, 147, 203-206
Gray-headed	203-206
Mexican	189, 206
Nevada	204
Oregon	203-204
Pink-sided	203-205
Red-backed	203-206
Shufeldt's	203-205
Sierra	203-205
Slate-colored	203-205
Thurber's	203-205
White-winged	203-205
Yellow-eyed	189, 206
KESTREL, American	27
Keys: to *Empidonax* Flycatchers	85
to Hummingbirds	60-61
KILLDEER	32-33, 34
KINGBIRD	70, 77, 78, 80, 84
Arkansas	77-78
Cassin's	53, 56, 77, 78-79, 81
Couch's	79-80
Eastern	77
Thick-billed	x, 79
Tropical	79-80, 82
Western	77-87, 79, 80, 181
West Mexican	79-80
KINGFISHER, Belted	4, 67
Green	67
Texas	67
KINGLET, Golden-crowned	134-135
Ruby-crowned	xi, 66, 110, 135-136, 142
KITE, White-tailed	19
KNOT	35
Lakes	xv, xvi-xvii
Lampornis	65
Laniidae	140
Lanius	140-141
Laridae	37
LARK, Horned	xiii, 33, 93-95, 127, 141
Larus	38-39
Laterallus	31
Law, "Phillips' "	xi
Leucophoyx (see *Ardea thula*)	6
Leucosticte	186
Life Zones	ix
Limnodromus	35-36
Limosa	36
LINNET	185
Lobipes	37
LONGSPUR	xiii, 93
Chestnut-collared, Lapland, Smith's	212
McCown's	93, 211-212
LOON, Arctic, Common, Pacific, Red-throated	1
Lophodytes (see *Mergus cucullatus*)	16-17
Loxia	183-184
MAGPIE, Black-billed	xi, 106
MALLARD	12-13
MAN-O-WAR BIRD	4
MARTIN, House	12
Purple	58, 95, 100-102, 141
Sand	96
MEADOWLARK	141, 162, 163-165
Arizona, Eastern, Texas	163-164
Western	121, 127, 156, 163, 164-165, 168

Megaceryle	67
Melanerpes	70-71
Melanitta	16
Meleagrididae	30
Meleagris	30
Melospiza	235-238
MERGANSER	4, 16
American, Common, Red-breasted	17
Hooded	16-17
Mergus	16-17
MERLIN	26, *27*
Micrathene	52-53
Micropalama	36
Migration (*see also* Catastrophes and Frost flight)	.x, xi,xiii, 24, 45, 58, 80, 83-84, 91, 93, 98, 100, 103, 131, 139-140, 150-151, 154, 176, 178, 181, 184, 189, 201, 205, 206
Mimidae	120
Mimus	121
Mniotilta	146
MOCKINGBIRD	*121*, 133, 141
Molothrus	143, 172-173
Mormon Lake	x
Motacillidae	137
MUD-HEN	32
Muscivora	80
Museums	viii, 17
Myadestes	133, 156
Mycteria	7-8
Myiarchus	17, 80-83, 141
Myioborus	161
Myiodynastes	80
Nesting (*see* Breeding seasons)	
Nests	62
NIGHTHAWK	55
Booming	56-57
Common	*56-57*, 91
Lesser	57-58
Texas	55, *57-58*
Trilling	57-58
Western	56-57
NIGHT HERON (*see* Heron)	
NIGHTJAR, Buff-collared	x, 56
Nomenclature (*see also* Taxonomy)	11, 208
Nucifraga	108-109
Numbers	x, xiii, 162
Numenius	33-34
NUTCRACKER, Clark's	103, *108-109*
NUTHATCH	71, 113
Black-eared	114-115
Pine	87, *114-115*
Pygmy	87, *114-115*
Red-breasted	114
Rocky Mountain	113-114
White-breasted	x, 110, *113-114*
Nyctanassa (see *Nycticorax violaceus*)	7
Nycticorax	6-7
OLDSQUAW	16
Olor	9
Oporornis	156-157
Oreortyx	29
Oreoscoptes	125-126
ORIOLE	162, 167, 173
Baltimore	169-170
Bullock's	167, *169-170*
Hooded	*167-168*, 170, 173, 174
Northern	167, *169-170*, 173, 176
Orchard	167
Scarlet-headed	169, *170*
Scott's	127, *186-169*
Ortstreue	12
OSPREY	4, 25
Otus	46-51
OUZEL, Water	116
OVENBIRD	155
Overgrazing	xiii, xvii, 28, 200
OWL (*see also* Pygmy-Owl and Screech-Owl)	46, 102, 141
Barn	46
Billy	53
Burrowing	53
Elf	x, 47, 50, *52-53*, 200
Ferruginous (*see* Pygmy-Owl, Ferruginous)	
Flammulated (*see* Screech-Owl, Flammulated)	
Great Horned	*51*, 53, 54
Ground	53
Horned (*see* Owl, Great Horned)	
Long-eared	54
Monkey-faced	46
Saw-whet	50, *54-55*
Scops	46, 47, *50-51*
Short-eared	54
Spotted	53-54
Whiskered	47, *49-50*
Pachyramphus	76
Pandion	25
Pandionidae	25
Panyptila	58
Parabuteo	23-24
Paridae	109
PARROT	102
Thick-billed	44-45
PARULA (Warbler)	149
Parulidae	146
Parus	109-111
Passer	162
Passerculus (see *Ammodramus sandwichensis*)	192-193
Passerella	210-211
Passeriformes	76
Passerina	179-181
Pelecanidae	3
Pelecanus	3
PELICAN, Brown	3
White	3
Perisoreus	103
Petrochelidon (see *Hirundo pyrrhonota*)	99-100
Peucedramus	136-137
Phaethon	3
Phaethontidae	3
PHAINOPEPLA, *Phainopepla*	x, 139-140
Phalacrocoracidae	4
Phalacrocorax	4
Phalaenoptilus	56
PHALAROPE, Northern, Red, Wilson's	37
Phalaropodidae	37
Phalaropus	37
Phasianidae	28
Phasianus	30
PHEASANT, Ring-necked	*30*, 45
Pheucticus	178-179
"Phillips' Law" (*see* Law)	
PHOEBE, Black	65, *83-84*, 92
Eastern	*83*, 91
Say's	x, 78, 82, *84*, 133
Picidae	68
Picoïdes	75
PIGEON	40
Band-tailed	20, *40-41*
domestic	40
Pinicola	186
PINTAIL	13
Pipilo	188-191

Pipit, American	137-138
Sprague's	138
Water	xi, *137-138*, 155
Piranga	174-176
Platypsaris (see *Pachyramphus*)	76
Plegadis	8
Ploceidae	162
Plover	32, 33
American Golden	32, *33*
Black-bellied	32, *33*
Golden (see American Golden)	
Mountain	33
Ringed	32
Semipalmated	32
Snowy	32
Upland	34
Pluvialis	33
Podiceps	1-2
Podicipedidae	1
Podilymbus	2-3
Polioptila	133-134
Polyborus (see *Caracara*)	25-26
Polymorphism	112-113
Pooecetes	194-195
Poor-will	55-56
Nuttall's	56
Porphyrula	31-32
Porzana	31
Preste-me-tu-cuchillo	x, 55 *56*, 78
Progne	100-102
Protonotaria	146
Psaltriparus	111-113
Psilorhinus	106
Psitlacidae	44
Purple-finch	184
California	184-185
Cassin's	183, *185*, 186
Common	184-185
Pygmy-Owl	47, 52
Ferruginous	52
Mountain	47, *51-52*
Rocky Mountain	51
Pyrocephalus	91-92
Pyrrhuloxia	177-178
Pyrrhuloxia (see *Cardinalis sinuatus*)	
Quail, California	29
Desert	29
Fool (see Mearn's Quail)	
Gambel's	29, 52
Harlequin (see Mearn's Quail)	
Mearns'	xiii, 28, *29*, 30
Scaled	xiii, 29
Quiscalus	171
Races (see Subspecies)	
Rail, Black, Carolina, Clapper, Virginia, Yellow	31
Rallidae	30
Rallus	31
Rarity (see *Abundance categories*)	
Raven	127
American	106
Common	xi, *106*
White-necked	106-107
Recurvirostra	36-37
Recurvirostridae	36
Red color (see Diet)	
Redhead	15
Redstart, American	160-161
Painted, Slate-throated	161
Regulus	134-136
Rhynchophanes (see *Calcarius mccownii*)	211-212
Rhynchopsitta	44-45
Ridgwayia	127
Ring-bill (see Ring-necked Duck)	15
Riparia	96-97, 156
Rivers	xiii-xvi
Roadrunner	45-46
Robin, American	*126*, 127, 175
Rufous-backed	127
Rough-leg, American, Ferruginous	23
Saguaro	102
Sanderling	36
Sandpiper	17, 32, 33, 127
Baird's	35
Bartramian	34
Common	34
Least	*35*, 36
Pectoral	35
Red-backed	35
Semipalmated	36
Solitary	*34*, 35
Spotted	*34*, 36, 155
Stilt	36
Western	*35*, 36
Santa Cruz River valley	xvi
Sapsucker, Natalie's	73-74
Red-breasted	72-73
Red-naped	72-73
Williamson's	72, *73-74*
Yellow-bellied	72-73
Sayornis	83-84
Scardafella	43-44
Scaup	15, 16
Greater	9, *16*
Lesser	9, *15*, 16
Sceloporus	23
Scolopacidae	33
Scoter, Surf, White-winged	16
Screech-owl	46-47, 48, 53
Common	xi, *47-49*, 209
Eastern Common	48
Flammulated	x, 47, *50-51*
Mexican	47-49
Saguaro	47-49
Spotted	47, *49-50*, 53, 55
Western Common	48
Seiurus	155-156
Selasphorus	61, 63-64, 156
Setophaga	160-161
Shitepoke (see Green Heron)	5
Shoveler	14
Shrike	121, 140
Loggerhead, Northern, White-rumped	140-141
Sialia	131-133
Siskin, Pine	xi, 161, 184, *186-187*
Sitta	113-115
Sittidae	113
Smaragdolanius	142
Snipe, Common, Wilson's	33, 36
Solitaire, Townsend's	*133*, 156
Sora	31
Sparrow	93
Baird's	xiii, 194
Bell's	201-202
Black-chinned	197-198
Black-throated	202-203
Botteri's	xiii, 189, 200
Brewer's	156, 195, 196, *197*
Cassin's	x, 165, 189, *200-201*
Chipping	155, *196*, 197
Clay-colored	196-197

Desert	202-203
English	xi, 27, 119, *162*, 183
Five-striped	201
Fox	93, 189, *210-211*
Gambel's	206-207
Golden-crowned	207
Grasshopper	189, *193-194*
Harris'	207-208
House (see English Sparrow)	162
Large-billed	192-193
Lark	195
Lincoln's	189, *208*
Rufous-crowned	200
Rufous-winged	123, *198-199*
Sage	189, *201-202*
Savannah	156, *192-193*, 194
Scott's	200
Song	xi, 93, 189, *209-210*
Swamp	208-209
Tree	195-196
Vesper	189, *194-195*
White-crowned	x, 189, *206-207*, 207
White-throated	207
SPECKLE-BELLY (see White Fronted Goose)	10
Speotyto	53
Sphyrapicus	72-74
Spinus	186-188
Spiza	181
Spizella	195-198
SPOONBILL	14
Roseate	8
SPRIG	13
STARLING	xi, 141-142
Statistics	87
Steganopus	37
Stelgidopteryx (see *Riparia ruficollis*)	97
Stellula	64-65
Stercororiidae	37
Stercorarius	37
Sterna	39
STIFF-TAILS	17
STILT, Black-necked	37
STORK	7, 8
Strigidae	46
Strigiformes	46
Strix	53-54
Sturnella	156, 163-165
Sturnidae	141
Sturnus	141-142
Subspecies (see also Taxonomy)	x, 132, 183
Sula	4
Sulidae	4
SWALLOW	58, 62, 64, 95, 98
Bank	*96-97*, 97, 156
Barn	12, *97-99*
Cliff	xi, 12, 96, 98, *99-100*, 127
Rough-winged	95, 96, 97
Tree	96
Violet-green	xi, *95-96*
SWAN, Trumpeter, Whistling	9
SWIFT	58, 95, 101, 127
Black	58
Chimney	*59*, 95
Vaux's	59
White-throated	*59*, 95
Sylviidae	133
Tachycineta	95-96
TANAGER	174, 177
Cooper's	176
Hepatic	175
Louisiana	174-175
Scarlet	175
Summer	xv, 173, *176*
Western	xi, 167, *174-175*
Tangavius	169, 173-174
Taxonomy (see also Nomenclature)	117-118, 208
TEAL	26
Blue-winged	9, *14*
Cinnamon	9, *14*
Common	13
Green-winged	14
TERN	37-38, 127
Black, Caspian, Common, Forster's Gull-billed, Least	39, *40*
Tetraonidae	27
THRASHER	120
Bendire's	*122-123*
Brown	122
Crissal	124-125
Curve-billed	122-123, *123-124*, 125
LeConte's	124
Palmer's	123-124
Sage	125-126
Thraupidae	174
Threskiornithidae	8
THRUSH (see also Water-Thrush)	126, 127, 131
Aztec	127
Gray-cheeked	131
Hermit	xi, *127-130*, 175
Olive-backed	130-131
Russet-backed	130-131
Swainson's	127, *130-131*
Varied	127
Willow	131
Thryomanes	118-119
TITMOUSE	102, 109
Bridled	xi, 110-111
Gray	109, *111*
Plain	109, *111*, 144
Totanus	35
TOWHEE, Abert's, Brown, Canyon, Green-tailed, Mountain, Nevada, Rufous-sided, Spotted Spurred	188-191
Toxostoma	120, 122-125
Tringa (see also *Totanus*)	34
Trochilidae	59
Troglodytes	117-118
Troglodytidae	117
TROGON, Coppery-tailed	66-67
Trogonidae	66
TROPIC-BIRD, Red-billed	3
Tucson (see Santa Cruz River valley)	
Turdidae	126
Turdus	126-127
TURKEY	30
Water	4
TURNSTONE, Black, Ruddy	33
Tyrannidae	77
Tyrannus	77-80
Tyto	46
Tytonidae	46

Variation from year to year (see also breeding seasons)
9, 24, 44-45, 71, 72, 74, 80, 83-84, 88, 92, 95, 103-104, 108-109, 139-140, 166, 180, 182-183, 198-199

Variation, geographic (see also subspecies)	xi
VEERY	xi, 131
VERDIN	xiii, 109, *113*, 143
Vermivora (see *Helminthophila*)	146-149
Vireo	92, 142-145

VIREO, Arizona 142-143
    Bell's ....xv, 142-143, 173
    Cassin's 144
    Gray *143-144*, 176
    Hutton's 85, 110, 135, *142*
    Least 142-143
    Philadelphia 144-145
    Plumbeous 144
    Red-eyed ....xi, 143, *144*
    Solitary ....xi, 143, *144*, 161
    Stephens' 142
    Warbling ....xi, 144, *145*
    White-eyed 142
    Yellow-throated 144
Vireolanius 142
Vireonidae 142
VULTURE 18
    Black ....xi, *18-19*, 24
    Turkey *18*, 22, 24

WARBLER, WOOD 142, 146, 148, 206
WARBLER, Audubon's *151-153*, 154
    Black-and-white 146
    Black-throated Blue 151
    Black-throated Gray ....ix, 110, *153*
    Black-throated Green 153-154
    Blue-winged 146
    Calaveras 148
    Cape May 151
    Cerulean 155
    Chestnut-sided 155
    Colima 148
    Connecticut 156
    Fan-tailed 161
    Grace's ....ix, 154, 155
    Gray-headed 148
    Hermit 136, 153, *154-155*, 156
    Hooded 159
    Kentucky 156
    Lucy's ....xv, 148-149
    Lutescent 146-148
    MacGillivray's *156-157*, 181
    Magnolia 151
    Mourning 156, 157
    Myrtle *151-153*, 154
    Nashville 148
    Olive 133, *136-137*
    Orange-crowned ....ix, 110, *146-148*, 159
    Palm 155
    Parula 149
    Pileolated 117, 156, *159-160*
    Prairie 155
    Prothonotary 146
    Red-faced 159
    Rufous-capped 148
    Tennessee 146
    Townsend's *154*, 155
    Virginia's 148
    Wilson's 117, 156, *159-160*
    Worm-eating 146
    Yellow ....xv, 127, 147, *149-151*, 173
    Yellow-rumped ....xi, 110, *151-153*, 154, 196
WATERFOWL 9, 25
WATERHEN 32
WATER-THRUSH, Alaska 155-156
    Grinnell's 155-156

    Louisiana 156
    Northern 155-156
WAXWING, Bohemian 138
    Cedar 138-139
WHIMBREL 34
WHIP-POOR-WILL 55-56
    Ridgway's ....xi, *56*, 78
    Stephens' 55-56
WHISKEY-JACK (*see* Gray Jay) 103
Whistler (*see* Common Goldeneye) 16
WIDGEON, American, European 13
WILLET *34*, 35
Wilsonia 156, 159-160
Wing measurements xi
WOODPECKER 68, 102, 141
    Acorn 68, *70-71*, 184
    Alpine Three-toed 75
    Ant-eating 70-71
    Arizona 75
    Batchelder's 74-75
    Cactus 75
    California 70-71
    Chihuahua 74
    Downy 68, *74-75*
    Gila *69-70*, 73, 75
    Hairy *74*, 75
    Ladder-backed 68, *75*
    Lewis' 68, *71-72*
    Mearns' 70-71
    Northern Three-toed 75
    Nuttall's 75
    Pileated 69
    Red-bellied 69
    Red-headed 70
    White-breasted 74
WOOD-PEWEE 92
    Eastern 90-91
    Western 77, *91*
Wood Warbler (*see* Warbler, Wood)
WREN 117
    Apache 117-118
    Baird's 118-119
    Bewick's ....xv, *118-119*, 148
    Brown-throated 117-118
    Cactus 117, *119*
    Cañon 120
    House *117-118*, 119
    Long-billed Marsh 119-120
    Parkman's 117-118
    Rock 120
    Western Marsh 119-120
    Winter 117

Xanthocephalus 165, 171
Xema 39

Yellow color (see Diet)
YELLOW-HAMMER 68
YELLOWLEGS, Greater, Lesser 35
YELLOW-THROAT, Common 157-159

Zenaida 41-42
Zenaidura 42-43
Zones, Life ix
ZONOTRICHIA 206-208